NATURAL SINKS OF CO_2

AF793842

NATURAL SINKS OF CO_2

Palmas Del Mar, Puerto Rico, 24–27 February 1992

Edited by

JOE WISNIEWSKI

and

ARIEL E. LUGO

Reprinted from *Water, Air, and Soil Pollution*,
Volume 64, Nos. 1–2, 1992

SPRINGER SCIENCE+BUSINESS MEDIA, B.V.

Library of Congress Cataloging-in-Publication Data

ISBN 978-94-010-5240-5 ISBN 978-94-011-2793-6 (eBook)
DOI 10.1007/978-94-011-2793-6

Printed on acid-free paper

All Rights Reserved
© 1992 Springer Science+Business Media Dordrecht
Originally published by Kluwer Academic Publishers in 1992
Softcover reprint of the hardcover 1st edition 1992
No part of the material protected by this copyright notice may be reproduced or utilized in any form or by any means, electronic or mechanical, including photocopying, recording or by any information storage and retrieval system, without written permission from the copyright owner.

Table of Contents

PART V

NATURAL SINKS OF CO_2 WORKSHOP SUMMARY – TECHNICAL SYNTHESIS, CONCLUSIONS, KEY FINDINGS AND RESEARCH RECOMMENDATIONS

Foreword

Humans have obviously altered the global carbon cycle in a major and fundamental way. The results of these changes potentially have enormous consequences for the functioning of the earth system and the well being of human endeavors. Yet, it is clear that we have, as yet, inadequate information of the full consequences of the altered carbon cycle. We cannot, for example, balance the carbon cycle nor do we have unequivocal information that would let us predict the effects of enhanced CO_2 on natural systems. Our uncertainty is evidenced by the ferocity of debate on these fundamental issues. This uncertainty has placed developing a better understanding of the carbon cycle as the highest priority of a large number of national and international programs dealing with global change issues.

This book addresses one of the intriguing, and major, issues of the carbon cycle, that of natural carbon sinks. This area, in particular, has been the subject of considerable debate and controversy. This book will most likely fuel these discussions even further since new dimensions are introduced, and conclusions are given that will not be universally accepted. In this sense this book will have served an important function – that of fostering new work and discussion to resolve the uncertainties that exist. To further aid this natural scientific progression research priorities are proposed. We cannot afford the luxury of debate on these issues too long however. We must resolve them quickly to foster a more rational management of our biospheric resources that will help us regulate the concentration of carbon in our atmosphere.

Dr. Harold Mooney
Department of Biological Sciences
Stanford University

Foreword

Preface

Scientific interest in the greenhouse effect and the global carbon cycle is not new. Jones and Henderson-Sellers (1990) traced the history of the greenhouse effect as far back as 100 years ago. In the early 1900's, Clarke (1908) suggested that the biota were the mitigating atmospheric carbon sink for carbon production in cities such as Paris. We can safely say, however, that beginning in the late 1970's the scientific literature on both subjects experienced unprecedented growth, leading to important summary studies such as the U.S. Department of Energy's (1985) state-of-the-art reports and the Intergovernmental Panel on Climate Change (IPCC) recent scientific assessment (Houghton et al. 1990). In spite of the voluminous attention to the greenhouse effect and the global carbon cycle, the treatment of the biosphere has been incomplete because most of earth's ecosystems have been assumed to be in carbon steady state with the atmosphere. Perhaps because of incomplete attention to the carbon cycle, scientists have been unable to balance the earth's carbon budget. More human generated carbon is accumulated in oceans and terrestrial ecosystems than the leading carbon models used to study the carbon cycle predict.

The 'Natural Sinks of CO_2' workshop held in Palmas Del Mar, Puerto Rico during 24–27 February 1992 had an objective to focus on the issues of the carbon cycle and carbon sinks. For far too long, carbon sources have been emphasized, while carbon sinks have been comparatively neglected. Consideration of carbon sinks is a vitally important component of any actions taken to manage the carbon cycle.

A key finding of the workshop was the unanimous decision to discard the steady state assumption and explore the implications of such a departure for the understanding of the global carbon budget. During the workshop, numerous potential carbon sinks that to date had not been considered in the modeling of the carbon cycle were identified. These included the coastal zone, biomass-accreting mature forests, soils, and other sectors of the biota such as arid lands. In a sense, this workshop's conclusion closes the historical loop in the analysis of the carbon cycle by validating Clarke's (1908) initial assumption that the biota was the sink for the excess CO_2 produced by burning of organic fuels.

A second key finding of the workshop was that substantial potential exists for capturing and sequestering carbon in natural systems. It is economically feasible to manage existing and additional terrestrial (forests, soils, saline lands) and aquatic (coastal, wetland and, possibly, ocean) ecosystems to substantially increases the level of carbon sequestration. The prospect of managing natural systems to absorb additional carbon should begin to change the mind-set under which scientists, policy makers and society deal with the issue of further greenhouse gas increases.

This book consists of 25 peer-reviewed chapters in five sections: Importance of Carbon Sinks, Effects of CO_2 on Natural Systems, Managing Natural Sinks of CO_2, Modeling Carbon Fluxes, and Natural Sinks of CO_2 Workshop Summary – Technical Synthesis, Conclusions, Key Findings and Research Recommendations.

Water, Air and Soil Pollution **64:** ix-x, 1992.

We found this workshop to be exciting and stimulating and offer these proceedings with the hope that they will stimulate further research in the areas of carbon sinks and non-steady state ecosystems, which we believe include most of the world's ecosystems.

Dr. Joe Wisniewski
Workshop Coordinator

Dr. Ariel E. Lugo
Workshop Chairman

References

Clarke, F.W. 1908, Data on biogeochemistry, U.S. Geological Survey Bulletin No. 330.

Houghton, J.T., G.J. Jenkins, and J.J. Ephraums (editors), 1990. Climate change: The IPCC scientific assessment, Cambridge University Press, Cambridge, England, 385 p.

Jones, M.D.H. and A. Henderson-Sellers, 1990, History of the greenhouse effect. Progress in Physical Geography 14:1-18.

U.S. Department of Energy, 1985. Atmospheric carbon dioxide and the global carbon cycle. J.R. Trabalka (ed.), DOE/ER-0239.

Acknowledgments

We acknowledge the input from all authors who contributed papers for this workshop and book. We acknowledge the support of the U.S. Environmental Protection Agency (EPA), Office of Research and Development; the U.S. Department of Energy (DOE), Office of Energy Research; the Edison Electric Institute (EEI), Environmental Affairs Department; and the Umweltbundesamt (UBA) Forestry Program. With respect to funding organizations, we specifically thank the following individuals: Dr. Courtney Riordan of EPA who continually offered encouragement, Dr. Roger Dahlman of DOE who provided invaluable technical insight, Robert Beck and John Kinsman of EEI for providing project direction, and Dr. Heinz Gregor of the UBA who provided a European perspective. We thank Battelle Pacific Northwest Laboratories (PNL) for providing administrative and technical input throughout the project. Particularly, we thank Dr. Peter Van Voris for considerable advice and guidance; Dr. John Clarke for moral and technical support; Dr. Gerry Stokes who enthusiastically approved PNL's participation; Kim Hough for administrative and clerical support; and Drs. Dominic Cataldo and John Downing, who helped synthesize a wide variety of workshop ideas. We acknowledge Dr. Billy McCormac, Editor of *Water, Air and Soil Pollution*, and Dee McCormac, the copy editor, for technical guidance necessary to produce the book. Finally, thanks are extended to Diane Loiello for personal and moral support throughout this effort.

Dr. Joe Wisniewski
Workshop Coordinator

Dr. Ariel E. Lugo
Workshop Chairman

Acknowledgments

[illegible]

THE IMPORTANCE OF CARBON SINKS

PART I

THE SEARCH FOR CARBON SINKS IN THE TROPICS

Ariel E. Lugo
Institute of Tropical Forestry
USDA Forest Service
Southern Forest Experiment Station
Call Box 25000
Rio Piedras, P.R. 00928-2500 USA

ABSTRACT. The study and modeling of the global C cycle have been dominated by the assumption that the atmosphere and the biota were in C steady state prior to the industrial revolution. This view led to the perception that most of the terrestrial biota was neutral with regards to the C concentration of the atmosphere. Recent evidence suggests that neither the atmosphere nor the biota were in C steady state prior to, or since, the industrial revolution. Therefore, it is now necessary to re-visit the role of natural processes in the global C cycle, study the C cycle in its totality, and focus attention on the magnitude of potential C sinks in ecosystems previously thought to be neutral with respect to atmospheric C.

"An improved understanding of the CO_2 cycle is essential to predict the future rate of any atmospheric CO_2 increase and to plan eventually for an international CO_2 management strategy" Tans et al. (1990).

1. Introduction

Most of the recent scientific activity concerning the global C cycle has focused on the anthropological sources of CO_2 (Melillo et al., 1990; Watson et al., 1990). Concern with the sources of atmospheric C is based on the fact that human activities such as burning of fossil fuels, put in circulation about 6 Pg yr^{-1} of which 3.4 Pg yr^{-1} accumulates in the atmosphere. This rate of C accumulation is sufficient to eventually cause global warming and other environmental changes (Houghton et al., 1990). While many of the future consequences of increased atmospheric CO_2 concentration are hard to predict, no one can argue the fact that people have changed the gaseous composition of the atmosphere (Watson et al., 1990).

Intensive study of atmospheric C sources has not led to a comprehensive understanding of the global C cycle. In fact, scientists are still unable to balance the global C budget. For example, the ocean and terrestrial biota must sequester the difference between the accumulation of C in the atmosphere (3.4 Pg yr^{-1}) and the C produced by the combustion of fossil fuels (6 Pg yr^{-1}). Models suggest a C uptake by oceans of about 1.6 Pg yr^{-1} which means that about 1.0 Pg yr^{-1} must be sequestered in the terrestrial biota. However, models of the C flux in terrestrial systems suggest they are a **net source** of C to the atmosphere of an additional 0.3 to 2.5 Pg yr^{-1} (Houghton et al., 1990;

Water, Air, and Soil Pollution **64**: 3–9, 1992.
© 1992 *Kluwer Academic Publishers.*

Tans et al., 1990). This causes an imbalance in the C budget and explains the euphemism of the "missing carbon" or that fraction of the anthropologically produced C that cannot be accounted for in C budgets.

The time has come to focus substantial scientific attention on the identification and quantification of global C sinks. Moreover, all aspects of the C cycle require study to balance the global C budget. I will review some of the historical obstacles to progress in the understanding of the terrestrial aspects of the global C cycle with emphasis on tropical forests (c.f., Lugo and Brown, 1980).

2. Identification of Carbon Sinks in Tropical Forests

2.1 CONCEPTUAL ISSUES

The perception that the only C fluxes deserving study are those that have "changed drastically in recent times" (Houghton et al., 1983) underscores most of the research effort since the 1970's when the Carbon Dioxide Program of the U.S. Department of Energy began to support research in the C cycle. The concept that has guided this important research program is that the global C cycle has been modified by human activity, particularly since the industrial revolution (about 1860). Before this time, it was assumed that the world was in C steady state and the atmosphere maintained homeostasis of its gaseous composition because the inputs of the various gases were balanced by their respective outputs. Scientists reasoned that to facilitate the analysis of the problem, it was best to ignore all C fluxes and storages that had not changed since 1860, no matter how large they were. This decision basically limited the study of C to a small fraction of its global cycle because it considered only changes in land use (Detwiler and Hall, 1988; Houghton et al., 1983; Post et al., 1990).

Moreover, if the world was in steady state, it was possible to divide any C pool by its estimated age to obtain its annual flux and thus an idea of its potential importance. For example, Woodwell et al. (1983) divided the C storage in soil, muskeg, and forests of the boreal and tundra zones by the elapsed time since the retreat of the ice some 5,000 to 15,000 yr ago, to estimate the annual C flux of this part of the world. They concluded that this region was not relevant as a C sink because it was such a small flux that it could be assumed to be at steady state.

Calculations such as the one by Woodwell et al. (1983) ignored two alternative considerations. (1) If the systems were not in steady state, what appeared to be a small sink could in fact have been a very large one if the C accumulated in a shorter time interval that was assumed, and (2) many small C sinks can add up to a single large one. For example, Brown et al. (1986) found that tropical tree plantations constituted a small sink of C (0.1 Pg yr^{-1} or similar to the upper level of the estimate of Woodwell et al. [1983] for the boreal and tundra region) which nevertheless was sufficient to offset the C source associated with deforestation rate in the temperate zone. In fact, should tropical forests not be in C steady state, it would only take a small C sink of 20 to 30 g C m^{-2} yr^{-1} to add up to 1 Pg yr^{-1}, or the same order of magnitude of the missing C (Lugo and Brown, 1986)

The hypotheses that guided the early analysis of the global C cycle have been identified by Lugo and Brown (1992) as the steady state atmosphere (SSA) and the steady state forest (SSF) assumptions. Steady state assumptions are very convenient because they simplify modeling and reduce the need for data, as large sectors of the C budget can be declared neutral in terms of their effect on the atmospheric CO_2 concentration.

When ice core and tree ring data were used to develop time series of CO_2 concentration in past atmospheres (Oeschger and Stauffer, 1986; Barnola et al., 1987), they showed oscillations in atmospheric CO_2. Even before the onset of the industrial revolution, atmospheric CO_2 concentrations were increasing (Fraser et al., 1986). Therefore, even though the recent increase in atmospheric CO_2 concentration has been steeper than "normal" and due to human activity, it is nevertheless very difficult to identify a period of atmospheric CO_2 steady state, unless one recognizes an oscillating steady state. Moreover, for the terminology to be consistently applied, it would be necessary that the period of time used in the simulations of C flux be in synchrony with the period of time required to establish a C steady state. The implication of the history of oscillation of atmospheric CO2 to the study of the C cycle is that one cannot ignore any of the C fluxes and pools, particularly those associated with "natural" biotic processes. For unlike physico-chemical processes, biotic processes are dynamic, respond to change, and are able to adapt.

Regarding the SSF assumption, recent research and analysis have shown that tropical forests cannot be assumed to be in C steady state either (Lugo and Brown, 1986; Brown and Lugo, 1990a). Biomass and rates of biomass production for tropical forests have been changing constantly over time as a result of past disturbances by people and natural factors (Flint and Richards, 1991; Brown et al., 1991, 1992; Brown and Lugo, 1992). It is unrealistic to assume steady state in the C cycle of tropical forests.

2.2 RESEARCH ORIENTATION ISSUES

The constraints imposed by the SSA and SSF assumptions not only assure that C models based on land-use change and steady state assumptions will always predict net releases of C to the atmosphere (because the predominant human activity is generally associated with sources of C), but they have also influenced research orientation. Research activity has focused almost exclusively on terrestrial and deep water oceanic ecosystems, and ignored the coastal interface where C metabolism is very active. In the tropics, coastal wetlands, coral reefs, sea grass beds, coastal shelf, coastal waters, and estuaries have high rates of C cycling and high C accumulation but are usually ignored in global C budgets (Twilley, this volume). They too are assumed to be in C steady state and neutral with regards to the C cycle of the world.

Today, geographically referenced data have become the preferred strategy for modeling the global C cycle (Dale, 1992). This is an important technological advance which should help overcome the initial breakdown of the large tropical biome into two forest types (rain and seasonal forests). One reason for using geographic references to model the C cycle is the wide diversity of tropical forest life zones (33 forested tropical life zones vs 19 in the rest of the world; Lugo and Brown, 1991), and the fact that each appears to have its own unique biodiversity (Lugo and Brown, 1991) and set of C fluxes and stocks (Brown and Lugo, 1982). The initial simplification of life zone diversity retarded progress because it lumped too much variety into two or at most six categories (Detwiler, 1986). The fewer the number of categories, the less precise are the estimates of C flux, because the subtleties of C dynamics are averaged out without proper weighting. The recent report by the Intergovernmental Panel on Climate Change (IPCC) used a two biome concept to represent all tropical forests (Melillo et al., 1990).

Finally, the research focus on the C enrichment question took a "potted plant" approach and ignored the response of whole ecosystems to C enrichment (Melillo et al., 1990). Only now are experiments under way to study whole ecosystem response to C enrichment (even if in small microcosms; Drake, this volume), and almost a decade has been lost in a species by species approach to study plant photosynthesis without sufficient attention to heterotrophic respiration and C cycling under elevaated CO_2 concentrations. Thus, after much research, we still don't know how the full biota will respond to greater CO_2 availability.

2.3 DATA BASE ISSUES

The way data have been made available to those modeling the global C budget has also retarded progress in the understanding of the C cycle. Five examples from the study of tropical forests will be sufficient to make the point.

(1) Rates of tropical deforestation have proved to be very difficult to establish (Dembner, 1991). Well documented and geographically referenced deforestation rate information is now required to improve the accuracy of C model predictions. Studies by Brown et al. (1992) show the importance of these approaches, but they must be expanded to all terrestrial biomes.

(2) Accurate estimates of tropical forest biomass have also been elusive. Initially values that were high by a factor of two were used because the only available information was ecological data from small plots (< 1 ha). These data were later shown to be biased towards high biomass values (Brown and Lugo, 1984). Data from large forest inventories (thousands of hectares) with sound statistical sampling for whole landscapes yield lower biomass values (Brown et al., 1991, 1992).

(3) Soils were considered to be large sources of C (Woodwell et al., 1983; Houghton et al., 1983, 1991; Detwiler, 1986; Houghton, 1990) before it was realized that the conversion of tropical forests to pastures had little effect on soil organic carbon (SOC) pools regardless of life zone, and that pastures often had more SOC than the original forest (Brown et al., 1984; Lugo et al., 1986; Brown and Lugo, 1990b). Soil C recovery under certain uses was faster than expected (Lugo et al., 1986).

(4) The influence of humans on "primary forests" was underestimated with the result that forests believed to be in C steady state with the atmosphere were in fact accumulating small amounts of C (Lugo and Brown, 1986, 1992). Natural perturbations disrupt C cycles with periods of <300 yr while C models limit consideration to 120 yr, a short and inadequate time interval for realistic assumptions about steady states in the C cycle of forests (Lugo and Brown, 1986). Such important aboveground C sinks as coarse woody debris and growth of large trees have been ignored in C cycle models (Lugo and Brown, 1992; Brown and Lugo, this volume). Root information is yet to be considered in a meaningful way as the data base is extremely limited (Brown and Lugo, 1982).

(5) Carbon fluxes in rivers were estimated but the data were not used to correct estimates of atmospheric C sources because the flux was assumed to be in steady state. Yet, a fraction of this flux originated as an atmospheric C sink through plant photosynthesis followed by leaching to streams. Carbon export by tropical rivers was underestimated because the data base favored the lower C export values associated with temperate rivers (Schlesinger and Melack, 1981). More recent data show higher exports of C from wet tropical areas (Lugo and Quiñones, 1983; Frangi and Lugo, 1985; Lewis et al., 1990; Paolini, 1990). The IPCC report doubles the estimate for river C export to the oceans from 0.4 Pg yr^{-1} (Schlesinger and Melack, 1981) to 0.8 Pg yr^{-1} (Watson et al., 1990). This means that our perception of the importance of this C sink continues to increase to a magnitude that approaches that of the missing C.

3. Conclusion

The disproportionate attention to the problem of anthropogenic C sources, although initially justified, clouded the C cycle issue, because it gave impetus to the exclusion of large sectors as being irrelevant to the function of the whole. Moreover, limitations imposed by poor data availability in the terrestrial biota, instead of stimulating the modeling of a range of values and conditions, led to the use of a limited, and eventually shown to be biased, set of data and assumptions. When modeled, these data and assumptions resulted in unrealistic predictions of sources of atmospheric C. In spite of the limitations, there has been progress in the solution of the missing C issue. Between 1978 and 1991 the release of C from the biota has been revised downward from a high of 8 Pg yr^{-1} (Woodwell et al., 1978) to <2 Pg yr^{-1} (Detwiler and Hall, 1988; Tans et al., 1990).

The search for C sinks is now an important and priority activity because such a search will: (1) shift the focus of the C cycle program towards a more comprehensive analysis of the issue, (2) result in a quicker resolution of the missing C issue, (3) lead to more credible recommendations to world leaders with regards to the management of the C budget and the biosphere, and (4) result in a better understanding of the function of the global ecosystem.

In addition to the advantages of the above, a greater emphasis on C sinks changes the mind set under which scientists and society attack the greenhouse issue. The focus on C sources results in a problem definition mind set while the focus on C sinks shifts thinking towards problem solving. The attention on the function of the biota, which is certainly not neutral with respect to the C cycle, has numerous social and economic benefits as well because it leads to conservation of forests, to reforestation, and the proper recognition of hidden natural subsidies to human activity. We cannot underestimate the importance of these considerations.

4. References

Barnola, J.M., D. Raynaud, Y.S. Korotkevich, and C. Lorius. 1987. Vostok ice core provides 160,000-year record of atmospheric CO_2. Nature 329:408-414.

Brown, S. and A.E. Lugo. 1982. The storage and production of organic matter in tropical forests and their role in the global carbon cycle. Biotropica 14:161-187.

Brown, S. and A.E. Lugo. 1984. Biomass of tropical forests: a new estimate based on volumes. Science 223:1290-1293.

Brown, S. and A.E. Lugo. 1990a. Tropical secondary forests. Journal of Tropical Ecology 6:1-32.

Brown, S. and A.E. Lugo. 1990b. Effects of forest clearing and succession on the carbon and nitrogen content of soils in Puerto Rico and US Virgin Islands. Plant and Soil 124:53-64.

Brown, S. and A.E. Lugo. 1992. Aboveground biomass estimates for tropical moist forests of the Brazilian Amazon. Interciencia 17:8-18.

Brown, S., A. Glubczynski, and A.E. Lugo. 1984. Effects of land use and climate on the organic content of tropical forest soils in Puerto Rico. Pages 204-209 in New forests for a changing world. Society of American Foresters, Bethesda, Md.

Brown, S., A.J.R. Gillespie, and A.E. Lugo. 1991. Biomass of tropical forests of South and Southeast Asia. Canadian Journal of Forest Research 21:111-117.

Brown, S., L. Iverson, and A.E. Lugo. 1992. Land use and biomass changes in Peninsular Malaysia during 1972-1982: a GIS approach. 1992. In V.H. Dale (editor). Effects of land use change on atmospheric carbon dioxide concentration: Southeast Asia as a case study. Springer-Verlag, New York. (in press).

Brown, S., A.E. Lugo, and J. Chapman. 1986. Biomass of tropical tree plantations and its implications for the global carbon budget. Canadian Journal of Forest Research 16:390-394.

Dale, V.H. (editor). 1992. Effects of land use change on atmospheric carbon dioxide concentration: Southeast Asia as a case study. Springer-Verlag, New York, (in press).

Dembner, S.A. 1991. Provisional data from the forest resources assessment 1990 project. Unasylva 164(42):40-44.

Detwiler, R.P. 1986. Land use change and the global carbon cycle: the role of tropical soils. Biogeochemistry 2:67-93.

Detwiler, R.P. and C.A.S. Hall. 1988. Tropical forests and the global carbon cycle. Science 239:42-47.

Flint, E.P. and J.F. Richards. 1991. Historical analysis of changes in land use and carbon stocks of vegetation in South and Southeast Asia. Canadian Journal of Forest Research 21:91-110.

Frangi, J.L. and A.E. Lugo. 1985. Ecosystem dynamics of a subtropical floodplain forest. Ecological Monographs 55:351-369.

Fraser, P.J., W.P. Elliott, and L.S. Waterman. 1986. Atmospheric CO_2 record from direct chemical measurements during the 19th century. Pages 66-88 in J.R. Trabalka and D.E. Reichle (editors). The changing carbon cycle: a global analysis. Springer-Verlag, New York.

Houghton, J.T., G.J. Jenkins, and J.J. Ephraums (editors). 1990. Climate change, the IPCC scientific assessment. Cambridge University Press, Cambridge. 365 p.

Houghton, R.A. 1990. The global effect of tropical deforestation. Environmental Science and Technology 24:414-422.

Houghton, R.A., J.E. Hobbie, J.M. Melillo, B. Moore, B.J. Peterson, G.R. Shaver, and G.M. Woodwell. 1983. Changes in the carbon content of terrestrial biota and soils between 1860 and 1980: a net release of CO_2 to the atmosphere. Ecological Monographs 53:235-262.

Houghton, R.A., D.L. Skole, and D.S. Lefkowitz. 1991. Changes in the landscape of Latin America between 1850 and 1985 II. net release of CO_2 to the atmosphere. Forest Ecology and Management 38:173-199.

Lewis, W.M., F.H. Weibezahn, J.F. Saunders III, and S.K. Hamilton. 1990. The Orinoco River as an ecological system. Interciencia 15:346-357.

Lugo, A.E. and S. Brown. 1980. Tropical forest ecosystems: sources or sinks of atmospheric carbon? Unasylva 32(129):8-13.

Lugo, A.E. and S. Brown. 1986. Steady state terrestrial ecosystems and the global carbon cycle. Vegetatio 68:83-90.

Lugo, A.E. and S. Brown. 1991. Comparing tropical and temperate forests. Pages 319-330 in J. Cole, G. Lovett, and S. Findlay (editors). Comparative analyses of ecosystems: patterns mechanisms, and theories. Springer-Verlag, New York.

Lugo, A.E. and S. Brown. 1992. Tropical forests as sinks of atmospheric carbon. Forest Ecology and Management (in press).

Lugo, A.E. and F. Quiñones. 1983. Organic carbon export from intensively used watersheds in Puerto Rico. Pages 237-241 in E.T. Degens (editor). Transport of carbon and minerals in mayor world rivers. Part 2 SCOPE/UNEP Sonderband, Heft 52 Hanburg, Germany.

Lugo, A.E., M.J. Sanchez, and S. Brown. 1986. Land use and organic carbon content of some subtropical soils. Plant and Soil 96:185-196.

Melillo, J.M., T.V. Callaghan, F.I. Woodward, E. Salati, and S.K. Sinha. 1990. Effects on ecosystems. Pages 283-310 in J.T. Houghton, G.J. Jenkins, and J.J. Ephraums (editors). Climate change, the IPCC scientific assessment. Cambridge University Press, Cambridge. 365 p.

Oeschger, H. and B. Stauffer. 1986. Review of the history of atmospheric CO2 recorded in ice cores. Pages 89-108 in J.R. Trabalka and D.E. Reichle (editors). The changing carbon cycle: a global analysis. Springer-Verlag, New York.

Paolini, J. 1990. Carbono organico disuelto y particulado en grandes rios de la America del Sur. Interciencia 15:358-366.

Post, W.M., T.H. Peng, W.R. Emanuel, A.W. King, V.H. Dale, and D.L. DeAngelis. 1990. The global carbon cycle. American Scientist 78:310-326.

Schlesinger W.H. and J.M. Melack. 1981. Transport of organic carbon in the world's rivers. Tellus 33:171-187.

Tans, P.P., I. Fung, and T. Takahashi. 1990. Observational constraints on the global atmospheric CO_2 budget. Science 247:1431-1438.

Watson, R.T., H. Rodhe, H. Oeschger, and U. Siegenthaler. 1990. Greenhouse gases and aerosols. Pages 1-40 in J.T. Houghton, G.J. Jenkins, and J.J. Ephraums (editors). Climate change, the IPCC scientific assessment. Cambridge University Press, Cambridge. 365 p.

Woodwell, G.M., J.E. Hobbie, R.A. Houghton, J.M. Melillo, B. Moore, B.J. Peterson, and G.R. Shaver. 1983. Global deforestation: contribution to atmospheric carbon dioxide. Science 222:1081-1086.

Woodwell, G.M., R.H. Whittaker, W.A. Reiners, G.W. Likens, C.S. Delwiche, and D.B. Botkin. 1978. The biota and world carbon budget. Science 199:141-146.

WHY UNDERSTANDING THE NATURAL SINKS AND SOURCES OF CO_2 IS IMPORTANT: A POLICY ANALYSIS PERSPECTIVE

Jae Edmonds
Pacific Northwest Laboratory
901 D St., S.W., Suite 900
Washington, D.C. 20024-2115

ABSTRACT: The mechanisms that regulate the concentration of carbon dioxide in the atmosphere, the carbon cycle, is an integral part of the analysis of the greenhouse issue. The present understanding of the carbon cycle is inadequate to the purpose of assessing the relationship between future anthropogenic emissions and concentrations of atmospheric CO_2. The most important problem is that natural science cannot presently explain the relationship between present and past anthropogenetic emissions and concentrations. Sinks for CO_2 are inadequate to explain present and past dispositions of emissions. This deficiency in scientific understanding leads to uncertainty in the analysis of potential future emissions and atmospheric CO_2 accumulation, and to uncertainty in the specification of other policy analysis instruments such as global warming potential coefficients.

1. Carbon Dioxide, Climate, and Policy

There may be very serious implications for the way humans conduct their affairs that hinge on a proper understanding of the carbon cycle. The urgency that surrounds improved understanding of the natural science of carbon sinks in the carbon cycle stems not so much from the usual source, scientific curiosity, as from more practical considerations. It is well known that carbon dioxide, CO_2, is a radioactively active gas. It along with other radioactively active gases such as water vapor, methane, nitrous oxides, and ozone, are effectively transparent to incoming solar radiation, but trap infrared radiation reradiated to space. These gases are important determinants of the Earth's present mean global surface temperature of approximately 15°C.

The modern record of atmospheric carbon dioxide concentrations dates from 1958 when David Keeling began systematic observations of atmospheric concentrations at Mauna Loa, Hawaii. Since that time Keeling's station and others subsequently established, have shown a persistent tendency toward increasing concentrations.

Water, Air, and Soil Pollution **64**: 11–21, 1992.
© 1992 *Kluwer Academic Publishers.*

Estimate of the implication of doubling the concentration of atmospheric CO_2 from pre-industrial levels, using general circulation models (GCMs), indicate that the direct and indirect effects on mean global surface temperature would be 1.5 to 4.5°C, although even greater ranges of estimates have emerged from GCM calculations.

In the period since the end of World War II, there has been an increasing trend toward the use of fossil fuels. As fossil fuels represent carbon reservoirs that were laid down more than a million years before the present, the oxidation of fossil fuels represents an anthropogenic perturbation of the carbon cycle of potential significance. The increase in human activities has also resulted in changes to land-use over time. Replacement of forest ecosystems with other less carbon dense ecosystems can also result in an anthropogenic release of significant quantities of carbon to the atmosphere.

2. Overview of Present Understanding of Carbon Cycle

The carbon cycle describes the carbon reservoirs and exchanges between various Earth systems elements. A summary of present understanding is given in Figure 2. This figure reflects a present understanding of the system as a general ocean-atmosphere-terrestrial system with relatively strong earth-atmosphere and ocean-atmosphere linkages and relatively weak earth-ocean interchanges. Gross fluxes within the carbon cycle are dominated by natural systems, ocean-atmosphere and earth-atmosphere exchanges, (IPCC,1990; IPCC,1992; Wuebbles and Edmonds, 1991; Post et al., 1990; Bolin, 1986; Trabalka, 1985). Estimates of the gross flux between various components of the carbon cycle are subject to great uncertainty.

2.1 Carbon Emissions Sources

Human activities generate emissions of carbon from stocks of natural carbon through the oxidation of fossil fuels and land-use change. Current rates of emission from fossil fuel burning are approximately 6 PgC/yr (IPCC,1992; Marland,1992). This estimate is known with relatively good precision, plus or minus 15% (Wuebbles and Edmonds,1991). Current rates of emission from land-use change, predominantly tropical deforestation, are estimated to be 1.6 ± 1.0 PgC/yr (IPCC,1990; IPCC,1992).

Nineteenth century rates of fossil fuel emissions were small relative to present rates, and relative to estimated rates of carbon release from land-use change. Fossil fuel carbon emissions have grown from 0.1 PgC/yr in 1860 (Boden et al.,1990; Trabalka, 1985), Figure 3. Emissions of carbon from fossil fuel use are generally believed to be smaller than those from land-use change before the twentieth century. Houghton and Skole (1990) estimate that nineteenth century emissions from land use change were in the neighborhood of 0.5 PgC/yr.

Concentrations have risen from approximately 315 parts per million volume (ppm) in 1959 to their present 350 ppm. Preindustrial concentrations have been estimated using various procedures. Ice core measurements indicate that the concentration in 1800 was approximately 280 ppm (IPCC,1990), Figure 1.

Figure 1: Atmospheric CO2 Concentrations: 1800 through Present

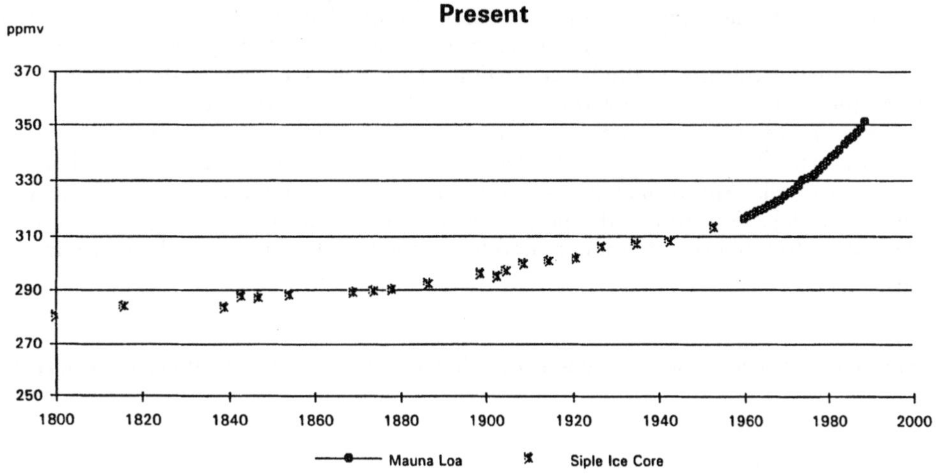

While anthropogenic releases of carbon to the atmosphere are small relative to the gross fluxes of carbon from natural systems, they are generally taken to be predominantly responsible for the observed accumulation of CO_2 in the atmosphere. The causality of human emissions is based on a set of circumstantial evidence including ice core measurements, isotopic carbon ratios, differences between northern and southern hemisphere CO_2 concentrations, and the rate of change of atmospheric CO_2 concentrations (IPCC,1990; Wuebbles and Edmonds,1991).

2.2 CARBON REMOVAL MECHANISMS

Over the period 1850 to 1986 approximately 195±20 PgC were introduced into the atmosphere by fossil fuel use. Over the same period approximately 117±35 PgC were introduced into the atmosphere by net deforestation. Net deforestation was the principal anthropogenic source of carbon release in the earlier part of the period, while fossil fuel use dominates releases in the latter portion. The average carbon content of the atmosphere increased by approximately 125 PgC over this same period. Thus the increase in atmospheric carbon corresponds to approximately 40% of the carbon released over the period from 1860 to 1986. This phenomenon can be observed over shorter time periods as well. Approximately 3.4 PgC/yr accumulate in the atmosphere, Table 1. The annual net injection from anthropogenic activities is thought to be range from 6 to 9 PgC/yr. Atmospheric retention is therefore approximately half of anthropogenic emissions.

Figure 2: Carbon Flow Throughout the Global System (PgC/yr - Rource: Moore 1988.

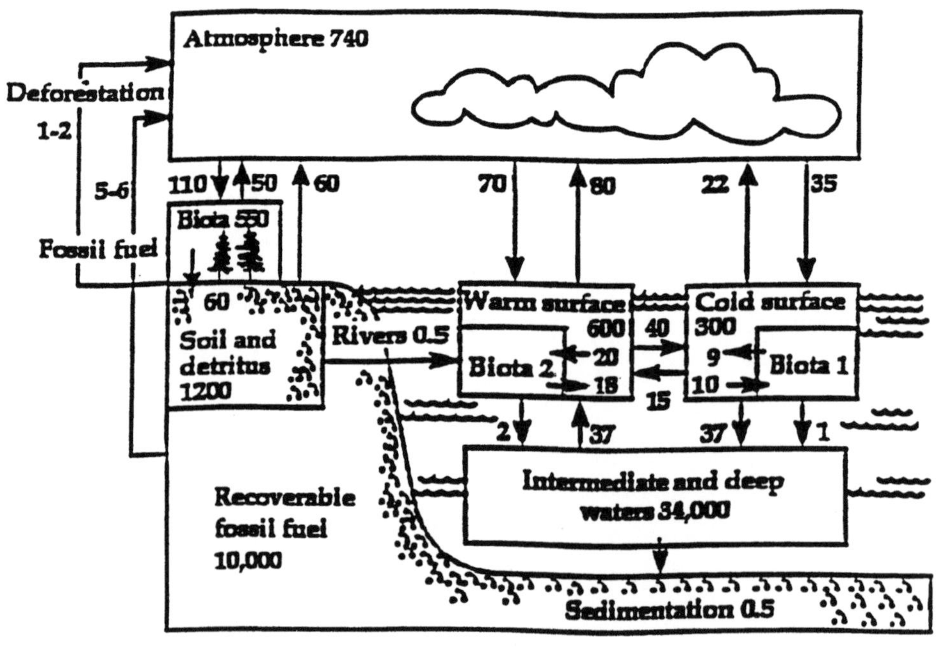

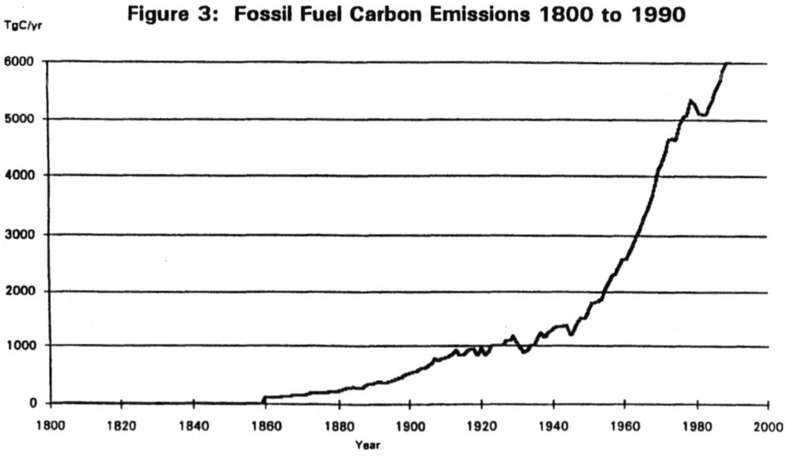

Figure 3: Fossil Fuel Carbon Emissions 1800 to 1990

Table 1: Average Global Carbon Fluxes for the 1980s

SOURCES AND SINKS	PgC/yr
Emissions from fossil fuels into the atmosphere	5.4±0.5
Emissions from net deforestation and land-use change	1.6±1.0
Total Net Emissions	**7.0±1.5**
Accumulation in the atmosphere	3.4±0.2
Uptake by the ocean	2.0±0.8
Total Disposition	**5.4±1.0**
Net Imbalance (the "missing carbon" sink)	***1.6±1.4***

Source: IPCC (1990), p.1.14.

The only generally agreed upon permanent sink for carbon are deep ocean and soils. The net buildup of carbon in soils is presently thought to proceed at a slower pace than the removal of carbon to oceans. At present, soils, may be a net source of carbon to the atmosphere as a result of land-use changes. Ocean models calibrated to observed removal rates, can account form only part of the 1 to 3 PgC/year removal of carbon from the atmosphere associated with human activities. The disposition of the so called "missing carbon," cannot yet be adequately explained by present research models. Nevertheless approximately 1.6 PgC/yr appear to be removed from the atmosphere in a way that cannot yet be adequately explained. This estimate is predicated on the assumption that, if it were not for human activities, that the carbon cycle would be in balance. That is, without human activities injecting carbon into the atmosphere, the total of all fluxes into and out of the atmosphere would balance. This may or may not be the case.

The terrestrial system is frequently treated as if it were in rough equilibrium with the atmosphere. CO_2 uptake by plants, 110 PgC/yr, is just balanced by releases from the biosphere to the atmosphere, 50 PgC/yr, and from soils and detritus to the atmosphere, 60 PgC/yr, Figure 2. Carbon leaving the biota to soils, 60 PgC/yr, leave both the biota and soils as neutral carbon sources (sinks). Uncertainty surrounds this conventional view of the terrestrial system. The proposition that either the carbon cycle, or terrestrial component of the carbon cycle would be in balance were it not for human activities is open to question, (Lugo and Brown, 1986; Wuebbles and Edmonds, 1991). Because CO_2 in a fertilizer for plants, the net increase in atmospheric CO_2 should in principle stimulate plant growth and add to the net carbon stock in the biota. This is referred to as the CO_2 fertilization effect. While the CO_2 fertilization effect is a well established phenomenon, the extent to which it affects the carbon cycle is a matter of heated debate. As noted earlier, estimates of up to 2.5 PgC/yr net removal have been estimated either by inference or by direct calculation, Goudriaan (1987,1989,1991), Esser (1991), Tans (1991), Tans et al. (1990). Empirical evidence provides an inconclusive basis for resolving the question at present.

3. Implications of the "Missing Carbon" Problem

The failure of researchers to adequately explain the carbon cycle leads directly to problems in the analysis of potential future concentrations of CO_2, and in the evaluation of the relative global warming potential of CO_2 as compared with other gases.

3.1 FUTURE ATMOSPHERIC COMPOSITION

In evaluating potential future concentrations of greenhouse gases it is common practice to assume a "neutral biosphere." A neutral biosphere is one in which the emissions from land-use change are exactly balanced by uptake from the "missing sink." Of the five analyses averaged to obtain future atmospheric CO_2 concentrations in the IPCC (1990) four employed this assumption (Wigley, 1992). See also for example Trabalka and Reichle (1986).

The assumption of a "neutral biosphere" is justified on the grounds of rough parity between the present "missing sink" and estimated emissions from land-use change, Table 1. Deconvolution studies show this rough parity over approximately the period 1940 through 1980, (Post et al.,1990). Such studies infer net land-use change emissions as a residual when ocean carbon uptake models are run against fossil fuel emissions and atmospheric records from Mauna Loa and the Siple ice core[1] only. Interestingly, the biosphere appears not to have been neutral historically. The residual roughly matches reconstructed land-use change emissions over the period 1860 to 1920, but departs sharply in the following decades. Thus a net source beyond fossil fuel emissions are needed prior to 1920, and a sink in 1980, (IPCC,1990; Post et al.1990). The transition occurs about the time when the reconstructed land-use carbon emissions are of approximately equal magnitude with fossil fuel carbon emissions, (Houghton and Skole,1991; Post et al.,1990).

One of the many problems with assuming a "neutral biosphere" for the analysis of potential future atmospheric concentrations is the implied assumption that increases and decreases in land-use change emissions do not matter. Implicitly every tree comes with its own sink. If more deforestation occurs, a greater sink is implicitly postulated. If deforestation declines the sink contracts. Such an approach is unacceptable for useful policy analysis. Furthermore, even if land-use change were prescribed and were anticipated to remain near present rates, the relative contributions of fossil fuels and land-use emissions cannot be expected to remain constant. As the relative contribution of fossil fuel carbon emissions increases, the need for realistic treatment of the entire carbon cycle becomes increasingly acute. The present set of alternatives for dealing with future atmospheric accumulation of carbon are not very attractive. They include the airborne fraction model[2], the addition of an arbitrary sink to an ocean model, or non-mechanistic removal mechanism.[3] A more attractive alternative is the use of a process oriented reduced form carbon cycle model which employs coefficient estimates capable of closing the carbon cycle. Such models include those by Goudriaan and Ketner (1984), Emanuel et al. (1984), and Keepin (1986). But until a scientific understanding of carbon cycle is significantly improved, even balanced carbon cycle models are subjecto to the uncertainty that surrounds the closure mechanism.

---

[1] See Boden et al. 1990 for these records, and primary sources.

[2] An airborne fraction model hypothesizes that a constant fraction of emissions is removed instantaneously to an unspecified sink, but that the remaining fraction accumulates in the atmosphere indefinitely. See Trabalka (1985), Bolin (1986), and Wuebbles and Edmonds (1991) for discussions of airborne fraction models.

[3] See for example, Ruether and Smith (1991).

3.2 GLOBAL WARMING POTENTIAL COEFFICIENTS

To facilitate the comparison of a marginal release into the atmosphere of different radiatively important gases, the IPCC (1990), building on work by such researchers as Lashof and Ahuja (1990), Derwent (1990), Derwent, Rodhe, and Wuebbles (1990), and Nordhaus (1990), developed a set of coefficients which the they referred to as Global Warming Potential (GWP) coefficients. These coefficients were developed by estimating the atmospheric decay pattern over time of the release into the atmosphere of one kilogram of each radiatively important gas. In each year the resulting change in radiative forcing was calculated. The amount of change in radiative forcing by each gas remaining in the atmosphere in each year was summed over a specific number of years. Three Integration Time Horizons (ITH) were chosen: 20, 100, and 500 years. The integrated change in radiative forcing of CO_2 was assigned the value of unity and other gases compared to it.

The role of CO_2 is particularly important. The GWP for all gases are stated in terms of carbon dioxide equivalent emissions. The use of CO_2 as the numeraire reflects its importance in overall emissions analysis. CO_2 has the greatest share of total global and United States GWP weighted emissions. Getting the change in cumulative radiative forcing for a one kilogram release of CO_2 correct therefore is crucial to getting the entire set of GWP indices correct. But carbon dioxide, unlike most of the other radiatively important gases has no sink in the atmosphere. Its removal mechanisms are complex and poorly understood. The IPCC (1990) used a decay function which reproduced the behavior of an ocean carbon cycle model.

This behavior is given by the equation:

$$f(t) = 0.30036\, e^{(-t/6.9993)} + 0.34278\, e^{(-t/71.109)} + 0.35686\, e^{(-t/815.727)},$$

where f(t) is the fraction of a kilogram of CO_2 remaining in the atmosphere a period of time t after release. This equation can be shown to reproduce the Mauna Loa record of carbon dioxide concentrations for the period 1959 through 1988 if the atmospheric injection is assumed to be given by the sum of the fossil fuel record as per Boden et al. (1990) and land-use change is assumed to proceed at the rate of 600 TgC/yr between 1800 and 1939 and then ceases thereafter. The problem with this assumption is that cumulative loading from land-use change is only 84 PgC, less than the lower bound given by Trabalka (1985) for cumulative carbon emission between 1800 and 1980, and inconsistent with the IPCC (1990) itself in its estimate of contemporary emissions from land-use change, 0.6 to 2.6 PgC/yr.

The use of an ocean uptake model in conjunction with a "neutral biosphere" assumption in the computation of GWP coefficients leads directly to a bias, (Edmonds,1991; IPCC,1992). The amount of this bias has been estimated by Wigley (1991) and Wuebbles (1992) to be in the range of 15 percent for the 100 year ITH. The extent of the bias depends on identifying the mechanism postulated as the "missing sink" of the carbon cycle.

4. Conclusions

There was a time in which the global climate change problem was referred to as the CO_2 problem. At that time one of the central research issues was the carbon cycle. Understanding and quantifying the relationship between emissions of anthropogenic carbon and atmospheric accumulation was considered a high priority. In recent years the recognition of multiple greenhouse gases and their associated processes has shifted attention to other areas. But the carbon cycle has never been adequately understood. Until such time as research succeeds in actually closing the carbon cycle, there will be unresolvable uncertainties which hinder the decision making process.

REFERENCES

Boden, T.A., Kanciruk, P., Farrell, M.P. 1990. ***Trends '90: A Compendium of Data on Global Change***, ORNL/CDIAC-36, Oak Ridge National Laboratory, Oak Ridge, TN.

Bolin, B. 1986. "How Much CO_2 Will Remain in the Atmosphere?" in ***The Greenhouse Effect, Climatic Change, and Ecosystem***, Bolin, B., Doos, B.R., Jaeger, J., and Warrick, R.A. (eds.), SCOPE 29, John Wiley & Sons. New York.

Derwent, R.G. 1990. ***Trace Gases and their Relative Contribution to the Greenhouse Effect***, AERE-R 13716, United Kingdom Atomic Energy Authority, Harwell, Harwell Laboratory, Oxfordshire OX11 ORA.

Derwent, R., Rodhe, H., and Wuebbles, D. 1990. "Global Warming Potential of Greenhouse Gases," special report for the United Nations Environmental Programme.

Edmonds, J.A. 1991. "A Review of Policy Implications of Greenhouse Warming: Report of the Mitigation Panel, National Academy of Sciences," submitted to ***Climatic Change***.

Emanuel, W.R., Killough, G.G., Post, W.M., and Shugart, H.H. 1984. ***Computer Implementation of a Globally Averaged Model of the World Carbon Cycle***, DOE/NBB-0062, United States Department of Energy (DOE), Washington DC.

Esser, G. 1991. "Uncertainties in the Dynamics of Biosphere Developments with the Accent on Deforestation," ***Pure and Applied Chemistry, 65***(5):775-778.

Goudriaan, J. 1991. "Uncertainties in Biosphere/Atmosphere Exchanges, CO_2 Enhanced Growth," ***Pure and Applied Chemistry, 65***(5):772-775.

Goudriaan, J. 1989. "Modeling Biospheric Control of Carbon Fluxes Between atmosphere, Ocean and Land in View of Climatic Change," in Berger, A., Schneider, S., and Duplessy, J.C., (eds.); ***Climate and Geo-Sciences***. NATO-ASI Series C, Vol 285. Kluwer Academic Publishers, Dordrecht, NE.

Goudriaan, J. 1987. "The Biosphere as a Driving Force in the Global Carbon Cycle," ***Netherlands Journal of Agricultural Science, 35***:177-187.

Goudriaan, J. and Ketner, P. 1984. "A Simulation Study for the Global Carbon Cycle Including Man's Impact on the Biosphere," ***Climatic Change 6***:167-192.

Houghton, R.A., and D.L. Skole. 1990. "Carbon," in ***The Earth Transformed by Human Action***, B.L. Turner, W.C. Clare, R.W. Kates, J.F. Richards, J.T. Mathews, and W.B. Meyer, (eds.), Cambridge University Press with Clark University, New York. Chapter 23:393-408.

IPCC (Intergovernmental Panel on Climate Change), 1992. ***IPCC 1992 Science Assessment.*** Report to IPCC from Working Group 1, prepared by the IPCC Group at the Meteorological Office, Bracknell, UK.

IPCC (Intergovernmental Panel on Climate Change), 1990. ***Scientific Assessment of Climate Change.*** World Meteorological Organization/United Nations Environmental Program. Cambridge University Press, New York.

Keepin, W. 1986. ***Review of Global Energy and Carbon Dioxide Projections***. Annual Reviews, Inc.

Lashof, D.A. and Ahuja, D.R. 1990. "Relative Global Warming Potentials of Greenhouse Gas Emissions," ***Nature, 344***:529-531.

Lugo, A.E. and S. Brown. 1986. "Steady state terrestrial ecosystems and the global carbon cycle," ***Vegetatio 68***:83-90.

Marland, G. 1992. Personal communication based on preliminary calculations for the year 1991.

Nordhaus, W.D. 1990. ***Contribution of Different Greenhouse Gases to Global Warming: A New Technique for Measuring Impact***. Department of Economics, Yale University, New Haven CT.

Post, W.M., Peng, T.-H., Emanuel, W.R., King, A.W., Dale, V.H., and DeAngelis, D.L. 1990. "The Global Carbon Cycle," ***American Scientist, 78***:310-326.

Ruether, J.A. and D.N. Smith. 1991. ***Average Atmospheric Lifetime of Carbon Dioxide in the Premodern Steady State and Present Era***, Pittsburgh Energy Technology Center, United States Department of Energy, P.O. Box 10940, Pittsburgh, PA 15236.

Tans, P.P., Fung, I.F., Takahashi, T., 1990. "Observational Constraints on the Global Atmospheric CO2 Budget," ***Science, 247***:1431-1438 (March 23, 1990).

Tans, P.P. 1991. "Uncertainties in the Global Carbon Cycle," ***Pure and Applied Chemistry, 65***(5):766-768.

Trabalka, J. (ed.) 1985. ***Atmospheric Carbon Dioxide and the Global Carbon Cycle***. DOE/ER-0239. National Technical Information Service, U.S. Department of Commerce, Springfield Virginia 22161.

Trabalka, J.R. and Reichle, D.E., eds. 1986. ***The Changing Carbon Cycle, A Global Analysis***. Springer-Verlag, New York.

Wigley, T.M.L. 1992. Personal communication.

Wigley, T.M.L. 1991. ***Global Warming Potentials***. Climate Research Unit, University of East Anglia, Norwich, NR4 7TJ, UNITED KINGDOM.

Wuebbles, D.J. 1992. Personal communication.

Wuebbles, D.J. and J.A. Edmonds. 1991. ***Primer on Greenhouse Gases***, Lewis Publishers, Chelsea, Michigan.

EFFECTS OF CO_2 ON NATURAL SYSTEMS

PART II

THE IMPACT OF RISING CO_2 ON ECOSYSTEM PRODUCTION

BERT G. DRAKE
Smithsonian Environmental Research Center
P.O. Box 28
Edgewater, Maryland, 21037
U.S.A.

Abstract. A fundamental property of green plants is that the rate of photosynthesis is dependent in the ambient CO_2 concentration. There is overwhelming experimental evidence that this effect increases plant production in most C_3 plants: hundreds of experiments with many species show that plant growth increases an average 30% to 40% for a doubling of the present normal ambient CO_2 concentration (Kimball, 1986). External environmental factors, such as temperature and the availability of nutrients, modify this response. The greatest stimulation of photosynthesis and growth can be expected to occur at high temperatures and much smaller responses at low temperature. Factors which restrict growth, such as low nutrients, will reduce but usually do not eliminate the stimulation of production with increasing CO_2 even when nitrogen is severly limiting. There are also reports of direct effects of ambient CO_2 concentration on dark respiration which show that there is an immediate reduction in the rate of CO_2 efflux or O_2 consumption when the CO_2 around plant tissues is increased. There have been very few long-term field studies of the effects of increased CO_2 on whole plants and ecosystem processes but the data from these studies are consistent in showing an increase in plant production with an increase in CO_2 concentration of the ambient air.

1. Introduction

Interest in the direct effect of rising atmospheric CO_2 on plants centers on its potential to: 1. increase plant growth and ecosystem production, 2. reduce evapotranspiration and 3. mitigate the rate of increase in atmospheric CO_2. Although it has been known for a long time that increasing the concentration of CO_2 around plants increases their growth and water use efficiency, there has been considerable doubt cast on this conclusion by those who feel that native plant species are limited in

Water, Air, and Soil Pollution **64**: 25–44, 1992.
© 1992 *Kluwer Academic Publishers.*

their production by nutrient supply and not CO_2. Moreover, it is also believed by some that the direct effects of rising CO_2 on photosynthesis and growth would be overwhelmed by effects of rising temperature on ecosystem respiration, the result of which would be that terrestrial ecosystems will become a source, rather than a sink for CO_2. This viewpoint appears to have been accepted by many in the modeling community since many models of the effects of global climate change on plant distribution and production have ignored possible direct effects of carbon dioxide on photosynthesis, growth and plant production (e.g. Esser, 1987).

The evidence that elevated CO_2 increases plant production is overwhelming. Kimball (1983) surveyed over 450 citations of experiments with the effects of CO_2 on plant growth. He determined that the mean value for the increase in production and the decrease in transpiration was about 34%. Cure and Acock (1986) found that the stimulation of growth was about 40%. In these studies, there was wide variation in the responses to CO_2, which probably represents real variability in the capacity of different species to respond to elevated CO_2 as well as interaction between the response to elevated CO_2 and different environmental conditions (Cure, 1985, Kimball *et al.*, 1983). At least 30 different effects have been attributed to increased CO_2 (Strain, 1985), most of which may be a consequence of increased carbon assimilation or reduced transpiration. Recently, attention has focused on the possibility that dark respiration may be reduced in elevated CO_2 (Amthor,1992; Bunce, 1990). The response of stomata to elevated CO_2 is largely independent of the effects of CO_2 on photosynthesis as shown by the fact that elevated CO_2 increase water use efficiency about the same amount in both C_3 and C_4 plants. Reduction of stomatal aperture occurs in all species whether or not photosynthesis is increased and there is no evidence that stomata acclimate to elevated CO_2 (Morison, 1987). Thus, response of plants to elevated CO_2 is almost certain to occur with consequent effects on the carbon flow in terrestrial ecosystems.

This review will briefly mention several key points about the known plant responses to elevated CO_2. I will also summarize the responses of a wetland plant community to long-term elevated CO_2 exposure because this study offers a somewhat different perspective on what might be expected to occur in native ecosystems as the atmospheric concentration of CO_2 continues it's all but inexorable rise than is projected from laboratory and greenhouse studies.

2. The effects of elevated CO_2 on photosynthesis.

The effect of CO_2 on photosynthesis arises from the fact that the major carboxylating enzyme, rubisCO, catalyzes either the reduction or oxidation of rubp depending on whether the substrate is CO_2 or O_2. Carboxylation of rubP supports the flow of carbon through the photosynthetic carbon reduction pathway, a major product of which is the substrate rubp. Oxidation of rubp supports the flow of carbon through the photosynthetic carbon oxidation cycle resulting in the loss of CO_2 in a process called photorespiration. The two catalytic reactions are competitive so increasing CO_2 increases photosynthetic carbon reduction at the expense of photorespiration. The effects of increased CO_2 on photosynthesis of individual leaves approximate the growth of plants in elevated CO_2 (Table 1).

The effect of elevated CO_2 on photosynthesis is very temperature sensitive (Long, 1991). As temperature rises in a background CO_2 concentration equal to the present normal ambient level, the oxidative reaction increases more rapidly than the reductive reaction resulting in more rapid increases in photorespiration. The increase in the O_2/CO_2 specificity of rubisCO with increasing temperature accounts for the increased stimulation of photosynthesis by elevated CO_2 with increasing temperature as proportionately more carbon is diverted from the photorespiratory pathway to the carbon reduction pathway.

The data in Table 1 indicate that the increase in growth and yield are correlated with photosynthesis in elevated CO_2. A commonly observed effect of chronic CO_2 exposure is that the full potential for carbon assimilation is reduced after plants grown in normal ambient CO_2 are exposed to elevated CO_2 for several days to weeks.

Table 1. The relative effect of increasing CO_2 concentration to twice the present level on photosynthesis, biomass production and water loss. Data are the means of percent change in a category by comparing plants in elevated CO_2 with those grown in normal ambient CO_2. Source: Cure and Acock (1986).

Species	Photosynth short term	Photosynth acclimated	Water loss	Biomass	Yield
Wheat	41	27	-17	31	35
Rice	42	46	-16	27	15
Soybean	78	42	-23	39	29
Cotton	60	13	-18	84	209
Weighted mean	(52)	(29)	(-23)	(30)	(41)
Corn	26	4	-26	9	29
Sorghum	-3	6	-27	9	-

The fact that down regulation of photosynthesis is commonly observed in controlled environment studies (Stitt, 1991) has been taken to mean that long-term exposure to elevated CO_2 would eventually yield little additional carbon accumulation in terrestrial ecosystems as happened in a study of the Arctic tundra ecosystem (Grulke *et al.*, 1990). As will be seen shortly, the results of this particular study are consistent with the repsonses of photosynthesis to temperature and sink regulation.

2.1. ACCLIMATION AND SOURCE/SINK CONTROL

I will now discuss the acclimation responses of photosynthesis to elevated CO_2. The reader is also referred to the interesting review by Stitt (1991) for more detail and a broader discussion of this topic.Some have expressed the view that plant growth and crop production in a future high CO_2 atmosphere will be limited by environmental factors other than CO_2 concentration (Goudriaan *et al.*, 1985, Kramer, 1981). Although the absolute magnitude of the effect of increased CO_2 on plant growth and production is determined by the availability of nitrogen, water and other nutrients, the relative effects of increased CO_2 are often larger under conditions of stress. High temperature, drought, atmospheric pollutants and salt stress tend to amplify the responses to CO_2 (Cure, 1985). The relatively larger response to CO_2 during some stresses arises largely from the effects of CO_2 on stomata which reduces the loss of water, improves water balance, and thereby reduces the sensitivity to salt stress and the susceptibility to atmospheric pollutants such as ozone. Elevated CO_2 also increases the temperature optimum for photosynthesis (Berry & Bjorkman, 1980, Pearcy & Bjorkman, 1983) and reduces heat stress (Kriedemann, Sward & Downton, 1976).

The idea that the stimulation of growth by rising CO_2 is unlikely to be sustained is due primarily to the observation that the stimulation of photosynthesis initially observed when plants grown in normal ambient CO_2 are exposed to higher levels declines after a period of days to weeks as shown in Table 1 (DeLucia, Sasek & Strain, 1985, Sasek, DeLucia & Strain, 1985, Tissue & Oechel, 1987). In the most extreme cases of the so-called acclimation response, there is no difference between the rate of photosynthesis of plants grown and tested at elevated CO_2 and the rate for those grown and tested at normal ambient CO_2 (Tissue & Oechel, 1987) but these are the exception. In most studies, even those which show the reduction of photosynthesis with time, the rate of CO_2 assimilation in elevated CO_2 is greater than the rate in normal ambient CO_2 (Long & Drake, In press, Pearcy & Bjorkman, 1983). In many experiments, plants grown in elevated CO_2 for an extended period have a lower rate of photosynthesis at a common intercellular CO_2 concentration than do plants grown in normal ambient CO_2 (DeLucia *et al.*, 1985, Downton, Björkman & Pike, 1980, Ehret & Jolliffe, 1985, Havelka *et al.*, 1984, Jarvis & Mansfield, 1980, Kaiser,

Weber & Sauer, 1983, Kaushal, Guehl & Aussenac, 1989, Kriedemann & Wong, 1984, Sage, Sharkey & Seemann, 1988, Sage, Sharkey & Seemann, 1989, Tolley & Strain, 1985). The reduction in photosynthetic capacity of plants grown in elevated CO_2 occurs after a period of exposure of some days or weeks duration. These responses are interpretable in terms of the control of the source of reduced carbon, the chloroplast, by the sink (Herold, 1980).

A key factor in the response of plants to long-term CO_2 exposure is the rate of transport of carbon away from the chloroplast. This is largely determined by the capacity of the sink for carbon and the rate of translocation of carbon between them (Stitt, 1986, Wardlaw, 1990). The sink is concieved as any pool of carbon outside the chloroplast envelope into which the products of photosynthesis are fed. Reducing the amount of non-photosynthetic tissues (the sink for photosynthate) reduces leaf level photosynthesis rates (Herold, 1980).

But reduction of photosynthetic capacity after long-term CO_2 exposure is not universally observed. In a number of studies, photosynthetic capacity either increased or did not change after growth in elevated CO_2 (Arp & Drake, 1991, Chen & Sung, 1990, Conroy, 1989, Fetcher *et al.*, 1988, Hollinger, 1987, Radin *et al.*, 1988, Radin *et al.*, 1987, Valle *et al.*, 1985a, Ziska *et al.*, In press). Growth of soybean in the elevated CO_2 treatment did not lead to significant down regulation of photosynthesis (Jones *et al.*, 1985, Valle *et al.*, 1985b). Comparisons of photosynthesis at 330 and 800 ppm for soybean canopies grown and tested in the different treatments showed that the CO_2 concentration during exposure accounted for 80% of the effect of high CO_2 on photosynthesis while acclimation during growth accounted for only 10% of the difference between treatments . In this field study of soybean it was shown in measurements of single leaf photosynthesis within the canopy that elevated CO_2 did not cause down regulation of photosynthesis (Valle *et al.*, 1985b). Soybean is considered to be sink limited during most of its life cycle except during canopy development and flowering. Radin <u>et al.</u> (1987) showed that there was no effect of elevated CO_2 exposure on photosynthetic capacity of field grown cotton during the reproductive and boll fill stages of field grown cotton.

In soybean, it appears that elevated CO_2 can increase canopy photosynthesis throughout the life cycle and not just during the production or flowers or filling of pods. Acock and Allen (1985) reported that canopy photosynthesis of soybean grown in elevated CO_2 was always higher than canopy photosynthesis for soybean in normal ambient CO_2 and that the rate for varied up to threefold depending on the developmental stage of the crop. During the week when pods began filling, the ratio of photosynthesis in elevated CO_2 (= 800 ppm) to photosynthesis in normal ambient CO_2 rose from 1.25 to 1.85 and then returned to 1.25 within the following week. However, Jones et al. (1984) report that at 70 days after planting, well past the time of pod filling, maximum photosynthesis in soybean canopy grown in 800 ppm CO_2 was 1.70 times the photosynthesis in a canopy of plants grown in 300 ppm CO_2 at midday.

Experiments carried out under conditions that limited the sink capacity, such as growing plants in small pots, also limited the response of the plants to elevated CO_2. Following the suggestion of Herold, (Herold, 1980), Wardlaw (1990), Thomas and Strain (1991), Ziska (1990) and others, Arp (1991a) constructed a correlation between pot size and the relative response of root:shoot and photosynthesis to elevated CO_2. Photosynthetic capacity of plants exposed to elevated CO_2 while growing in small pots (< 10L) was less than in plants grown in normal ambient CO_2 while photosynthetic capacity of plants grown in large pots or in the field was the same or greater than in plants grown in normal ambient CO_2. Root:shoot and rubisCO content were down regulated by exposure to elevated CO_2 while the plant was growing in conditions which reduced the capacity of the sink for carbon.

The effects of pot size are poorly understood but they are consistent with the hypothesis that the photosynthetic capacity is in part regulated by the carbon storage capacity. Wardlaw, (1990) suggested that the effects of elevated CO_2 are consistent with the control of the sink by environmental factors: when growth is limited the response to elevated CO_2 will be small and photosynthesis will be down regulated. When the sink is small, as happens at low temperature, there is very little response to CO_2 (Tissue & Oechel, 1987). Conversely, at high temperature, during leaf expansion and during pod fill or tuber expansion, the response of photosynthesis and growth to elevated CO_2 is relatively large (Valle *et al.*, 1985a). Environmental control over the sink size and the response to CO_2 has probably confounded the summaries of the effects of CO_2 offered in previous reviews (Kimball, 1983).

Both the content and activity of rubisCO are reduced by growth in elevated CO_2 (Allen *et al.*, 1989, Campbell, Allen & Bowes, 1988, Chen & Sung, 1990, Downton *et al.*, 1980, Havelka, Wittenbach & Boyle, 1984, Kriedemann & Wong, 1984, Peet, Huber & Patterson, 1986, Sage *et al.*, 1989, Vu, Allen & Bowes, 1983, Yelle *et al.*, 1989). Yelle (1989) has suggested that the down regulation of leaf photosynthesis in elevated CO_2 is caused by reduction of the rubisCO activity but the evidence is that photosynthesis of plants in elevated CO_2 may be high even when the activity of rubisCO is reduced (Chen & Sung, 1990, Sage *et al.*, 1989). However, the control of the rate of photosynthesis by rubisCO occurs only at CO_2 concentrations below present ambient (Stitt, 1991) and even with a 50% reduction in the rubisCO content, there would be a significant increase in photosynthesis by elevated CO_2 at temperatures above 30C (Long, 1991).

Long and Drake, (In press) showed that the responses of the metabolite pool sizes important in CO_2 assimilation are clearly effected by long-term exposure to elevated CO_2 and that the rooting volume has a strong impact on this response.

3. Effects of elevated CO_2 on respiration.

Elevated CO_2 appears to reduce dark respiration in a number of species. Short-term responses have been reported by Kidd (1916); Nilovskaya and Razoryonova (1968); Bunce (1990); and Amthor (1992). Long-term responses measured by comparing dark respiration of plants grown in elevated CO_2 with plants grown in present normal ambient CO_2 are reported in the studies by Ludwig et al. (1975); Gifford et al. (1985); Bunce (1990); Reuveni and Gale (1985); Spencer and Bowes (1986); Drake et al. (1989); Baker et al. (1992); and Wullshleger, et al. (1992). Both of these effects of elevated CO_2 were observed on dark respiration measured on excised samples of leaves and stem tissues of the C_3 sedge *Scirpus olneyi* grown in elevated CO_2 in the field. Both an instantaneous effect in which the dark respiration of all tissues was reduced as CO_2 concentration increased, and a long-term effect in which the tissues grown in elevated CO_2 had a lower rate of respiration than tissues grown in normal ambient CO_2 when they were compared at common conditions were observed (Drake *et al.*, 1989).

We have also found that the long-term aspect of the effect of increased CO_2 on dark respiration is time dependent: tissues tested early in the season during rapid leaf expansion show no difference in rate between treatments when tested at the same CO_2 concentration. The differences developed after about eight weeks of growth (unpublished data). On the other hand, effects of elevated CO_2 on dark respiration of rice were observed throughout the growth cycle (Baker, et al. 1992). The effects of elevated CO_2 on dark respiration are also seen in the net ecosystem gas exchange measurements. Respiration in the community level measurements are subject to many more complications than are those determined using excised tissues including the differences in live and dead tissues, the involvement of soil and root respiration, and the total amount of green tissues in the chambers. Nevertheless, the data show that ecosystem dark respiration, expressed per unit of green dry matter, was reduced in the elevated CO_2 treatments (Drake *et al.*, 1989).

The implications of a reduction in dark respiration of green plants and/or a reduction of the respiration rate of decomposers for the carbon budgets of terrestrial ecosystems are significant and deserve careful study to determine the mechanisms of this response as well as the potential effects on ecosystem processes. These observations on respiration suggest that the generally held assumptions about the effects of rising CO_2 and climate change on ecosystem respiration need to be reevaluated and that modeling exercises should include scenarios which take this reduced rate of carbon loss into account.

4. Long-term CO_2 effects on carbon accumulation in field studies.

All of the field experiments to date have confirmed the basic findings of many studies conducted in greenhouses and controlled environments that elevated CO_2 stimulates carbon accumulation in all C_3 species. But laboratory and field studies differ in that the stimulation of carbon uptake by a doubling of atmospheric CO_2 is greater in field experiments (Arp, 1991a).

Table 2. The effect of doubling atmospheric $[CO_2]$ on growth and carbon accumulation in field studies. The numbers in the final column on the right are the ratio of the response in elevated CO_2 to that in normal ambient CO_2 concentration.

Author, date	Species	Property	Elevated CO_2 / Present CO_2
Havelka et al., (1984)	Soybean, two varieties	Biomass	2.24, 1.28
Havelka et al. (1984)	Winter Wheat	Biomass	1.11
Rogers et al., (1984)	Soybean	Biomass	1.63
Overdieck and Lieth (1986)	Lollium perenne Trifolium repens	Biomass	1.50 1.50
Idso, Kimball & Allen, (1991)	Orange trees, 2.5 years	Volume of wood in trunk and branches	2.79
Grulke, et al. (1990)	Arctic tussock tundra	Net ecosystem CO_2 exchange	1.10
Drake & Leadley, (1991)	Salt marsh, 4 years	Net ecosystem CO_2 exchange	1.55
		Mean	1.63

This difference is in part due to the longer duration of field studies, the use of perennial species and to the multiple secondary effects on plants and on soil metabolic processes which additional carbon may bring about. The values in table 2 indicate that elevated CO_2 increases biomass production in field studies by about 60% which is considerably greater than the stimulation reported by Kimball (1983) and Cure and Acock (1986). The range of values in Table 2 is from about 11% in winter wheat to 179% in sour oranges. The causes of this variability are due in large part to the environmental conditions of the experiments: the stimulation of photosynthesis by elevated CO_2 is temperature dependent with the greatest stimulation occurring at highest temperatures.

There have been two studies of soybean canopy photosynthesis (Acock *et al.*, 1985, Jones, Allen & Jones, 1985, Jones *et al.*, 1985, Jones *et al.*, 1984), a study of canopy photosynthesis in rice (Allen *et al.*, 1989); a study of NCE with "model-ecosystems" using *Lolium* and *Trifolium* (Overdieck & Lieth, 1986); and a study in which NCE was measured in the Arctic tussock tundra (Oechel & Strain, 1985). All except the open top chamber study used closed chambers in which temperature, humidity and CO_2 concentration were controlled in an outdoor setting and all tracked environmental temperature. In the two crop studies the rooting environment was not as great as in the field studies although there appears to be no evidence that this restricted the response to elevated CO_2. There have also been other field studies of the photosynthetic properties of soybean (Ackerson, Havelka & Boyle, 1984, Havelka *et al.*, 1984, Sionit *et al.*, 1984) and of cotton (Kimball *et al.*, 1986, Mauney *et al.*, 1979, Radin *et al.*, 1987) exposed to elevated CO_2 throughout the growing season.

Baker et al. (1992) studied the effect of three different day-night temperature regimes, 28-21 C, 34-27 C, and 40-33 C on the effects of elevated CO_2 on rice. Although the highest absolute values for canopy photosynthesis occurred in elevated CO_2 at the lower two temperature treatments, elevated CO_2 had the greatest relative effect on canopy photosynthesis at the highest growth temperatures (40-33 C). Similarly, the initial slope of the light response curve was increased in elevated CO_2 compared to normal ambient CO_2 and the greatest relative increase occurred at the highest temperature. In soybean, a 5 C temperature difference (28-33 C) was undetectable either at 330 ppm or at 800ppm: the light response curves for both temperature treatments during full day measurements were identical (Jones *et al.*, 1985).

The available evidence from ecosystem studies supports the conclusion that plants growing in widely different temperature regimes should have different responses to CO_2. There was a relatively small response of canopy photosynthesis to a doubling of the normal ambient CO_2 in a study of the Arctic tussock tundra (Oechel & Strain, 1985, Grulke et al. 1990) and plants quickly acclimated by down regulation of photosynthesis (Tissue & Oechel, 1987). Canopy photosynthesis in a mixture of perennial ryegrass (Lolium perenne) and white clover(Trifolium repens) in Osnabruck, Germany, increased 25-45% in 600 ppm CO_2 compared to 340 ppm (Overdieck & Lieth, 1986). Net ecosystem gas exchange per unit ground area in a plant community dominated by the C_3 sedge, *Scirpus olneyi*, was stimulated about 50%. Average daytime temperature in this community was 28-30 C. Rice grown in out-door controlled environment chambers at 40/33 C day/night temperature showed an average 54% stimulation of canopy photosynthesis (Allen *et al.*, 1989).

5. The effect of long-term elevated CO_2 exposure on a Chesapeake Bay wetland.

The longest running field experiment with elevated CO_2 is being conducted on three plant communities in a brackish Chesapeake bay wetland. Results from the first four years of this study have recently been reviewed (Drake In press) and only the highlights will be discussed here. A summary of the effects of elevated CO_2 on various plant and ecosystem processes is shown in Table 3. There were many interacting responses in the C_3 plant community to the elevated CO_2 treatment but very little response in the community composed entirely of the C_4 grass *Spartina patens*.

This study has provided no evidence after 4 years exposure that photosynthesis in the C_3 sedge *Scirpus olneyi* has down regulated (Arp & Drake, 1991, Arp *et al.*, In press, Drake *et al.*, In press, Ziska *et al.*, 1990). Data on quantum yield and photosynthetic capacity of *Scirpus olneyi* measured in plants after four years of growth in elevated CO_2 provide no evidence that down regulation of metabolic aspects of photosynthesis have occurred (Arp & Drake, 1991, Long & Drake, In press). In fact, data in August and September 1990 indicated that photosynthetic capacity increased after acclimation to high CO_2 (Arp & Drake, 1991).

Ecosystem level carbon balance depended almost entirely on the photosynthetic responses to elevated CO_2 in the dominant species. During the first four years of exposure, elevated CO_2 had the following effects on mono-specific stands of the C_3 sedge, *Scirpus olneyi*: increased quantum yield and photosynthetic capacity, reduced dark respiration, increased numbers of shoots, roots and rhizomes, reduced nitrogen concentration of all tissues, increased nitrogen fixation and increased ecosystem carbon accumulation. In a mixed community of the sedge and C_4 grass species, *Spartina patens* and *Distichlis spicata*, biomass of the C_3 component increased over 100% and this was accompanied by decreased biomass in the C_4 component of the community. Elevated CO_2 reduced water loss, increased water potential and delayed senescence in all three species.

Many factors responded to elevated CO_2 to increase carbon accumulation in the pure C_3 community dominated by *Scirpus olneyi* including sustained, high photosynthetic capacity, decreased respiration, delayed senescence and increased allocation of carbon to roots and rhizomes. The complex interaction of these diverse responses suggests that the direct effects of rising atmospheric CO_2 on ecosystem processes may be greater than previously anticipated.

Table 3. The effect of elevated CO_2 on plant and ecosystem processes in a long-term study of the effects of elevated CO_2 concentration on a Chesapeake Bay wetland. Symbols mean: elevated CO_2 treatment decreases (-), increases (+), or has no effect (0). Source: Drake (In press).

Category	Scirpus C_3	Spartina C_4
Evapotranspiration	-	-
Water potential	+	+
Photosynthesis	+	0
Shoot number	+	0
Relative root growth	+	0
Nitrogenase activity	+	0
Nitrogen fixation	+	0
Nitrogen concentration	-	0
Senescence	-	0
Decomposition	-	0
Herbivory	-	0
Fungal infection	-	+
Methane production	+	0

Many of the results of this long-term ecosystem study might be predicted from the known responses of plants to elevated CO_2. Evapotranspiration was reduced resulting in increased water potential and WUE in both types of species; photosynthesis was increased in the C_3 plant but not in the C_4 species; both root and shoot growth increased in the elevated CO_2 treatment in the C_3 species and the relative effect on root growth was greater than the relative effect on shoot growth; in the mixed community the C_3 plants increased relative to the C_4 plants; and nitrogen concentration decreased and the C/N increased in the C_3 but not the C_4 plant. These effects essentially confirm the basis for the expected response of vegetation to rising atmospheric CO_2 as determined by a great many studies from greenhouse and controlled environment experiments.

However, some results obtained in this study differ from those from controlled environments in the following unexpected ways: photosynthetic capacity increased and there was no evidence of down regulation (Arp & Drake, 1991); there was evidence of real reduction in water loss owing to the relatively small increase in above-ground biomass production; dark respiration of one of the three species was reduced (Drake *et al.*, 1989); nitrogen fixation increased in the C_3 but not the C_4 species (Dakora, unpublished data); and herbivory by insects and fungi decreased (Thompson & Drake, In review).

6. What we need to know

Most of the data on the effects of elevated CO_2 was obtained in relatively short-term experiments using crop species. The greatest need is for studies in the field using native species of trees and which last for enough time to observe the effects of elevated CO_2 on all aspects of ecosystem carbon balance as well as on plant growth and photosynthesis.

It is necessary to model the responses of ecosystems to rising CO_2 but this effort will continue to be rather anemic without a more complete knowledge of acclimation of photosynthesis and the responsess of dark respiration to elevated CO_2. As illustrated in this paper, there is clear evidence that both of these processes are altered by prolonged growth in elevated CO_2, but variation among the different experimental approaches, methods of measurement and the basis for expressing the results has led to much speculation regarding the occurrence and meaning of acclimation of these two fundamental processes. It is not yet clear whether the control of acclimation in the field occurs to the same extent as it does under laboratory and controlled environment conditions.

Other important but unresolved issues of the responses of crops and ecosystems to rising CO_2 in nutrient limited conditions. Specific issues are: growth requirement of nutrients in altered CO_2; the possibility that extra carbon may stimulate N fixation of either free living N fixers or those associated with non-leguminous plants; the capacity of plants with additional rooting volume to mine extra N from the soil; the absorption efficiency for N in a nutrient deficient condition. The available data on the interaction between elevated CO_2 and nutrient deficiency may be complicated by experiments in which many nutrients were varied along with the one in question as for example when N is studied by varying the strength of Hoaglands solution which necessarily varies all other nutrients.

The changes in nitrogen content of plant tissues as the

concentration of CO_2 in the air rises cannot be simply a dilution of the nitrogen content because this phenomenon occurs even when nitrogen is supplied in abundance (Conroy, In press). The reduction in nitrogen content during growth in elevated CO_2 appears to reflect a profound change in nitrogen metabolism and should not be viewed necessarily as an indication of nitrogen deficiency.

7. Summary

The case for the stimulation of vegetative growth in elevated CO_2 has been made in the following ways:

7.1. STUDIES IN CONTROLLED ENVIRONMENTS AND GREENHOUSES

The record of experiments since the 19th century overwhelming demonstrates that under virtually all experiments and with many species of C_3 plants, **elevated CO_2 increases growth in most** C_3 plants. C_3 plants comprise over 95% of the species of higher plants and 100% of the trees. While there is evidence of down-regulation of photosynthesis in many species, this effect may be in part an artifact of experimental approach in which the growth environment constrains the available sink for carbon. Moreover, it is not clear that down-regulation of photosynthesis occurs under natural conditions or that it would occur with slowly increasing atmospheric CO_2. In any case elevated CO_2 almost always leads to an increase in growth and yield in crops in the range of 34%-42% for a doubling of the present normal ambient CO_2 (Kimball, 1983; Cure and Acock, 1986).

7.2. FIELD STUDIES

In studies of whole plants in the field including collectively crops, wild perennial, and trees the result is qualitatively the same: elevated CO_2 increased growth. However, the growth enhancement in the field is greater than from laboratory studies, primarily because plants in the field have unrestricted soil volumes within which to accumulate the additional carbon. The greatest increases occurred at highest temperatures and the range of responses was from about 10% in the Arctic study (Grulke *et al.*, 1990) to over 200% in a three year study of sour orange trees in the desert (Idso *et al.*, 1991). This record is consistent with the responses of photosynthesis to the combined effects of elevated CO_2 and temperature. The available data from the few field studies that have been done do not give clear support the view that the stimulation of growth by elevated CO_2 is transitory as was shown in the Arctic study (Grulke et al., 1990). In fact, the Arctic study is the only one so far accomplished in which complete homeostatic adjustment to elevated CO_2 occurred. Results from the few field studies tend to suggest that the effects of elevated CO_2 may be greater than has been estimated from the responses of plants in controlled environments.

7.3. ECOSYSTEM CARBON DIOXIDE EXCHANGE.

The longest running ecosystem experiment to date is a five year study of ecosystem carbon accumulation in three Chesapeake Bay wetland communities. Carbon accumulation, measured by ecosystem gas exchange, in the mono-specific stands of the C_3 sedge, *Scirpus olneyi*, has increased 50-100% throughout the course of the study. This increased carbon accumulation resulted from a number of factors including the sustained stimulation of photosynthesis, an increase in aboveground production, a decrease in dark respiration, delayed senescence, and increased allocation of carbon to more slowly decomposing belowground plant parts.

The direct effects of rising CO_2 on vegetation may turn out to have impacts on crop and ecosystem production more significant than the effects of climate change. The effect of rising CO_2 on photosynthesis is very temperature sensitive: in single leaves the effect varies from about 10% increase in photosynthesis at 10C to over 100% at 40 C for an increase in CO_2 from the pre-industrial level of 270 ppm to about 800 ppm. Experimental evidence suggests that rising CO_2 will increase ecosystem production most in warm climates.

It has recently been shown that there is a significant artifact in the data on the effects of elevated CO_2 in experiments in which plants grown in small pots showed a rapid acclimation response reflected in down-regulation of photosynthesis. Plants grown in large pots or in the field tend to show either no regulation of photosynthesis or an increase in the capacity to assimilate CO_2. The few field studies which have been done also suggest that there are many important unanswered questions regarding the role of rising CO_2 in plant production including regulation of evapotranspiration, photosynthesis, respiration and carbon partitioning.

8. References

Ackerson, R.C., Havelka, U.D. & Boyle, M.G. (1984) CO_2-enrichment effects on soybean physiology. II. Effects of stage-specific CO_2 exposure. *Crop Science* 24, 1150-1154.

Acock, B. & Allen, L.H., Jr. (1985) Crop responses to elevated carbon dioxide concentrations. In *Direct Effects of Increasing Carbon Dioxide on Vegetation* (ed. B.R. Strain & J.D. Cure), pp. 53-97. United States Department of Energy, Washington, D.C.

Allen, L.H., Jr., Boote, K.J., Jones, P.H., Jones, J.W., Rowland-Bamford, A.J., Bowes, G., Graetz, D.A. & Reddy, K.R. (1989) Temperature and CO_2 effects on rice: 1988. United States Department of Energy, Office of Energy Research, Carbon Dioxide Research Division, Washington, DC.

Amthor, J.S., Koch, G.W., and A.J.Bloom. (1992) CO_2 inhibits respiration in leaves of Rumex crispus L. Plant Physiology 98:

Arp, W.J. (1991a) Effects of source-sink relations on photosynthetic acclimation to elevated CO_2. *Plant, Cell and Environment* **14**, 869-875.

Arp, W.J. (1991b) Vegetation of a North American salt marsh and elevated atmospheric carbon dioxide. Doctoral thesis, Free University of Amsterdam,

Arp, W.J. & Drake, B.G. (1991) Increased photosynthetic capacity of *Scirpus olneyi* after 4 years of exposure to elevated CO_2. *Plant, Cell and Environment* **14**, 1003-1006.

Arp, W.J., Drake, B.G., Pockman, W.T., Curtis, P.S. & Whigham, D.F. (In press) Effects of four years exposure to elevated atmospheric CO_2 on competition between C_3 and C_4 salt marsh plant species. *Vegetatio*

Baker, J.T., Laugel, F., Boote, K.J. and L.H. Allen, Jr. 1992. Effects of daytime carbon dioxide concentration on dark respiration in rice. Plant Cell and Environment 15:231-239.

Berry, J. & Bjorkman, O. (1980) Photosynthetic response and adaptation to temperature in higher plants. **31**, 491-543.

Bunce, J.A. (1990) Short and long term inhibition of respiratory carbon dioxide efflux by elevated carbon dioxide. *Annals of Botany* **65**, 637-642.

Campbell, W.J., Allen, L.H., Jr. & Bowes, G. (1988) Effects of CO_2 concentration on rubisCO activity, amount, and photosynthesis in soybean leaves. *Plant Physiology* **88**, 1310-1316.

Chen, J.J. & Sung, J.M. (1990) Gas exchange rate and yield responses of Virginia-type peanut to carbon dioxide enrichment. *Crop Science* **30**, 1085-1089.

Conroy, J. (1989) Influence of high CO_2 on *Pinus radiata*. Ph.D. Thesis, Macquarie Unversity, School of Biological Sciences, Canberra, Australia

Conroy, J.D. In press. Influence of elevated atmospheric CO_2 concentrations on plant nutrition. Australian Journal of Biological Sciences.

Cure, J.D. (1985) Carbon dioxide doubling responses: a crop survey. In *Direct effects of increasing carbon dioxide on vegetation* (ed. B.R. Strain & J.D. Cure), pp. 99-116. United States Department of Energy, Office of Energy Research, Carbon Dioxide Research Division, Washington, DC.

Cure, J.D. & Acock, B. (1986) Crop responses to carbon dioxide doubling: a literature survey. *Agricultural and Forest Meteorology* **38**, 127-145.

DeLucia, E.H., Sasek, T.W. & Strain, B.R. (1985) Photosynthetic inhibition after long-term exposure to elevated levels of atmospheric carbon dioxide. *Photosynthesis Research* **7**, 175-184.

Downton, W.J.S., Björkman, O. & Pike, C.S. (1980) Consequences of increased atmospheric concentrations of carbon dioxide for growth and photosynthesis of higher plants. In *Carbon Dioxide and Climate: Australian Research* (ed. G.I. Pearman), pp. 143-151. Australian Academy of Science, Canberra.

Drake, B.G. In press. A field study of the effects of elevated CO_2 on ecosystem processes in a Chesapeake Bay wetland. Australian Journal of Biological Sciences.

Drake, B.G., Arp, W.J., Balduman, L., Curtis, P.S., Johnson, J., Kabara, D., Leadley, P.W., Pockman, W.T., Seliskar, D., Sutton, M.L., Whigham, D., Ziska, L. & (1989) Effects of elevated CO_2 on Chesapeake Bay wetlands. IV. Ecosystem and whole plant responses. April-November 1988. United States Department of Energy, Office of Energy Research, Carbon Dioxide Research Division, Washington, DC.

Drake, B.G., Arp, W.J., Long, S.P. & Lawlor, D.W. (in press) Photosynthesis of the C_3 sedge, *Scirpus olneyi*, after long-term exposure to elevated CO_2 in open top chambers in the field. In *Trends in Photosynthesis Research* (ed. J.B. Barber, H. Medrano & M.G. Guerrero), Intercept, LTD, Andover, UK.

Drake, B.G. & Leadley, P.W. (1991) Canopy photosynthesis of crops and native plant communities exposed to long-term elevated CO_2. *Plant, Cell and Environment* **14**, 853-860.

Drake, B.G., Leadley, P.W., Arp, W.J., Nassiry, D. & Curtis, P.S. (1989) An open top chamber for field studies of elevated atmospheric CO_2 concentration on saltmarsh vegetation. *Functional Ecology* **3**, 363-371.

Ehret, D.L. & Jolliffe, P.A. (1985) Photosynthetic carbon dioxide exchange of bean plants grown at elevated carbon dioxide concentrations. *Canadian Journal of Botany* **63**, 2026-2030.

Esser, G. (1987) Sensitivity of global carbon pools and fluxes to human and potential climatic impacts. Tellus, 39B,No.3.

Fetcher, N., Jaeger, C.H., Strain, B.R. & Sionit, N. (1988) Long-term elevation of atmospheric CO_2 concentration and the carbon exchange rates of saplings of *Pinus taeda* L. and *Liquidambar styraciflua* L. *Tree Physiology* **4**, 255-262.

Gifford, R.M., Lambers, H. & Morison, J.I.L. (1985) Respiration of crop species under CO_2 enrichment. *Physiologia Plantarum* **63**, 351-356.

Goudriaan, J. & Ketner, P. (1984) A simulation study for the global carbon cycle, including man's impact on the biosphere. **6**, 167-192.

Goudriaan, J., van Laar, H.H., van Keulen, H. & Louwerse, W. (1985) Photosynthesis, CO_2 and Plant Production. In *Wheat Growth and Modeling* (ed. W. Day & R.K. Atkin), pp. 107-122.

Grulke, N.E., Riechers, G.H., Oechel, W.C., Hjelm, U. & Jaeger, C. (1990) Carbon balance in tussock tundra under ambient and elevated atmospheric CO_2. *Oecologia* **83**, 485-494.

Havelka, U.D., Ackerson, R.C., Boyle, M.G. & Wittenbach, V.A. (1984) CO_2-enrichment effects on soybean physiology. I. Effects of long-term CO_2 exposure. *Crop Science* **24**, 1146-1150.

Havelka, U.D., Wittenbach, V.A. & Boyle, M.G. (1984) CO_2-enrichment effects on wheat yield and physiology. *Crop Science* **24**, 1163-1168.

Herold, A. (1980) Regulation of photosynthesis by sink activity - the missing link. *New Phytologist* **86**, 131-144.

Hollinger, D.Y. (1987) Gas exchange and dry matter allocation responses to elevation of atmospheric CO_2 concentration in seedlings of three species. *Tree Physiology* **3**, 193-202.

Houghton, R.A. (1987) Terrestrial metabolism and atmospheric CO_2 concentrations. Independent geophysical and ecological estimates of seasonal carbon flux address global change. *BioScience* **37**, 672-678.

Idso, S.B., Kimball, B.A. & Allen, S.G. (1991) Net photosynthesis of sour orange trees maintained in atmospheres of ambient and elevated CO_2 concentration. *Agricultural and Forest Meteorology* **54**, 95-101.

Jarvis, R.G. & Mansfield, T.A. (1980) Reduced stomatal responses to light,carbon dioxide and abscisic acid in the presence of sodium ions. **3**, 279-283.

Jones, P., Allen, L.H., Jr. & Jones, J.W. (1985) Responses of soybean canopy photosynthesis and transpiration to whole day temperature changes in different CO_2 environments. *Agronomy Journal* **77**, 242-249.

Jones, P., Allen, L.H., Jr., Jones, J.W. & Valle, R. (1985) Photosynthesis and transpiration responses of soybean canopies to short-and long-term CO_2 treatments. *Agronomy Journal* **77**, 119-126.

Jones, P.H., Allen, L.H., Jr., Jones, J.W., Boote, K.J. & Campbell, W.J. (1984) Soybean canopy growth, photosynthesis, and transpiration responses to whole-season carbon dioxide enrichment. *Agronomy Journal* **76**, 633-637.

Kaiser, W.M., Weber, H. & Sauer, M. (1983) Photosynthetic capacity, Osmotic response and solute content of leaves and chloroplasts from Spinacia oleracea under salt stress. **113**, 15-27.

Kaushal, P., Guehl, J.M. & Aussenac, G. (1989) Differential growth response to atmospheric carbon dioxide enrichment in seedlings of *Cedrus atlantica* and *Pinus nigra* ssp. *Laricio* var. *Corsicana*. *Canadian Journal of Botany* **19**, 1351-1358.

Kidd, F. (1916) The controlling influence of carbon dioxide. Part III. The retarding effect of carbon dioxide on respiration. **89**, 136-156.

Kimball, B.A. (1983) Carbon dioxide and agricultural yield: an assemblage and analysis of 430 prior observations. *Agronomy Journal* **75**, 779-788.

Kimball, B.A., Mauney, J.R., Guinn, G., Nakayama, F.S., Pinter, P.J., Jr. Clawson, K.L., Reginata, R.J., Idso, S.B., Mitchell, S.T., Macdonald, R., Seay, R., Schnell, S., Brummet, D., Pros, H., Savoy, B., Martinex, J., Tyrell, G. & Anderson, M. (1983) Effects of increasing atmospheric CO_2 on yield and water use of crops. Annual Report, U.S. Water Conservation Laboratory

Kimball, B.A., Mauney, J.R., Radin, J.W., Nakayama, F.S., Idso, S.B., Hendrix, D.L. & others (1986) Effects of increasing atmospheric CO_2 on the growth, water relations, and physiology of plants grown under optimal and limiting levels of water and nitrogen. In *Direct Effects of Increasing Carbon Dioxide on Vegetation* (ed. B.R. Strain & J.D. Cure), pp. 187-204. United States Department of Energy, Office of Energy Research, Carbon Dioxide Research Division , Washington, DC.

Kramer, P.J. (1981) Carbon dioxide concentration, photosynthesis, and dry matter production. *BioScience* **31**, 29-33.
Kriedemann, P.E., Sward, R.J. & Downton, W.J.S. (1976) Vine response to carbon dioxide enrichment during heat therapy. *Australian Journal of Plant Physiology* 3, 605-618.
Kriedemann, P.E. & Wong, S.C. (1984) Growth response and photosynthetic acclimation to CO_2: Comparitive behavior in two C_3 crop species. *Acta Horticulturae* **162**, 113-120.
Long, S.P. (1991) Modification of the response of photosynthetic productivity to rising temperature by atmospheric CO_2 concentrations: Has its importance been underestimated? *Plant, Cell and Environment* 14, 729-739.
Long, S.P. & Drake, B.G. (1991) Effect of the long-term elevation of CO_2 concentration in the field on the quantum yield of photosynthesis of the C_3 sedge, *Scirpus olneyi*. *Plant Physiology* **96**, 221-226.
Long, S.P. & Drake, B.G. (In press) Photosynthetic CO_2 assimilation and rising atmospheric CO_2 concentrations. In *Crop Photosynthesis: Spatial and Temporal Determinants* (ed. N.R. Baker & H. Thomas), Elsevier Science Publishers, Amsterdam.
Ludwig, R., Charles-Edwards, D.A. & Withers, A.C. (1975) Tomato leaf photosynthesis and respiration in various light and carbon dioxide environments. In *Environmental and Biological Control of Photosynthesis* (ed. R. Marcelle), pp. 29-36. W. Junk, The Hague.
Mauney, J.R., Guinn, G., Fry, K.E. & Hesketh, J.D. (1979) Correlation of photosynthesis carbon dioxide uptake and carbohydrate accumulation in cotton, soybean, sunflower and sorghum. *Photosynthetica* 13, 260-266.
Morison, J.I.L. (1987) Intercellular CO_2 concentration and stomatal response to CO_2. In *Stomatal Function* (ed. E. Zeiger, G.D. Farquhar & I.R. Cowan), pp. 229-251. Stanford University Press, Stanford, California.
Nilovskaya, N.T. & Razoryonova, T.A. (1968) Respiration rate of vegetable plants at various partial pressures of carbon dioxide. **5**, 876-883.
Oechel, W.C. & Strain, B.R. (1985) Native species responses to increased atmospheric carbon dioxide concentration. In *Direct Effects of Increasing Carbon Dioxide on Vegetation* (ed. B.R. Strain & J.D. Cure), pp. 117-154. United States Department of Energy, Office of Energy Research, Carbon Dioxide Research Division, Washington, DC.
Overdieck, D. & Lieth, H. (1986) Final report of the project: Long-term effects of increased atmospheric CO_2 concentration level on terrestrial plants in model-ecosystems. Universitat Osnabruck, Fachbereich Biologie/Chemie, Arbeitsgruppe Okologie, Osnabruck.
Pearcy, R.W. & Bjorkman, O. (1983) Physiological effects. In *CO_2 and Plants:The Response of Plants to Rising Levels of Atmospheric Carbon Dioxide* (ed.E.R. Lemon), pp. 65-105. Westview Press, Boulder, CO.
Peet, M.M., Huber, S.C. & Patterson, D.T. (1986) Acclimation to high CO_2 in monoecious cucumbers II. Carbon exchange rates, enzyme activities, and starch and nutrient concentrations. *Plant Physiology* **80**, 63-67.

Radin, J.W., Hartung, W., Kimball, B.A. & Mauney, J.R. (1988) Correlation of stomatal conductance with photosynthetic capacity of cotton only in a CO_2-enriched atmosphere: Mediation by abscisic acid? **88**, 1058-1062.

Radin, J.W., Kimball, B.A., Hendrix, D.L. & Mauney, J.R. (1987) Photosynthesis of cotton plants exposed to elevated levels of carbon dioxide in the field. *Photosynthesis Research* **12**, 191-203.

Reuveni, J. & Gale, J. (1985) The effect of high levels of carbon dioxide on dark respiration and growth of plants. *Plant, Cell and Environment* **8**, 623-628.

Rogers, H.H., Cure, J.D., Thomas, J.F. & Smith, J.M. (1984) Influence of elevated CO_2 on growth of soybean plants. *Crop Science* **24**, 361-366.

Sage, R.F., Sharkey, T.D. & Seemann, J.R. (1988) The in-vivo response of the ribulose-1,5-bisphosphate carboxylase activation state and the pool sizes of photosynthetic metabolites to elevated CO_2 in Phaseolus vulgaris L. *Planta* **174**, 407-416.

Sage, R.F., Sharkey, T.D. & Seemann, J.R. (1989) Acclimation of photosynthesis to elevated CO_2 in five C_3 species. *Plant Physiology* **89**, 590-596.

Sasek, T.W., DeLucia, E.H. & Strain, B.R. (1985) Reversibility of photosynthetic inhibition in cotton after long-term exposure to elevated CO_2 concentrations. *Plant Physiology* **78**, 619-622.

Sionit, N., Rogers, H.H., Bingham, G.E. & Strain, B.R. (1984) Photosynthesis and stomatal conductance with CO_2-enrichment of container-and field-grown soybeans. *Agronomy Journal* **76**, 447-451.

Spencer, W. & Bowes, G. (1986) Photosynthesis and growth of water hyacinth under CO_2 enrichment. *Plant Physiology* **82**, 528-533.

Stitt, M. (1986) Limitation of photosynthesis by carbon metabolism, I. evidence for excess electron trasport capacity in leaves carrying out photosynthesis in saturating light and CO_2. **81**, 1115-1122.

Stitt, M. (1991) Rising CO_2 levels and their potential significance for carbon flow in photosynthetic cells. *Plant, Cell and Environment* **14**, 741-762.

Strain, B.R. (1985) Physiological and ecological controls on carbon sequestering in terrestrial ecosystems. *Biogeochemistry* 1, 219-232.

Thomas, R.B. & Strain, B.R. (1991) Root restriction as a factor in photosynthetic acclimation of cotton seedlings grown in elevated carbon dioxide . *Plant Physiology* **96**, 627-634.

Thompson, G.B. and B.G. Drake. In review. Herbivory and fungal infection of a C_3 plant are reduced and the fungal infection of a C_4 plany is increased by elevated CO_2 in open top chambers in the field.

Tissue, D.T. & Oechel, W.C. (1987) Response of *Eriophorum vaginatum* to elevated CO_2 and temperature in the Alaskan tussock tundra. *Ecology* **68**, 401-410.

Tolley, L.C. & Strain, B.R. (1985) Effects of CO_2 enrichment and water stress on gas exchange of *Liquidambar styraciflua* and *Pinus taeda* seedlings grown under different irradiance levels. *Oecologia* **65**, 166-172.

Valle, R., Mishoe, J.W., Campbell, W.J., Jones, J.W. & Allen, L.H., Jr. (1985a) Photosynthetic responses of 'Bragg' soybean leaves adapted to different CO_2 environments. *Crop Science* **25**, 333-339.
Valle, R., Mishoe, J.W., Jones, J.W. & Allen, L.H., Jr. (1985b) Transpiration rate and water use efficiency of soybean leaves adapted to different CO_2 environments. *Crop Science* **25**, 477-482.
Vu, C.V., Allen, L.H., Jr. & Bowes, G. (1983) Effects of light and elevated atmospheric CO_2 on the ribulose bisphosphate carboxylase activity and ribulose bisphosphate level of soybean leaves. *Plant Physiology* **73**, 729-734.
Wardlaw, I.F. (1990) Tansley Review No. 27. The control of carbon partitioning in plants. *New Phytologist* **116**, 341-381.
Woodwell, G.M. (1987) Forests and climate: Surprises in store. *Oceanus* 29, 71-75.
Wullschleger, S.D., Norby, R.J. and D.L.Hendrix. 1992. Carbon exchange rates, chlorophyll content, and carbohydrate status of two forest tree species exposed to carbon dioxide enrichment. Tree Physiology 10:21-31.
Yelle, S., Beeson, R.C., Jr., Trudel, M.J. & Gosselin, A. (1989) Acclimation of two tomato species to high atmospheric CO_2. I. Sugar and starch concentrations. *Plant Physiology* **90**, 1465-1472.
Ziska, L.H., Drake, B.G. & Chamberlain, S. (1990) Long-term photosynthetic response in single leaves of a C_3 and C_4 salt marsh species grown at elevated atmospheric CO_2 *in situ*. *Oecologia* **83**, 469-472.
Ziska, L.H., Hogan, K.P., Smith, A.P. & Drake, B.G. (In press) Growth and photosynthetic response of nine tropical species with long-term exposure to elevated carbon dioxide. *Oecologia*

FIELD MEASUREMENTS OF CO_2 ENHANCEMENT AND CLIMATE CHANGE IN NATURAL VEGETATION

BOYD R. STRAIN AND RICHARD B. THOMAS
Department of Botany
Duke University
Durham NC 27706 USA

ABSTRACT: It is generally assumed that healthy, natural ecosystems have the potential to sequester carbon under favorable environmental conditions. There is also evidence that CO_2 acts as a plant fertilizer. It is of interest to know if these assumptions are valid and how natural systems might respond under future scenarios of CO_2 increase and possible climate changes. Few measurements of the effects of CO_2 and global climate change have been made on "natural" ecosystems under realistic field conditions. Most measurements have been conducted in the synthetic environments of totally controlled greenhouses and growth chambers. Several lines of evidence indicate that controlled environment studies using plants growing in pots induce experimental artifacts that reduce confidence in the use of results for prediction of future global responses. Open top chambers are being used in several autecological field studies in an attempt to obtain more realistic field environments. A few field microcosm studies have been completed and a system for the free air release of CO_2 has been applied in cotton fields. Unfortunately, the requirement of large amounts of CO_2 and financial restrictions have precluded the initiation of larger scale field studies in natural vegetation. This paper lists and summarizes the best field studies available but draws heavily on studies from artificial environments and conditions in an attempt to summarize knowledge of global environmental change on forests and other non-agricultural ecosystems. Finally the paper concludes that there is a need for the development and application of equipment for field measurements in several representative natural ecosystems and makes specific recommendation of the creation of a tropical research center.

1. Introduction

There is irrefutable evidence that atmospheric carbon dioxide is limiting to plant productivity at the current mean global concentration of 355 ppm (Strain and Cure 1985, Cure and Acock 1986, Strain 1992, Poorter 1992). There is considerable uncertainty, however, about the response of natural, "unmanaged" ecosystems to CO_2 increase and climate change (Sionit and Kramer 1985, Kramer and Sionit 1987, Jarvis 1989, Bazzaz 1990, Arp and Drake 1991, Oechel and Billings 1992). Although several attempts have

Water, Air, and Soil Pollution **64**: 45–60, 1992.
© 1992 *Kluwer Academic Publishers.*

been made to model CO_2 and climate change effects on forests (Botkin 1977, Shugart et al. 1986, Solomon 1986, Pastor and Post 1988), shortcomings of the application of empirical stand models for prediction into uncertain futures have been described in detail and a recommendation made for mechanistic modeling using modular, generic models (Reynolds and Acock 1985). These problems and other issues specific to forests were considered at a workshop in 1984 (Shands and Hoffman 1986). Most of the uncertainty considering natural ecosystem responses derives from the fact that few realistic field experiments have been attempted. Most evidence available has been obtained from the synthetic environments of controlled chambers using plants grown and studied in an autecological fashion (Rogers et al. 1983a, Williams et al. 1986, Norby et al. 1986a, 1987, Fetcher et al. 1988, Conroy et al. 1990, Ziska et al. 1991). Recent studies, however, indicate that the restrictive conditions of potted plants growing in controlled environment chambers or greenhouses induce artifacts in plant response that prevent the extrapolation of results to real time field conditions (Thomas and Strain 1991, Arp 1991).

This paper will review the information available from the few field studies that have been published and will draw on selected autecological studies that provide evidence that can be extrapolated to open field conditions. The review will reveal that the primary producers of unmanaged ecosystems have the potential to absorb and store carbon from the atmospheric reservoir of CO_2. Resource limitations (e.g. soil fertility, drought, air and soil temperature), however place important restrictions on the magnitude of carbon sequestration to be realized in net primary productivity (NPP). In addition, the energy and carbon requirements of animals and microorganisms reduce the realized net ecosystem productivity (NEP) to values signigificantly below potential NPP. Known responses of primary producers to CO_2 increase and climate change will be itemized. A conclusion will be drawn that extensive field experimentation will be required to obtain reliable information on carbon sequestering and on the natural carbon cycle of "unmanaged" ecosystems. Available and percieved technologies appropriate for application to the problem are summarized from the more extensive recommendations made in Strain (1991b).

2. Summary of Studies Completed in Model Ecosystems (Microcosms)

2.1. ALASKAN TUNDRA

A microcosm approach was applied by W.D. Billings and his colleagues in a series of studies beginning in 1980. They removed cores from the permafrosted tundra of Point Barrow in March when the ground was still ice covered. The cores were air frieghted to the Phytotron at Duke University were they were placed in controlled environments and then thawed in simulated spring environments. Environmental conditions were established representing 1980 and a future year in which CO_2 was doubled and mean July air temperature warmed from 4 to 8 °C. In the warmer condition, water table was lowered simulating deeper and more rapid draw down with climate warming. Both warming and increased depth to permafrost and the water table decreased carbon storage in the cores (Billings et al. 1982, 1983). In their last experiment of the series they controlled temperature and water table but also increased soil N. Productivity increased with N enrichment but there was no CO_2 enhancement of primary productivity (Billings et al. 1984). They concluded that warming and drying would increase system respiration and carbon loss whereas N additions would stimulate net primary productivity. There were no direct CO_2 effects observed in their studies.

2.2. PASTURE

In 1982, Overdieck and Reining (1986) constructed a model pasture system using white clover and perennial ryegrass (*Trifolium repens* and *Lolium perenne*), respectively. He established a given density of each species in small, refrigerated greenhouses where temperature could be controlled and CO_2 provided at ambient and 2X ambient for 13 mo. Selected physiological, growth, and ecological measurements showed that the 2 species responded differently to the CO_2 enrichment with the clover increasing in number of seed produced while the rye grass produced less seed but increased in the number of vegetative tillers produced. Total standing crop was greater under CO_2 enrichment at the end of the study and there were differential species responses in several growth parameters such as leaf area index, petiole and leaf length. Overdieck concluded that future similar meadows would have a larger standing crop but that vegetation would change in floristic makeup. Although this study was of a synthetic system it was the first field demonstration of CO_2 effects in native vegetation.

2.3. SHORT GRASS PRAIRIE

William Hunt, James Detling and their colleagues at Colorado State University worked with 2 species of the Colorado short grass prairie in a simulated ecosystem study in the Duke Phytotron. They used the C_4 grass *Bouteloua gracilis* and the C_3 *Andropogon smithii* . In addition to doubling the CO_2 they increased temperature 4 °C and applied soil water and nutrient stress. The data are still being analyzed but it appears that the C_3 grass was enhanced by high CO_2 in all treatments. Contrary to theoretical predictions made by the investigators, a high percentage of the C_4 plants did not survive the high temperature and drought treatments at current CO_2 levels. CO_2 enrichment, however, ameliorated that loss as has also been observed for salinty (Bowman and Strain 1987, high (Kriedeman et al. 1976) and low (Sionit et al. 1981, Potvin 1985) temperatures, drought (Sionit et al. 1980, Marks and Strain 1989) and low soil fertility (Sionit et al.1981, Norby et al. 1986a).

2.4. TEXAS RANGELAND

A fourth totally synthetic system was established in 1990 at Temple, Texas by Hyrum Johnson. Representative shrub and grass species from a semi-arid rangeland were established in controlled mixtures in a long plastic enclosure within a standard greenhouse. Ambient air is drawn in one end of the tube and becomes depleted of CO_2 as plants photosynthesize. Thus, Johnson is studying the effect of CO_2 depletion in an attempt to learn how the rangeland ecosystem may have functioned in preindustrial times when atmospheric CO_2 was below 300 ppm. CO_2 is the only factor controlled in this experiment and it is too early to determine the outcome or validity of the approach. Preindustrial levels of CO_2 have been shown to seriously limit plant productivity (Milner and Hiesey 1964, Overdieck et al. 1988) and the Johnson study and evidence from tree-ring studies (LaMarche et al. 1985, Graybill 1987) indicate that natural ecosystems have responded positively to the 27% increase observed in global atmospheric CO_2 concentrations since preindustrial times.

3. Field Studies

Only four field studies have been initiated in "natural" ecosystems as of this writing. These studies are briefly summarized below.

3.1. ALASKAN TUNDRA

Oechel initiated the first field ecosystem study on CO_2 and thermally enriched plots of *Eriophorum* tussock tundra on the Alaskan North Slope in 1982 (Oechel and Strain 1985, Oechel et al. 1991). The tundra project, with supplementary controlled environment studies in the Duke Phytotron, has focused on 6 vascular species (*Eriophorum vaginatum, Carex bigelowii, Vacinium vitis-idaea, V. ulignìosum, Betula nana* and *Ledum palustre*). Basically, Oechel et al. have demonstrated that an early stimulation of carbon fixation by the entire system occured in yr 1 but declined in yr 2 and was unchanged or slightly negative in yr 3 (Oechel and Riechers 1986, Grulke et al. 1990, Oberbauer et al. 1986a and b). Physiological homeostasis (Tissue and Oechel 1987) returned the ecosystem to a similar net primary productivity within 3 yrs (Oechel et al. 1991). There were significant species shifts observed by yr 3, however, with the tillering graminoid species increasing in plant frequency and density at the expense of the shrubs (Oechel et al. 1991). As the new environments select species and genotypes from individuals surviving the change, genecological differentiation and gene frequency changes in populations can be expected (Strain 1991a). The field CO_2 enrichment studies were concluded in 1987. Oechel is now using gas exchange chambers to measure net ecosystem CO_2 flux. During the 1990 and 1991 seasons net losses of CO_2 were measured throughout the dry upland tundra and smaller loses were found in the wetter, lowland tundra (Oechel et al. 1992). These latter studies are the first documented feedbacks between climate, unmanaged ecosystems, and the global atmospheric CO_2 concentration. The feedbacks are positive; climate warming releases CO_2 from permafrosted peatlands which adds to the atmospheric CO_2 load perhaps intensifying greenhouse warming.

3.2. COASTAL SALT MARSH

Drake initiated the second study in a coastal salt marsh in 1987 (Drake et al. 1989, Drake and Leadley 1991). His study was explained in detail in the proceeding chapter. The coastal study compared monospecific habitats of a C_4 grass (*Spartina patens*) with a C_3 sedge (*Scirpus olneyi*). As predicted from an earlier study (Patterson and Flint 1980) and from theoretical considerations (Pearcy and Bjorkman 1983), Drake's group has shown a larger response of the C_3 system to CO_2 enrichment (Curtis et al. 1989a, b; Arp et al. 1991). Unlike the tundra study of Oechel, however, no homeostasis has been observed in the salt marsh. Both the C_3 and C_4 types are still showing significant biomass increases after 5 yrs of CO_2 doubling. At the ecotone between the monospecific populations of the 2 species, high CO_2 is contributing to a replacement of the C_4 sedge by the C_3 grass. Drake has received funding to continue these enrichment studies through 1994 at which time the plants will have been treated for 8 yrs. This will be the longest running field study of native vegetation available. The conclusion of the Drake study is opposite to the Oechel study in that the feedback between the vegetation and the climate is negative; increasing atmospheric CO_2 increases carbon sequestering in the marsh decreasing the atmospheric CO_2 concentrations and perhaps decreasing greenhouse warming.

3.3. TALL GRASS PRAIRIE

Clinton Owensby began a study of pristine tall grass prairie in Kansas in 1989. He placed open top chambers over representative species and has maintained ambient and 2X ambient for three growing seasons. The vegetation is a mixture of C_3 and C_4 species and is dominated by *Andropogon gerardii* and *Sorghastrum nutans* (C_4's) with several subdominant species including the C_4 *Bouteloua curtipendula* and the C_3's *Poa pratens* and several *Carex* species and several species of forbs including the C_3's *Sporobolus asper* and *Ambrosia psilostachya*. After the first 2 yrs, growth stimulation by CO_2 doubling was evident but unlike the saltmarsh results the largest stimulation was in the C_4 grass *Andropogon gerardi*. Changes are occurring in floristic makeup of the chambers but there is extreme variability in this parameter and Owensby has not yet been able to establish a significant pattern (Owensby, personal communication). In yrs 1 and 2 Owensby tried grazing sheep in controlled sections of the open tops to examined the effects of possible changes in nutrient quality on grazers such as observed in entomological studies by Lincoln and his colleagues (Lincoln et al. 1984, 1986, 1989). Lincoln has shown that insects decline in feeding efficiency when feeding on leaves grown under CO_2 enrichment. I understand that the sheep were too destructive to the ecosystem and that Owensby has dropped the study of herbivory for the time being.

3.4. CALIFORNIA ANNUAL GRASSLAND

The fourth system level study is currently being initiated by Harold Mooney and Christopher Field in serpentine annual grassland near Stanford, California. Equipment was installed in 1991 and 1992 will be the first complete growing season of measurement. The investigators hypothesize increasing carbon fixation and rapid CO_2 concentration and climate equilibration because of the annual nature and low carbon stock of this vegetation type.

4. Ecosystem Oriented Autecological Studies

In addition to the system studies described above, some system oriented autecological studies have been conducted on other ecosystem types. It is unfortunate that the only system level studies intitiated under realistic field conditions to date are those decribed in sections 3 above on graminoid systems. It is understandable, however, when one considers ecosystem stature and current technical limitations and the expense of setting up to work on larger stature systems. These limitations are well understood and were recently presented in a workshop report on field technologies for global change research edited by Strain (1991b, see Appendix A).

Currently, attempts are being made to study trees in temperate coniferous and deciduous trees (e.g. Wong and Dunin 1987, Norby 1987, Norby et al.1986a,b, 1987, see also Table 1 below) and in tropical trees (Oberauer et al. 1985, Reekie and Bazzaz 1989) but all of the studies are autecological in approach with the plants being grown in pots or soil and singly without competition or species interaction. In some cases ecological interactions of herbivores, mycorrhizae, and competition are underway but the starting points are synthetic mixtures and not the existing natural ecosystems as decribed in section 3 above.

No one has yet obtained the funding to establish a system level study of CO_2 enrichment in a forest. There are, however, several ecological studies of temperate forest species underway at the autecological level. Most of these studies are using open top chambers (Heagle et al. 1973, Rogers et al. 1983a) in an attempt to grow trees under partially controlled but field-like conditions. These studies are listed in Table 1 to identify the projects and to indicate the minimal level of research activity completed and underway.

Table 1. Current and terminated studies of CO_2 effects on trees. All of these studies employed open top chambers in an attempt to control CO_2 without greatly changing other environmental factors.

Principal Investigator	Location of Study	Species Studied
Current Studies	Underway	
Richard Norby	Oak Ridge, Tennessee	*Quercus alba* *Liriodendron tulipifera*
Bert Drake	Cheasepeake Bay, MD	Sub canopy shrubs
Boyd Strain, R. Thomas	Duke University, NC	*Pinus taeda*
Tim Ball, Dale Johnson	Desert Research Instit. Reno, NV	*Pinus ponderosa*
S.C. Wong	Australian National University	*Eucalyptus* spp.
S. Idso, B. Kimball	Water Conservation Laboratory, Tempe, AZ	Valencia orange
Peter Curtis T. Jurik	Michigan Biol. Station	Various northern tree species
Older Studies	Now Terminated	
Hugo Rogers	NC State University, Raleigh, NC	*L. styraciflua. Pinus taeda, Liriodendron tuliperifera*
J. Shinn, J. Houpis, and K. Surano	Livermore National Laboratory, CA	*Pinus ponderosa*

5. Comparative Studies of Pine Forests

In 1990 a comparative study was initiated of the 2 most widely distributed pine forests in North America. *Pinus taeda* (loblolly pine) and *Pinus ponderosa* (ponderosa pine) forests are being studied in open top chambers to determine the effects of long-term CO_2 enrichment interacting with soil fertility and water status. It is planned to incorporate other environmental factors (e.g. temperature, air humidity, climatic variance) into the study as the study progresses and additional funds and staff become available. This study is being conducted in a loblolly pine forest by Boyd Strain and colleagues at Duke University and by Tim Ball and Dale Johnson and their colleagues of the Desert Research Institute in a ponderosa pine forest at Placerville, California . The study is jointly funded by the Electric Power Research Institute and by the Department of Energy with technological support from the U.S. Forest Service.

One reason for picking these 2 species for detailed, long-term studies is the relatively large amount of research that has already been completed on the response of these 2 species to atmospheric CO_2. Loblolly pine has been examined by Rogers et al. (1983b), Thomas and Harvey (1983), Teskey and Shasta (1985), Sionit et al. 1985, Tolley and Strain (1984a,b, 1985),Teskey et al. 1986, Miller et al. 1986, Fetcher et al. (1988). Ponderosa pine has been investigated by Tinus (1972), Green and Wright (1977), Houpis et al. (1986, 1988) and Surano et al. (1986).

The overriding hypothesis linking the 2 studies is that CO_2 will induce greater net primary productivity in both forests if soil fertility and available soil water are adequate. As soil fertility (specifically N and P) and soil water decline, we expect that the fertilization effect of CO_2 enhancement will decline. Factor interaction with some amelioration of stresses are expected. At both sites, soil microbiological interactions are being examined with the hypothesis being that mycorrhizal interactions will improve P status but that N limitation will become more stringent as CO_2 increases.

Results 2 yrs into the study at the loblolly site indicate that total biomass produced on plants grown under optimal conditions of soil water, N and P, saturating photosynthetically active irradiance, no biological competition, and no herbivory or disease is enhanced by CO_2 enrichment. Data are shown at the end of the first year of the study in the lefthand graph of Figure 1 and the results of the second year are shown in the righthand graph. The total biomass of plants grown at three CO_2 concentrations and at reduced soil P and soil N levels were not significantly different from each other at any

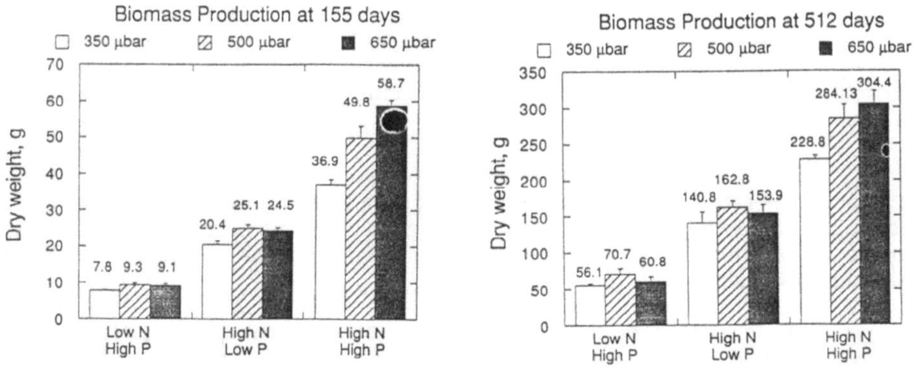

Figure 1. Biomass production of loblolly pine after 1 growing season (left graph) and 2 growing seasons (right graph) under variable CO_2, N and P levels. Note the difference in the dry weight scale for the 2 yrs.

time in the 2 yrs of the study. In other words, no homeostasis and no amelioration of nutrient deficiency stress has been observed, contrary to our hypotheses. Results from the ponderosa site are similar except that the CO_2 stimulation under the most fertile conditions peaks at 550 ppm CO_2 and declines at higher CO_2. If validated this evidence for CO_2 saturation is critically important because it implies that any CO_2 fertilization effect will be removed within a few decades of continuing atmospheric CO_2 enhancement. These studies will proceed through 1995 so conclusions concerning long term effects will not be possible until then.

6. Summary of Knowledge Applicable to Unmanaged Ecosystems: Reponse to CO_2 Enrichment and Climate Change

From all of the controlled environment studies itemized above and the studies reviewed by Drake in the previous paper, we know the following:

6.1. CO_2 has the properties of traditional plant fertilizers. If other environmental resources are present at required levels, CO_2 enrichment will induce increased photosynthesis and plant growth.

6.2. CO_2 interacts with all other environmental factors in true holocoenotic fashion. The general effect is that CO_2 ameliorates stress effects of other factors at moderate stress levels but the CO_2 enhancement effect is completely removed at higher stress conditions.

6.3. Plants stringently limited by resource deficiencies (e.g. N and P) or climate stress will respond slightly or not at all to CO_2 enrichment. Interactions with symbotic organisms will, however ameliorate soil fertility limitations.

6.4. CO_2 reduces transpiration and improves plant water status. Water-use-efficiency increases significantly because of increasing photosynthesis and decreasing water loss. Thus, drought stresses will be partially alleviated in arid and semi-arid ecosystems.

6.5. CO_2 enhancement leads to increasing C to N ratios (and C to other essential element ratios) in plant leaves and perhaps in litter. This increase in C:N affects the biological activity of associated leaf-feeding herbivores and may inhibit microbial decomposers of litter. These responses may impede decomposition and slow C cycling back to the atmosphere.

6.6. CO_2 and all other environmental factors of global change, affect species in a differential manner. Differences in species responses through time will generate ecological changes in ecosystem flora and fauna. Variance in response has also been confirmed within species indicating the opportunity for human selection of adaptive characters and the possibility of increased ecotypic fitness due to natural gene frequency changes.

7. Candidate Ecosystems and Available Technologies

The need for studies in several representative ecosystem types has been stated previously (Strain and Bazzaz 1983, Oechel and Strain 1985, Strain 1985, 1987, Strain 1991b, Mooney et al. 1991a) but to little avail. If we are to develop an understanding of ecosystem response to CO_2 increase and other aspects of global environmental change, it will be necessary to obtain a broad sampling of the major ecosystems of the earth. Figure 2 presents a classification scheme that considers functional aspects of a range of ecosystem diversity based on water availability and soil fertility factors. These environmental factors were shown in this paper to be critical to

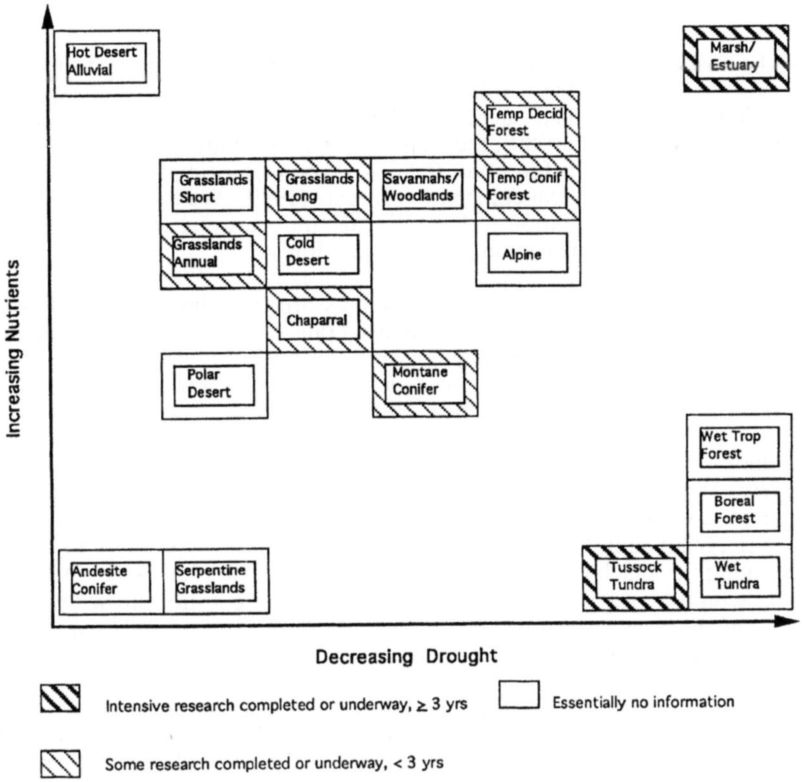

Figure 2. Ecosystem types representing the major biomes of the earth arranged on a matrix of increasing gradients of nutrient and drought stress. The largest effects are expected to occur in the most mesic but most nutrient-poor ecosystems. Some ecosystem-level information exists for the marshes and the Arctic tussock tundra systems. Autecological studes of varying intensity have been completed or are underway in the other systems as indicated by the degree of cross hatching. (Modified from Oechel and Strain 1985).

the determination of carbon cycling dynamics in plants and ecosystems. In addition to selecting ecosystems that represent the arrary shown in Figure 2, ecosystem site selection will be dependent upon the technology available to accomplish the degree of environmental control desired. That explains why the ecosystems analyzed to date are all of small stature. In fact, the only ecosystems measured have all been graminoid in structure as described earlier in this paper.

Facilities that have been used to control atmospheric gases (e.g. toxic air pollutants, CO_2, water vapor), include leaf chambers, sunlit controlled environment facilities, mobile greenhouses, larger more permanent greenhouses, and open-top chambers. A system of gas lines that deliver CO_2 for release over open fields (FACE) has recently been used in cotton fields and shows great promise for use in natural ecosystems as well (Strain 1991b).

8. Conclusions

Ecosystems to be analyzed and appropriate technologies to be applied have been adequately discussed elsewhere (Strain 1991b). Although high latitude ecosystems will experience the greatest changes in temperature and drought according to the Manabe scenario of global climate change (Manabe and Wetherald (1980), the tropical zone will undergo important ecosystem modification associated with differential species responses directly to carbon dioxide enrichment and subsequent biological competition Bazzaz (1990. We conclude this paper with a special plea for immediate attention to the tropical rainforest and support the associated section that describes a field experiment in a tropical rain forest.

Tropical rainforests are under extreme deforestation pressure. Continuing destruction of these forests is removing biological diversity at an increasing rate. As presented earlier in this book, however, recovery of deforested lands may be sequestering carbon at rates greater than we assumed. It is critical to determine if this is so, to learn how it might occur, and to quantify the effect on ecosystem carbon content and on the global carbon cycle.

Forests with canopy trees in excess of 30 m height generate the requirement for large scale control facilities. The minimum functional size of a rain forest at La Selva, Costa Rica, is estimated to be in excesss of 1 ha. A system has to be maintained under CO_2 control for several seasons to obtain equilibrium with the new environment. No ecosystem has yet been enriched long enough to reach carbon equilibrium under high CO_2 levels. In fact, we do not know how long would be required for an ecosystem to reach equilibrium with an enriched atmosphere in a square-wave enrichment experiment. Tropical forests might require a century while annual grasslands would presumably require only decades.

Discussion has been underway for several years on the feasibility of using a regeneration approach to the study of tropical rainforest response to global change. Assuming that increasing population and development pressures in tropical areas will generate large tracts of regenerating ecosystems, it would seem to be desirable to study early stages of ecosystem regeneration when plants are becoming established and are small enough to be contained in FACE arrays or ecocosms larger than 200 m^2. Complete ecosystems could be analyzed in long-term studies in large-scale Astrodome-type facilities as suggested in Strain (1991b) and in the associated section in this book on a large-scale field experiment in a tropical rainforest.

References

Arp, W.J. 1991. Effects of source-sink relations on photosynthetic acclimation to elevated CO_2. Pl., Cell, Environ. 14:869-875.

Arp, W.J. and B. Drake. 1991. Increased photosynthetic capacity of *Scirpus olneyi* after four years exposure to elevated atmospheric CO_2. Pl., Cell, Environ. 14:1004-1008.

Arp, W.J., B.G. Drake, W.T. Pockman, W.T. Curtis, and D.F. Whigham. 1991. Effects of four years exposure to elevated atmospheric CO_2 on competition between C_3 and C_4 salt marsh species. Vegetatio (In Press).

Bazzaz, F.A. 1990. The response of natural ecosystems to the rising global CO_2 levels. Ann. Rev. Ecol. Sys. 21:167-196.

Billings, W.D., J.O. Luken, D.A. Mortensen, K.M Peterson. 1982. Arctic tundra: A source or sink for atmospheric carbon dioxide in a changing environment? Oecologia 53:7-11.

Billings, W.D., J.O. Luken, D.A. Mortensen, K.M. Peterson. 1983. Increasing atmospheric carbon dioxide: Possible effects on Arctic tundra. Oecologia 58:286-289.

Billings, W.D., K.M. Peterson, J.O. Luken, D.A. Mortensen. 1984. Interaction of increasing atmospheric carbon dioxide and soil nitrogen on the carbon balance of tundra microcosoms. Oecologia 65:26-29.

Botkin, D.B. 1977. Forests, lakes, and the anthropogenic production of carbon dioxide. BioScience 27:325-331.

Bowman, W.D. and B.R. Strain. 1987. Interaction between CO_2 enrichment and salinity stress in the C_4 non-halophyte *Andropogon scoparius* (Walter) BSP. Pl., Cell, Environ. 10:267-270.

Conroy, J.P., P.J. Milham. M.L. Reed and E.W. Barlow. 1990. Increases in phosphorus requirements for CO_2-enriched pine species. Pl. Physiol. 92:977-982.

Cure, J.D. and B. Acock. 1986. Crop responses to carbon dioxide doubling: A literature survey. Agric. For. Meteorol. 38:127-145.

Curtis, P. S., B. G. Drake, P. W. Leadly, W. Arp and D. Whigham. 1989a. Growth and senescence of plant communities exposed to elevated CO_2 concentrations on an estuarine marsh. Oecologia 78:20-26.

Curtis, P. S., B. G. Drake and D. F. Whigham. 1989b. Nitrogen and carbon dynamics in C_3 and C_4 estuarine marsh plants grown under elevated CO_2 *in situ.* Oecologia 78:297-301.

Drake, B.G. and P.W. Leadley. 1991. Canopy photosynthesis of crops and native plant communities exposed to long-term elevated CO_2. Pl., Cell, Environ. 14:853-860.

Drake, B. G., P. W. Leadley, W. J. Arp, D. Nassiry and P. S. Curtis. 1989. An open top chamber for field studies of elevated atmospheric CO_2 concentration on saltmarsh vegetation. Func. Ecol. 3: 363-371.

Fetcher, N., C. H. Jaeger, B. R. Strain and N. Sionit. 1988. Long-term elevation of atmospheric CO_2 concentration and the carbon exchange rates of saplings of *Pinus taeda* L. and *Liquidambar styraciflua* L. Tree Physiol. 4:255-262.

Graybill, D.A. 1987. A network of high elevation conifers in the western U.S. for detection of tree-ring growth response to increasing atmospheric carbon dioxide. In: Jacoby, G.C.J. and J.W. Hornbeck (eds.). Proceedings of the International Symposium on Ecological Aspects of Tree-Ring Analysis. N. T. I. S. Springfield VA. pp 463-474.

Green, K. and R. Wright. 1977. Field response of photosynthesis to CO_2 enhancement in ponderosa pine. Ecology 58:687-692.

Grulke, N.E., G. Richers, W.C. Oechel, U. Hjelm, and C. Jaeger. 1990. Carbon balance in tussock tundra under ambient and elevated atmospheric CO_2. Oecologia 83:485-494.

Heagle, A.S., D.E. Body, W.W. Heck. 1973. An open-top field chamber to assess the impact of air pollution on plants. J. Environ. Qual. 2:365-368.

Houpis, J.L., K.A. Surano, S. Cowles and J.H. Shinn. 1988. Chlorophyll and carotenoid concentrations in two varieties of *Pinus ponderosa* seedlings subjected to long-term elevated carbon dioxide. Tree Physiol. 4:187-193.

Houpis, J. L. J., K. A. Surano, P. F. Daley and J. H. Shinn. 1986. Growth and morphology of *Pinus ponderosa* seedlings fumigated for two years with elevated CO_2. In: Proc. Physiological and Genetic Basis of Forest Decline, Ninth North American Forest Biology Workshop, 19-26.

Jarvis P. G. 1989 Forests, Weather and Climate. Philosophical Transactions of the Royal SocietyAtmospheric carbon dioxide and forests. In: Jarvis, P.G. (ed). Forests, Weather and Climate. Philosophical Transactions of the Royal Society Monteith J L, Unsworth M H and Shuttleworth J Series B.

Kramer P. J. and N. Sionit. 1987. Effects of increasing carbon dioxide concentration on the physiology and growth of forest trees. In: Shands. W.E. and J.S. Hoffman (eds). The Greenhouse Effect, Climate Change and U.S. Forests. Washington, D.C.: The Conservation Foundation. pp 219-246.

Kriedemann, P.E., R.J. Sward, W.J.S. Downtown. 1976. Vine response to carbon dioxide enrichment during heat therapy. Aust. J. Pl. Physiol. 3:605-618.

LaMarche, J. V., D. A. Graybill, H. C. Fritts and M. R. Rose. 1986. Carbon dioxide enhancement of tree growth at high elevations. Science 231:1019-1021.

Lincoln, D. E. and D. Couvet. 1989. The effect of carbon supply on allocation to allelochemicals and caterpillar consumption of peppermint. Oecologia 78:112-114.

Lincoln, D. E., D. Couvet and N. Sionit. 1986. Response of an insect herbivore to host plants grown in carbon dioxide enriched atmospheres. Oecologia 69:556-560.

Lincoln, D.E., N. Sionit, and B.R. Strain. 1984. Growth and feeding responses of *Pseudoplusia includens* (Lepidoptera: Noctuidae) to host plants grown in controlled carbon dioxide atmospheres. Environ. Entomol. 13:1527-1530.

Manabe, S. and R.T. Wetherald. 1980. On the distribution of climatic change resulting from an increase in CO_2-content of the atmosphere. J. Atmos. Sci. 37:99-118.

Marks, S. and B.R. Strain. 1989. Effects of drought and CO_2 enrichment on competition between two old-field perennials. New Phytol. 111:181-186.

Miller W. F., Dougherty P. M. and Switzer G. L. 1986. Rising CO_2 and changing climate: Major southern forest management implications. In: Shands, W.E. and J.S. Hoffman (eds). CO_2, Climate Change,and Forest Management in the United States. Washington, D.C. Conservation Foundation. pp 157-187.

Milner, and Hiesey. 1964. Photosynthesis in climaatic races of *Mimulus*. II. Effect of time and CO_2 concentration on rate. Plant Physiol. 39:746-750.

Mooney, H.A., B.G. Drake, R.J. Luxmore, W.C. Oechel, and L.F. Pitelka. 1991a. Predicting ecosystem responses to elevated CO_2 concentration. BioScience 41:96-104.

Mooney, H.A., E. Medina, D.W. Schindler, E-D. Schulze and B. Walker (eds). 1991b. Ecosystem Experiments. SCOPE 45, John Wiley & Sons, New York. 268 pp.

Norby, R. J. 1987. Nodulation and nitrogenase activity in nitrogen-fixing woody plants stimulated by CO_2 enrichment of the atmosphere. Physiol. Planta. 71:77-82.

Norby, R. J., E. G. O'Neill, W. G. Hood and R. J. Luxmoore. 1987. Carbon allocation, root exudation and mycorrhizal colonization of *Pinus echinata* seedlings grown under CO_2 enrichment. Tree Physiol. 3:203-210.

Norby, R. J., E. G. O'Neill and R. J. Luxmoore. 1986a. Effects of atmospheric CO_2 enrichment on the growth and mineral nutrition of *Quercus alba* seedlings in nurtient poor soil. Pl. Physiol. 82:83-89.

Norby, R. J., J. Pastor and J. Melilo. 1986b. Carbon-nitrogen interactions in CO_2-enriched white oak: Physiological and long-term perspectives. Tree Physiol. 2:233-241.

Oberbauer, S. F., W. C. Oechel and G. H. Riechers. 1986a. Soil respiration of Alaskan tussock tundra at elevated atmospheric carbon dioxide concentration. Plant and Soil 96:145-148.

Oberbauer, S. F., N. Sionit, S. J. Hastings and W. R. Oechel. 1986b. Effects of CO_2 enrichment and nutrition on growth, photosynthesis and nutrient concentrations of Alaskan tundra species. Can, J. Bot. 64:2993-2998.

Oberbauer, S. F., B. R. Strain and N. Fetcher. 1985. Effect of CO_2 enrichment on physiology and growth of seedlings of two tropical tree species. Physiol. Planta. 65:352-356.

Oechel, W.C. and W.D. Billings. 1992. Effects of global change on the carbon balance of arctic plants and ecosystems. In: Arctic Ecosystems in a Changing Climate. Academic Press, New York. pp 139-168.

Oechel, W. C. and G. H. Riechers. 1986. Impacts of increasing CO_2 on natural vegetation, particularly the tundra. Proceedings of Climate-Vegetation Workshop, NASA/Goddard Space Flight Center, Greenbelt MD. pp. 36-42.

Oechel, W.C. and Strain B.R. 1985. Native species responses to increased carbon dioxide. In: Strain B.R. and J.D. Cure (eds). Direct Effects of Increasing Carbon Dioxide on Vegetation. Nat. Tech. Infor. Ser. U.S. Dept. Comm. Springfield, VA. pp 117-154.

Oechel, W.C., M Jenkins, S.Hastings, G. Riechers, N. Grulke and G. Vourlitis. 1992. The changing carbon balance of Arctic tundra ecosystems: First evidence of climatic feedbacks on the global carbon cycle. (Submitted).

Oechel, W.C., G. Riechers, W.T. Lawrence, T.T. Prudhome, N. Grulke, and S.J. Hastings. 1991. Long-term *in situ* manipulation and measurement of CO_2 and temperature. Func. Ecol. 6: 86-100.

Overdieck, D., D. Reid and B.R. Strain. 1988. The effects of preindustrial and future CO_2 concentrations on growth, dry matter production and the C/N relationship in plants at low nutrient supply:*Vigna unguiculata* (Cowpea), *Abelmoschus esculentus* (Okra) and *Raphanus sativus* (Radish). Angew. Botanik 62:119-134.

Overdieck, D. and E. Reining. 1986. Effect of atmospheric CO_2 enrichment on perennial ryegrass (*Lolium perenne*) and white clover (*Trifolium repens* L.) competing in managed model-ecosystems. I. Phytomass production. Acta Oecologia 7:26-35.

Pastor, J. and W. M. Post. 1988. Response of northern forests to CO_2-induced climate change. Nature 334:55-58.

Patterson, E.P. and E. Flint. 1980. Potential effects of global atmospheric carbon dioxide enrichment on the growth and competitiveness of carbon 3 and carbon 4 weed and crop plants. Weed Sci. 28:71-75.

Pearcy, R. and O. Bjorkman. 1983. Physiological effects. In: E.R. Lemon (ed.). CO_2 and Plants: The Response of Plants to Rising Levels of Atmospheric Carbon Dioxide. AAAS Selected Symposium Vol 84. Westview Press Inc., Boulder, CO. pp 65-105.

Poorter, H. 1992. Interspecific variation in the growth response of plants to an elevated ambient CO_2 concentration. Vegetatio (In press).

Potvin, C. 1985. Amelioration of chilling effects by CO_2 enrichment. Physiol. Veg. 23:345-352.

Reekie, E.G. and F.A. Bazzaz. 1989. Competition and patterns of resource use among seedlings of five tropical trees grown at ambient and elevated CO_2. Oecologia 79:212-222.

Reynolds, J.F. and B. Acock. 1985. Predicting the response of plants to increasing carbon dioxide: A critique of plant growth models. Ecol. Model. 29:107-129.

Rogers, H., W. Heck, and A.S. Heagle. 1983a. A field technique for the study of plant responses to elevated carbon dioxide concentrations. J. Air Pollut.Cont. Assoc. 33:42-44.

Rogers, H., J.. Thomas and G. Bingham. . 1983b. Response of agronomic and forest species to elevated atmospheric carbon dioxide. Science. 220:428-429.

Shands, W.E. and J.S. Hoffman (eds.). 1986. CO_2, Climate Change, and Forest Management in the United States. Washington D.C. Conservation Foundation.

Shugart, H.H., M.Y. Antonovsky,P.G. Jarvis and A.P.Sandford. 1986. CO_2, Climatic Change and Forest Ecosystems. Assessing the response of global forests to the direct effects of increasing CO_2 and climatic change. In: The Greenhouse Effect, Climatic Change, and Ecosystems. John Wiley and Sons, New York, N.Y. 475 pp.

Sionit, N., H. Hellmers and B.R. Strain. 1980. Growth and yeild of wheat under carbon dioxide enrichment and water stress conditions. Crop Sci. 20:687-690.

Sionit, N. and P.J. Kramer. 1985. Woody plant reactions to CO_2 enrichment. In: Enoch, H.Z. and B.A. Kimball (eds). CO_2 Enrichment of Greenhouse Crops. CRC Press, Boca Raton FL.

Sionit, N., D.A. Mortensen, B.R. Strain and H.Hellmers. 1981. Growth responses of wheat to carbon dioxide enrichment with different levels of mineral nutrition. Agron. J. 73:1023-1027.

Sionit, N., B.R. Strain, H. Hellmers, C.W. Jaeger. 1985. Long-term atmospheric CO_2 enrichment affects the growth and development of *Liquidambar stryaciflua* and *Pinus taeda* seedlings. Can. J. For. Res. 15:468-471.

Solomon, A.M. 1986. Transient response of forests to CO_2-induced climate change: Simulation modeling experiments in eastern North America. Oecologia 68:567-579.

Strain, B.R. 1985. Physiological and ecological controls on carbon sequestering in terrestrial ecosystems. Biogeochemistry 1:219-232.

Strain, B.R. 1987. Direct effects of increasing atmospheric CO_2 on plants and ecosystems. Trends in Ecol. Evol. 2:18-21.

Strain, B.R. 1991a. Possible genetic effects of continually increasing atmospheric CO_2. In: Taylor Jr., G.E., L.F. Pitelka and M.T. Clegg (Eds). Ecological Genetics and Air Pollution. Springer-Verlag, New York. pp 237-244.

Strain, B.R. (ed.). 1991b. Available technologies for field experimentation with elevated CO_2 in global change research. In: Mooney, H.A. et al. (eds.). Ecosystem Experiments. SCOPE 45, John Wiley & Sons, New York. pp 245-261.

Strain, B.R. 1992. Atmospheric carbon dioxide: A plant fertilizer? The New Biologist 4:87-89.

Strain, B.R. and F. Bazzaz (eds). 1983. Terrestrial Communities. In CO_2 and Plants: The response of plants to rising levels of carbon dioxide. E. Lemon (ed.). AAAS Selected Symposium Vol 84. Westview Press, Boulder, CO. 280 pp.

Strain, B.R. and J.D. Cure (eds.). 1985. Direct Effects of Increasing Carbon Dioxide on Vegetation. Nat. Tech. Infor. Ser. U.S. Dept. Comm. Springfield, VA.

Surano, K. A., P. F. Daley, J. L. J. Houpis, J. H. Shinn, J. A. Helms, R.J. Palassou and M. P. Costella. 1986. Growth and physiological responses of *Pinus ponderosa* Dougl. ex P. Laws. to long-term elevated CO_2 concentrations. Tree Physiol. 2:243-259.

Teskey, R.O. and R.B. Shrestha. 1985. A relationship between carbon dioxide, photosynthetic efficiency and shade tolerance. Physiol. Planta. 63:126-132.

Teskey, R.O., J.A. Fites, L.J. Samuelson and B.C. Bongarten. 1986. Stomatal and non-stomatal limitations to net photosynthesis in *Pinus taeda* under different environmental conditions. Tree Physiol. 2:131-142.

Thomas, J.F. and C.N. Harvey. 1983. Leaf anatomy of four species grown under continuous CO_2 enrichment. Bot. Gaz. 144:303-309.

Thomas, R.B. and B.R. Strain. 1991. Root restriction as a factor in photosynthetic acclimation of cotton seedlings grown in elevated carbon dioxide. Pl. Physiol. 96:627-634.

Tinus, R.W. 1972. Carbon dioxide enriched atmosphere speeds growth of ponderosa pine and blue spruce seedlings. Tree Planters' Notes 23:12-17.

Tissue, D.T. and W.C. Oechel. 1987. Physiological response of *Eriophorum vaginatum* to elevated CO_2 and temperature in the Alaskan tussock tundra. Ecology 68:401-410.

Tolley, L.C. and B.R. Strain. 1984a. Effects of CO_2 enrichment on growth of *Liquidambar styraciflua* and *Pinus taeda* seedlings under different irradiance levels. Can. J. For. Res. 14:343-350.

Tolley, L.C. and B.R. Strain. 1984b. Effects of CO_2 enrichment and water stress on growth of *Liquidambar styraciflua* and *Pinus taeda* seedlings. Can. J. Bot. 62:2135-2139.

Tolley, L.C. and B.R. Strain. 1985. Effects of CO_2 enrichment and water stress on gas exchange of *Liquidambar stryaciflua* and *Pinus taeda* seedlings grown under different irradiance levels. Oecologia 65:166-172.

Williams, W.E., K. Garbutt, F.A. Bazzaz and P.M. Vitousek. 1986. The response of plants to elevated CO_2 IV. Two deciduous-forest tree communities. Oecologia 69:454-459.

Wong, S.C. and F.X. Dunin. 1987. Photosynthesis and transpiration of trees in a eucalypt forest stand: CO_2, light and humidity responses. Aus. J. of Pl. Physiol.14:619-632.

Ziska, L.H., K.P. Hogan, A.P. Smith and B.G. Drake. 1991. Growth and photosynthetic response of nine tropical species with long-term exposure to elevated carbon dioxide. Oecologia 86:383-389.

POTENTIAL EFFECTS OF THE INCREASE IN CARBON DIOXIDE AND CLIMATE CHANGE ON THE DYNAMICS OF VEGETATION

DIETER MUELLER-DOMBOIS
Department of Botany
University of Hawaii at Manoa
3190 Maile Way
Honolulu, Hawaii 96822
U.S.A.

ABSTRACT. The continued CO_2 loading of the atmosphere appears to be responsible for inducing three new force factors controlling dynamic changes in the world's vegetation. They come from (1) enhanced fertilization with the single most important plant nutrient, (2) the widely expected global temperature increase, and (3) aggravated weather disturbances. Increased CO_2 absorption may enhance plant growth but it may also increase soil-nutrient limitations. It surely will enhance the metabolism of forest trees similarly as global warming will enhance plant metabolism, but both factors may also shorten the lifespan of perennial plants. Increased weather disturbances can be expected to produce new physiological stresses on the standing vegetation, particularly on habitats with poor soils. Since wide-spread forest decline has been reported from both the Atlantic and Pacific region, it seems possible that the roughly synchronic mass mortality of trees during the past two decades is related to the global increase in CO_2. The paper gives an overview of forest decline and dieback as known from past and present research and suggests how the changing atmospheric environment may interact in this widely observed contemporary phenomenon of vegetation dynamics.

1. Introduction

The continued increase in CO_2, since the middle of the last century, has by now certainly resulted in a changed atmospheric environment for the world's vegetation cover. It thus appears necessary to consider changes in the CO_2 enriched environment for at least those aspects of vegetation dynamics that occur at the landscape level.

Several such aspects of vegetation dynamics have been predicted under the climate change scenario. They include spatial shifts of the major biomes or life zones in accordance with climate change models (Shugart, 1990), migration of individual species rather than whole biomes or communities based on paleo-ecological evidence (Graham and Grimm, 1990), fragmentation and isolation of populations through climate-related environmental barriers as have occurred in the past (Cronin and Schneider, 1990), and survival of stress-tolerant

Water, Air, and Soil Pollution **64**: 61–79, 1992.
© 1992 *Kluwer Academic Publishers.*

species adapted to harsh environments versus extinction of species adapted to more favorable environments based on metabolic cost considerations (Parsons, 1990).

None of these predictions have as yet become evident in the real environment. However, landscape-level aspects of vegetation dynamics which have been documented over the past two decades include several major forest declines and diebacks in the Pacific and Atlantic regions.

Landscape-level forest decline and tree mortality are perhaps something to be expected with climate change. However, thus far there is no substantiated claim that the predicted global warming has transgressed the temperature tolerance level of any tree species.

So, what is going on? Does the widely observed forest decline have anything to do with the changing atmospheric environment or is it caused by so many other factors that its complexity escapes any analytical inquiry?

In providing an answer to these questions it is necessary to first recall briefly what is now known about forest declines. Thereafter, I will discuss the possible role of climate change in the contemporary forest decline syndrome.

2. Forest Decline and Dieback: An Overview

2.1 HISTORIC DECLINES AND DIEBACKS

2.1.1 Definitions: Historic declines and diebacks are those that were observed and researched long before climate change became an issue, i.e. prior to the 1960's. **Declines** are easily observed losses of vigor affecting many trees in a stand. They can be visually compared to the autumn stage in deciduous forests prior to leaf drop but occur in evergreen species as well as deciduous species and typically out of season. **Diebacks** are the next step, namely drop of crown foliage followed usually by death. Since crown dieback in forests occurs at the stand-level I have referred to this phenomenon as canopy dieback (Mueller-Dombois 1983). The process can be rapid or very slow, i.e. drawn out over several decades. In the latter case barren crowns often develop associated with adventitious resprouting from the stem. Trees may even regain vigor with more leaf area but usually decline in foliar height before they finally die. The terms decline and dieback are often used interchangeably. Another characteristic of diebacks and declines is that they cannot be related to a single cause such as fire, flooding, insect pests, disease causing agents, or air pollution (Manion, 1981).

2.1.2 The Maui Dieback: A major forest decline occurred in the early 1900's in the lowland rainforest on the island of Maui, Hawaii. The forest, dominated by *Metrosideros polymorpha*, a native evergreen tropical hardwood species, displayed a remarkable loss of vigor. The canopy foliage turned yellow and started to drop similarly as in a deciduous forest at the onset of the unfavorable season. The affected forest was a lowland vegetation belt about 30 km wide and 5 km upslope, between 300 and 700 m altitude in an area of very high rainfall (4000 to 6000 mm/yr). Following initial observations, the affected forest was named the "Maui Forest Disease" (Lyon, 1909, Lewton-Brain, 1909). A long and intensive search for a biotic disease agent ended without success. Instead it was discovered that the soils had become paludified. No specific biotic disease agent was identified but soil aging with concomitant development of reducing Fe and increases in H_2S were considered to have resulted in toxifying the standing vegetation. Lyon (1918, 1919) concluded that the native Metrosideros forest was doomed as a pioneer vegetation, unable to adapt to aging tropical soils. Today, Metrosideros is still in the area, but it recovered only partially in variously clustered scrub islands or dwarf forest stands. Accordingly we have called this type of dieback "stand-reduction dieback" (Holt, 1988, Mueller-Dombois, 1988a).

2.1.3 Forest Declines in the USA and Canada: Other forest declines were discovered in the 1910's through 1950's on the U.S. mainland. They include the western white pine pole blight (first observed in Idaho 1929; Wright and Graham, 1952), the little leaf disease of southern pines (first receiving attention in the late 1930's; Roth, 1954), the birch dieback (first recognized in 1927 as spreading through the maritime provinces of Canada and the northwestern U.S.; Spaulding and Macaloney, 1931), the maple decline (which received much attention in the early 1960's but had been observed repeatedly since before 1915; Westing, 1966), and the oak decline (which occurred in three periods since 1912 in the central Appalachians; Staley, 1965), and several others (Manion, 1981). A literature review of these earlier declines and diebacks showed (Mueller-Dombois et al., 1983) that they all were approached first as biotic diseases. When this failed to give a satisfactory explanation, they were considered as caused by abiotic stresses, usually involving climatic change in combination with poorly buffered soils. Western white pine pole blight, for example, was explained as occurring in stands on shallow soils that could become flooded in heavy rains or completely dry up in periods of drought. Birch dieback was explained as primarily resulting from climatic change during a 30 yr warming period in the Northeast that raised surface soil temperatures by a few degrees during the summers. Redmond (1955) provided experimental evidence by heating the root zone of birch trees with copper wire by 1 to 2°C over normal summer temperatures causing feeder root mortality in mature birches and lowering vigor in seedlings. Another important characteristic of these historic declines is that they were restricted largely to mature trees (Walker et al., 1990).

Manion (1981), in his textbook on forest disease concepts, defined three types of diseases: biotic, abiotic and decline diseases. The first being the classical type of disease whereby a single disease agent, a pathogen or pest insect, is identified by adherence to Koch's postulates. Examples are the dutch elm disease, white pine blister rust, and the chestnut blight. The second type, abiotic diseases, are those that can be identified by a single abiotic cause, such as a mineral deficiency. The third type, decline diseases, are those that have no obvious single cause but instead display an interaction of predisposing, inciting and contributing causes, a three-stage cause-factor sequence originally proposed by Sinclair (1967).

2.2. CONTEMPORARY DECLINES: CONSIDERED PRIMARILY NATURAL

2.2.1 The Hawaiian Rainforest Decline: In 1965 another major forest decline was discovered in the *Metrosideros* rainforest biome, this time on the wet windward side of the island of Hawaii in an elevational range between 600 and 1,500 m. It rapidly expanded from a small core area of 120 ha (on the 1954 air photos) in interrupted patches to a combined area of 50,000 ha out of a total of 100,000 ha in 1977, when its spread seems to have come to a halt (Mueller-Dombois, 1985, Hodges *et al.*, 1986, Jacobi *et al.*, 1988). Initially, the decline was thought to be caused by a biotic disease agent, and it took almost 10 yr of intensive pathological research to clarify that pathogens played only a minor contributing role (Papp *et al.*, 1979). Instead a number of facts were established (Mueller-Dombois 1985). Taken together, they relate to edaphic stresses that increase with the maturation of forest stands of simplified cohort structure. When these stands reach growth-stagnation and the decline stage, they eventually are triggered into dieback by fluctuating site factors, considered to be extreme weather disturbances.

2.2.2 Other Primarily Natural Pacific Forest Declines: The Hawaiian rainforest is clearly not the only Pacific forest biome with contemporary stand or canopy dieback showing up in mass mortality at the landscape level. Diebacks of this sort occur in several New Zealand forest biomes, particularly in the southern mountain beech (*Nothofagus solandri* var. *cliffordioides*) forest (Wardle and Allen, 1983, Skipworth, 1983, Ogden, 1988), but also in New Zealand's *Metrosideros* -*Weinmannia* forest (Stewart and Veblen, 1983, Batcheler, 1983) and the *Ixerba brexiodes* cloud forest in the Kaimai Ranges on the North Island (Jane and Green, 1983). Further diebacks have been reported for native *Araucaria heterophylla* stands on Norfolk Island (Holloway, 1977) and for *Nothofagus pullei* stands in the montane forests of Papua New Guinea (Arentz, 1983, Ash, 1988), the *Scalesia pedunculata* moist tropical forest in the Galápagos Islands (Itow and Mueller-Dombois, 1988, Lawesson, 1988), the montane tropical forest in Sri Lanka away from the Pacific Rim (Werner, 1988), the subalpine fir forests of Japan (Kohyama, 1988), and for *Chamaecyparis nootkatensis* in southeast Alaska (Hennon *et al.*, 1989). Surely, there will be more such forest declines, if thorough literature and field surveys are made.

The principal characteristic of the above mentioned diebacks is that they have been judged to be natural, i.e. not caused by anthropogenic disturbances. There is still an ongoing argument, however, whether the New Zealand Metrosideros dieback is caused mainly by the introduced Australian possum (Batcheler, 1983) or by factors associated with stand dynamics (Stewart and Veblen, 1983), i.e. related to the advanced maturation of declining cohort stands.

2.3 CONTEMPORARY DECLINES: CONSIDERED PRIMARILY ANTHROPOGENIC

2.3.1 The Australian Rural Dieback: Native eucalypt diebacks and declines have become of great national concern in Australia since the late 1970's (Old et al., 1980). Initially, much emphasis was placed on disease causing agents (pests and pathogens). More recently, abiotic environmental stresses are implicated as critical even in the well-known jarrah (Eucalyptus marginata) forest dieback of western Australia (Davison, 1988), which earlier was considered a classical biotic disease caused by the root pathogen Phytophtora cinnamomi (Podger, 1972). Of wide-spread occurrence in eastern Australia is the so-called rural eucalypt (Landsberg and Wylie, 1988) or New England dieback (White, 1986, Heatwole and Lowman, 1986). It occurs from south of Canberra to north of Brisbane on the native eucalypts left as remnant trees in pastures and as forest fragments in the otherwise pastoral agricultural landscape.

A number of causes have been implicated, among which insect pests and droughts have been emphasized. It has become clear, however, that there is a strong anthropogenic component that overrides the natural stresses which in these structurally and biologically simplified forests can also lead to natural dieback (White, 1986).

Development of the Australian pasture industry required introduction of grazing resistent grasses from other continents. With this, N fixing legumes, such as clover, were introduced. Moreover, the ancient Australian soils are very low in P. This required heavy doses of superphosphate, which are applied often from airplanes. Both the pastures and the remnant eucalypts responded favorably. But the latter started to become negatively affected in a species sequence documented by Williams and Nadolny (1980). Briefly stated, the eucalypt foliage became more protinaceous and its chemical defenses (tannins) were diluted (Landsberg, 1990). Herbivorous insects became overwhelming: in the spring the larvae of chrysomelid beetles and the adult beetles feed on new shoots, in the summer, mature leaves are eaten by Christmas beetles (Scarabaeidae) whose larvae feed on the grass roots, in late summer leaves are again fed upon by a new generation of chrysomelid beetles (Heatwole and Lowman, 1986). Thus, there is a precarious balance between herbivorous insects and eucalypts that has been broken by agricultural development.

2.3.2 The European Forest Decline: This much publicised phenomenon has been explained as caused by industrial pollution, whereby acid precipitation, N_2O, O_3, ammonia, and pollution with heavy metals have been emphasized (Schulze et al., 1989). Lately, however, Mg deficiency has been found to be the cause of needle yellowing in Norway spruce and other species, and fertilization with Mg helped to eliminate this mineral deficiency (Zoettl and Huettl, 1991). However, the so-called "new type damage" in form of needle yellowing is not the only symptom of European forest decline. Needle and leaf browning associated with loss of crown foliage is another wide-spread phenomenon (Kandler et al., 1990, Landmann, 1991). Both types of symptoms are not considered "new", but certainly the wide-spread occurrence throughout much of Europe is considered new, and likewise the sequential synchrony which started with silver fir, then Norway spruce and then included the principal hardwoods, beech and oak (Bauer, 1988).

Most of the European decline research has been done in restricted areas. A recent landscape-level analysis including all of former West Germany (FRG) has come up with new findings based on the spatial dynamics of the European forest decline (Neuland et al., 1990). This statistical analysis revealed that (1) forest decline on nutrient-poor soils is much greater than on nutrient-rich soils; (2) that it is also much more prevalent on soils with low waterholding capacities. This led to the conclusion that drought periods were involved; (3) that forests with normal stocking density were less affected than open or over-stocked forest stands; (4) that in all affected tree species there was a highly significant correlation with tree age. Trees under 40 yr showed much less damage than those older than 40 yr and further, that young and old trees of the same species responded very differently to the stress factors analysed; (5) wind-exposure was found to be correlated with damage only in Pinus sylvestris and (6) no relationship could be established with the emission of SO_2 or the deposition of pollutants such as SO_4 and NO_2.

While the wide-spread "leaf yellowing" is now clearly identified as Mg deficiency, and in some cases as K deficiency (Zoettl et al., 1989, Huettl et al., 1990) "leaf browning and out-of-season leaf abortion" is not apparently reduceable to a single factor, but instead to an interaction of the factors 1 through 4 mentioned above. Moreover, since the Mg and K deficiencies are so wide-spread that they are recognized as "new type" forest damage, they still must be considered declines by definition, because the causes of the soil-mineral deficiencies are as yet unclarified.

3. Conditions of Ecosystem Dynamics Leading to Forest Decline

It is clear that forest declines are brought about by several causes which interact in a chain reaction. A single cause would not escape detection for very long. If industrial pollution was the only cause of the European forest decline, it would by now be clarified. Instead, the decline still remains largely a mystery. Three aspects, however, are well established; firstly, that the European forest decline and the Australian rural dieback have a strong anthropogenic component although of different forms; secondly, that forest decline is not restricted to the industrially polluted part of the world but instead is a global phenomenon; and thirdly, that decline and canopy dieback also occur naturally under certain conditions. Canopy dieback (cd) can be explained as an integral function (f) of five generic factors: cd = **f** (**s**, **e**, **p**, **b**, **t**). The five factors leading to canopy dieback are as follows:

3.1 SIMPLIFIED STRUCTURE (**s**)

This is expressed through simplified age-and/or size-structure, and low diversity of canopy species. Simplified structure applies to forests in which many individuals of the dominant canopy species form even-sized cohorts that belong to the same age-state or life stage, similarly as in plantation stands. Under natural conditions such low-diversity biomes have often developed on islands due to their isolation where monospecific canopy stands are common even in benign environments. Monospecific or other low-diversity canopy stands are typical for harsh environments such as found in mountain areas. A forest biome made up of a spatial mosaic of canopy cohorts can be expected to support stand segments that have reached the senescing life stage after a period of peak maturity. Synchrony in life stages of cohorts can be initiated through catastrophic disturbances, such as fires, hurricanes, volcanic explosions, or lava flows that offer open areas for species with opportunistic strategies for invasion. Where such catastrophic disturbances are recurring only at long intervals in the same location, senescing stands are bound to develop in the absence of fire or commercial exploitation.

3.2 EDAPHIC EXTREMES (**e**)

In islands, dominant species are often distributed over a wide spectrum of habitats because of the absence of competing species, a phenomenon known as "ecological release" (Grant, 1966, Williams, 1969). Opportunistic, stress-tolerant tree species, thereby become established in edaphically extreme habitats, where chronic stresses can develop with advancing stand age. Stands on such sites may be forced into premature senescence on account of poor soil water and nutrient relations to which such "ecologically released" species are not well adapted. To a lesser degree, this can also occur naturally in continental areas. The question relates to the availability of adapted species. Incomplete evolutionary adaptation is the natural counterpart to poor site matching or to planting a tree species or variety on the "wrong" site.

3.3 PERTURBATIONS RESULTING IN PHYSIOLOGICAL SHOCKS (**p**)

These are fluctuating site factors such as extreme weather events, that occur repeatedly throughout the life of a tree or cohort population in most forest biomes. When young and programmed primarily for vegetative growth, and also in the life phases of early and peak maturity, trees are generally equipped to recover after being exposed to adverse weather conditions. Their response will be different when they enter the senescing life phase, which is generally indicated by a negative photosynthesis/respiration ratio, an unfavourable leaf area/respiring wood area and by a stagnation in diameter growth. At this senescing stage, the trees will have difficulties recovering from the same strength of perturbation that during their earlier life stages amounted to only a transitory set-back.

3.4 BIOTIC AGENTS (**b**)

Unless they are infectuous diseases for which the vigor of the host tree is no barrier, biotic agents (pathogens or insect pests) become epidemic only when the vigor of their host tree declines (Manion, 1981, Mueller-Dombois et al., 1983, Sinclair and Hudler, 1988). Usually this may be related to a chemical imbalance that is physiologically conditioned. Maintenance of chemical defences requires metabolic costs that can become too high when trees are under stress from other causes. It is therefore typical for biotic agents to enter into the decline or dieback phases after these phases have been initiated by other factors. In an ecosystem context, the function of these weak biotic disease agents can be of importance in hastening the dieback, which in turn can be beneficial for the next generation of cohorts which follows a dying cohort.

3.5 TIME (**t**)

Time provides the demographic dimension for the process of stand development and decline. Time is a function of the life span of the affected species, which follows a specific genetic and tree-architectural program. After a period of peak development, trees will naturally decline from endogenous age-related stresses also under environmentally favorable conditions. However, then their decline can be expected to become desynchronized. Under favorable conditions, environmental stresses accumulate additively to the endogenous stresses and trees do not achieve their potentially possible life span. This, together with the cohort structure, uniformity of habitat and stand history conditions, will lead to a certain synchronization of tree decline which then is expressed in stand-level or canopy dieback.

4. The Cohort Senescence Theory

Unless tree mortality is clearly the result of a catastrophic disturbance, a decline phase will preceed dieback and death. Decline and dieback under natural conditions appears to be reduceable to the four generic causes (**s**, **e**, **p**, **b**) explained above, which interact in the time-frame of stand demography (**t**) in a chain reaction as explained through the cohort senescence theory (Mueller-Dombois, 1983, 1988b). This theory is based on a series of natural events that occur in the course of stand demography in certain low-diversity biomes with specific disturbance regimes.

The dieback chain reaction can be stated as follows:

- **t** - demographic time from stand origin, following a disturbance, to dieback
- **s e p** - predisposing factors that lead to synchronized cohort senescence
- **p** - trigger factor, a repeat perturbation, resulting in stand-level dieback
- **b** - contributing factor that may hasten and further synchronize tree mortality

The factors **s**, **e**, and **p** are seen as predisposing forest stands with similar histories and site factor relationships to dieback, which then is initiated by a recurrence of **p** and thereafter may conclude with the contributary role of **b**. However, **b** may also become a predisposing or trigger factor of dieback. This depends on the evolution and composition of pathogen/host relationships which are expected to vary among biomes.

Biomes in which stand- and landscape level diebacks occur without apparent anthropogenic influences are considered to be naturally destabilized or vulnerable. Anthropogenic destabilization is considered a new element now becoming apparent in Europe and Australia. In both the Australian rural and European declines, anthropogenic stresses add to those found under natural conditions.

5. Forest Decline and Climate Change

One of the remarkable aspects of the contemporary forest decline is that it is not restricted to the industrially polluted areas of Europe and eastern North America. Unless one takes the view that the Atlantic and Pacific declines have nothing in common, one cannot escape the conclusion that climate change may play a certain role. However, at the present state of knowledge, this can only be stated as a hypothesis. Other things in common are the four generic factors (structural simplification, edaphic stresses, periodic physiological shocks, and biotic disease agents) that are part of the cohort senescence theory. Climate change may now be considered as driving forest dynamics in three ways: by CO_2 fertilization, global warming, and increases in weather extremes. How does this relate to dieback?

5.1 CARBON DIOXIDE FERTILIZATION

The most obvious effect of the rising level of atmospheric CO_2 should be a stimulation of plant growth and an increase in the net primary production of the global vegetation. Earlier greenhouse studies have shown that CO_2 is an important limiting factor (Walter, 1960). Increases in the seasonal drawdown of atmospheric CO_2 have been documented from Mauna Loa by Bacastow et al. (1985), suggesting an increased rate of CO_2 fixation in the northern hemisphere. Experiments in agricultural systems have generally supported the fertilization effect of increasing CO_2 levels (Gifford, 1977, Acock and Allen, 1985, Overdieck and Reining, 1986, Woodward, 1990), but important limitations apply.

Plants with the C4 pathway of photosynthesis (mostly tropical grasses such as sugarcane, sorghum, maize and numerous savanna grasses) are near CO_2 saturation at ambient levels. They have a CO_2 concentration mechanism that makes these plants less responsive to additional supplies of CO_2. Carter and Perterson (1983), however, found that high levels of CO_2 may delay the C4 maturation process. Thus, there seems to be an effect in delaying senescence in C4 plants.

Plants with the C3 pathway of photosynthesis (which include all tree species mentioned above that are involved in the forest decline syndrome) have generally a great capacity for fixing additional CO_2 (Woodward et al., 1991). However, there are other limitations depending on species and habitats. Oechel and Strain (1985), working with native tundra plants in Alaska, found that by experimentally elevating the level of atmospheric CO_2, there was a stimulating effect for only the first 3 to 4 weeks, and thereafter a downward acclimation. This was interpreted as an emerging soil-nutrient deficiency. Wong (1979), experimenting with cotton (a C3 plant) and maize (a C4 plant) in a greenhouse with elevated CO_2 and N fertilization found that maize was insensitive to increased CO_2 but cotton grew much more with high CO_2 when combined with additional N. The higher the N level, the greater the growth enhancement with CO_2. In addition to soil-nutrient limitations which may significantly lessen the response of C3 plants to elevated carbon dioxide, there are a number of physiological limitations (Oechel and Strain, 1985, Woodward et al., 1991). They largely boil down to C allocation problems. For example, starch may accumulate in the chloroplast when the additionally acquired photosynthates are not readily transported into the appropriate C sinks of the plant such as the roots, stems, flowers or fruits. Damage by large starch granules has been suggested as a cause of photosynthetic inhibition. In extreme cases starch grains have been observed to disrupt the grana of the chloroplasts. Such carbon allocation problems have been described as symptoms of a "source-to-sink" imbalance by Acock and Allen (1985).

Other important plant responses noted were partial closing of stomata under elevated levels of CO_2 (Woodward et al., 1991). Partial closing does not interfere with CO_2 absorption but has an effect on lowering conductivity. While this may improve the water use efficiency, it also elevates leaf temperatures as a consequence of decreased evaporative cooling through reduced transpiration. This in turn increases leaf respiration. Leaf respiration may not increase significantly relative to the increase in photosynthesis, and certainly not in actively growing plants (Drake, pers. comm.). However, Woodward et al. (1991) point out that the reponses of respiration to elevated CO_2 are as yet poorly understood. These authors state in their latest review that "... dark respiration is the negative side of net primary production and so is an important issue in any consideration of ecosystem responses to global change..."

Thus while the increase in atmospheric CO_2 can definitely be expected to boost primary production generating also leafier canopies in the world's forests, increases in respiration clearly point to an overall increase in the rate of plant metabolism. An accelerated metabolism translates into faster growth rates and probably also shorter lifespans. When trees have reached peak maturity, senescence is likely to be hastened by increased atmospheric CO_2. This has already been suggested as a longer-term effect of exposure to elevated CO_2 by Oechel and Strain (1985). Experimental evidence for accelerated whole-plant senescence under elevated CO_2 has been given by St. Ohmer and Horvath (1983) for C3 annuals and by Chang (1975) for cotton, a C3 perennial.

An accelerated rate of maturation and senescence through atmospheric CO_2 fertilization can be considered an additional stress factor in forest decline at the global level. This hypothesis is consistent with the fact that in all forest decline and dieback cases cited above the tree species in dieback condition also have actively growing young and mature populations coexisting next to the dying ones in the same biomes. Fertilized trees and those on nutritionally better habitats in Europe are said to be growing better than ever before (Huettl, pers. comm.), while some stands formerly reported as declining are regaining vigor without fertilizer additions and others are continuing to decline. The CO_2 fertilizing and senescing scenario explains this apparent contradiction.

5.2 GLOBAL WARMING

Although considered the most important consequence of the rising level of CO_2 on the world's climate, the evidence for global warming is as yet not very conclusive, but it is getting stronger. Henderson-Sellers (1990) provides data from all 'acceptable' surface-air temperature measurements taken worldwide from 1900 to 1987. The resulting curve displays a warming trend of 0.7°C over the last 88 yr. This value still includes some climatic records from urban "heat islands." When these are removed Henderson-Sellers estimates a 'real' global temperature increase of half a degree (0.5°C) for this period. The curve shows two warming periods, the first from 1900 to 1940, coinciding with the historic forest declines discussed above, and a second from 1975 to 1987, more or less coinciding with the contemporary decline records. Certainly one cannot expect a direct relationship to forest decline. Warming by such a small amount should encourage forest growth rather than depress it. But the warming effect has recently been found by Karl et al. (1991) to apply mostly to an increase in night temperatures, while day temperatures have largely remained at the same level. This is said to be related possibly to increased cloudiness and to reflectance of incoming short-wave radiation by microscopic sulfate aerosols, which, at the same time, may intecept outgoing radiation. Night time warming will contribute to an increase in plant metabolism and thus may have a synergistic effect together with CO_2 fertilization on shortening the lifespan of C3 plants. Tropical to warm-temperate ecosystems are likely to respond to global warming with a greater increase in respiration relative to their increase in photosynthesis (Townsend et al., 1992). This would accelerate the maturation process leading to an earlier senescence.

5.3 WEATHER EXTREMES

Global warming cannot be expected to result in a spatially uniform temperature increase. General climate circulation models predict regional differences. For example, according to Henderson-Sellers (1990), the 1988 GISS (Goddard Institute of Space Studies) model of Hansen et al. predicts the earliest notable warming in ocean areas near Antarctica and China and the interior of Asia. The latest upswing of the global average temperature from 1975-1987 coincided with two major El Niño southern oscillation (ENSO) events in 1982/83 and 1986/87 (Henderson-Sellers, 1990). This Pacific-wide phenomenon brings extreme weather changes primarily because of the warm equatorial current reversing its normal direction (Ramage, 1986, Kerr, 1987). Normally the southwest Pacific is very cloudy and rainy with the low pressure system centered over the Solomon Islands, Bougainville, and the Bismarck Archipelago, while Easter Island and the Galápagos Islands are sunny and dry. During an ENSO event this pattern is reversed; drought may be experienced in the southwest Pacific as far as Indonesia and extreme wetness over the eastern Pacific including California. So far, there is no real evidence for increases in EL Niño events. Quinn et al. (1987) made a careful analysis of past EL Niño events going back to the year 1525. Table 1 summarizes their data on strong and moderate EL Niño events.

In spite of the fact that one cannot yet conclude an increase in frequency and intensity of EL Niño events with the current state of global warming, it seems rather likely that changes in moisture availability will develop with the CO_2 increase and further global warming. Overpeck et al. (1990) predicted increased storminess and more frequent weather disturbances under increased CO_2 conditions using a climate model which also predicted accelerated tree mortality and forest dieback. An analysis of extreme weather events in conjunction with a search for dieback triggers in the Hawaiian forest decline revealed that the frequency of extreme storm-floods

Table 1. EL Niño events during the last five centuries (based on data by Quinn and Neal 1987)

	Number of EL Niños strong	moderate	Global temperature
1525-1625	12	nd	"little
1625-1725	11	nd	ice age"
1725-1825	8	nd	
1825-1925	10	18	warming
1925-1987	6	8	trend

nd = no data available

and extreme drought events had increased on windward Hawaii from 1900 to 1984 (Mueller-Dombois, 1986, 1992). For the first 50 yrs, the record showed four extreme storm-flood periods and five for the next 35 yr. Likewise, extreme drought periods increased from two in the first 50 yrs to three in the following 35 yrs. The data were too scanty for attaching any statistical significance to them. But a trend is indicated, which may become significant through additional biome-specific analyses of weather extremes.

6. Conclusions

Forest decline and dieback is a multiple cause phenomenon of vegetation dynamics which occurs in a number of low-diversity biomes. Unlike other predicted vegetation changes expected to appear in response to climate change, forest decline and dieback is now clearly evident as a landscape-level pattern in both the industrially polluted Atlantic region and in relatively "clean-air" environments of the Pacific region. Therefore, it has become expedient to search for underlying causes of canopy dieback that are common to both hemispheres which can be inferred globally as generic causes.

The cohort senescence theory provides four such causes, namely (1) structural simplification (**s**) associated with cohort stand development followed later by synchronized senescence. (The structural simplification is being promoted by biotic impoverishment and long-interval catastrophic disturbances), (2) edaphically extreme situations (**e**) associated with poor species adaptations, (3) periodic weather disturbances associated with physiological shocks (**p**), and (4) biotic disease agents (**b**) associated with host-tree stands weakened from prior stresses. These four causes are implied to interact in form of a chain reaction involving them as predisposing, inciting, and contributory stresses and disturbances that appear in the demographic time frame (t) of the affected stands.

Climate change adds a new dimension to that theory in so far as it impinges on all four causes. CO_2 fertilization in addition to increasing net primary production in C3 trees is likely to shorten their lifespan. The principal reason is the enhanced metabolism, which may accelerate development in all life stages prior to and past peak maturity. Past peak maturity, the senescing process, which is a function of the changing photosynthesis/ respiration ratio, is likely to become more synchronized among physiologically similar cohort individuals. Global warming will have a synergistic effect.

In edaphically extreme or nutrient-poor habitats, CO_2 fertilization may have no positive effect, because it will only raise the plant metabolism, again promoting together with global warming an accelerated senescence.

Periodic weather disturbances, extreme enough to be associated with physiological shocks, are suggested to eventually push the declining stands into dieback. Such disturbances are probably increasing in frequency and intensity with global warming, but so far, the evidence is rather scanty.

Biotic disease agents will respond to declining stands by assuming their role as hasteners of dieback. Whether CO_2 fertilization will promote or discourage insect herbivory is still an open question. Lincoln and Couvet (1989) found that allelochemicals are not diluted under elevated CO_2, while soil nutrients were diluted resulting in increased leaf C/N ratios. In their study this prompted increased caterpillar feeding activity on peppermint, Mentha perita). In the Australian rural eucalypt dieback, the opposite, namely foliar enrichment with N and P promotes herbivory by insects.

Thus far, the effects of climate change on forest decline are largely unknown. That they interact with forest decline, however, can hardly be doubted. Among the atmospheric changes, CO_2 fertilization has probably an important direct effect. It should not be difficult to design a research program on forest decline and climate change that builds on the questions and hypotheses raised in this paper.

Acknowledgments

I thank my wife, Annette Mueller-Dombois, for wordprocessing and Guillermo H. Goldstein, Boyd R. Strain and an anonymous reviewer for critical and constructive comments on the manuscript. Financial support from the National Science Foundation for Research Grant BSR 8918526 "Ecosystem Dynamics in Hawaii" is gratefully acknowledged.

References

Acock, B. and Allen L. H. Jr. (1985) Crop responses to elevated carbon dioxide concentrations. Pages 53-97 in "Direct Effects of Increasing Carbon Dioxide on Vegetation". B. R. Strain and J. D. Cure (eds.) DOE/ER-0238, U.S. Dept. of Energy Rept., 286 p.
Arentz, F. (1983) Nothofagus dieback on Mt. Giluwe, Papua New Guinea. Pac. Sci. 37(4):453-458.
Ash, J. (1988) Nothofagus (Fagaceae) forest on Mt. Giluwe, New Guinea. New Zealand J. Bot. 26:245-258.
Bacastow, R. B., Keeling, C. D., and Whort T. C. (1985) Seasonal amplitude increase in CO_2 concentration at Mauna Loa, Hawaii. 1959-1982. J. Geophys. Research 90 (D6):10,529-10,540.
Batcheler, C. L. (1983) The possum and rata-kamahi dieback in New Zealand: a review. Pac. Sci. 37(4):415-426.
Bauer, F. (1988) Zur Waldschadenssituation in der Bundesrepublik Deutschland. Allgem. Forstzeitschrift 51-53:1417-1420.
Carter, D. R., and Peterson R. M. (1983) Effects of a CO_2-enriched atmosphere on the growth and competitive interaction of a C3 and C4 grass. Oecologia 58:188-193.
Chang, C. W. (1975) Carbon dioxide and senescence in cotton plants. Plant Physiology 55:515-519.
Cronin, T. M. and Schneider C. E. (1990) Climatic influences on species: evidence from fossil record. Trends Ecol. &. Evol. 5(9):275-279.
Davison, E. M. (1988) The role of waterlogging and Phytophthora cinnamomi in the decline and death of Eucalyptus marginata in Western Australia. GeoJournal 17:239-244.
Gifford, R. M. (1977) Growth pattern, CO_2 exchange and dry weight distribution in wheat growing under different photosynthetic environments. Australian J. of Plant Physiology 4:99-110.
Graham, R. W. and Grimm, E. C. (1990) Effects of global climate change on the patterns of terrestrial biological communities. Trends Ecol. & Evol. 5(9):289-292.
Grant, P. R. (1966) Ecological compatability of bird species. Am. Nat. 100:451-462.
Heatwole, H. and Lowman, M. (1986) Dieback, Death of an Australian Landscape. Reed Books Pty Ltd, Frenchs Forests, NSW, Australia, 150 p.
Henderson-Sellers, A. (1990) Modelling and monitoring "greenhouse" warming. Trends Ecol. & Evol. 5(9):270-275.
Hennon, R. E., Hanson, E. M. and Shaw C. G. III. (1990) Dynamics of decline and mortality of Chamaecyparis nootkatensis in southeast Alaska. Canadian J. Bot. 68:651-662.
Hodges, C. S., Adee, D. T., Stein, J. D., Wood, G. B., and Doty, R. D. (1986) Decline of 'ōhi'a (Metrosideros polymorpha)in Hawaii: a review. USDA Forest Service, Pac. SW. For. and Range Expt. St. Gen. Tech. Report PSW-86, 22p.

Holloway, J. D. (1977) The Lepidoptera of Norfolk Island. Dr. W. Junk, The Hague. 291 p.
Holt, R. A. (1988) The Maui Forest Trouble: Reassessment of an Historic Forest Dieback. M.Sc. Thesis. Univ. of Hawaii,Honolulu.
Huettl, R. F., Fink, S., Lutz, H.-J., Poth, M., and Wisniewski, J. (1990) Forest decline, nutrient supply and diagnostic fertilization in southwestern Germany and in southern California. Forest Ecol. and Managmt. 30:341-350.
Itow, S. and Mueller-Dombois, D. (1988) Population structure, stand-level dieback and recovery of Scalesia pedunculata forests in the Galápagos Islands. Ecol. Research 3:333-339.
Jane, G. T. and Green, T. G. A. (1983) Episodic forest mortality in the Kamai Ranges, North Island, N.Z. New Zealand J. Botany 21:21-31.
Jacobi, J. D., Gerrish, G., Mueller-Dombois, D., and Whiteaker, L. (1988) Stand-level dieback and Metrosideros regeneration on the montane rainforest of Hawaii. GeoJournal 17:193-200.
Kandler, O., Senser, M., and Miller, W. (1990) Vergiftung und Wiederergruenung der Fichten. Pages 113-137 in Neuartige Waldschaeden - Erkenntnisse und Folgerungen. J. Jositz (ed.), Hans-Seidel-Stiftung e.V., Munich, Agrarpolitik Serie 56.
Karl, T. R., Kukla, G., Razuvajev, V. N., Changery, M. J., Quayle, R. G., Heim, Jr., R. R., Easterling, B. R., and Fu, C. B. (1991) Global warming: evidence for asymmetric diurnal temperature change. Geophys. Res. Letters 18(12):2253-2256.
Kerr, R. A. (1987) El Niño and winter weather. Science 238: 1508.
Kohyama, T. (1988) Etiology of "Shimagare" dieback and regeneration in subalpine Abies forests of Japan. GeoJournal 17:201-208.
Landmann, G. (ed.) (1991) French research into forest decline. DEFORPA Programme. ENGREF, 14, rue Girardet, F-54042 Nancy Cedex. 120 p.
Landsberg, J. and Wylie, F. R. (1988) Dieback of rural trees in Australia. GeoJournal 17:231-237.
Landsberg, J. (1990) Dieback of rural eucalypts: Response of foliar dietary quality and herbivory to defoliation. Austr. J. Ecol. 15:89-96.
Lawesson, J. E. (1988) Stand-level dieback and regeneration of forests in the Galápagos Islands. Vegetatio 77:87-93.
Lewton-Brain, L. (1909) The Maui forest trouble. Hawaiian Planter's Record 1:92-95.
Lincoln, D. E. and Couvet, D. (1989) The effect of carbon supply on allocation of allelochemicals and caterpillar consumption of peppermint. Oecologia 78:112-114
Lyon, H. L. (1909) The forest disease on Maui. Hawaiian Planter's Record 1:151-159.
Lyon, H. L. (1918) The forests of Hawaii. Hawaiian Planter's Record 20:276-281.

Lyon, H. L. (1919) Some observations on the forest problems of Hawaii. Hawaiian Planter's Record 21:289-300.
Manion, P. D. (1981) Tree disease concepts. Prentice-Hall, Englewod Cliffs, New Jersey, USA. 399 p.
Mueller-Dombois, D. (1983) Canopy dieback and successional processes in Pacific forests. Pac. Sci. 37:317-325.
Mueller-Dombois, D. (1985) 'Ōhi'a dieback in Hawaii: 1984 synthesis and evaluation. Pac. Sci. 39:150-170.
Mueller-Dombois, D. (1986) Perspectives for an etiology of stand-level dieback. Ann. Rev. Ecol. Syst. 17:221-243.
Mueller-Dombois, D. (1988a) Vegetation dynamics and slope management on the mountains of the Hawaiian Islands. Env. Conserv. 15:255-260.
Mueller-Dombois, D. (1988b) Towards a unifying theory for stand-level dieback. GeoJournal 17:249-251.
Mueller-Dombois, D. (1992) Biotic impoverishment and climate change: Global causes of forest decline? In "Forest Decline in the Atlantic and Pacific Region". R. F. Huettl & D. Mueller-Dombois (eds.), Ecol. Series, Springer-Verlag. In press.
Mueller-Dombois, D., Canfield, J. E., Holt, R. A., and Buelow, G. P. (1983) Tree-group death in North American and Hawaiian forests: a pathological problem or a new problem for vegetation ecology? Phytocoenologia 11:117-137.
Neuland, H., Hank, H., Tenhagen, P., Mueller, D., Paus, L., and Schaefer, T. (1990) Regionalstatistische Analyse des Zusammenhangs zwischen Standortsbedingungen und Waldschaeden. Bundesministerium f. Forschung & Technologie. Dornier GmbH, PO Box 1420, 7990 Friedrichshafen 1, 198 p.
Oechel, W. C. and Strain, B. R. (1985) Native species response to increased atmospheric carbon dioxide concentration. Pages 117-150 in "Direct Effects of Increasing Carbon Dioxide on Vegetation". B. R. Strain and J. D. Cure (eds.) DOE/ER-0238, US Dept. of Energy Report, 286 p.
Ogden, J. (1988) Forest dynamics and stand-level dieback in New Zealand's Nothofagus forests. GeoJournal 17:225-230.
Old, K. M., Kile, G. A., and Ohmart, C. P. (1981) Eucalypt dieback in forests and woodlands. Proceed. Conf. CSIR Div. of For. Research, Canberra, ACT., 185 p.
Overdieck, D. and Reining F. (1986) Effect of atmospheric CO_2 enrichment on perennial rye grass (Lolium perenne L.) and white clover (Trifolium repens L.) competing in managed model-ecosystems I. Phytomass production. Acta Oecologia 7:357-366.
Overpeck, J. T., Rind, D., and Goldberg, R. (1990) Climate-induced changes in forest disturbance and vegetation. Nature 343:51-53.
Papp, R. P., Kliejunas, J. T., Smith, Jr., R. S. and Scharpf, R. F. (1979) Association of Plagithmysus bilineatus (Coleoptera: Cerambycidae) and Phytophthora cinnamomi with the decline of 'ōhi'a-lehua forests on the island of Hawaii. Forest Science 25:187-196.
Parsons, P. (1990) The metabolic cost of multiple environmental stresses: implications for climatic change and conservation. Trends Ecol. &. Evol. 5(9):315-317.

Podger, F. D. (1972) Phytphthora cinnamomi a cause of lethal disease in indigenous plant communities in Western Australia. Phytopathology 62:976-981.
Quinn, W. H., Neal, V. T., and de Mayolo, S. E. A. (1987) El Niño occurrences over the past four and a half centuries. J. Geogr. Res. 92(C13):14,449-14,461.
Ramage, C. S. (1986) El Niño. Scientific American 254(6):77-83.
Redmond, D. R. (1955) Studies in forest pathology XV. Rootlets, mycorrhiza, and soil temperatures in relation to birch dieback. Canadian J. of Bot. 33:595-627.
Roth, E. R. (1954) Spread and intensification of little leaf disease of pine. J. For. SL:592-596.
Schulze, E.-D., Lange, O. L., and Oren, R. (1989) Forest decline and air pollution. Springer-Verlag, Heidelberg. Ecol. Studies 77, 475 p.
Shugart, H. H. (1990) Using ecosystem models to assess potential consequences of global climatic change. Trends Ecol. &. Evol. 5(9):303-307.
Sinclair, W. A. (1967) Decline of hardwoods: possible causes. Proc. 42nd Intern. Shade Tree Conf. (1966):17-32.
Sinclair, W. A. and Hudler, G. W. (1988) Tree declines: four concepts of causality. J. Arboric. 14(2):29-36.
Skipworth, J. P. (1983) Canopy dieback in a New Zealand mountain beech forest. Pac. Sci. 37:391-395.
Spaulding, P. and Macaloney, H. J. (1931) A study of organic factors concerned in the decadence of birch on cut-over lands in northern New England. J. of Forestry 29:1134-1149.
Staley, J. M. (1965) Decline and mortality of red and scarlet oaks. Forest Science 11:2-17.
Stewart, G. H. and Veblen, T. T. (1983) Forest instability and canopy tree mortality in Westland, New Zealand. Pac. Sci. 37:427-431.
St. Omer, L. and Horvath, S. M. (1983) Elevated carbon dioxide concentration and whole plant senescence. Ecology 64:1311-1314.
Townsend, A. R., Vitousek, P. M., and Holland, E. A. (1992) Tropical soils dominate the short-term carbon cycle feedbacks to atmospheric carbon dioxide. Climatic Change. In press.
Walker, S. L., Auclair, A. N., and Martin, H. (1990) History of crown dieback and deterioration symptoms of hardwoods in eastern Canada. Atmospheric Environm. Service, 4905 Dufferin St., Downsview, Ontario, 62 p.
Walter, H. (1960) Phytology III: Grundlagen der Pflanzenverbreitung I. Teil: Standortslehre. Verlag Eugen Ulmer, Stuttgart, 566 p.
Wardle, J. A. and Allen R. B. (1983) Dieback in New Zealand Nothofagus forests. Pac. Sci. 37:397-404.
Werner, W. (1988) Canopy dieback in the upper montane rain forests of Sri Lanka. GeoJournal 17:245-248.
Westing, A. H. (1966) Sugar maple decline: an evaluation. Economic Botany 20:196-212.
Williams, E. E. (1969) The ecology of colonization as seen in

Williams, E. E. (1969) The ecology of colonization as seen in the zoography of anoline lizards on small islands. Quarterly Rev. Biol. 44:345-389.

Williams, J. B. and Nadolny, C. (1981) A survey of eucalypt species involved in New England dieback. Pages 74-86 in Eucalypt Dieback in Forests and Woodlands. K. M. Old, G. A. Kile and C. P. Ohmart (eds.), For. Res. CIRO Melbourne, 285 p.

White, T. C. R. (1986) Weather, Eucalyptus dieback in New England, and a general hypothesis of the cause of dieback. Pac. Sci. 40:58-78.

Wong, S. C. (1979) Elevated atmospheric partial pressure of CO_2 and plant growth I. Interaction of nitrogen nutrition and photosynthetic capacity in C3 and C4 plants. Oecologia 44:68-74.

Woodward, F. I. (1990) Global change: translating plant ecophysiological responses to ecosystems. Trends Ecol. &. Evol. 5(9):308-315.

Woodward, F. I., Thompson, G. B., and McKee, I. F. (1991) The effects of elevated concentrations of carbon dioxide on individual plants, populations, communities and ecosystems. Annals of Botany 67 (supplement):23-38.

Wright, E. and Graham, D. P. (1952) Surveying for pole blight. J. Forestry 50:680-682.

Zoettl, H. W., Huettl, R. F. Fink, F., Tomlinson, G. H., and Wisniewski, J. (1989) Nutritional disturbances and histological changes in declining forests. Water, Air, and Soil Pollution 48:87-109.

Zoettl, H. W. and R. F. Huettl (eds.) (1991) Management of nutrition in forests under stress. Kluwer Acad. Publishers, Dordrecht, 668 p.

MANAGING NATURAL SINKS OF CO_2

PART III

EFFECTS OF FOREST MANAGEMENT ON SOIL CARBON STORAGE

Dale W. Johnson
Desert Research Institute
P.O. Box 60220
Reno, NV 89506 USA
and
Department of Range, Wildlife, and Forestry
College of Agriculture
University of Nevada, Reno
Reno, NV 89512 USA

ABSTRACT. The literature on soil C change with forest harvesting, cultivation, site preparation, burning, fertilization, N fixation, and species change is reviewed. No general trend toward lower soil C with forest harvesting was apparent, unless harvesting is followed by intense burning or cultivation. Most studies show no significant change (± 10%) with harvesting only, a few studies show large net losses, and a few studies show a net gain following harvesting. Cultivation, on the other hand, results in a large (up to 50%) loss in soil C in most (but not all) cases. Low-intensity rescribed fire usually results in little change in soil C, but intense presribed fire or wildfire can result in a large loss of soil C. Species change can have either no effect or large effects on soil C, depending primarily upon rooting patterns. Fertilization and (especially) nitrogen fixation cause increases in soil C in the majority of cases, and represent an opportunity for sequestering soil C and causing long-term improvements in site fertility.

1. Introduction

Forest soil scientists have,until very recently, paid little attention to soil C relative to other nutrients, even though the role of soil organic matter in soil fertility (cation exchange capacity, structure, bulk density, N, P, S, and water status) is well recognized (Jurgensen et al 1989). Evaluations of the effects of harvesting, burning, and site preparation on forest productivity have concentrated upon nutrient losses and gains with much less attention given to soil organic matter (e.g., Marion, 1979; Boyle et al 1973; Johnson et al 1982, 1988b).

Water, Air, and Soil Pollution **64**: 83–120, 1992.
© 1992 *Kluwer Academic Publishers.*

With the recent concerns over increases in atmospheric CO_2 levels and global warming, forest soil scientists have an additional reason to consider changes in soil C as affected by management practices.

Even a cursory review of published global C budgets reveals that soils could be either a major source or sink for C. For example, the global C budget in Figure 1 indicates that net annual release of C as CO_2 from fossil fuel combustion is 5.3 x 10^{15} g yr-1, whereas detrital inputs to the soil are estimated at about 60 x 10^{15} g yr^{-1} and decomposition at 50 to 60 x 10^{15} g yr^{-1} (Harrington, 1977; Post et al 1990). Clearly, a slight imbalance (10%) between detrital production and decomposition could either equal (if negative) or offset (if positive) the fossil fuel contribution. There is no a priori reason to suspect that litter and soil organic matter are in a steady-state condition or even within 10% of such a condition at present, especially in view of the large changes in land use patterns that continue to occur at a global scale.

A recent analysis by Tans et al (1990) suggests that the terrestrial ecosystems of the Northern Hemisphere are absorbing 2 to 3.4 x 10^{15} g yr^{-1} of C. This amount of C is not otherwise accounted for, and could easily be sequestered within either vegetation or soils. Thus, it is important to gain a better understanding as to whether soils are a net source or a net sink of C, and, if possible make some estimate as to the magnitude of the imbalance between inputs and outputs of C to the soil. While calculations of the latter have been made (e.g., Mann, 1986; Schlesinger 1990), little is known about the error bounds surrounding such estimates; uncertainties may well be so large as to make even the overall direction of change questionable.

There have been several papers suggesting that aboveground biomass in forest ecosystems has been either a significant sink (Delcourt and Harris, 1980) or source (Houghton et al 1983; Harmon et al 1990) of C. Much less information is available on changes in C content of forest soils, and the existing literature is contradictory. Delcourt and Harris (1980) assumed that clearing and cultivation caused a 40% reduction in soil C in the southeastern U.S. In their global C model, Houghton et al (1983) assume 35, 50, and 15% losses of litter and soil C after forest clearing in tropical, temperate, and boreal forests, respectively, and a further delayed loss to 50% of original C content with cultivation. In a later paper on the C balance of Latin American forests, Houghton et al (1991) assumed that cultivation resulted in a 25% loss of soil C, whereas "Logging was assumed not to change the storage of organic carbon in soil." (p. 183). Based upon a review of the literature on tropical forest clearing, Detwiler (1986) concluded that clearing and burning alone do not cause a loss of soil C, and in some cases, may cause a gain. He notes that while clearing followed by cultivation or pasturing cause losses of soil C, "... the decrease in soil carbon is a result of the soil's use, not its clearing" (p.

75). In the absence of hard information on changes in soil C following harvesting, Harmon et al (1990) apparently took a conservative approach by assuming no change; however, they indicate that this assumption is likely false and that, "soil organic matter ... will most likely decrease under intensive management". Schlesinger (1990) analyzed chronosequence studies for soil C accumulation rate, and concluded that, on a worldwide basis, only approximately 0.4 x 10^{15} g C yr^{-1}, or 2.4 g $m^{-2}yr^{-1}$ is stored. Musselman and Fox (1991) quote estimates of 52.5 x 10^{15} g of C stored in U.S. forests, with 59% of this (31 x 10^{15} g) stored in the soil. After accounting for long-term storage of harvested wood, they estimate C lost to the atmosphere from trees to be 6.7 x 10^{15} g of C. For soils, they assume a 25 to 50% loss of C after harvest, which converts to 8 to 15 x 10^{15} g of soil C.

The objective of this paper is to review the literature on forest management with special attention to effects on soil C. It was assumed that, in addition to those few papers which actually highlight soil C changes, a great deal of information on soil C change with management was contained within publications and reports that focused on other nutrients. With the assistance of the International Energy Agency's IEA T6/A6 project ("Impacts of Forest Harvesting on Long-term Site Productivity", W.J. Dyck, Project Leader) letters were sent to foresters, soil scientists and ecologists throughout the world asking for such information. Over 100 reprints were received, most of which are summarized in this document. The reprints were divided into seven categories for the purpose of this report: 1) Harvesting, 2) Cultivation 3) Site Preparation, 4) Burning, 5) Species Conversion, 6) Reforestation and Succession, and 7) Fertilization (including N-fixation).

2. Methods and Materials

Several significant difficulties and uncertainties were encountered while trying to synthesize the literature on forest management and soil C. The question of how to report the data arose immediately: the use of absolute values of soil C would tend to minimize the importance of changes in soils with inherently low C whereas the changes in soils with large soil C would be emphasized. Changes in soils with low C may be of less importance to global C budgets than changes in soils with high C, but low C soils are of equal if not greater interest in terms of changes in site fertility. Converting the results to percentages results in the opposite effect, i.e., emphasizing the results in low C soils, perhaps unduly, in view of their significance to global C budgets. However, the advantages of converting percentages were deemed to outweigh the disadvantages; also, much of the literature on global C deals with changes in soil C on a percentage basis (e.g., Houghton et al 1983; Mann 1986; Musselman and Fox 1991).

One of the most serious problems encountered in summarizing the literature was the differences in sampling protocols, both in space and time. For example, how does one legitimately compare percent changes in C in the top 2 to 5 cm of soil reported by some authors (e.g., Ellis and Graley 1983) with percent changes in the top meter of soil reported by other authors (e.g., Johnson et al 1991)? In order to try to standardize the depth effect somewhat, the percentage changes for the entire soil profile sampled were calculated when not reported.

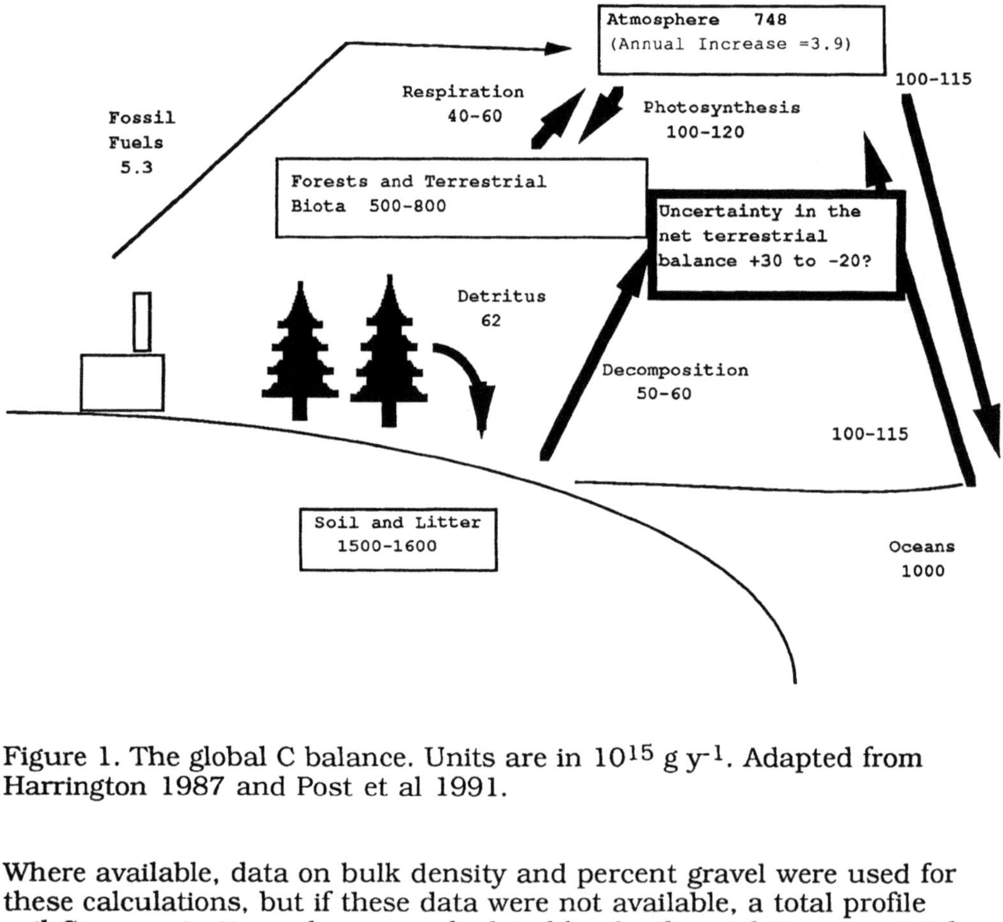

Figure 1. The global C balance. Units are in 10^{15} g y^{-1}. Adapted from Harrington 1987 and Post et al 1991.

Where available, data on bulk density and percent gravel were used for these calculations, but if these data were not available, a total profile soil C concentration value was calculated by depth weighting, i.e., total profile C = Σ(%C x depth)/ Σdepth. This procedure biases the values toward surface horizons, which generally have lower bulk densities and would therefore be given a lower weighting if bulk density data were available for the calculations. Additional sources of bias and error include differences in total soil sampling depth and unrecorded changes in bulk density (Detwiler 1986).

The intensity of sampling varied considerably among these studies; the intensity of sampling in the Hubbard Brook studies reported by Johnson et al (1991) was sufficient to detect statistical significance in a mere 8% change in soil C, whereas in other cases, differences of 20 to 50% were not statistically significant (e.g., Edmonds and McColl 1989; Mattson and Swank 1989). All changes that could be calculated are reported here, along with statistical significance or lack of it, in an attempt to detect overall patterns across sites. The alternative - to assume that no changes in soil C occurred - was deemed to be potentially more misleading than reporting actual changes, in that differences which were significant at any level up to 89% (or more, depending upon the level of significance selected) would be ignored. While the absence of statistical analyses normally negates the value of any particular study, it was reasoned that a collection of data points showing an overall trend would provide useful information, even though each data point in itself may not be particularly significant. An appropriate statistical analogy would be a regression equation, where no statistics are normally available for each point, but the collection of several points may show a meaningful trend.

Finally, many comparisons are confounded by temporal differences. Sampling intervals varied from as short as 1 mo to as long as 83 yr after treatment, and several of the long-term and chronosequence studies indicated significant temporal trends in soil C (e.g., Jenkinson 1970; Durgin 1980; Gholz and Fisher 1982). At this stage, no attempt has been made to stratify the results temporally, given the paucity of data, but potential and documented temporal variations add a significant caveat to the following summaries.

3. Results and Discussion

3.1 HARVESTING

Several studies reported soil C changes with harvesting, either alone or in combination with other treatments. For our purposes here, harvesting alone will be considered separately from harvesting with other treatments in order to avoid confusion as to what the actual effects of harvesting versus other treatments are. This is an especially important consideration when harvesting is followed by cultivation, as will be shown later.

The results of thirteen studies which considered harvesting alone are summarized in Table 1. As one would expect, harvesting alone had a significant effect upon forest floor mass, causing either increases or decreases, depending upon how much slash was left behind. The effects of harvesting on mineral soil C varied from site to site, with reports of increases, decreases, or no effects. However, the majority of the studies reported either no effects or very small changes (<10%; Table 1). The two exceptions occurred in the studies of a tropical rain forest in Ghana (Cunningham 1963) and those in a Eucalypt forest in Tasmania (Ellis and Graley 1983). Cunningham reported soil C reductions that related to the degree of shading and, presumably, soil temperature, after harvesting a tropical rain forest in Ghana. Three yr after harvesting, he noted 57, 49, and 25% decreases in soil C in the 0 to 5 cm depth with full exposure, half-exposure, and full shading, respectively. The same basic pattern held in the 5 to15 cm depth, also (30, 25, and 17% decreases in full exposure, half-exposure, and full shading, respectively). Ellis and Graley (1983) report data indicating a statistically significant (i.e., non overlapping 95% confidence intervals) 23% difference in 0 to 2 cm layer soil C between harvested and unharvested Eucalyptus sites in Tasmania. However, no significant differences were noted in deeper horizons, and the authors did not regard the surface soil differences to be of any real consequence. Thus, the overall effects of harvesting on soil C in this case, also, were actually quite small.

3.2 CULTIVATION

Mann (1985, 1986), Detwiler (1986), and Schlesinger (1986) have provided excellent reviews of the effects of cultivation on soil C. In one of her two papers, Mann (1985) utilized data from 303 loess-derived soil profiles, mostly Alfisols and Mollisols, in the central U.S., and in another paper (Mann 1986) she reviewed the results of 50 studies in the literature involving comparisons of 625 profiles. These comparisons involved both forested and prairie ecosystems, and results were not separated by these two categories. In both cases, she noted that cultivation resulted in a substantial net loss (at least 20%, mostly in the plough layer) in soils that were initially relatively high in C, but a slight net gain (e.g., 11% in Udolls of the central U.S.) in soils that were initially low in C. Using a computer model of land use changes in the tropics, Detwiler (1986) estimated that clearing followed by cultivation results in an average 40% loss of soil C, and clearing followed by pasturing results in an average 20% loss of soil C, each within 5 yr. Schlesinger summarized the effects of clearing and cultivation on soil C from several studies in the literature, and obtained an average loss of 21%, with a range of +1 to -69%.

Table 2 summarizes the reviews by Mann (1985, 1986), Detwiler (1986), and Schlesinger (1986), as well as several other studies that

Table 1. Effects of harvesting on soil C.

Location and species	Treatments	Results	Reference
Hubbard Brook, NH (Northern hardwoods)	Whole -tree harvesting	At 3 yr, 16 to19% loss in LOI in O horizons, 8% loss in E horizon, no change in B. There was a loss in Oi mass, but only 1% change (increase) in total profile organic matter.	Johnson et al 1991
La Selva, Costa Rica (tropical rain forest)	Mature forest vs secondary forest cut 1, 6, and 17 yr previously	Soil C was 8% lower in secondary forest	Reich 1983
Challenge Forest, CA (Mixed conifer)	Chronosequence 100 yr-old, clearcut 6 and 18 yr ago	6-yr-old site had 23 and 35% more total C than 100-yr-old in 0-70 and 70-140 cm, resp., 18-yr-old had 12% less in 0-70 cm, no difference deeper. (No statistics.)	Frazer et al 1990
Klamath Mts. OR (Pseudotsuga menziesii)	Clearcut 20 years previously	1% greater OM in clearcut	Aztet et al 1989
Ghana (Tropical rain forest)	Clearcut, including different degrees of artificial shading	After 3 yr, 57 and 30% decrease in 0-5 and 5-15 cm (resp.) in full exposure, 49 and 25% in half-exposure, 25 and 17% in shade. N losses slightly less.	Cunningham 1963
Hubbard Brook, NH (Northern hardwood)	Whole-tree harvest	30% increase in forest floor, 8% increase in soil after three years (non-significant)	Huntington and Ryan 1990 (same study as Johnson et al 1991)
Petawawa, Ontario (Mixed conifer, hardwood)	Whole-tree (WTH) vs conventional harvest (CH)	Increase in both forest floor (20%) and top 20 cm of soil (18%) with CH; no significant change in forest floor (-2%), but increase in top 20 cm of soil (14%) with WTH	Hendrickson et al 1989

Coweeta, NC (Mixed deciduous)	Whole-tree (WTH) vs bole only (BO)	20% greater C in BO soil, 74% greater C in WTH soil, but may be due to natural, pre-treatment variation	Mattson and Swank 1989
Michigan (Northern hardwood)	Whole-tree harvest on sites of varying quality	40-70% loss in forest floor. No soil data, but soil N changes from +50% to -60%. No statistics	Mroz et al 1985
Minnesota (Populus tremuloides)	Stem-only and whole-tree harvesting	1-2 % increases (non-significant)	Alban and Perla 1990
Tasmania (Mixed Eucalyptus, rainforest)	Clearcut	20% loss in surface 10 cm; not considered significant to site productivity	Ellis and Graley 1983
Australia (Pinus radiata)	Comparison of 4-mo 5 yr-old, and 45-yr-old clearcut	Compared to 45-yr-old, 4-mo-old site had 22% greater soil OM; 5-yr-old site had 20% greater soil OM (not stastically significant)	Edmonds and McColl 1982, 1989
Maine (Mixed hardwood, conifer)	Biomass harvest	At well-drained site, average change of -3% (not significant); at poorly-drained site, average change of -39% (significant in one of 4 horizons only)	Fernandez et al; 1989

Table 2. Effects of cultivation on soil C.

Location	Treatments	Results	Reference
303 sites in central US	Cultivation of either prairie or mature forest	28% overall decease in Udalfs, 11% increase in Udolls	Mann 1985
625 sites worldwide	Cultivation	20% decreases in 20 yr in sites initially in C, slight gain in low C sites	Mann 1986
USA, Central and South America, Africa, India, Thailand, New Guinea	Clearing and Cultivation	Average of 21% loss of soil C, range of +1 to -69%.	Schlesinger 1986
General Tropical Forests	Clearing followed by cultivation or pasturing (estimated from model)	Average loss of 40% soil C with cultivation, 20% with pasturing. No change with logging.	Detwiler 1986
Nigeria (20 sites)	Clearing and cultivation	In most cases, cropping caused decreases in soil C (10 to 50%); increases (10-40%) when crop residues returned to soil	Ayanaba et al 1976
British Columbia Washington (60 sites)	Clearing and cultivation	20% soil C decrease in 35 yr, mostly in first 15 yr.; C:N ratio narrowed from 15:1 to 12:1	Goldin and Lavkulich 1990
Brazil	Clearing and cultivation	Changes after 1 yr difficult to determine because of variability (>50%)	Tiesson and Santos 1989
North Carolina	Tillage; replacement of grasses by annual herbs	24% reduction in total C, mostly (76%) due to reduction in roots.	Richter et al 1990

Indonesia	Shifting cult., conversion to 2° forest, pine plantation and grassland	13-yr-old pine plantation was ~15% lower, grasslands were ~20% higher than natural forest. 2° forest the same as natural forest.	Arimatsu 1983
U.S.A. and tropics	Forest clearing, cropping	0-50% decrease, most in high C soils. 50% greater loss in old soils in tropics than in temperate or young tropical soils	Allen 1985
Peru	Slash and burn, bulldozer clearing with straight and shear blade then cropped with and without fertilizer	No change with slash and burn, approx. 20% reduction with bulldozing at 3 mo. Bedding and fertilization resulted in approx. 15% increase	Alegre and Cassel 1986
Brazil	Pasturing (Chronosequence)	Apparent 20-25% loss of soil C during first 1-2 years followed by increase to forest levels after 8 years as grass roots incorporate into humus.	Cerri et al 1991
Brazil	Pasturing (Chronosequence)	No clear pattern; rapid loss during first year, then gain during 2-6 years, then loss again after 12 years. Spatial variability may confuse results.	Eden et al 1991
Brazil	Cultivation	10-20% loss of soil C; also mobilization of fulvic acids and attendant degradation of soil structure and permeability.	Martins et al 1991
Ghana	Clearing, burning different types of cultivation including "local practice"	Soil C was 13 to 30% higher after clearing and burning in experimental treatments but not in "local practice". After 2 yr, soil C declined by 6% with "local practice" and 20 to 35% in other treatments	Nye and Greenland 1964

were not included in these reviews. The additional papers generally confirm earlier conclusions as to the effects of cultivation: soil C changes ranged from a slight gain to over 50% loss. One study worthy of particular mention is that of Richter et al (1990) who point out the importance of accounting for root biomass when evaluating soil C change. They found a 24% overall decrease in soil after 7 yr of annual tillage of a Udalf in Michigan, and that 76% of this decrease was due to a reduction in root biomass associated with a transition from grasses to annual herbs.

3.3 SITE PREPARATION

The effects of site preparation prior to establishment of a new forest plantation on soil C can be quite considerable, as seen in Table 3. However, the implications for global C budgets are often unclear, in that it is frequently not possible to separate soil C lost by displacement (e.g., bulldozed into slash piles and therefore not necessarily lost to the atmosphere as CO_2) and that which is lost due to decompositon. In general, there is a net loss of soil C with site preparation, the magnitude of which is dependent upon the severity of the disturbance. In that site preparation occurs only once during a forest rotation, one would expect that its overall effects on soil C loss would be less than that of continuous cultivation, however. In cases where site preparation involved incorporating logging residues into the soil, soil C values can obviously be expected to increase (e.g., Smethurst and Nambiar, 1990a). Thus, the effects of site preparation on soil C varied not only with site but with treatment. For instance, Morris and Pritchett (1983) found that only slight changes in mineral soil C due to site preparation (chopping, burning, KG-blade, disking, and bedding) in one Florida slash pine site, whereas Burger and Pritchett (1984) found significant (20-40%) reductions in soil C following site preparation (burning followed by chopping and burning followed by windrowing, disking, and bedding) in another Florida slash pine site (Table 3).

Finally, in a more unusual study, Laine and Vasander (1991) evaluated the effects of drainage and forest establishment upon the C balance of a peat bog in Finland. They found an overall ecosystem C increase of 9% due to increases in tree, litter, and peat C. They concluded that the effects of the increased productivity due to forest establishment more than compensated for any loss of peat C due to increased decomposition rate.

3.4 BURNING

The literature on burning included both prescribed burning and wildfire. The effects of burning upon both forest floor and soil C were very dependent upon fire intensity, as is to be expected. Prescribed

Table 3. Effects of forest site preparation on soil C.

Location and species	Treatments	Results	Reference
Golden Triangle, ME (Mixed hardwood)	Chopped (CH), raked (RK) windrow burned (WB)	Consistent decreases in O horizons; increases of 139, 38, and 69% with CH, RK, and WB in well-drained site, increases of 39, 17, and 11% in poorly-drained site. Changes stastically significant in only 3 of 14 horizons.	Fernandez et a; 1989
Finland (Pinus sylvestris)	Peatland drainage	9% increase overall in drained land due to increase in bulk density, %C and trees.	Laine and Vasander 1991
Mississippi (Pinus spp.)	Skidding vs skyline logging	OM up to 50% lower with severe disturbance, due to displacement. Real losses not known	Miller and Sirois 1986
Florida (P. eliottii)	Burning, low- and high-intensity site prep. at 2 sites (Cary and Bradford)	At Cary, 27% increase with minimal disturbance 20% decrease with intensive disturbance. At Bradford, no change with minimal disturbance 15% decrease with intensive disturbance	Morris and Pritchett 1983
New Zealand (Pinus radiata)	Harvesting, root-raking, burning	Displacement of considerable N, and, presumably C, by soil removal (3-18 cm) during root raking. No idea of actual total C lost to atmosphere.	Sims et al 1988
New Zealand (P. radiata)	Litter raking	No significant differences in 0-5 cm, but 23 and 15% lower in raked plots in 5-10 and 10-10-20 cm levels after 16 yr	Ballard and Will 1981
Australia (P. radiata)	Slash and litter intact (SL), slash removed (L), litter ploughed (LP), and slash and litter removed (SLR)	At 4 mo, 0-15 cm: relative to SL: L= -5%, LP= +53%, SLR = +8%. At 48 mo, all treatments decreased by 14-25% , and relative to SL: L = -7%, LP = +33%, SLR = +7%. No significant effects in 15-30 cm depth.	Smethurst and Nambiar 1990a

Florida (P. eliottii)	Clearcut and burn, then chop vs windrow and bed	At 2 yr, 54% lower C concentration with burning and chopping, 68% lower C with windrow and bedding. Contents (Mg/ha) were 21 and 40% lower, resp.	Burger and Pritchett 1984
Australia (P. radiata)	Windrow and plough	Continuous monitoring of surface soil (unspecified depth) showed decrease of approx. 40% over 42 mo.	Smethurst and Nambiar 1990b

fire usually caused a reduction in 0 horizon weight (Table 4), but either no change or an increase in mineral soil C. Often, the invasion of N-fixing species after burning caused an increase in soil C over the long-term. Kraemer and Hermann (1979) found no significant differences in soil organic matter 25 yr after broadcast burning in 34 plot pairs in western Washington and Oregon. They did, however, find a significant increase in soil C in sites occupied by N-fixing Ceanothus. Wells (1971) reported the results of a 20-yr study of prescribed burning at the Santee forest in South Carolina. Treatments included annual summer burning (AS), annual winter burning (AW), periodic (4 times) summer burning (PS) and periodic winter burning (PW). He found forest floor reductions to be as follows: AS>AW>PS>PW. However, there was a tendency for the forest floor to regain this organic matter over time and approach the control condition in the periodically-burned plots. He also found that during the first 10 yr of the study organic matter and N increased in the top 5 cm of soil in approximately the same order as forest floor was lost. Thus, "the principal effect of burning was the redistribution of the organic matter in the profile and not in any reduction" (p. 88). One treatment (annually-burned plots) showed especially large increases in soil N (550 to 990 kg/ha) during the second 10 yr of the study, which were attributed to increased activity of N-fixers.

McKee (1982) summarized the results of several prescribed burning studies throughout the southeast (including Wells' study) and concluded that burning generally resulted in a decrease in forest floor but an increase in soil C in the top 5 to 10 cm within the first 10 yr, the result being a small net overall system C loss. The causes of the increase in surface soil following prescribed burning likely include incorporation of charcoal and partially burned organic matter into the mineral soil and, in some cases, the increase in the presence of N-fixing species following burning.

In contrast to these studies of low-intensity prescribed burning, other studies of the effects of high-intensity burning (either prescribed or wildfire) show significant soil C loss. Sands (1983) reported that 24 yr after an intense broadcast burn in a Radiata pine site in Australia, soil C was approximately 40 to 50% lower throughout the profile (to a 60 cm depth) than in an unburned plot. Grier (1975) noted a 40% loss of litter and soil N after an intense fire on the eastern slope of the Cascade Mountains of Washington. Neither organic matter nor C changes were reported by Grier, but were presumably quite high, also.

Not all wildfires result in a reduction in soil C, however. Fernandez et al reported large losses of 0 horizon but no significant change in mineral soil C 1 yr after a wildfire in Maine. Dyrness et al (1989) sampled soils within one week of a wildfire in interior Alaska and found that organic matter losses from the forest floor (assessed by comparing to unburned areas) varied from 5 to 80% depending upon

Table 4. Effects of burning on soil C.

Location and Species	Treatments	Results	Reference
Australia (Eucalyptus)	Clearcut and burn (CF2); regular fire over 40 yrs (RB40); and control (C)	No significant difference between RB40 and C; CF2 was 25-43% lower than C	O'Connell 1987
Tasmania (Mixed Eucalyptus, rainforest)	Broadcast burning	Approximately 50% loss in top 10 cm with harvesting and burning, mostly in top 2 cm. Overall effect of burning seen as beneficial	Ellis and Graley 1983
Australia (Eucalyptus)	Prescribed fire once only, 1 month 3 yr, and 19 yr previously	No differences in sites newly burned or burned 3- or 19 yr previously	Edmonds and McColl 1982
South Carolina (P. plaustris)	Prescribed, 1,2,3, and 4-yr intervals	Reduction in O horizons; no effect in mineral soil	Binkley et al in press
Australia (P. radiata)	Intense broadcast burning	40-50 % reduction to 60 cm depth	Sands 1982
Washington (Mixed conifers)	Wildfire	40% loss of forest floor and soil N (No C data given)	Grier 1975
Alaska (Picea glauca, Picea mariana, Betula papyrifera, Populus tremuloides)	Wildfires of varying intensity	Loss of forest floor increased with intensity; losses (up to 15%) gains, (up to 15%), or no change in mineral soil, depending upon intensity and forest type.	Dyrness et al 1989
Maine (Mixed hardwoods, conifers)	Wildfire	Large reduction in O horizon, no effects in mineral soil one year after fire.	Fernandez et al 1989

Santee, SC (Pinus taeda)	20-yr results of annual and periodic (7 yr) burns	Little effect of periodic burns on either O or mineral soil. Annual burning decreased O and increased surface mineral soil C (30%)	Wells 1971
Oregon and Washington (Conifers)	Broadcast burning (Morris plots)	Burned plots 26% higer in OM in north Cascades, 2% lower in south Cascades (Not stastically significant). Increased with Ceanothus noted	Kraemer and Hermann 1979
British Columbia (P. contorta, Picea glauca x engelmannii)	Broadcast burning	Slight decrease (20-30%) at 9 mos, but increase again (40-70%) at 21 mos	Macadam 1987
Brewerton, AL (Pinus palustris)	Biennial winter prescribed fire	After 5 fires, burned = 4% greater than control (not significant)	McKee 1982
Olustee, FL (Pinus elliottii)	Periodic (4-yr) winter (PW), annnual winter (AW) prescribed fire	At 20 years, PW = 17% greater (significant at 95%) AW = 16% greater (not significant) than control	McKee 1982
Roberts, LA (Pinus palustris)	Annual winter prescribed fire	After 65 years. burned was 7% greater than control overall, but lower in surface 5 cm (not significant)	McKee 1982
Santee, SC (Pinus taeda)	Annual winter (AW) , annual summer (AS) periodic (7 yr) winter (PW) and periodic summer (PS) prescribed fire. (Same site as Wells 1971)	PW = -16%, PS = +6%, AW = +11%, AS = +28% relative to control at 30 yr. Only AS was significant, and only at 30-50 cm depths	McKee 1982

fire intensity. Changes in the top 5 cm of mineral soil ranged from +16 to -18% depending upon fire intensity.

3.5 SPECIES CHANGE OR COMPARISONS

Many of the studies dealing with species changes or comparisons involved N-fixers. These results have been lumped together with fertilization for the purposes of this review and are discussed below. Species change studies involving non-N fixers are summarized in Table 5.

The effects of tree species on soil C was often significant but inconsistent. For instance, Turner and Kelly (1985) and Turner and Lambert (1988) compared soil properties beneath planted radiata pine (Pinus radiata) and native Eucalyptus forest at various sites in New South Wales (NSW), Australia. In some cases, they noted greater (average of 35 to 57%) soil C in pine plantation than in the native Eucalyptus forest, in some cases they noted the reverse, and in other cases they found no differences (Table 5). The authors noted that organic matter was the main soil property that was affected by plantation establishment in their study and other related studies.

Gilmore and Rolfe (1980) reported the results of a very careful, statistically sound comparison of loblolly and shortleaf pine plantations at various spacings on soil properties over a period of 25 yr. Results showed no effect of spacing, but significant differences between species: mineral soil organic matter was higher but 0 horizon weight was lower in the shortleaf pine than in the loblolly pine stand after 25 yr. Lane (1989) reported no differences in soil C after conversion of native hardwoods to loblolly pine in South Carolina. Alban (1982) compared soil properties in adjacent stands of trembling aspen (Populus tremuloides), white spruce (Picea glauca), jack pine (Pinus banksiana), and red pine (P. resinosa) at two sites in northern Minnesota. There were no differences in total forest floor weight among the stands at either site, but the aspen soil had significantly lower surface soil organic matter (10 to 40%) than the other species at both sites. One study of peripheral interest was that of Amendinger (1990) which indicated a large (>50%) loss of soil C with the invasion of jack pine in prairie during the Holocene inferred from a chronosequence study.

Feger et al (1990) reported the nutrient budgets of two contrasting watersheds in Germany which are relevent to the subject of species effects on soil C change. No soil C or N data were presented to document the actual decline in soil C, N, and S contents, and thus the study is not summarized in Table 5; however, some of the results are worth summarizing here. The sites were: Schluchsee, which has granitic bedrock and within which Norway spruce (Picea abies) are experiencing Mg deficiency, and Villingen, which has sandstone bedrock and within which Scots pine (Pinus sylvestris) and silver fir

Table 5. Effects of species change on soil C.

Location	Treatments	Results	Reference
Australia (Pinus radiata, Eucalyptus)	Radiata pine vs Eucalyptus	Soil C was greater (by 45% in surface and 27% in subsurface horizons) beneath pines in one site, no difference in another site.	Turner and Kelly 1985
Australia (P. radiata, Eucalyptus)	Radiata pine vs Eucalyptus	Pine had greater soil C in low fertility site, lower soil C in high fertility site.	Turner and Lambert 1988
Illinois (P. echintata, P. taeda)	Loblolly vs shortleaf pine plantations on former agricultural land	Greater soil C (20%) but lower O horizon in shortleaf plantations at age 25. No effect of spacing.	Gilmore and Rolfe 1980
Minnesota (P. banksiana)	Chronosequence of prairiet o jack pine	Large loss (>50%) inferred in jack pine invasion during the Holocene	Almendinger 1990
Clemson, SC (P. taeda)	Conversion of hardwood to loblolly pine	No change after 23 yr	Lane 1989
Minnesota (Pinus resinosa, P. banksiana, Picea glauca, Populus tremuloides)	Comparison of four species at 40 yr of age	Significantly less soil C under aspen (10-40%) than other species	Alban 1982

(Abies alba) are experiencing K deficiency. Of interest to this review are high rates of N and S leaching from the Schlucsee watershed, an effect attributable to the mineralization of organic matter left in subsoils from a deeper-rooted beech stand (Fagus sylvatica) which occupied the site 150 yr previously.

3.6 REFORESTATION AND SUCCESSION

Examining the changes in soil C during reforestation and succession is one way of gaining some insight into the long-term effects of harvesting, cultivation, etc., on soil C reserves (Table 6). In cases where former agricultural land is reverted to forest or where newly developing soil undergoes afforestation, soil C usually increases substantially.

In a study involving resampling of soils over time, Johnson et al (1988) noted either increases (30 to 100%) or no significant change in surface soil (0 to 15 cm) C over an 11-yr period in aggrading forests growing on former agricultural land on Walker Branch Watershed, Tennessee. Jenkinson (1970, 1991) reported the results of the Rothamsted studies of organic matter and nutrient changes in soils left uncultivated since the early 1880's. One site (Broadbalk) was on a calcareous soil that had been limed sometime during the 18th or early 19th century, the effects of which were still evident in the pH of samples taken in 1964 to 1965. The other site (Geescroft) received N and P fertilizers but no lime and consequently experienced significant acidification (pH 7.1 to 4.5) from 1883 to 1965. Differences in acidification were thought to have resulted in substantial differences in soil organic C, N, S, and P, all of which were greater at Broadbalk. Soil organic C in Broadbalk increased by 80% over the 83-yr period, whereas Geescroft increased by only half as much (Jenkinson 1970). Rates of N, S, and P accumulation were also considerably greater at Broadbalk than Geescroft. Of special interest was the finding that the rates of organic N accumulation (65 and 23 kg N ha^{-1} yr^{-1}) were greater than could be accounted for by either atmospheric deposition or N fixation by legumes.

Chronosequence studies have also shown significant soil C accumulation when former agricultural land is reforested or afforested. Wilde (1964) examined soil organic matter in 100 red pine (Pinus resinosa) plots of varying age (from 10 50 yr) planted on former agricultural land in Wisconsin. He found a linear increase in soil organic matter in the top 15 cm of soil with stand age, with the overall increase being 300 to 400% over 40 yr. Lugo et al (1986) assessed the affects of conversion of former agricultural land to either forest or pasture in Puerto Rico. They were motivated to test some of the assumptions used in global climate models that the top 1 m of soil loses 65% of its C after deforestation, and that it can return to within

Table 6. Effects of reforestation and succession on soil C.

Location	Treatments	Results	Reference
Wisconsin (Pinus banksiana)	100 plantations on old fields	Linear increase (+300 to 400%) in soil C in top 15 cm up through age 50	Wilde 1964
Virgnina (P. taeda, P. virginiana)	Old field chronosequence	35% increase in soil C over 50 yr	Schiffman and Johnson 1990
Puerto Rico (Subtropical wet and moist forests)	Coffee shade plantations vs secondary forest	50% greater soil C in secondary forest	Weaver et al 1987
Puerto Rico (Subtropical forests)	Conversion of agricultural land to forest or pasture	Approximate doubling of soil C within 50 yr, to within 90% of mature forest.	Lugo et al 1986
Florida (P. eliottii)	Chronosequence after forest harvest	Initially high A horizon soil C due to bedding of slash decreases by 50% to pre-harvest levels by age 5. No trend from ages 5 to 35.	Gholz and Fisher 1982
Rothamsted, UK (Mixed hardwoods)	Agricultural land reverted to forest	Approximate doubling of soil C over 83 years	Jenkinson 1970, 1991
Hubbard Brook, NH (Northern hardwood)	Old field chrono-sequence	Decrease (-20%) in Ap over 75 years. Increase in O horizon was greater than decrease in Ap.	Hamburg 1984
Oregon (Tsuga mertensiana)	Mt. hemlock waves	Decrease in forest floor in regrowth, no change in soil C	Boone et al 1988
Hawaii (Metrosideros collina ssp. polymorpha)	ohi'a forest on ash and lava flow	380 kg ha^{-1} yr^{-1} accumulation over first 191 yr; slower thereafter up to 4000 yr	Vitousek et al 1983

Tennessee (Mixed deciduous)	11-year interval sampling	Either no change or increases in soil C (20-100%)	Johnson et al 1988
Six Rivers, CA (Mixed conifer, Sequoia sempervirens)	Chronosequences following clearcutting	No change in fir after 25 yr; slight decrease (25%) in redwood until age 25, then increase Likely logging residue effect.	Durgin 1980

75% of its original value within 50 yr of abandonment of agriculture. They evaluated the effects of abandonment of agriculture in several life zones (subtropical wet, dry, and moist forest) through resampling of specific sites through time and chronosequence studies. They found that recovery of soil C was much more rapid than generally assumed in models: for instance, chronosequence studies indicated that abandoned agricultural soils in the wet and moist forest life zones regained 90% of the soil C in mature forests within 50 yr, and in the dry zone, recovery of soil C to levels in mature forests occurred within 30 yr.

Schiffman and Johnson (1990) compared C contents of two loblolly pine chronosequences, one growing on a former agricultural field, and one on a site converted from naturally-regenerated Virginia pine (Pinus virginiana). There was a large increase in ecosystem C content (235%), mostly due to phytomass in the old field chronosequence but only a 24% gain in the natural forest conversion chronosequence. Similarly, there was a large increase in soil C (57%) in the old field chronosequence and no significant change in the forest conversion chronosequence. The authors drew two important conclusions as to the effects of reforestation in the southeastern U.S.: 1) there were "negligible oxidative losses of carbon from soils after harvest and site preparation" of natural forests (p. 69), and 2) "the conversion of natural forests to plantations is no substitute for the farm to forest conversion" in terms of C storage. The latter is of significance, in view of the marked reductions in the rate of old field reforestation in this region.

An exception to the general pattern of increased soil C after reforestation of formerly cultivated soils is the study by Hamburg (1984) at Hubbard Brook, NH. In this case, Ap horizon soil C decreased over the 75-yr chronosequence. Even in this case, however, increases in 0 horizon C more than offset the decreases in Ap horizon C.

In cases where reforestation and succession follow previous forests, there is no clear pattern. As noted above, Schiffman and Johnson (1990) found negligible effects of forest conversion from native Virginia pine to loblolly pine. Boone et al (1988) found no change in soil C but a decrease in 0 horizon C in regrowth following hemlock waves in Oregon. Durgin (1980) found no changes in fir forests 25 yr after clearcutting and slight decreases followed by increases in redwood forests following clearcutting in California. In a chronosequence study in slash pine (Pinus elliottii), Gholz and Fisher (1982) found that the A horizon of a 2-yr-old stand contained approximately twice as much soil C as the other stands (up to age 34) in the chronosequence), which was attributed to bedded slash. Assuming that the chronsequence truly represented trends with time, the effect of slash bedding was very short-lived: soil C decreased by 50% to approximately pre-harvest levels by age 5.

3.7 NITROGEN FIXATION AND FERTILIZATION

For the purposes of this review, the effects of N-fixing species and fertilization on soil C were combined into one category (Table 7). For the most part, the presence of N-fixers caused substantial (20 to 100%) increases in soil C and N (Table 7). The one exception to this general rule was the study by Paschke et al (1989) in Illinois, which evaluated the effects of black alder (Alnus glutinosa) and autumn olive (Elaegnus umbellata) interplantings with black walnut (Juglans nigra). In this case, interplanting with both alder and autumn olive resulted in significantly greater mineral N levels and N mineralization rates than in walnut alone, but there were no increases in either soil total C or N with interplanting after 18 yr. Indeed, there was a clear and significant trend toward lower soil C and N in the autumn olive interplanted plantations than in walnut only plantations. Reasons for this were not known.

Another seeming exception to this general rule was the Cascade Head site in Oregon, where there was only an 11% difference in soil C between red alder (Alnus rubra) and Douglas-fir (Pseudotsuga menziesii) soils. In this case, however, the Douglas-fir soil was quite high in both C and N, and the 11% difference was actually about equal in magnitude to larger percentage increases at other sites (Binkley and Sollins 1990).

Fertilization generally caused an increase in soil C, as one would expect given its expected effect upon primary productivity. However, the increases in soil C with fertilization were generally not as large as those due to the presence of N fixers. Nohrstedt et al (1989) evaluated soil C and microbial activity in two sites in Sweden: Kroksbo, which was treated with ammonium nitrate and urea at 150 and 600 kg N ha^{-1} 11 yr previously, and Nissafors, which was treated with 150 kg N ha^{-1} as ammonium nitrate at both 8-yr and 1 yr previously (for a total of 300 kg N ha^{-1}). They found an increase of 16 to 25% in litter plus soil C in the Kroksbo site, and an increase of 10% overall in the Nissafors site. The effect was more pronounced at higher fertilization levels and more pronounced with ammonium nitrate than with urea. Interestingly, they could not account for the increased C with increased litterfall, and attribute the effect to reduced microbial activity per unit organic C.

Van Cleve and Moore (1978) noted increases in soil C of 13 to 17% with N (ammonium nitrate) and P (triple super P) fertilization of aspen sites in central Alaska. Turner and Lambert (1986) noted up to 22% increase in soil C 30 yr after a single superphosphate fertilization in a P-deficient Radiata pine plantation in New South Wales, Australia. In contrast, McCarthy (1983) found only a slight (5%) increase in soil C in slash pine plantations in Florida 20 yr after P fertilization. Gilmore and Boggess (1963) and Gilmore (1977, 1980) reported on studies

Table 7. Effects of nitrogen fixation and fertilization on soil C.

Location	Treatments	Results	Reference
British Columbia, Washington (Pseudotsuga menziesii, Alnus rubra)	N fixation	Increases by 30-100%	Binkley 1983
Wind River and Cascade Head, OR (P. menziesii A. rubra)	N fixation	36% increase (+28,020 kg ha^{-1}, p=90%) at Wind River, non-significant 11% increase of about the same absolute magnitude (+25,020 kg ha^{-1}) at Cascade Head	Binkley and Sollins 1990
Oregon (P. menziesii, Ceanothus spp.)	N fixation	40 to 60% increase with Ceanothus	Binkley et al 1982
Thompson, WA (P. Menziesii, A. rubra)	N fixation	50% more C in red alder soil	Brozek 1990
Wind River, OR (P. menziesii, A. rubra)	Interplanting with N fixers	Greater in surface soils in interplanting vs fir alone (+30-40%)	Tarrant and Miller 1963
Illinois (Juglans nigra, Alnus glutinosa, Elaeagnus umbellata)	Interplanting with N fixers	Increase (+22%, , not significant) with Alnus ; decrease (-15%, not significant) with Elaeagnus	Paschke et al 1989
Illinois (Platanus occidentalis, Fraxinus pennsylvanica, Liriodendron tulipifera)	Plantations on agricultural land previously treated with manure, crop residue, rock phosphate, and lime	Previously limed plots were 20-30% higher (p=95%) throughout the study. Manuring added increase, and yellow-poplar had higher C, also	Gilmore 1977

Illinois (Pinus taeda, P. strobus, P. resinosa, P. echinata)	Plantations on agricultural land previously treated with manure, crop residue, rock phosphate, and lime	Soil C was 2x greater in limed plots in 1955 but by 1978, differed by only 10-20% (p=95%). Soil C in unlimed plots increased to near limed plot levels.	Gilmore 1980
Florida (Pinus elliottii)	P fertilization	Slight increase (5%) with highest application (157-314 kg ha^{-1}. Very large increase in forest floor. (No statistics)	McCarthy 1983
Sweden (Pinus sylvestris)	Urea and ammonium nitrate at 150 and 600 kg N ha^{-1}	10-26% greater C in fertilized plots most pronounced at higher fertilizer levels and with ammonium nitrate. Effect due to reduced microbial activity rather than greater litterfall. (Not significant at 95% level)	Nohrstedt et al 1989
New Zealand (P. radiata)	N,P,K,S,Ca, and Mg fertilizer and lupines on sand dunes	9% increase with lupine (p=95%), 6% increase with fertilization (not sign.) 17% increase with both (not sign.)	Baker et al 1986
Australia (P. radiata)	Superphosphate	Up to 22% increase with 100 kg P ha^{-1}.	Turner and Lambert 1986
Alaska (Populus tremuloides)	Ammonium nitrate (777 kg ha^{-1}), triple super P (385 kg ha^{-1}), KCl (111 kg ha^{-1}.as K) over a 6-yr period	Increase in surface soil with N (+30%, p=95%) and P (+34%, p=95%)	Van Cleve and Moore 1978

where various tree species (both pines and hardwoods) were planted on agricultural soils previously treated with manure, crop residue, P (rock phosphate) and limestone, both singly and in combination. In the hardwood plantations, previosusly limed plots had significantly greater (20 to 30%) soil organic matter at the start of the study, and the effect carried through the 18th yr. In the pine plantations (P. taeda, P. echinata, P. resinosa, and P. strobus), previously limed soils had twice as much organic matter (OM) initially (in 1955). Soil OM increased in both limed and unlimed plots, but at a greater rate in the unlimed plots, so that at age 24, the limed plots were only 20 to 25% higher in soil OM.

Baker et al (1986) compared the effects of mixed fertilizer (960, 410, 410, 140, 200, and 290 kg ha^{-1} of N, P, K, S, Ca, and Mg, respectively), lupine (Lupinus arboreus), and lupine plus fertilizer on planted radiata pine on sand dune sites in New Zealand. They documented a statistically significant effect of fertilization on soil C (115% in the top 5 cm). Lupine and lupine plus fertilizer also caused increases (47 and 89%, respectively), but these were not statistically significant. To a 1 m soil depth, the effects of treatments were somewhat different, however: only the lupine treatment was significantly different from control (9% greater), although fertilizer and lupine plus fertilizer treatments alone also resulted in increases (6 and 17% greater, respectively).

4. Conclusions

Despite the numerous uncertainties and caveats noted in Section 2, the results of this literature review reveal some reasonably consistent results and trends in soil C under various forest management scenarios. It has long been established and remains clear that cultivation leads to substantial decreases in soil C in all but the most C-poor soils (Figure 2; Mann 1985, 1986; Schlesinger 1986; Detwiler 1986). However, the assumption that soil C decreases on the order of 30 to 40% following forest harvesting (e.g., Houghton et al 1983; Musselman and Fox 1991) is not supported by the literature reviewed here. Rather, it appears as if losses of soil C after harvesting and reforestation are negligible in most cases. The effects of harvesting, site preparation, and burning on soil C (not including litter) are summarized in Figure 3a and b. In Figure 3a, only statistically significant results are reported whereas in Figure 3b, all results are reported.

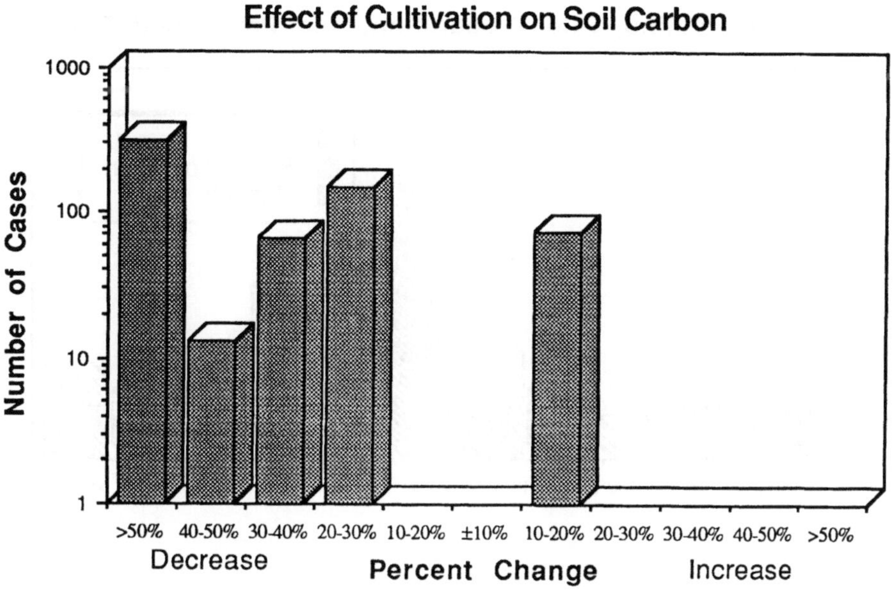

Figure 2. Frequency distribution of the percentage change in soil C with cultivation (see Table 2 for data sources).

Regardless of whether only statistically significant differences or overall trends are considered, the majority of studies reviewed here indicate little or no change in soil C (i.e., ± 10%) following harvesting and reforestation. The exceptions to this are primarily in the tropics, where recovery to original levels after reforestation is apparently quite rapid, and in cases where harvesting is followed by intense broadcast burning (e.g., Sands 1983). However, there are also instances where soil C increased after harvesting, probably due to the additions of slash, increased decomposition rates, and incorporation of organic matter into the mineral soil (e.g., Gholz and Fisher 1982; Henricksen et al 1989).

It is important to recognize that cultivation for crops differs substantially from harvesting and site preparation in new forest plantations. Crop cultivation typically involves much more severe and prolonged disturbance than harvesting, even with intensive site preparation. Crop cultivation also very likely leads to long-term increases in soil temperature, whereas soil temperatures are likely return to near pre-harvest levels rapidly after the development of a new forest canopy. Thus, it is not at all surprising that soil C losses following harvesting and reforestation are substantially less than with harvesting followed by cultivation, and these differences must be taken

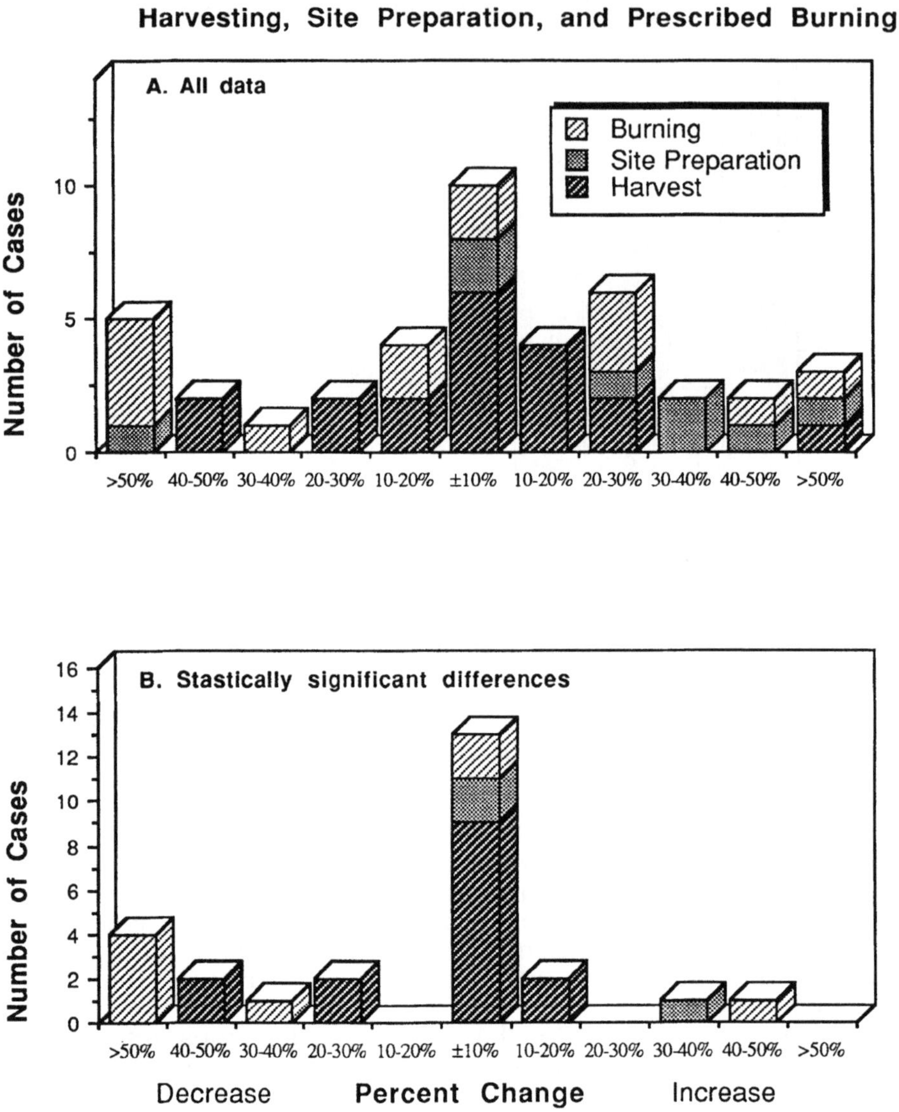

Figure 3. Frequency distribution of the percentage change in soil C with forest harvesting, site preparation, and burning. A. Statistically significant differences only shown (non-significant differences included in the ± 10% category), B. All differences shown. (See Tables 1,3, and 4 for data sources).

into account when evaluating the effects of forest harvesting in general on global C balances. It is likely true that harvesting and cultivation result in large changes in soil C on the order of 30 to 50% over a period of several decades. However, there is nothing in the literature to suggest that such changes occur when harvesting is followed by forest replanting.

It is clear that the effect of fire upon soil C is a function of fire intensity. A light or moderate burn causes a mobilization of nutrients, and may be beneficial to the growth of the subsequent forest. Figure 4 shows that the effects of regular prescribed fire on soil C is heavily

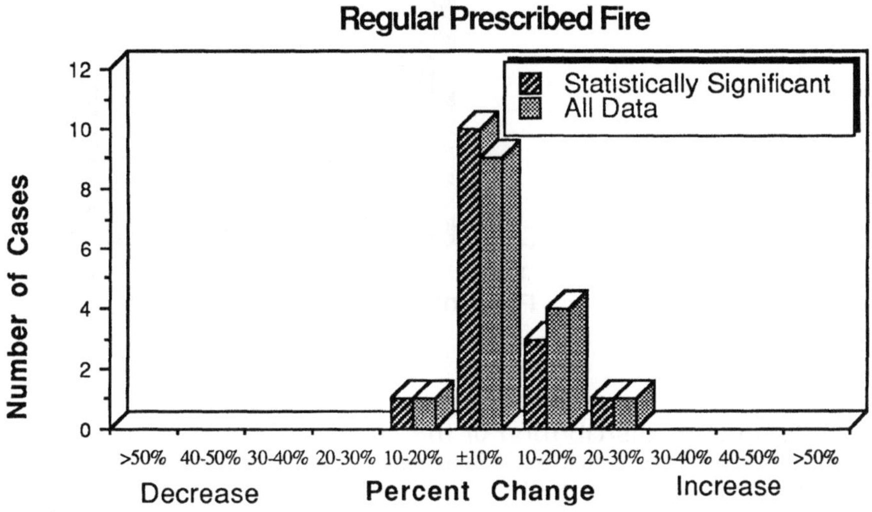

Figure 4. Frequency distribution of the percentage change in soil C with regular prescribed fire. S=Statistically significant differences only shown (non-significant differences included in the ± 10% category), A=All differences shown. (See Table 6 for data sources).

weighted toward the center (negligible effect) but somewhat skewed right, indicating a positive effect. An intense burn, on the other hand, may deplete the soil of volatile nutrients (including N, S and P; Raison et al 1985), causing a long-term decrease in forest productivity and C sequestration.

There are clearly opportunities for increasing soil fertility and the fixation of C in forest ecosystems through the management of forest nutrition: there is a marked, clear trend toward greater soil C with the introduction of N-fixers as well as with fertilization (Figure 5). The benefits of this must be weighed against the cost of fertilization or the cost of allowing N-fixing species of low economic value to inhabit the sites in question.

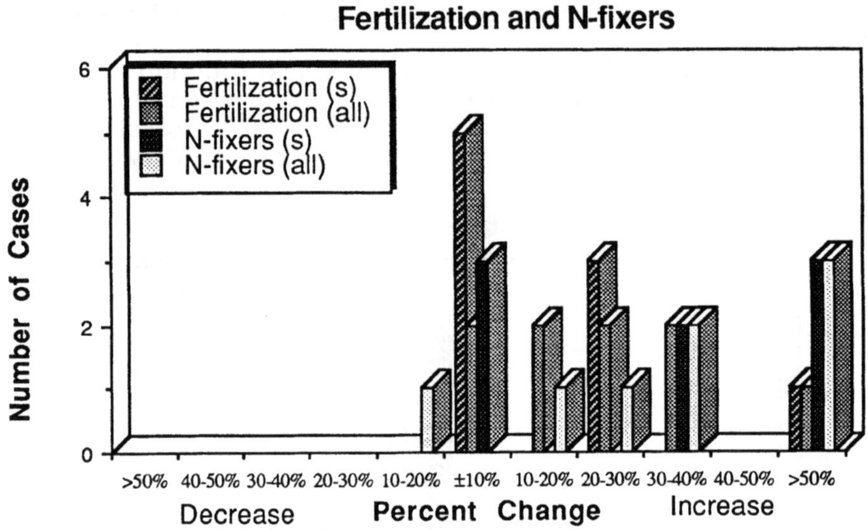

Figure 5. Frequency distribution of the percentage change in soil C with N fixers and fertilization S=Statistically significant differences only shown (non-significant differences included in the ± 10% category), A=All differences shown. (See Table 7 for data sources).

There are two possible reasons for the increase in soil C following N fixation and fertilization: 1) increased productivity and, therefore, increased organic matter input to soils, and 2) stabilization of soil organic matter. In the case of N, non-biological condensation reactions of phenols with ammonium are important in the production of humus (Mortland and Wolcott, 1965; Paul and Clark 1989). These reactions are enhanced by high pH (because NH_3 is the reactive form of N) and high NH_3 and/or NH_4^+ concentrations. Both of these conditions occur following urea fertilization, which is known to cause non-biological NH_4^+ fixation (Foster et al 1985). In the case of Ca and other polyvalent cations, cation bridging of organic colloids causes condensation and stabilization of organic matter (Oades 1988). Oades (1988) suggests that the well-documented positive relationship between soil clay and organic matter content may actually be the result of greater polyvalent cation availability (either Ca or Al) in clay rich soils. Because Ca is rarely limiting to tree growth, the positive effects of liming on soil C noted by Gilmore (1977, 1980) and Jenkinson (1970, 1991) are likely due to these reactions rather than a direct effect upon plant primary productivity.

5. Research Needs

As noted in the Objectives and Methods section, there are numerous inconsistencies in the way data was collected and summarized in the studies reviewed here. This is certainly not meant as a criticism of these studies, each of which was designed to test a specific hypothesis or answer a specific question. However, there is a clear need for a coordinated, regional study on the effects of forest management on soil C dynamics such as has been done for nutrient effects (e.g., Mann et al 1988) and such as that proposed by Powers et al (1990). Such a study should control for both management practices (e.g., degree of residue removed, burning, bedding, etc.) and establish sampling protocols that eliminate the current uncertainties arising from unknown effects of spatial and temporal variation among study sites.

In addition to (or in conjunction with) a coordinated regional study, more research is needed on the processes controlling soil C accumulation and loss. What are the roles of temperature versus moisture on decomposition? What are the effects of extremes versus changes in mean values of temperature and moisture? What role do nutrients play in stabilization or loss of soil organic matter? Such process studies should include not only changes in litter decomposition, which has been extensively studied, but also the incorporation of litter into soil organic matter and, ultimately, humus. Investigations of soil C fractions, even on existing study sites, may be a meaningful first step in obtaining insight into important processes.

Acknowledgments

Research supported by the National Council of the Paper Industry for Air and Stream Improvement, Inc. (NCASI). An earlier version of this paper was issued as a NCASI report. Many thanks are due to Bill Schlesinger and Kim Mattson for their comments.

References

Ayanaba, A. S.B. Tuckwell, and D.S. Jenkinson. 1976. The effects of clearing and cropping on the organic reserves and biomass of tropical forest soils. Soil Biol. Biochem. 8: 519-525.

Aztet, T., R.F. Powers, D.H. McNabb, M.P. Amaranthus, and E.R. Gross. 1989. Chapter 10. Maintaining long-term forest productivity in southwest Oregon and Northern California. pp. 185-201 IN: D.A. Perry, R. Meurisse, B. Thomas, R. Miller, J. Boyle, J. Means, D.R. Perry, and R.F. Powers (eds.) Maintaining long-term productivity of Pacific Northwest Forest Ecosystems. Timber Press, Inc. Portland, OR.

Baker, T.G., G.R. Oliver, and P.D. Hodgkiss. 1986. Distribution and cycling of nutrients in Pinus radiata as affected by past lupin growth and fertiliser. For. Ecol. Managem. 17:169-187.

Ballard, R., and G.M. Will. 1981. Removal of logging waste, thinning debris, and litter from a Pinus radiata pumice soil site. New Zealand Journal of Forestry Science 11:152-163.

Binkley, D. 1983. Ecosystem production in Douglas-fir plantations: Interactions of red alder and site fertility. For. Ecol. Manage. 5:215-227

Binkley, D., and N.L. Christensen. Chapter 8. The effects of canopy fire on nutrient cycles and plant productivity. IN: R. Laven and P. Omi (eds.) Pattern and process in crown fire ecosystems. Princeton University Press (in press).

Binkley, D., K. Cromack, and R.L. Fredriksen. 1982. Nitrogen accretion and availability in some snowbrush ecosystems. Forest Sci. 28:720-724.

Binkley, D., D. Richter, M.B. David, and B. Caldwell. Soil chemistry in a loblolly/longleaf pine forest with interval burning. Ecol Appl. (in press).

Binkley, D,, and P. Sollins. 1990. Factors determining differences in soil pH in adjacent conifer and alder-conifer stands. Soil Sci. Soc. Amer. J. 54:1427-1433.

Boone, R.D., P. Sollins, and K. Cromack. 1988. Stand and soil change along a mountain hemlock death and regrowth sequence. Ecology 69:714-722.

Boyle, J.R., J.J. Philiips, and A.R. Ek 1973 "Whole-tree" harvesting: Nutrient budget evaluation J. For. 71:760-762

Brozek, S. 1990. Effect of soil changes caused by red alder (Alnus rubra) on biomass and nutrient status of Douglas-fir (Pseudotsuga menziesii) seedlings. Can. J. For. Res. 20: 1320- 1325.

Burger, J.A., and W.L.Pritchett.1984. Effects of clearfelling and site preparation on nitrogen mineralization in a southern pine stand. Soil Sci. Soc. Amer. J. 48: 1432-1437.

Cunningham, W. 1963. The effect of clearing on a tropical forest soil. J. Soil Sci. 14: 334-345.
Delcourt, H.R., and W.F. Harris. 1980. Carbon budget of the southeastern U.S. Biota: Analysis of historical change in trend from source to sink. Science 210:321-323.
Detwiler, R.P. 1986. Land use change and the global carbon cycle: the role of tropical soils. Biogeochemistry 2: 67-93.
Durgin, P.B. 1980. Organic matter content of soil after logging of fir and redwood forests. USDA Forest Service Res. Note PSW-346. pacific Southwest Forest and Range Experiment Station, Berkeley, CA.
Dyck, W.J., P.D. Hodgkiss, G.R. Oliver, and C.A. Mees. 1991. Harvesting impacts on second-rotation productivity on sand dune forests. Rep. No. 2710, Forest Research Institute, Rotorua, New Zealand.
Dyrness, C.T., K. Van Cleve, and J.D. Levison. 1989. The effect of wildfire on soil chemistry in four forest types in interior Alaska. Can. J. For. Res. 19:1389-1396.
Eden, M.J., P.A. Furley, D.F. M. McGregor, W. Milliken, and J.A. Ratter. 1991. Effect of forest clearance and burning on soil properties in northern Roraima, Brazil. For. Ecol. Manage. 38: 283-290.
Edmonds, R.L., and J. G. McColl. 1983. Forest management effects on soil nitrogen in Eucalyptus pauciflora and Pinus radiata stands in the Australian Capital Territory, Australia. pp. 259-263 IN: R. Ballard and S. P. Gessel (eds.), IUFRO Symposium on Forest Site and Continuous Productivity. USDA Forest Service Gen. Tech. Rep. PNW-163, Portland, Oregon.
Edmonds, R.L., and J.G. McColl. 1989. Effects of forest management on soil nitrogen in Pinus radiata stands in the Australian Capital Territory. For. Ecol. Manage. 29:199-212.
Ellis, R.C., and A.M. Graley. 1983. Gains and losses in soil nutrients associated with harvesting and burning eucalypt rainforest. Plant. Soil. 74:437-450.
Feger, K.H., G. Brahmer, and H.W. Zoettl. 1990. Element budgets of two contrasting catchments the Black Forest (Federal Republic of Germany). J. Hydrol. 116:85-99.
Fernandez, I.J., J. Logan, and C.J. Spencer. 1989. The effects of site disturbance on the mobilization and distribution of nutrients and trace metals in forest soils. Environmental Studies Center, University of Maine, Orono, ME
Foster, N.W., E.G. Beauchamp, and C.T. Corke. 1985. Reactions of ^{15}N-labelled urea with Jack pine forest floor materials. Soil Biol. Biochem. 17:699-703
Frazer, D.W., J.G. McColl, and R.F. Powers. 1990. Soil nitrogen mineralization in a clearcutting chronosequence in a northern California conifer forest. Soil Sci. Soc. Amer. J. 54:1145-1152.

Gholz, H.L., and R.F. Fisher. 1982. Organic matter production and distribution in slash pine (Pinus elliottii) plantations. Ecology 63:1827-1839.
Gilmore, A.R., and W.R. Boggess. 1963. Effects of past agricultural practices on the survival and growth of planted trees. Soil Sci. Soc. Amer. Proc. 27: 98-102.
Gilmore, A.R. 1 977. Changes in a reforested soil associated with tree species and time. II. Soil organic content and pH in hardwood plantations. Univ. Illinois Agr. Exp. Sta. For. Res. Rep. No. 77-1.
Gilmore, A.R. 1980. Changes in a reforested soil associated withtree species and time. IV. Soil organic content and pH in pine plantations after 24 years. Univ. Illinois Agr. Exp, Sta. For. Res. Rep. No. 80-3.
Gilmore, A.R., and G.L, Rolfe. 1980. Variation in soil organic matter in shortleaf pine and loblolly pine plantations at different tree spacings. Univ. Illinois Agr. Exp. Sta. For. Res. Rep. No. 80-2.
Goldin, A. and L.M. Lavkulich. 1990. Effects of historical land clearing on organic matter and nitrogen levels in soils of the Fraser lowland of British Columbia. Can S. Soil Sci. 70: 583- 592.
Greenland, D.J., and P.H. Nye. 1959. Increases in the carbon and nitrogen contents of tropical soils under natural fallows. J. Soil Sci. 10: 284-299.
Grier, C.L. 1975. Wildfire effects on nutrient distribution and leaching in a coniferous ecosystem. Can. J. For. Res. 5:599-607.
Hamburg, S.P. 1984. Effects of forest growth on soil nitrogen and organic matter pools following release from subsistence farming. pp. 145-158 IN E. L. Stone (ed.), Forest Soils and Treatment Impacts, Proceedings of the Sixth North American Forest Soils Conference, University of Tennessee, Knoxville, TN.
Harmon, M.E., W.K. Ferrel, and J.F. Franklin. 1990. Effects on ca@on storage of conversion of old-growth forests to young forests. Science 247:699-702
Harrington, J.B. 1987. Climatic change: A review of the causes. Can. J. For. Res. 17: 1313-1339.
Henricksen, O.Q., L. Chatarpaul, and D. Burgess. 1989. Nutrient cycling following whole-tree and conventional harvest in a northern mixed forest. Can J. For. Res. 19:725-735.
Houghton, R.A., J.E. Hobbie, J.M. Melillo, B. Moore, B.J. Peterson, G.R. Shaver, and G.M. Woodwell. 1983. Changes in the carbon content of terrestrial biota and soils between 1860 and 1980: a net release of CO_2 to the atmosphere. Ecol. Monogr. 53: 235-262.
Houghton, R.A., D.L. Skole, and D.S. Lefkowitz. 1991. Changes in the landscape of Latin America between 1850 and 1985. II. Net release of CO_2 to the atmosphere. For. Ecol. Manage. 38: 173-199.
Huntington, T.G., and D.F. Ryan. 1990. Whole-tree-harvesting effects on soil nitrogen and carbon. For. Ecol. Manage. 31:193-204.

Jenkinson, D.S. 1970. The accumulation of organic matter in soil left uncultivated. Comm. Bur. Soil. Sci. Rothamstad Exp. Sta. Rep. for 1970.
Jenkinson, D.S. 1991. The Rothamsted long-term experiments: Are they still of use? Agron. J. 83: 2-10
Johnson, C.E., A.H. Johnson, T.G. Huntington, and T.G. Siccama. 1991. Whole-tree clear-cutting effects on soil horizons and organic matter pools. Soil Sci. Soc. Amer. J. 55:497-502.
Johnson, D. W., G. S. Henderson, and D. E. Todd. 1988. Changes in nutrient distribution in forests and soils of Walker Branch Watershed, Tennessee, over an eleven-year period. Biogeochemistry 5:275-293.
Johnson, D. W., D. C. West, D. E. Todd, and L. K. Mann. 1982. Effects of sawlog vs whole-tree harvesting on the nitrogen, phosphorus, potassium, and calcium budgets of an upland mixed oak forest. Soil Sci. Soc. Am. J. 46:1304-1309.
Jurgensen, M.F., A.E. Harvey, R.T. Graham, M.J. Larson, J.R. Tonn, and D.S. Page-Dumroes. 1989. Soil organic matter, timber harvesting, and forest productivity in the inland Northwest. pp. 392-415 IN: S.P. Gessel, D.S. LcCate, G.F. Weetman, and R.F. Powers eds.). Sustained productivity of forest soils. 7th North American Forest Soils Conference, University of British Columbia, Vancouver, B.C.
Kinerson, R.S., C.W. Ralston, and C,G. Wells. 1977. Carbon cycling in a loblolly pine plantation. Oecologia 29:1-10
Kraemer, J.F., and R.K. Hermann. 1979. Broadcast burning: 25-year effects on forest soils in the western flanks of the Cascade Mountains. For. Sci. 25:427-439.
Laine, J., and H. Vasander. 1991. Effect of forest drainage on the carbon balance of sa sedge fen ecosystem. Proc. Symp. "The changing face of fenlands and implications for their future use. April 9-11, 1991, Cambridge, England.
Lane, C.L. 1989. Forest stand conversion from Hardwoods to pine: Twenty-three years later. pp 253-256 IN: S.P. Gessel, D.S. LcCate, G.F. Weetman, and R.F. Powers eds.). Sustained productivity of forest soils. 7th North American Forest Soils Conference, University of British Columbia, Vancouver, B.C.
Lugo, A.E., A.J. Sanchez, and S. Brown. 1986. Land use and organic carbon content of some subtropical soils . Plant Soil 96: 185-196.
Macadam, A.M. 1987. Effects of broadcast slash burning on fuels and soil chemical properties in the Sub-boreal spruce Zone of central British Columbia. Can J. For. Res.17:1577-1584.
Mann, L.K. 1985. A regional comparison of carbon in cultivated and uncultivated alfisols and mollisols in the central United States. Geoderma 36:241-253.
Mann, L.K. 1 986. Changes in soil carbon storage after cultivation. Soil Sci. 142:279-288.

Mann, L. K., D. W. Johnson, D. C. West, D. W. Cole, J. W. Hornbeck, C. W. Martin, H. Riekerk, C. T. Smith, W. T. Swank, L. M. Tritton, and D. H. Van Lear. 1988. Effects of whole-tree and stem-only clear-cutting on post harvest hydrologic losses, nutrient capital, and regrowth. Forest Science 42:412-428.

Marion,G.K. 1979. Biomass and nutrient removal in long-rotation stands pp 98-110 IN:A. L. Leaf (ed.), Impact of Intensive Harvesting on Forest Nutrient Cycling. State University of New York, Syracuse.

Martins, P. F. da S., C.C. Cerri, B. Volkoff, F. Andreux, and A. Chauvel. 1991. Consequences of clearing and tillage on the soil of a natural Amazonian ecosystem. For. Ecol. Manage. 38: 273-282.

Mattson, K.G., and W.T. Swank. 1989. Soil detrital carbon dynamics following forest cutting in the southern Appalachians. Biol. Fert. Soils. 7:247-253.

McCarthy, J.W. 1983. Soil changes resulting from phosphorus fertilization of slash pine (Pinus elliottii Engem. var elliottii). Ph.D. thesis, University of Florida, Gainesville.

McKee, W.H. 1982. Changes in soil fertility following prescribed burning on coastal plain pine sites. Southeastern Forest Experiment Station, USDA Forest Service Res. Paper SE-234.

Miller, J.H., and D.L. Sirois. 1986. Soil disturbance by skyline yarding vs. skidding in a loamy hill forest. Soil Sci. Soc. Amer. J. 50:1579-1583.

Morris, L.A., and W.L. Pritchett. 1983. Effects of site preparation on Pinus elliottii - P. palustris flatwoods forest soil properties. pp. 243-251 IN: R. Ballard and S. P. Gessel (eds.), IUFRO Symposium on Forest Site and Continuous Productivity. USDA Forest Service Gen. Tech. Rep. PNW-163, Portland, Oregon.

Mortland, M.M., and A.R. Wolcott. 1965. Sorption of inorganic nitrogen compounds by soil materials. pp. 150-197 IN: W.V. Bartholomew and F.E. Clark (eds.) Soil Nitrogen. Agronomy 10. Amer. Soc. Agron., Madison, Wisconsin.

Mroz, G.D., M.F. Jurgensen, and D.J. Frederick. 1985. Soil nutrient changes following whole tree harvesting on three northern hardwood sites. Soil Sci. Soc. Amer. J. 49:1552-1557.

Musselman, R.C., and D.G. Fox. 1991 . A review of the role of temperate forests in the global CO_2 balance. J. Air Waste Manage. Assoc. 41:798-807.

Nohrstedt, H,-O., K. Arnebrant, E. Baath, and B. Soderstrom. 1989. Changes in carbon content, respiration rate, ATP content, and microbial biomass in nitrogen-fertilized pine forest soils in Sweden. Can. J. For. Res. 19:323-328.

Nye, P.H., and D.J. Greenland. 1964. Changes in the soil after clearing tropical forest. Plant Soil. 21:101-12.

Oades, J.M. 1988. The retention of organic matter in soils. Biogeochemistry 5: 35-70.

O'Connell. A.M. 1 987. Litter decomposition, soil respiration and soil chemical and biochemical properties at three contrasting sites in kari (Eucalyptus diversicolor F. Muell.) forests of south-western Australia. Australian J. For. Res. 12:31-40,

Paschke, M.W., J.O. Dawson, and M.B. David. 1989. Soil nitrogen mineralization in plantations of Juglans nigra interplanted with actinorhizal Elaegnus umbellata or Alnus glutinosa. Plant Soil 118:33-42.

Paul, E.A., and F.E. Clark. 1989. Soil Microbiology and Biochemistry. Academic Press, New York.

Post, W.M., T-H Peng, W.R. Emmanuel, A.W. King, V.H. Dale, and D. L. DeAngelis. 1990. The global carbon cycle. American Scientist 78:310-326.

Powers, R.S., D.H. Alban, G.A. Ruark, and A.E. Tiarks. 1990. A soils research approach to evaluating management impacts on long-term productivity. pp. 127-145 IN: W.J. Dyck and C.A. Mees (eds.) Impact of intensive harvesting on forest site productivity. proc. IE/BE A3 Workshop, Furneaux Lodge, Marlborough Sounds, New Zealand, March 1989. IE/BE T6/A6 Rep. No.2, Ministry of Forestry, Forest Research Institute, Rotorua, New Zealand.

Raich, J.W. 1 983, Effects of forest conversion on the carbon budget of a tropical soil. Biotropica 15:177-184.

Raison, R.J,, P.K. Khanna, and P.V. Woods. 1 985. Mechanisms of element transfer to the atmosphere during vegetation fires. Can. J. For. Res. 15:132-140.

Richter, D.D., L.I. Barbar, M.A. Huston, and M. Jaeger. 1990. Effects of annual tillage on organic carbon in a fine-textured Udalf: The importance of root dynamics to soil carbon storage. Soil Sci. 149:78-83.

Sands, R. 1983. Physical changes to sandy soils planted to radiata pine. pp. 146-152 IN: R. Ballard and S. P. Gessel (eds.), IUFRO Symposium on Forest Site and Continuous Productivity. USDA Forest Service Gen. Tech. Rep. PNW-163, Portland, Oregon.

Schiffman, P.M., and W.C. Johnson. 1990. Phytomass and detrital storage during forest regrowth in the southeastern United States Piedmont. Can. J. For. Res. 19: 69-78.

Schlesinger, W.H. 1986. Changes in soil carbon storage and associated properties with disturbance and recovery. pp. 194-220 IN: J.R. Trabalka and D.E. Reichle (eds.) The Changing Carbon Cycle: A Global Analysis. Springer-Verlag, New York.

Schlesinger, W.H. 1990. Evidence from chronosequence studies for a low carbon-storage potential of soils. Nature 348:232-234.

Sims, A. G.R. Oliver, and W.J. Dyck. 1988. Effects of forest management on the long-term productivity of Nelson soils - Estimate of nutrient removal. Proj. Rec. No.2034, Forest Research Institute, Rotorua, New Zealand.

Smethurst, P.J., and E.K.S. Nambiar. 1990a. Distribution of carbon and nutrients and fluxes of mineral nitrogen after clear-felling a Pinus radiata plantation. Can. J. For. Res. 20: 1490-1497.
Smethurst, P.J., and E.K.S. Nambiar. 1990b. Effects of slash and litter management on fluxes of nitrogen and tree growth in a young Pinus radiata plantation. Can. J. For. Res. 20: 1 498-507.
Smethurst, P.J., and E.K.S, Nambiar. 1990c. Effects of contrasting silvicultural practices on nitrogen supply to young Radiata pine. pp. 85-96 IN: W.J. Dyck and C.A. Mees (eds.) Impact of intensive harvesting on forest site productivity. proc. IEA/BE A3 Workshop, Furneaux Lodge, Marlborough Sounds, New Zealand, March 1 989. IE/BE T6/A6 Rep. No.2, Ministry of Forestry, Forest Research Institute, Rotorua, New Zealand.
Tans, P.P., I.Y. Fung, and T. Takahashi. 1990 Observational constraints on the global atmospheric C02 budget- Science 247:1431-1438.
Tarrant, R.F., and R.E. Miller. 1963. Accumulation of organic matter and soil nitrogen beneath a plantation of red alder and Douglas-fir. Soil Sci. Soc. Amer. Proc. 27:231-234.
Tiesson, H., and M.C.D. Santos. 1989. Variability of C, N, and P content of tropical and semiarid soil as affected by soil genesis, erosion, and land clearing. Plant Soil 119: 337-341.
Turner, J. and J. Kelly. 1985. Effect of radiata pine on soil chemical characteristics. For. Ecol. Manage. 11:257-270.
Turner, J. and M.J. Lambert. 1986. Fate of applied nutrients in a long-term superphosphate trial in Pinus radiata. Plant Soil 93: 373-382.
Turner, J. and M.J. Lambert. 1988. Soil properties as affected by Pinus radiata plantations. New Zealand J. For. Res. 18:77-91.
Van Cleve, K. and T.A. Moore. 1978. Cumulative effects of nitrogen, phosphorus, and potassium fertilizer additions on soil respiration, pH , and organic matter content. Soil Sci. Soc. Amer. J. 42:121-124.
Vitousek, P.M., K. Van Cleve, N. Balakrishnan, and D. Mueller-Dombois. 1983. Soil development and nitrogen turnover in montane rainforest soils on Hawaii. Biotropica 15:268-274.
Weaver, P.L., R.A. Birdsey, and A.L. Lugo. 1987. Soil organic matter in secondary forests of Puerto Rico. Biotropica 19: 17-23
Wells, C.G. 1971. Effects of prescribed burning on soil chemical properties and nutrient availability. pp. 86-99 IN: Prescribed Burning Symposium, Proceedings. USDA Forest Service Southeastern Forest Experiment Station, Asheville, N.C.
Wilde, S.A. 1964. Changes in soil productivity induced by pine plantations. Soil Sci. 97:276-278.

ON ESTIMATING AGRICULTURE'S NET CONTRIBUTION TO ATMOSPHERIC CARBON

R. B. JACKSON, IV
U.S. Environmental Protection Agency
960 College Station Road
Athens, Georgia 30613 USA

ABSTRACT. Fossil fuel combustion, chlorofluorocarbon releases, and agricultural activities (including deforestation) are the primary anthropogenic sources of greenhouse gases. Of the three sources, agriculture is the only one that also has a sink capacity. Thus, an accounting of the net carbon (C) flux is required to properly evaluate agriculture's contribution and to determine the opportunities for emissions mitigation through changes in agricultural practices. Common data sets and a standard accounting method are required to perform country-by-country net C analyses. This research used agricultural census data to determine that U.S. agriculture removed 1.3 Pg of CO_2 from the atmosphere, in 1987, in the plants that it produced. The turnover times and the fate of this C were not ascertained. The research also showed that 6.4 Tg of CH_4 was emitted from live U.S. agricultural animals. A net C flux was computed, but is incomplete, because rice CH_4, plant and animal waste CH_4 and CO_2, and soil-atmosphere C fluxes could not be estimated from the census data. Additionally, agriculture's net contribution to atmospheric C was found to depend critically on the boundaries of the analysis.

1.0 INTRODUCTION

1.1. Carbon Flux Estimates

In the interest of science assessments or greenhouse gas flux estimates, a country's greenhouse gas emissions must be calculated accurately by means that allow comparisons among different countries. This requirement dictates the use of a standard method that uses data provided by all countries. Such methods exist for calculating carbon (C) gas emissions from the production and use of fossil fuels and cement [Marland et al., 1989], and chlorofluorocarbons (CFC's) [IPCC, 1990], but not from agriculture.

Agriculture is believed to rank third, behind fossil fuel use and CFC releases, in its contribution to greenhouse gas emissions [White, 1990]. The components that contribute to the atmospheric C flux are: methane (CH_4) from live animals; CH_4 and carbon dioxide (CO_2) from animal waste; CH_4 from rice paddies; and CO_2, CH_4, and carbon monoxide (CO) emissions from soils as a result of cultivation. Agriculture is unique in that it is

Water, Air, and Soil Pollution **64**: 121–137, 1992.
© 1992 *Kluwer Academic Publishers.*

also a sink for these atmospheric gases. The plant crops that agriculture produces cycle CO_2, retaining some of the C. Soils also sequester C through fixation of atmospheric CH_4 (15-45 Tg $CH_4 \cdot yr^{-1}$ [IPCC, 1990]), CO_2 [Schimmel, 1987] and CO [Conrad and Seiler, 1985], although the global magnitude of the CO_2 and CO sinks are unknown. Fluxes of C containing gases (hereafter referred to as C gases) are also altered as a result of deforestation, reforestation, conversion of agricultural land to other uses, or conversion of other lands to agricultural use. These are land-use change fluxes, and are not addressed in this study. Other elements of the U.S. Environmental Protection Agency's Global Change Program are addressing these issues.

1.2. Study Objectives

The objectives of this analysis were to: 1) estimate to the extent currently possible, the net C flux from the U.S. agricultural sector using production statistics, and 2) highlight ongoing research that would facilitate 1), and 3) suggest future research and data collection strategies. The fossil fuel C pool p_4, and flux, ϕ_7, were not estimated in this study, since they are included in energy sector emissions computations much more accurately than would be possible here.

1.3. Data

Production and consumption data for fossil fuels, cement, and CFC's are readily available and are well suited to emissions calculations. Unfortunately, agricultural production data are not applicable without some modification. Table I lists the agricultural data that are reported by most countries. While all of these data are not typically suited to C gas emissions calculations, existing (and perhaps future) research may allow them to be used to estimate the magnitudes of some agricultural C fluxes.

Table I Readily Available Agricultural Data

Crop Yields
Animal Types and Populations
Land Area in Crops
Land Area in Pasture
Land Area Irrigated
Land Area Fertilized

1.4. Carbon Flux Pathways

Agroecosystems primarily exchange three C gases -- CO_2, CH_4, and CO -- with the atmosphere. Figure 1 shows graphically the flows of C through the agricultural ecosystem, and Table III identifies the specific C flows involved and the practices that produce them. Figure 1 also shows that C pools -- p_1, p_2, and p_3 -- consist of C that has been fixed by plant crops. In the case of pastured animals, much of p_2 may have also come from natural vegetation, which would not appear in an accounting based solely on plant crops. The plant crop C pool (p_1) that is not fed to animals or does not become a waste product, and the C in the bodies of the animals (p_2) leaves the agricultural sector.

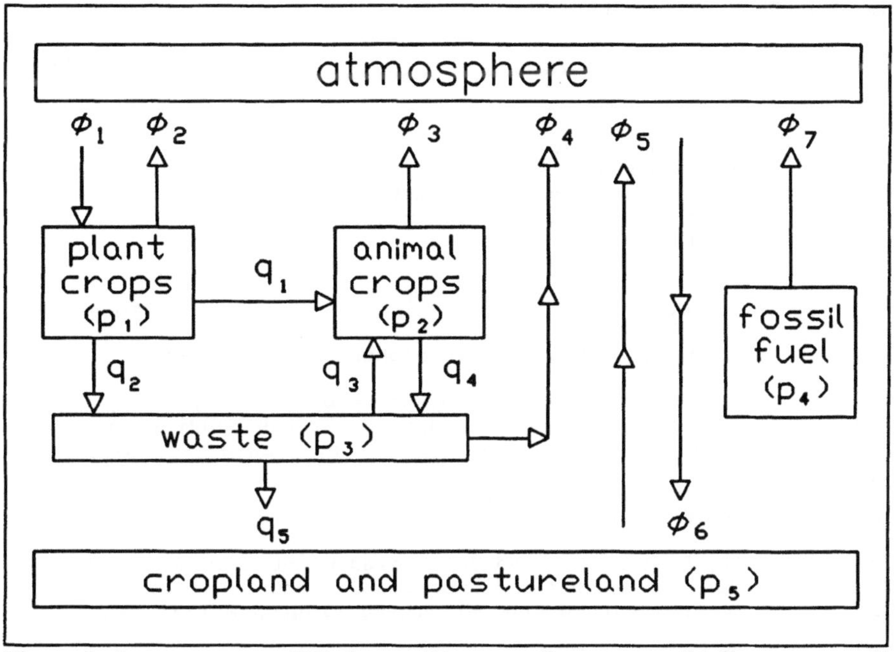

Figure 1 Agroecosystem C Pathways [ϕ_i = gas flux, q_i = solid flux, p_i = pool]

Its subsequent fate is not accountable to agriculture but to food processors, consumers, and waste facilities. Agricultural plant and animal waste (p_3) is processed and fed to animals; decomposes into CO_2 and CH_4; and also becomes part of the soil C pool. The distribution of C among p_3 sources and sinks depends on the management practices in use on the specific farm in question as well as other environmental factors. C in the fossil fuel pool (p_4) is converted to CO_2, CH_4, and CO during combustion for energy (ϕ_7).

The size of the agricultural soil C pool (p_5) depends on numerous factors (Table II), many of which humans control. Farmers constantly change the physical, chemical, and biological structure of cultivated land, in their labors to maintain and increase yields. Thus, p_5 is also a function of farm management. The physical and chemical properties

Table II Factors Affecting Agricultural Soil C Pool Size

Climate
Initial Organic Matter Amount
Initial Organic Matter Composition
Soil Physical Properties
Soil Chemical and Biological Properties
Crop Type and Cropping Frequency
Organic Matter Addition and Composition
Farm Management Practices

Table III Nomenclature for Figure 1

Symbol	Meaning	Units
	- Pools -	Tg C
p_1	C in harvested crops	
p_2	C in produced animals	
p_3	C in animal and plant waste	
p_4	C in fossil fuels used in agriculture	
p_5	C in agricultural soils	
	- Gaseous Fluxes -	Tg gas · yr^{-1}
$\phi_1[CO_2]$	CO_2 removed from the atmosphere by plant crops	
$\phi_2[CO_2]$	CO_2 returned to the atmosphere by plant crops	
$\phi_2[CH_4]$	CH_4 transported to atmosphere through paddy rice	
$\phi_2[CO]$	CO released to the atmosphere by plants [Bouwman, 1990]	
$\phi_3[CH_4]$	CH_4 released from live ag animals	
$\phi_3[CO_2]$	CO_2 released from live ag animals (respiration)	
$\phi_4[CH_4]$	CH_4 from burned and decomposed plant & animal waste	
$\phi_4[CO_2]$	CO_2 from burned and decomposed plant & animal waste	
$\phi_4[CO]$	CO from burned and decomposed plant & animal waste	
$\phi_5[CO_2]$	CO_2 emitted from ag soils	
$\phi_5[CH_4]$	CH_4 emitted from ag soils	
$\phi_5[CO]$	CO emitted from ag soils	
$\phi_6[CO_2]$	CO_2 removed from the atmosphere by soil processes	
$\phi_6[CH_4]$	CH_4 removed from the atmosphere by soil processes	
$\phi_6[CO]$	CO removed from the atmosphere by soil processes	
$\phi_7[CO_2]$	CO_2 released from ag fossil fuel use	
$\phi_7[CH_4]$	CH_4 released from ag fossil fuel use	
$\phi_7[CO]$	CO released from ag fossil fuel use	
	- Solid Fluxes -	Tg · yr^{-1}
$q_1[C]$	plant crop dry matter fed to animals	
$q_2[C]$	plant crop waste (includes litter & roots)	
$q_3[C]$	waste that is converted to animal feed	
$q_4[C]$	animal manure produced	
$q_5[C]$	waste applied to land + litterfall + animal droppings	

of pastures are altered, when forage or cover crops are planted and animals are maintained on the land. All of the other factors governing p_5 in Table II are controlled by the human-induced physical and chemical changes. Further complicating matters is the fact that the fluxes to and from the soil (ϕ_5 and ϕ_6) are affected by the factors in Table II. Thus p_5, ϕ_5, and ϕ_6 are difficult to quantify. This constraint presents significant barriers to computing agriculture's net emissions since there are only two fluxes of C gases from the atmosphere to agriculture, namely ϕ_1 and ϕ_6, and there are five C gas fluxes from agriculture to the atmosphere namely, ϕ_2, ϕ_3, ϕ_4, ϕ_5, and ϕ_7.

2.0 MATERIALS AND METHODS

2.1.1 The Data

The country chosen for this example is the United States of America (USA). Agencies of the USA regularly report agricultural information in electronic form, and agricultural practices in the USA are well known. To estimate values for the pools and fluxes in Figure 1, I collected agricultural production data at the state or county level from the 1987 U.S. Census of Agriculture [U.S. Department of Commerce (U.S. DOC), 1987]. Tables IV and V summarize animal and crop production respectively, in the USA [U.S. DOC, 1987].

Table IV Types and Populations of U.S. Agricultural Animals in 1987 *
Million Animals

Cattle and Calves	Dairy Cows	Hogs and Pigs	Chickens	Turkeys	Horses and Ponies	Sheep and Lambs
95.8	10.1	52.2	373	73.8	2.4	11.1

* These are U.S. totals -- state totals were used in all calculations and are available from U.S. DOC, 1987.

Table V Yield and Land Area of Selected U.S. Agricultural Crops in 1987 *

Crop	Yield unit 10^7	Desired Unit [Reported Unit]	Harvested Area unit 10^5 ha
Barley	45.8	kg [Bushels]	37.1
Corn for Grain	670.0	kg [Bushels (Dry)]	237.6
Corn for Silage	8.0	kg [Tons (Dry)]	23.4
Cotton	1.3	kg [Bales]	39.8
All Hay	12.9	kg [Tons (Dry)]	234.6
Oats	31.2	kg [Bushels]	24.2
Peanuts	340.0	kg [Pounds]	5.8
Irish Potatoes	36.7	kg [cwt]	5.3
Sweet Potatoes	1.5	kg [Bushels]	0.3
Rice	13.1	kg [cwt]	9.8
Wild Rice	1.3	kg [Pounds]	0.1
Rye	1.5	kg [Bushels]	2.2
Sugar Beets	5430.0	kg [Pounds]	5.1
Sugar Cane	6140.0	kg [Pounds]	3.7
Sorghum for Grain	63.3	kg [Bushels]	39.5
Sorghum for Silage	.6	kg [Tons]	2.0
Soybeans	184.0	kg [Bushels]	223.8
Tobacco	122.0	kg [Pounds]	2.6
Wheat for Grain	189.0	kg [Bushels]	215.4
Wheat for Grass	1.1	kg [Pounds]	0.5

* These are U.S. totals -- state totals were used in all calculations and are available from U.S. DOC, 1987.

2.2.0 First-Order Estimates

First-order estimates are defined here as C amounts that can be estimated as the product of values in Tables IV or V and conversion factors obtained from published research. In other words, the C estimate is not found through laborious calculation and requires only the most basic information, such as that provided by a comprehensive census.

2.2.1 Plant Crops

The plant crop C pool is p_1. It was computed for each state as the sum of the dry matter produced by each crop, multiplied by 0.45 g C / g dry matter [Starr and Taggart, 1989]. Many of the crop yields in Table V were given on a volume basis. Therefore, to obtain the dry matter values, the yields were multiplied by the necessary bulk density, and then by one minus their respective typical dried moisture contents. The conversion factors that were used are listed in Table VI. The yields also had to be increased by a weight equal to that portion of the crop that would not have been reported as yield (i.e., depending on the crop, the amount of roots, stems, leaves, seeds, or fruit). The factors used for that computation are given in Table VII.

Table VI Factors for Converting Reported Yield Units to Dry Yield (kg)

Crop	Reported Unit	Conversion‡ Factors	Wet basis* moisture content
Barley for Grain	Bushels	20.9 kg / bu	0.12
Corn for Grain	Bushels (Dry)	24.9 kg / bu	0.05
Corn for Silage	Tons (Dry)	907.2 kg / ton	0.05
Cotton	Bales	217.7 kg / bale	0.01
All Hay	Tons (Dry)	907.2 kg / ton	0.05
Oats	Bushels	14.9 kg / bu	0.11
Peanuts	Pounds	0.4536 kg / lb	0.15
Irish Potatoes	cwt	50.8 kg / cwt	0.67
Sweet Potatoes	Bushels	24.8 kg / bu	0.67
Rice	cwt	50.8 kg / cwt	0.12
Wild Rice	Pounds	0.4536 kg / lb	0.12
Rye	Bushels	24.9 kg / bu	0.11
Sugar Beets	Tons	907.2 kg / ton	0.77
	Pounds	0.4536 kg / lb	0.77
Sugar Cane	Tons	907.2 kg / ton	0.03
	Pounds	0.4536 kg / lb	0.03
Sorghum for Grain	Bushels	25.4 kg / bu	0.13
Sorghum for Silage	Tons	907.2 kg / ton	0.13
Soybeans	Bushels	27.1 kg / bu	0.10
Tobacco	Pounds	0.4536 kg / lb	0.19
Wheat for Grain	Bushels	28.1 kg / bu	0.14
Wheat for Grass	Pounds	0.4536 kg / lb	0.14

‡ [Henderson and Perry, 1982], * [Duke and Atchley, 1986; ASHRAE, 1985]

Table VII Multiplier for Estimating Total Biomass from Yield Data *

Crop	kg total biomass per kg dry yield
Barley	2.12
Corn	2.46
Cotton	5.00
All Hay	1.30‡
Oats	3.44
Peanuts	1.70●
Irish Potatoes	1.63
Sweet Potatoes	1.63
Rice	2.86
Wild Rice	2.86
Rye	2.65
Sugar Beets	1.32
Sugar Cane	1.44
Sorghum	3.44
Soybeans	2.46
Tobacco	2.64†
Wheat	2.15

* [Aselmann and Leith, 1983], ‡ [Sharpe, 1975], ● [Jones, 1967], † [Moore, 1977]

The plant crop C pool, p_1, is also affected by the gaseous and solid fluxes. The gaseous fluxes, $\phi_1[CO_2]$, $\phi_2[CO_2]$, and $\phi_2[CO]$ are balanced by the following equation $(\phi_1 - \phi_2) * \text{time} = p_1 + (q_1 + q_2)$ and are limited by the photosynthesis rate of the crop plant. The rate of atmospheric C removal by the plant crops then equals: $(\phi_1 - \phi_2) = p_1 / t$.

2.2.2 Agricultural Animals

The live animal pool of C is p_2. Its computation involved first estimating the live weight of animals from their numbers, and then the amount of C contained in their bodies from that weight. Typical average animal live weights (Table VIII) were computed using data from the American Society of Agricultural Engineers [ASAE, 1990], and the factor 0.15 g C / g animal [Starr and Taggart, 1989].

CH_4 from live agricultural animals was calculated from animal populations and types, using the emission factors shown in Table VIII. Since the U.S. census of agriculture reports the total number of cattle and calves, not the number of each, I assumed half of the total number were calves. Assuming that the animals in Table IV were alive the entire year of the data, the flux $\phi_3[CH_4]$ was estimated as total CH_4 released from the population distributed over the year. As animal respiration also produces CO_2, large numbers of animals contribute to the C flux in that way also; however, these fluxes have not been reported previously. Typical respiration rates are given in Table VIII. The animal waste production (q_4) was calculated using factors that give typical manure production per weight of animal (Table VIII).

Table VIII Factors Used To Relate Animal Types and Populations to Fluxes

Cattle and Calves	Dairy Cows	Hogs and Pigs	Chickens	Turkeys	Horses and Ponies	Sheep and Lambs
225.5†	640	61	0.9	6.8	450	27
55	90	1.5	ND	ND	18	8
0.085	0.095	0.095	0.389	0.328	0.110	0.195
60′	86	84	75	47	51	40

row 1: Typical animal live weights*, kg
row 2: Live animal CH_4 emission‡, kg CH_4 / animal·yr
row 3: Live animal CO_2 emission@ (respiration), $\times 10^{-3}$ ℓ CO_2 / g animal·hr
row 4: Typical kg manure produced per 1000 kg animal weight per day*

* [ASAE, 1990], †Average of 360 kg for cattle and 91 kg for calves, ‡ [Crutzen et. al., 1986]; ND - no data given, @ [Lasiewski, 1974]; using R = volume CO_2 produced per volume O_2 consumed = 0.76), ′ Average of 62 kg for cattle and 58 kg for calves.

2.3.0 Second-Order Estimates

Second-order estimates in this study are defined as those requiring extensive modelling, or computation, or those for which the gas producing or destroying mechanisms are not known or fully understood. Finding ways to quantify these C fluxes is part of ongoing research at the Athens Environmental Research Laboratory (AERL), and other EPA Environmental Research Laboratories.

2.3.1 Rice

CH_4 from rice paddies is currently a second-order estimate because the production and emission of the gas depends on numerous physical, biological, and chemical properties of soils as well as many plant characteristics. None of these variables are measured or reported regularly, and even if they were, there is no consensus on their influence and interactions relative to the production of CH_4 ($\phi_2[CH_4]$).

Rice production consists basically of four management types that are distinguished from each other on the basis of water use. In the United States each type occupies approximately the following percentage of total planted rice area: dryland, 61 %; irrigated, 27 %; rainfed, 5 %; and deepwater, 4 %. In 1987, rice was produced in only nine states (AR, CA, FL, LA, MN, MO, MS, TX, and WI) [U.S. DOC, 1987]. The sum of these rice areas accounted for only ≈ 1 M ha -- less than 1 % of the total rice area in the world.

At this time, an analysis of the available empirical information is the best way to extrapolate small-scale emissions data to larger areas (like the United States). If yield data are to be used effectively to estimate emissions, I argue that CH_4 emissions factors for rice must be given as a function of the type of rice production -- dryland, irrigated, rainfed, and deepwater -- (where the types are distinguished by the length of time inundated and the depth of water), the spatial location of the field (to include climate and soils), the type of rice plant, and the area of the rice field. These are the kinds of data that agriculture census will be capable of providing on an annual basis.

2.3.2 Animal and Plant Waste Emissions

The waste pool (p_3) consists of total animal and plant waste production, minus losses to the atmosphere, inputs to the land, and waste that is returned to animals as feed. The plant and animal waste fluxes (q_2 and q_4, respectively) are first order estimates. The flux of waste to land (q_5) and animals (q_3) could be first order as well, if sufficient data were collected and reported.

Often animal manure is composted, placed into aerobic or anaerobic lagoons to decompose, or left to decompose where it falls. Manure that falls on pasture or crops is part of q_5 (waste flux to the land), and much of the manure that is collected for lagooning or composting is eventually placed on the land as well.

The use and ultimate fate of plant crop wastes depend on what portion of the plant is harvested, where that portion resides, and the residue practices employed by the particular farmer. In lieu of direct information from the individual farms, I have identified a means of estimating plant matter inputs to the land from data compiled by the Conservation Technology Information Center (CTIC) [Conservation Technology Information Center, 1990]. CTIC defines conservation tillage as " any tillage and planting system that maintains at least 453.6 kg of flat small grain residue equivalent on the surface." CTIC compiles data at county resolutions for the entire United States annually. The procedure for computing the plant crop portion of q_5 is to: 1) determine the total agricultural production for a county (roots, stems, leaves, fruit) in units of Net Primary Production (NPP), 2) determine the percentage of that county area that is in conservation tillage as defined by CTIC, and 3) calculate the amount of plant crop residue reaching the land, based on 1) and 2). For counties that have no area in conservation tillage, perhaps amounts of residue applied to the land may be estimated from the crop type, harvest method, yield, and other studies of the region's management practices.

The flux of C gases from the waste pool ($\phi_4[CH_4]$, $\phi_4[CO_2]$, $\phi_4[CO]$) depends entirely on the size and composition of the pool, its ultimate use, and the environmental conditions to which it is subjected [Guenther et al., 1990]. Whether the animal manure is collected in a waste pit, lagooned, composted, or left in the field, some portion of it decomposes into CO_2, CH_4, and CO (represented by ϕ_4). The same is true of plant matter waste. However, in the case of lagoons and waste pits, CH_4, CO_2, and CO from animal waste could be estimated from the amount of waste, the lagoon or pit configuration, and the regional climate. Unfortunately, location and loading data are not currently available for lagoons and waste pits.

The plant waste that does not end up on the land is burned, composted, used as animal feed, or applied to other specialized uses, none of which easily lends itself to gas production estimates. Thus, the best current course of action for estimating C gas emissions from agricultural waste is to gather geographically-referenced information about the waste amount, composition, and management, in conjunction with efforts to define the factors that convert that data to $\phi_4[CH_4]$, $\phi_4[CO_2]$, $\phi_4[CO]$.

2.3.3 Agricultural Soil C Fluxes

The agricultural soil C pool is extremely difficult to quantify. Since C values obviously cannot be checked everywhere, one must (or does not) agree with the processes employed to compute them after carefully noting known errors and assumptions. Soil C pool estimates have been made by many researchers based primarily on climate-dependant modeling of either vegetation or soil types, or some combination of the two [Bouwman, 1990]. Most of the estimates are for world soils, not agricultural soils, and range from 700 to 3000 Pg C. One specific estimate of C in the agricultural soils of the U.S. (top 30 cm), placed the pool at between 5.3 and 8.6 Pg C [Kern and Johnson, 1991]. Agricultural soil C coverages are an essential component of AERL's modeling efforts to quantify the potential for C sequestration in agricultural soils.

3.0 RESULTS

3.1.0 First-Order Estimates

The plant crop C pool and its spatial distribution are shown in Figure 2 (multiply values in Figure 2 by 0.45 to convert to g of C). As expected, regions that had large agricultural yields sequestered larger amounts of atmospheric C. The total dry yield for the United States equaled 510.16 Tg, from which I estimated that 814.96 Tg of total biomass was produced. The NPP estimates presented in Figure 2 for some states are up to an order of magnitude larger than estimates of global agricultural productivity ($\approx$ 100 to 400 $g\ C \cdot m^{-2}$) from other researchers [Melillo et al., 1988; Olson et al. 1985]; while most are consistent with global and other U.S. estimates [Sharpe, 1975]. The total C contained in U.S. agricultural plant crops in 1987 was estimated to equal 367 Tg (814.96 * 0.45), which by the earlier definition equals the rate of atmospheric C removal by U.S. agricultural plant crops, that is, $p_1 / t = 367$ Tg C per year.

The C contained in the bodies of the animals and its spatial distribution are shown in Figure 3. The C amounts were divided by the area of the state to permit easier comparison with values from the plant crop NPP map. These numbers, however, are not NPP. The total C contained in the bodies of U.S. agricultural animals in 1987 was approximately 5 Tg (total weight * 0.15). CH_4 and CO_2 emissions from live agricultural animals are given in Table IX. Live animal CO_2 emissions from animal respiration equaled 52 Tg (Table IX) -- an almost insignificant figure in comparison to global CO_2 sources. However, any actual measurements of CO_2 production from live animals will likely be higher since the fermentation process that produces CH_4, produces CO_2 as well.

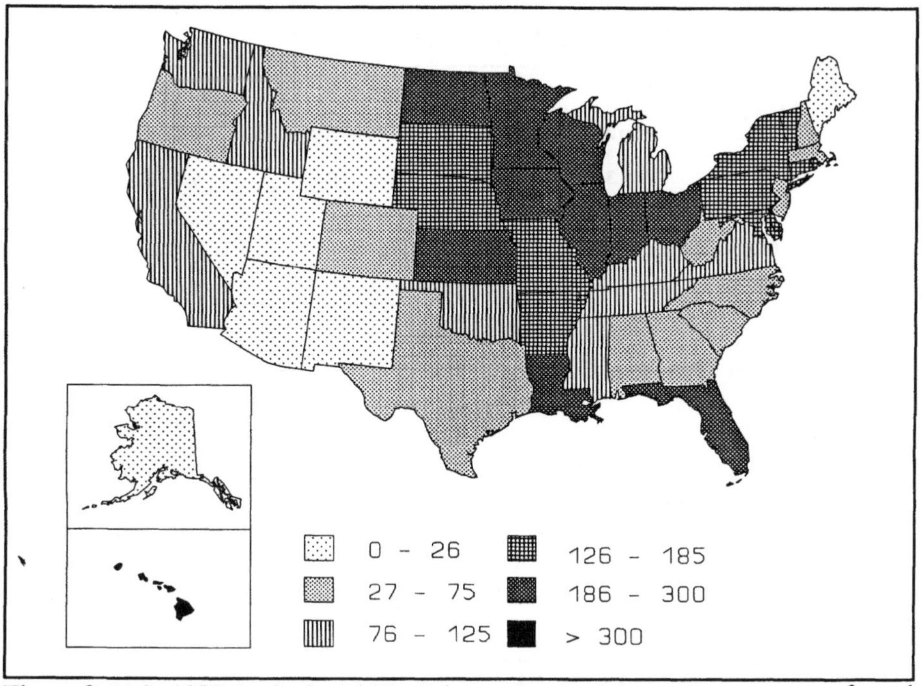

Figure 2 U.S. 1987 Agricultural Net Primary Productivity, g dry matter · m^{-2} · yr^{-1}

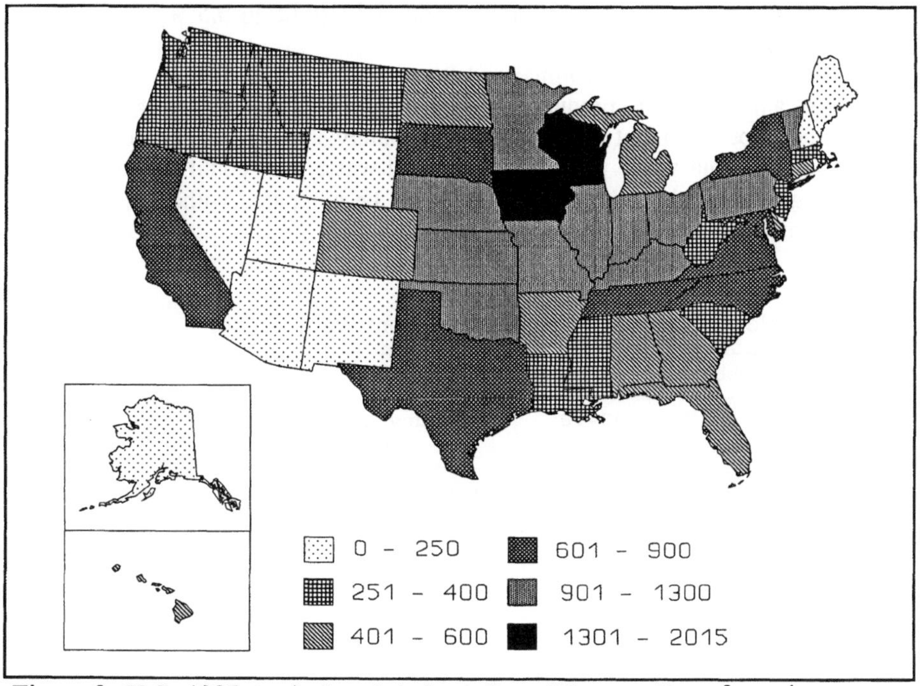

Figure 3 U.S. 1987 Agricultural Animal C, milligrams C · m^{-2} · yr^{-1}

Table IX Estimated CH_4 and CO_2 Emissions From Live U.S. Ag Animals in 1987*, Row 1: Gg $CH_4 \cdot year^{-1}$, Row 2: Tg $CO_2 \cdot year^{-1}$

Cattle and Calves	Dairy Cows	Hogs and Pigs	Chickens	Turkeys	Horses and Ponies	Sheep and Lambs
5269	909	78.3	ND	ND	43.2	88.8
29.6	9.9	4.9	2.1	2.7	1.9	0.9

* U.S. totals. The CO_2 values correspond to respiration only and were computed at 21 °C and 1 atmosphere.

3.2.0 Second-Order Estimates

3.2.1 CH_4 From Rice

One estimate of CH_4 emissions from inundated rice fields in the United States is between 3 and 28 g CH_4 / (m^2 * season) [U.S. EPA, 1990]. Deepwater rice fields are probably inundated all of the time (there are various definitions), and irrigated and rainfed fields may be inundated most or some of the time. Unfortunately, without more information, I cannot know for sure. Approximately 61 % of the rice in the United States is grown on dry land. Using the 3 to 28 g CH_4 / m^2 * season estimate, and assuming two 100 day growing seasons per year and 1 M ha of rice area in the United States, the $\phi_2[CH_4]$ emissions range is 0.06 to 0.56 Tg $CH_4 \cdot year^{-1}$ from U.S. rice fields. CH_4 emissions from rice fields worldwide, is estimated to be 110 Tg annually [IPCC, 1990], making the U.S. an almost insignificant contributor of rice CH_4 (0.05 to 0.5 %).

3.2.2 CO_2, CH_4, and CO from Animal and Plant Wastes

The amount of animal manure produced in the United States in 1987 was approximately 817 Tg and the amount of plant waste was approximately 304.8 Tg dry matter (Table X). Although both the plant waste pool and the animal waste pool could be estimated, the amount of C gas released by those pools could not be computed without more specific management knowledge or major assumptions.

Table X Agricultural Animal and Plant Waste Fluxes†

Symbol	Value	Units
q_2	304.8*	Tg dry plant matter $\cdot$ Yr^{-1}
q_4	2.2	Tg manure $\cdot$ Yr^{-1}

* Difference of Total Biomass (814.96 Tg) and Total Reported Yield (510.16 Tg)

† These are First Order Estimates.

3.2.2 Soil ↔ Atmosphere Carbon Fluxes

A portion of the C in agricultural soils changes on time scales of years to decades. This labile fraction is greatly influenced by gains from animal and plant waste (q_5), C gas uptake by agricultural soils ϕ_6, losses due to C gas emissions (ϕ_5), and C exports due to water movement. Climate, irrigation and drainage, tillage use, crop type, cropping frequency, soil type, and chemical use all change soil properties. Thus, they all regulate q_5, ϕ_5, and ϕ_6 to some extent. These interactions make agricultural soil processes extremely complex.

4.0 DISCUSSION

4.1.0 First- and Second-Order Estimates

In computing the first-order estimates, I used U.S. Census of Agriculture information [U.S. Department of Commerce, 1987]. The census is not the only source of this data however I believe their information to be highly reliable. The results show clearly that agriculture has both sinks and sources for C gases. The potential certainly exists to refine the calculations. In the case of the first-order estimates, it is the procedure that should be focused on, and in the case of the second-order estimates, the research directions should be emphasized. Table XI shows the result of net C gas computations using the first order estimates. The +18 for the animal crops is the sum of the C in the animal's respiratory CO_2 and fermentative CH_4 emissions.

Table XI Net U.S. Agroecosystem Carbon Fluxes From First-Order Estimates

	C [Tg]	CO_2 [Pg]	CH_4 [Tg]
Plant Crops	-367	-1.35	---
Animal Crops	+18	+0.05	+6.38
Balance	-349	-1.30	+6.38

+ Increased Atmospheric Burden, - Decreased Atmospheric Burden
Note: Animal crop body C was not included due to the lack of data about q_1.

CH_4 from rice is an extremely complex second-order estimate. Detailed current information about all of the important variables will probably never be available on a annual basis. However, rice yield, location, and area are reported now, and other pertinent variables such as rice plant type, management type (dryland, irrigated, rainfed, and deepwater), or management practice should be reported as well. EPA's Environmental Research Laboratory at Corvallis is working to refine rice CH_4 emissions estimates [Braatz and Hogan, 1991]. I would recommend that they develop specific factors for rice water management type that can be used with census information to estimate CH_4 emissions.

Plant and animal wastes constitute a large pool of managed C (animal excreta and plant residues), most of whose fate is determined by agriculture. Since the decomposition rates and amounts of gas evolved are highly variable, the C gas emissions may be controllable. Animal and plant wastes also represent a large resource and should be treated as such by agricultural census. Data about amounts and fate are fundamental to any estimates of gas production. Thus, our efforts in this area are currently focused on quantifying and characterizing agricultural waste, and determining how it is managed.

Soil C fluxes are indeed the most difficult to estimate. Consider that the natural soil processes are affected by cultivation, which is controlled by the farmer, who manages the crop and field conditions, and is directed in his choices by economics and the demand for specific crops, which is controlled by the structure of regional and global trade policies. This reasoning can be followed for all of animal and plant agriculture, once it has been established " which crop will grow where ". The Global Change Workgroup at AERL is working to produce accurate estimates of agriculture's net C flux due to animal and plant waste, but our largest effort concerns the relationships among soil C, agricultural management practices, and management's relationship to economics and policy. At AERL we are utilizing the CENTURY soil C model, the DNDC soil model, census data, CTIC data, climate data and numerous other geographically referenced databases to assess the impact of agricultural management on ϕ_5, ϕ_6, and q_5 over areas for which complete historical information on driving variables is available [Barnwell et al., 1991]. We anticipate assessments for the entire United States in 1 to 2 years.

4.2.0 Analysis Boundaries

Strategies to sequester C will ultimately have to consider all aspects of the agricultural system in order to compute an accurate net C balance. For instance, the first-order analysis results in Table XI show that U.S. agriculture removed 349 Tg of C from the atmosphere in 1987. However, since the rates at which the C pools are turning over, and even their ultimate fate are unknown, that C may be removed for an insignificant period of time. Thus, we do not know whether there is a net C sequestration.

In addition to the unknowns, agricultures' net contribution to atmospheric C varies with the boundaries of the analysis. For instance, if the proposed solution is to increase agricultural soil C with animal manure, will the transport, processing, and application of the manure to the land release more C than would eventually be stored in the soil ? Also, what are the turnover times of the pools involved and what then happens if the analysis is expanded to include the C costs in the manufacture of chemicals, implements, and machinery ? Clearly, stating the analysis boundaries, is a crucial aspect of determining the net balance.

4.3.0 Strategy Development and Research Directions

Debates to limit greenhouse gas emissions from agricultural systems assume that the emissions can be computed fairly and accurately. Currently, the only comprehensive global agricultural data upon which to base these estimates is census information; specifically the United Nations Food and Agriculture Yearbook [FAO Production, 1989]. Research should strive to develop those factors needed to convert that database to defensible estimates of net greenhouse gas emissions.

I argue that the factors used to compute the emissions should be the focus of strategy discussions in order that all countries might agree on a standard set, subject to scientific proof that an individual country should be judged using a different standard. Unfortunately, emissions factors cannot precede the science that would compute them -- speaking specifically in terms of second-order estimates. One suggestion then, would be to set the initial standard factors such that they produce emissions estimates on the high end of the range. Each country would then be encouraged to fund and conduct its own research (to be reviewed by all countries), if they believe their net emissions are indeed lower than the standard factors imply.

Perhaps, in the near future, a country-specific compendium of factors for converting census information to net greenhouse gas fluxes can be agreed upon. Only then can global warming potential, emissions credits, emission mitigation techniques, etcetera be more reliably and efficiently addressed.

ACKNOWLEDGMENTS

I gratefully acknowledge Angela Briley for her invaluable assistance in the preparation of graphics, and library literature searches. I also wish to thank the following persons for their guidance and critical reviews: Thomas Barnwell, David Brown, Lee Mulkey, Robert Swank, Robert Ryans, Kevin Weinrich, Allen Rowell.

REFERENCES

ASAE. 1990. *Manure Production and Characteristics - ASAE Data:ASAE D384.1* American Society of Agricultural Engineers. St. Joseph, Michigan.

Aselmann, I. and H. Lieth. 1983. The Implementation of Agricultural Productivity Into Existing Global Models of Primary Productivity. In: Degens, Kempe, Soliman (Eds), *Transport of Carbon and Minerals in Major World Rivers*, Part 2, Mitt. Geolog.-Paläontolog. Inst. Univ. Hamburg, SCOPE/UNEP Sonderband, Heft 55,107-118.

ASHRAE. 1985. *ASHRAE Handbook - 1985 Fundamentals.* American Society of Heating, Refrigerating, and Air Conditioning Engineers, Inc. Atlanta, GA.

Barnwell, T.O., R.B. Jackson, E.T. Elliott, E.A. Paul, K. Paustian, A.S. Donigian, A.S. Patwardhan, A.L. Rowell, and K.B. Weinrich. 1991. *Assessment of Methods, Models, and Databases for Soil Carbon Sequestration Potential for U.S. Agroecosystems.* U.S. Environmental Protection Agency (Internal Report), Environmental Research Laboratory, Athens, Georgia.

Bouwman, A.F. 1990. *Soils and the Greenhouse Effect.* Wiley Publishers. Chichester, New York.

Braatz, B.V. and K.B. Hogan. 1991. *Sustainable Rice Productivity and Methane Reduction Research Plan.* U.S. Environmental Protection Agency. Office of Air and Radiation. Washington, D.C. (Unpublished Report)

Conrad, R. and W. Seiler, 1985: Influence of Temperature, Moisture, and Organic Carbon on the Flux of H_2 and CO Between Soil and Atmosphere: Field Studies in Sub-tropical Regions, *J. Geophys. Res.*, 90, 5633 - 5709.

Conservation Technology Information Center. 1990. *1990 National Survey of Conservation Tillage Practices.* West Lafayette, Indiana.

Crutzen, P.J., I. Aselmann, and W. Seiler. 1986. Methane Production by Domestic Animals, Wild Ruminants, Other Herbivorous Fauna, and Humans. *Tellus* 38B(1986), 3-4, pages 271-284.

Duke, J.A. and A.A. Atchley. 1986. *CRC Handbook of Proximate Analysis of Higher Plants.* CRC Press, Inc. Boca Raton, Florida.

FAO Production. 1989. *FAO Production Yearbook*, v - 43. Food and Agriculture Organization of the United Nations. New York, New York.

Guenther, A., P. Zimmerman, A. Hills, C. Westburg and B. Lamb. 1990. *Flux Measurements of Methane from Cattle Feedlots.* National Center for Atmospheric Research; Atmospheric Chemistry Division. Boulder, Colorado.

Henderson, S.M. and R.L. Perry. 1982. *Agricultural Process Engineering.* AVI Publishing Company, Inc. Westport, Connecticut.

IPCC - Intergovernmental Panel on Climate Change. 1990. *Climate Change: The IPCC Scientific Assessment.* Cambridge University Press. Cambridge, Great Britain.

Jones, B.W. 1967. *The Peanut Plant: Its Cultivation and Uses.* Orange Judd Publishing Company. New York, New York.

Kern, J.S. and M.G. Johnson. 1991. *The Impact of Conservation Tillage Use on Soil and Atmospheric Carbon in the Contiguous United States.* A Report to the U.S. Environmental Protection Agency from: ManTech Environmental Technology, Inc. Corvallis, Oregon.

Lasiewski, R.C. 1974. *Vertebrate Oxygen Consumption:* In Altman, P.L. and D.S. Dittmer (Eds), Biology Data Book Volume II. Federation of American Societies for Experimental Biology. Bethesda, Maryland.

Marland, G., T.A. Boden, R.C. Griffin, S.F. Huang, P. Kanciruk, and T.R. Nelson. 1989. *Estimates of CO_2 emissions from fossil fuel burning and cement manufacturing based on the United Nations energy statistics and the U.S. Bureau of Mines cement manufacturing data.* ORNL/CDIAC-25, NDP-030, Oak Ridge National Laboratory, Oak Ridge, Tennessee.

Melillo, J.M., J.R. Fruci, R.A. Houghton, B. Moore, and D.L. Skole. 1988. Land-Use Change in the Soviet Union between 1850 and 1980: Causes of A Net Release of CO_2 to the Atmosphere. *Tellus* 40B (1988),2 page 118.

Moore, E.L., 1977., *Tobacco International and Tobacco Science.* Volume XXI, TS9, Lockwood Trade Journal Co., New York, New York.

Olson, J.S., J.A. Watts, and L.J. Allison. 1985. *Major World Ecosystem Complexes Ranked by Carbon in Live Vegetation: A Database.* NDP-017. Carbon Dioxide Information Center. Oak Ridge National Laboratory. Oak Ridge, Tennessee.

Schimmel, S.M. 1987. Dark Fixation of Carbon Dioxide in an Agricultural Soil. *Soil Science*. July 1987, 144(1).

Sharpe, D.M., 1975. *Methods of Assessing the Primary Production of Regions*. In-Lieth, H. and Whittaker, R.H. (Eds), Primary Productivity in the Biosphere. Springer Verlag., New York, New York.

Starr, C and R. Taggart. 1989. *Biology: The Unity and Diversity of Life*. Wadsworth Publishing Co. Belmont, California.

U.S. Department of Commerce - Bureau of the Census. *1987 Census of Agriculture* - United States Summary and Data Report. Vol. 1 - 2 Geographic Series. U.S. Government Printing Office. Washington, D.C.

U.S. EPA. 1990. U.S. Environmental Protection Agency, U.S. Department of Agriculture, and Inter-governmental Panel on Climate Change. 1990. *Greenhouse Gas Emissions From Agricultural Systems - v 1 and 2.* Government Printing Office. Washington, D.C.

White, R.M. 1990. The Great Climate Debate. *Scientific American* (July). New York, New York.

NOMENCLATURE

C	-	carbon
CFC	-	chlorofluorocarbon
g	-	gram
ha	-	hectare; 10,000 square meters
m	-	meter
NPP	-	Net Primary Production [dry matter per unit area & time]
p_i	-	carbon solid flux
ϕ_i	-	carbon gas flux
q_i	-	carbon pool
SOM	-	Soil Organic Matter
t	-	1,000,000 grams (tonne)
Ton	-	907.2 kg (2000 pounds)

k	-	kilo; 10^3
M	-	mega; 10^6
G	-	giga; 10^9
T	-	tera; 10^{12}
P	-	peta; 10^{15}
E	-	exa; 10^{18}

PROCESSES AND LANDS FOR SEQUESTERING CARBON IN THE TROPICAL FOREST LANDSCAPE

SANDRA BROWN
Department of Forestry
University of Illinois
110 Mumford Hall
1301 W. Gregory Dr
Urbana, IL 61801, USA

ARIEL E. LUGO
Institute of Tropical Forestry
USDA Forest Service
Southern Forest Experiment Station
Call Box 25000
Rio Piedras, Puerto Rico 00928

LOUIS R. IVERSON
Illinois Natural History Survey
607 E. Peabody
Champaign, IL 62820, USA

ABSTRACT. Balancing the C budget in the tropics has been hindered by the assumption that those forests not undergoing deforestation are in C steady state with respect to their C pools and thus with the atmosphere. The long history of human activity in tropical forests suggests otherwise. In this paper we discuss the forest compartments into which C can be stored, what the likely rates of storage are and for how long, and over which areas of the tropical landscape these processes occur. Results of our analysis suggest that tropical forests have the potential to sequester up to 2.5 Pg C yr^{-1} from the atmosphere if human pressure could be completely removed. Addition of agroecosystems and degraded lands could increase this estimate markedly.

Water, Air, and Soil Pollution **64**: 139–155, 1992.
© 1992 *Kluwer Academic Publishers.*

1. Introduction

The debate about the role of tropical forests as sources or sinks of atmospheric C is based mostly on different perceptions of how these forests function. What has hindered progress is the assumption that all tropical forest lands not undergoing active land-use change are mature or in C steady state with as much C being respired as is being fixed by photosynthesis, i.e., net ecosystem production is zero. What this assumption fails to account for is the long history of forest use by humans that has resulted in much of the tropical forest landscape being in less than a primary state. Biomass or C content of tropical forests has changed constantly over time as a result of past disturbances by people and natural catastrophes (Brown and Lugo, 1990a, 1992; Brown et al., 1991; 1992; Flint and Richards, 1991).

Before the wide-scale use of fossil fuels, and even today to a large extent, forest biomass has been used for a multitude of purposes including fuel and construction. When tropical countries relied solely on renewable resources (prior to the start of the industrial era), the net C flux from tropical forest lands could have already been in transient (i.e., not in steady state with respect to the atmosphere) because C in biomass per unit area could have been declining. Evidence suggests that this was the case and that the forests lands today contain less C per unit area than they could potentially store (Brown and Lugo, 1992; Iverson et al., 1992).

We suggest that essentially no forests are in C steady state, either with respect to all forest compartments or with the atmosphere, including secondary forests, logged forest, forest fallows, degraded forests (caused by biomass removal by illicit activities; Brown et al., 1991), naturally disturbed forests, and even mature forests (Lugo and Brown, 1986, 1992). We have argued that mature forests that are in C steady state with respect to all C pools in the forest cannot be in C steady state with the atmosphere (Lugo and Brown, 1986, 1992). All forests leach C into the soil profile and eventually rivers, therefore photosynthesis in the forest removes more C from the atmosphere than is returned. Therefore, if the C balance of tropical forests is not in steady state, how much C are they sequestering or can they potentially sequester if human pressure on them was reduced?

To address this question fully, we need to know in what ecosystem compartments C can be stored, what are the likely rates of storage and for how long, and over which areas of the landscape can these processes occur. This information has a practical value if serious consideration is going to be given to mitigating greenhouse gas emissions by sequestering C on the landscape. To meet this need, knowledge of the areas of land potentially and actually suitable for C sequestering is needed.

The objectives of this paper are to: (1) discuss the compartments and fluxes in tropical forest lands that lead to C sequestering, and their likely magnitude, (2) discuss briefly actual and potential forest land uses for sequestering carbon, and (3) present estimates of areas of land potentially suitable for C sequestering and the magnitude of the potential C sink for one tropical region.

2. Carbon Sequestering Processes in Tropical Forest Lands

The net balance of natural processes of the C cycle determine whether a tropical forest is a source or a sink of atmospheric C. Primary productivity fixes atmospheric C which then accumulates at various rates in above and below ground biomass, necromass (dead organic matter, excluding soil), and soil organic C. Fire and respiration pathways return C to the atmosphere. Depending upon the frequency of fires, C storage capacity can increase because of the production of charcoal which does not readily decompose. Carbon export (from soil and necromass) by leaching and the flow of goods (wood mostly) to the human economy transfer C to other systems where it can act as a sink.

Here we discuss the processes in tropical forest lands that have the potential to sequester C. We will focus on C accumulation in above ground biomass, necromass, and soil organic matter. Belowground processes have the potential to sequester C, but information needed for such an analysis is sorely lacking.

2.1. ACCUMULATION IN ABOVE GROUND BIOMASS

Observations of organic matter dynamics of long-term, small permanent tropical forest plots do not support the notion that they are in C steady state. Average net C accumulation in above ground biomass for 26 late secondary, slightly disturbed, or mature forest stands in Venezuela, grouped into five life zones, ranged between 1 to 2 Mg ha^{-1} yr^{-1} (Figure 1). The same is true of mature stands in Puerto Rico (Brown et al., 1983). In all cases, the balance (net wood

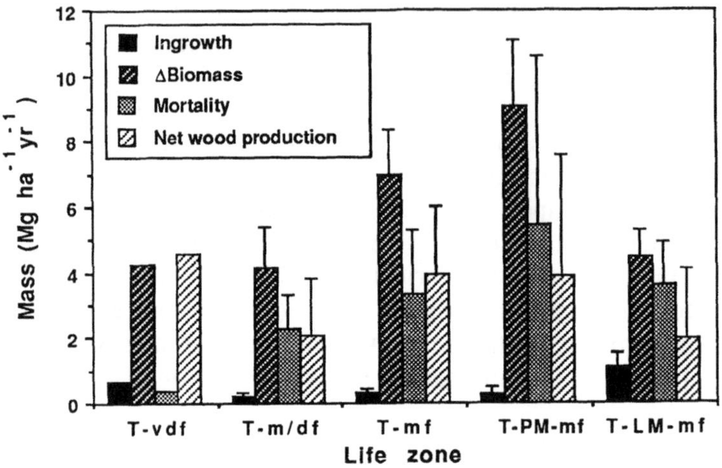

Figure 1. Organic matter budgets (values are means + 95% confidence interval) for 26 permanent plots in Venezuela, grouped into five life zones. T = tropical, vdf = very dry forest, m/df = moist transition to dry forest, mf = moist forest, Pm-mf = premontane moist forest, and LM-mf = lower montane moist forest.

production) between the sum of ingrowth (new trees entering the minimum diameter class of 10 cm) and biomass increment in live trees and the biomass lost by tree mortality was positive. Ingrowth into these stands tended not to be a significant C sink unless the stand was recovering from acute disturbance such as a hurricane.

Results of analysis of forest inventories done 10 yr apart in Peninsular Malaysia showed a similar trend (Brown et al., 1992). Carbon storage in vegetation of some inaccessible mature forests increased by 7 to 14%, while in recovering, previously logged forests, C storage increased by 7 to 152%. Accessibility to the forests explained the variation in the increases in C storage: forests with increased fragmentation (increase in the perimeter-to-area ratio and more accessible to humans) gained less C than forests with decreased fragmentation (less accessible to humans) in the 10 yr interval (Brown et al., 1992). Large increases in fragmentation led to biomass reductions and thus, potentially, to C sources to the atmosphere. In the case of the logged forests, a large proportion of the increased C storage was caused by ingrowth of the smaller trees (Brown et al., 1992).

Where is this C accumulating? We believe that a large fraction of the net C accumulation in above ground biomass occurs in the growth of older trees that get progressively larger with age. As a tree ages, its annual biomass increment continuously increases. Large trees (defined as those with a dbh > 70 cm and biomass of >5 Mg) generally occur at low densities in forests (less than 5 or so per ha), but they can account for more than 40% of the biomass of stands (Brown and Lugo, 1992; Brown et al., 1992). Such trees must be considered as long-term C sinks because they tend to survive for long periods. The presence and growth dynamics of large trees in forests may explain how C can continue to accumulate in forests for many decades or even centuries without any appearance of C excesses.

Change in species composition during forest succession is another way for biomass to continue to accumulate in a forest. Typically, mature forest species have higher density woods than early colonizing species (Smith, 1970; Whitmore and Silva, 1990). As a result, more C can be stored per unit wood volume produced. Weaver (1987) documented changes in the weighted density of montane forests in Puerto Rico from 0.61 Mg m^{-3} to 0.63 Mg m^{-3} or an increase of about 4% in mass per unit volume. Our biomass accumulation rates do not consider this process explicitly. These changes may be small per unit area of forest but can account for larger quantities when extrapolated over large forested landscapes.

The rate of above ground C accumulation in tropical forest lands ranges widely. Natural stands that are recovering from previous disturbances accumulate C in above ground biomass at the fastest rates in early stages (< 20 yr-old), i.e., up to 3.5 Mg $ha^{-1}yr^{-1}$ (Table 1). Older secondary forests accumulate biomass at rates of about half this amount (Brown and Lugo, 1990a). Fast growing plantations can accumulate C at rates as high as 15 Mg C $ha^{-1}yr^{-1}$ (Lugo et al., 1988).

2.2. Accumulation in Necromass

Most estimates of the quantity of organic matter in tropical forests have tended to ignore the amount of C in coarse woody debris (CWD) on the forest floor on the assumption that it is likely to be small relative to the C in biomass of live vegetation. Recent work

TABLE 1. Processes that create C sinks and their potential magnitude in the tropical forest landscape (from Lugo and Brown, 1992)

Process	Magnitude ($Mg\ C\ ha^{-1}yr^{-1}$)	Source
Biomass accumulation in forests >60-80 yr-old and logged forests	1-2	Brown and Lugo, unpublished data
Biomass accumulation in secondary forest fallows, 0-20 yr-old	2-3.5	Brown and Lugo, 1990a
Biomass accumulation in plantations[a]	1.4-4.8	Brown et al., 1986
Accumulation of coarse woody debris in:		
forests >60-80 yr-old	0.2-0.4	Brown and Lugo, unpublished data
forests 0-20 yr-old	0.2-0.3[b]	Brown and Lugo, 1990a
Accumulation of SOC:		
background rates	0.02-0.03	Schlesinger, 1990
forest succession	0.5-2.0	Brown and Lugo, 1990b
plantations <40 yr-old	0.5-2.0[c]	
conversion of cultivation to pasture/grassland	0.3-0.42	Lugo et al., 1986

[a] Weighted average rates across all species and age classes.

[b] Two studies given in Brown and Lugo (1990a) report an average amount of coarse woody debris at age about 20 yr of 8.5% of the aboveground biomass; we assumed this percent of the biomass accumulation rate goes into coarse woody debris during the 20 yr period.

[c] Assumed rates of SOC accumulation similar to secondary forests.

suggests otherwise. The few data available for late secondary and mature tropical forests suggest that the standing stock of CWD may be about 10 to 40% of the above ground biomass (Greenland and Kowal, 1960; Saldarriaga et al., 1986; Uhl et al., 1988; Uhl and Kauffman, 1990). Further, annual rates of production of CWD can range from 0.2 to 4 $Mg\ C\ ha^{-1}\ yr^{-1}$ (unpublished data of authors, see also Figure 1 for mean rates). This large pool, high rate of production, and its slow decomposition rate makes this compartment another potential long-term sink of C (Table 1). We have found that wood decomposition can be slower than the growth of new wood in a disturbed stand (unpublished data of authors). Thus, by retaining CWD *in situ*, a stand gains additional C (as above ground necromass) that can last for a little as 30 yr to as long as 200 yr depending upon the mix of, and the decomposition rates of, the wood.

2.3. Accumulation as Soil Organic Carbon

Soil organic C (SOC) is a long-term storage compartment for atmospheric C and may take on the order of millenia to reach steady state levels (Jenny, 1980). Schlesinger (1990) recently confirmed this trend when he showed that soils can continue to accumulate SOC over thousands of years (we define this continued accumulation as the background rate, Table 1).

There are large losses of SOC associated with permanent agriculture (Detwiler, 1986) but not with short-term use by shifting agriculture (Ewel et al., 1981; Ramakrishnan and Toky, 1981)). Soils recover their SOC following abandonment of agriculture to forest succession, with the rate depending upon life zone (Figure 2a, Table 1). Forests in moist or wet life zones recover SOC faster than in dry life zones, but all appear to reach levels of SOC approaching those of nearby forests in about the same length of time (about 50 yr).

Conversion of forests to pastures often results in no loss or gains in SOC compared to nearby, native forests (Figure 2b). Furthermore, resampling of cultivated sites that had been converted to pastures showed increases in SOC over several decades (Table 1). In other words, soils under pasture can be C sinks, regaining some of the C that was released during clearing. High rates of root production by grasses may explain why pastures accumulate SOC (Cerri et al., 1991).

Soil organic matter will recover under forest plantations at rates similar to or faster than secondary forests (Figure 2c). Differences in rates of SOC accumulation are caused by differences in species and environmental factors (Lugo et al., 1990). Some species produce more litter and roots than others, thus producing more organic inputs which eventually influences SOC. Clearly, this species effect must be considered in plans for enhancing C sequestering.

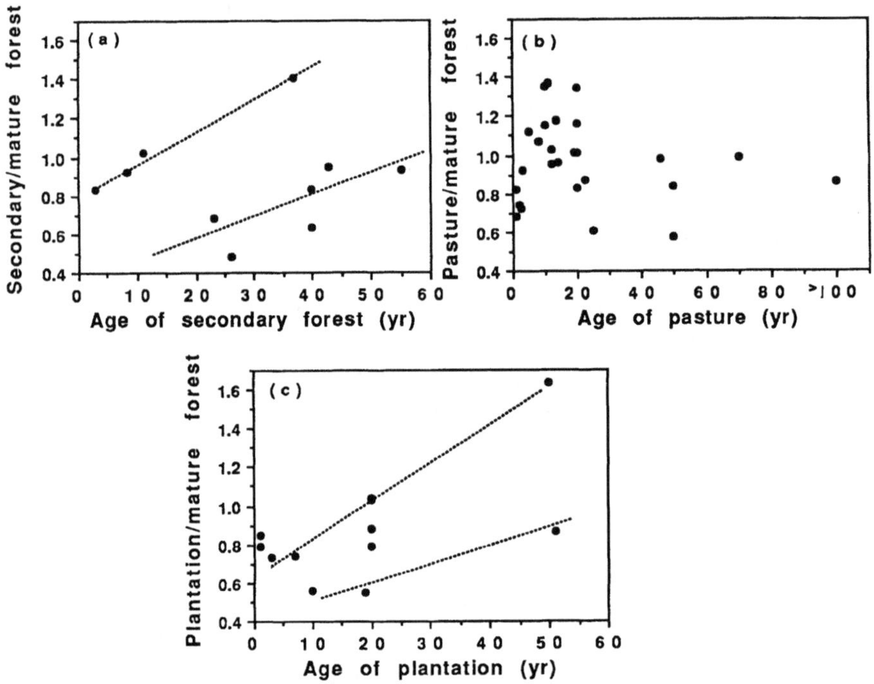

Figure 2. The ratio of soil organic C in (a) secondary forests, (b) pastures, and (c) plantations to paired mature forests versus age. Lines in (a) and (c) are drawn in by hand to illustrate range of likely rates of soil C sequestering. Data are from Lugo and Brown (unpublished paper).

3. Forest Land Uses for Sequestering Carbon

Land-use change in the tropics can be brought about by human and natural events. These changes in forest land use are complex, with forests undergoing degradation, logging, and conversion to shifting cultivation/forest fallow, permanent agriculture, and urban lands, as well as the converse.

In addition to outright forest clearing, large areas of tropical forests experience some measure of human degradation that affects their biomass and forest composition (Brown et al., 1991, 1992; Flint and Richards, 1991; Brown and Lugo, 1992; Iverson et al., 1992). The intensity of this biomass degradation varies from the removal of one or two valuable large trees per hectare, often illicitly, to clearing of small areas (Brown et al., 1991).

All these land-use changes (discussed in detail by Lugo and Brown, 1992) have numerous implications for the C cycle because each land use has a particular potential and actual C density and functions at a particular speed of C accumulation and export. During the time interval between events that reduce the biomass of a stand, C accumulates at various rates by regrowth, retention of necromass or wood products, and/or increased SOC. All these processes must be considered when doing a complete accounting of C fluxes in the tropical landscape.

Because forest lands undergo reductions in their C pools when they are converted to other uses, they have the potential to regain this C if human pressure is removed or reduced. Logged forests are a clear example; most logging operations extract a few trees per hectare of a given diameter and then the forest is allowed to recover for many decades before it is logged again.

Clearly lands used for food production have little chance of being abandoned (except for those lands under shifting cultivation). However, alternative soil organic matter management or conversion to agroforestry systems, permanent tree crop plantations (rubber, oil palm, cacao, etc.) or pastures could improve the sequestering capabilities of these lands as well as make them more sustainable. These possibilities and their effect on C sequestering will not be discussed here. Instead we will concentrate on present forest lands (mature, logged, and plantations) because they have the potential to sequester significant amounts of C to help balance the net flux of C to the atmosphere from tropical deforestation (Lugo and Brown, 1992).

4. Estimating Lands Potentially Suitable for Carbon Sequestering: South/Southeast Asia as a Case Study

Here we will discuss preliminary results of current research designed to identify forest lands in continental South/Southeast Asia suitable for C sequestering. As important as the results being presented here is the methodology that we have developed, which we believe is the approach needed to better estimate C sequestering on the landscape. Land areas alone are not enough, as we have shown, because C sequestering rates are dependent on environmental factors, as are the potential steady state C pools.

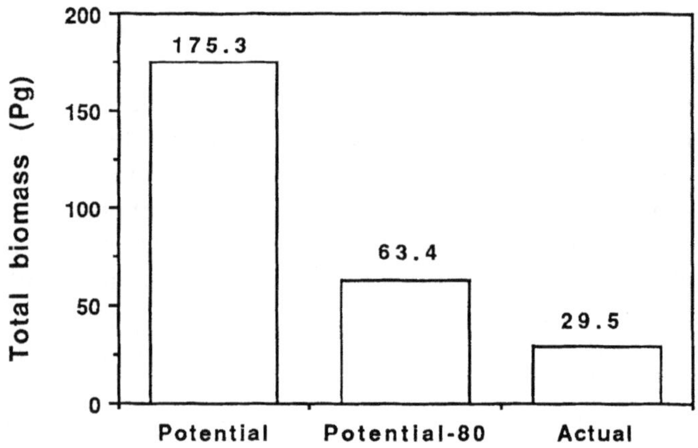

Figure 3. Estimates of potential forest biomass, and potential and actual forest biomass as of about 1980 for continental South/ Southeast Asia (from Iverson et al., 1992).

The amount of C that can ultimately be sequestered in forest vegetation (potential C pool) on the landscape is related to the amount of biomass that the biophysical factors of climate, soils, topography, and elevation can support. Using geographic information systems technology (GIS) and digital maps of these biophysical factors, estimates of the potential above ground biomass (or C pool), in digital map form, have been made for South/Southeast Asia (Iverson et al., 1992). This digital map was then overlain with a forest map as of about 1980 to produce a potential forest biomass map for this time period (potential biomass-80; Iverson et al., 1992). The final step was to then add the influence of humans that cause biomass reductions through many types of activities, discussed above, to produce a digital map of actual forest biomass. This biomass reduction was done by taking into consideration human population densities, presence/absence of shifting cultivators, historical accounts of the area, forest inventories, and the advice of experts on the region (Iverson et al., 1992).

This approach resulted in three maps of forest biomass and three estimates of the total biomass for these three conditions (Figure 3). The overall decrease in above ground biomass caused by the long history of human impacts on the forest lands of continental South/Southeast Asia was 83%. The first decrease, between the potential biomass and potential biomass-80 (64%

reduction), represents the cumulative effect of deforestation only in the region. The further decrease of an additional 53% represents the impact of biomass degradation on present forest lands, including planned logging operations.

Iverson et al's. (1992) analysis considered C in above ground biomass only. We have added estimates for C in below ground biomass and coarse woody debris to complete the picture (Table 2). Clearly, achieving the potential C is unrealistic because humans cannot be removed from the landscape completely. However, achievement of the potential C pool in forest lands as of about 1980 is more realistic (although the forest areas in this region have declined by at least another 9% during the last decade [Food and Agriculture Organization, 1991]), particularly if economically sustainable practices were implemented on all agricultural lands. Thus the potential amount of C that can be sequestered in forest vegetation in the region is 28 Pg (Table 2). Accumulation of organic C in any degraded soils of the forest lands is not included here which could increase the potential amount.

Although we refer to the amount of C that can be sequestered as "potential", some of this is actually being sequestered by those processes described in Table 1 in previously logged forests, inaccessible forests, or forests now under protection as described above for Peninsular Malaysia, for example. At present we have not separated these forests from other forest lands still undergoing degradation, thus we refer to the whole amount as potential.

Knowing the total amount of C that potentially can be sequestered in continental South/Southeast Asia is only part of the issue; annual rates and over what time interval are also important. The range in magnitude of C sequestering rates in forests is very much a function of climate or life zone, thus the need to add this level of detail to the analysis.

In the absence of a reliable digital life zone map, we reclassified an ecofloristic zone (EFZ) map produced by the Food and Agriculture Organization (1989). This map distinguished 36 zones based on climate (rainfall and its regime, length of the dry season, temperature, and relative humidity), physiography, and edaphic factors (Food and Agriculture Organization, 1989). We reclassified these 36 zones for continental South/Southeast Asia into six that approximated a simple life zone classification (based on the groupings given in Brown and Lugo, 1982) (Figure 4).

TABLE 2. Potential and actual C pools in forest vegetation of continental South/Southeast Asia and potential magnitude of C sequestering

	---Pg C---			
	Aboveground vegetation[a]	Below-ground[b]	Coarse woody debris[c]	Total
Potential (without humans)	88	18	26	132
Potential as of 1980	32	6.3	9.5	48
Actual as of 1980	15	3.0	2.2	20
Potential amount of C sequestering[d]	17	3.3	7.3	28
Potential rate of C sequestering[e]		**---0.25 Pg yr^{-1}---**		

[a] From Figure 3, and using 1 Pg of biomass = 0.5 Pg C
[b] Assumed belowground mass = 20% of aboveground (Brown and Lugo, 1982)
[c] Assumed coarse woody debris was 30% of potential aboveground biomass (mature forests) and 15% of actual biomass (disturbed forests) (see text).
[d] Estimated by difference between potential as of 1980 and actual C pool as of 1980
[e] Estimated as follows: first, the distribution of forest areas, based on their ranking to sequester C by ecofloristic zone (EFZ, Figure 5) was used to estimate a weighted rank for each EFZ. Then the rates in Table 1 were applied differentially, based on their C sequestering rank and EFZ, to the areas of each EFZ shown in Figure 4.

Clearly, the lowland seasonal forest zone dominates the region (53% of the total area), due in the most part to the common occurrence of this zone in India, by far the largest country in the region. This large area of the lowland seasonal zone will dominate the rates of C sequestering. The lowland moist zone, the one likely to have the highest C sequestering rates, occupies about 25% of the region.

The next important issue is how much of the potential C sequestering (Table 2) occurs in which EFZ? To address this question, we subtracted the actual forest biomass map from the

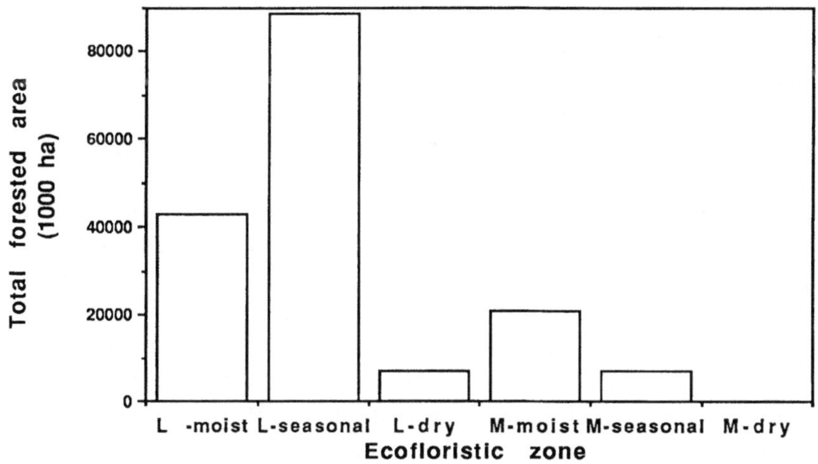

Figure 4. Areas of forest in continental South/Southeast Asia, as of about 1980, grouped into six ecofloristic zones (reclassified ecofloristic zone map produced by the Food and Agriculture Organization, 1989). L= lowland zones and M= montane zones.

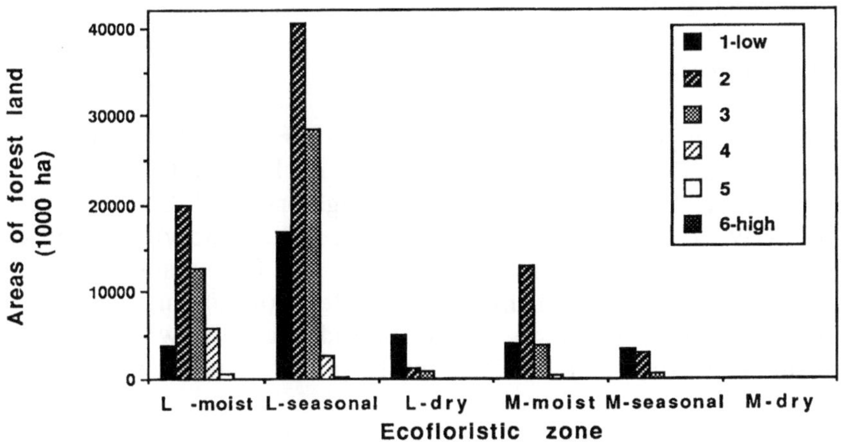

Figure 5. Areas of forest in South/Southeast Asia, as of about 1980, ranked according to their ability to sequester new C in vegetation, and grouped into six ecofloristic zones (see legend to Figure 4 for source of ecofloristic zones and key to symbols).

potential biomass-80 map and, using GIS techniques, overlaid this with the reclassified EFZ map. This resulted in the distribution of areas, classed by their ability to sequester C from low to high, by EFZ shown in Figure 5. Very little forested area has the ability to sequester high amounts of C; most of the total area can sequester low amounts (rank of 1 to 2).

Using the information in Figure 5 and in Table 2, the annual rate of potential C sequestering in forests of the region was estimated to be 0.25 Pg yr^{-1}, which could occur for more than 100 yr (annual rate divided into the total amount given in Table 2). This seems a small amount compared to the present emissions of C from fossil fuels (about 6.0 Pg yr^{-1}). However, the forest area in the region is only about 10% of the total tropical forest area. If rates similar to these were assumed to occur over the whole tropics, then the total could be 2.5 Pg yr^{-1}, or a little under half that produced from fossil fuel. And, as C sequestering in agroecosystems and degraded lands is not included, the total amount could be even more significant. An analysis by Lugo and Brown (1992), without the geographical component, produced an estimated C sink of 1.5 to 3.2 Pg yr^{-1} for all tropical forest lands.

Acknowledgements

This study was supported by grant DEFGO2-90ER61081 between the Department of Energy and the University of Illinois (S. Brown, L. Iverson, and A. E. Lugo, principal investigators) and Cooperative Agreement PNW 91-0115 between the USDA Forest Service, Pacific Northwest Region and the US EPA (S. Brown and L. Iverson, principal investigators). We thank J. P. Veillon for making his permanent plot data available to us, A. Prasad fpr help with GIS analysis, and C. A. Hall and W. Post for helpful comments on the manuscript.

References

Brown, S. and Lugo, A. E. (1982) 'The storage and production of organic matter in tropical forests and their role in the global carbon cycle', Biotropica, 14, 161-167.

Brown, S. and Lugo, A. E. (1990a) 'Tropical secondary forests', Journal of Ecology, 6, 1-32.

Brown, S. and Lugo, A. E. (1990b) 'Effects of forest clearing and succession on the carbon and nitrogen content of soils in Puerto Rico and US Virgin Islands', Plant and Soil, 124, 53-64.

Brown, S and Lugo, A. E. (1992) 'Aboveground biomass estimates for tropical moist forests of the Brazilian Amazon', Interciencia (in press).

Brown, S., Gillespie, A. J. R. and Lugo, A.E. (1991) 'Biomass of tropical forests of south and southeast Asia', Canadian Journal of Forest Research, 21, 111-117.

Brown, S., Iverson, L., and Lugo, A. E. (1992) 'Land use and biomass changes in Peninsular Malaysia during 1972-82: a GIS approach', in V. H. Dale (ed.), Effects of Land Use Change on Atmospheric Carbon Dioxide Concentrations: Southeast Asia as a Case Study, Springer-Verlag, New York, in press.

Brown, S, Lugo, A. E., Silander, S., and Liegel, L. (1983) 'Research history and opportunities in the Luquillo Experimental Forest', General Technical Report SO-44, USDA Forest Service, Southern Forest Experiment Station, New Orleans, LA.

Brown, S, Lugo, A. E., and Chapman, J. (1986) 'Biomass of tropical tree plantations and its implications for the global carbon budget', Canadian Journal of Forest Research, 16, 390-394.

Cerri, C., Volkoff, B., and Andreaux, F. (19910 'Nature and behaviour of organic matter in soils under natural forest, and after deforestation, burning and cultivation, near Manaus', Forest Ecology and Management, 38, 247-257.

Delcourt, H. R. and W. F. Harris. (1980) 'Carbon budget of the southeastern US biota: analysis of historical change in trend from source to sink', Science, 210,321-322.

Detwiler, R.P. (1986) 'Land use change and the global carbon cycle: the role of tropical soils', Biogeochemistry, 2, 67-93.

Ewel, J., Berish, C., Brown, B., Price, N., and Raich, J. (1981) 'Slash and burn impacts on a Costa Rican wet forest site', Ecology, 62, 816-829.

Flint, E.P. and Richards J.F. (1991) 'Historical analysis of changes in land use and carbon stocks of vegetation in South and Southeast Asia', Canadian Journal of Forest Research, 21, 91-110.

Food and Agriculture Organization. (1989) Classification and Mapping of Vegetation Types in Tropical Asia, FAO, Rome.

Food and Agriculture Organization. (1991) 'Second interim report on the state of tropical forests', Forest Resource Assessment 1990 Project, FAO, Rome.

Greenland, D. J. and Kowal, J. M. J. (1960) 'Nutrient content of the moist tropical forest of Ghana', Plant and Soil, 12, 154-174.
Iverson, L. R., Brown, S., Prasad, A, Mitasova, H., Gillespie, A. J. R., and Lugo, A. E. (1992) 'Use of GIS for estimating potential and actual forest biomass for continental South and Southeast Asia', in V. H. Dale (ed.), Effects of Land Use Change on Atmospheric Carbon Dioxide Concentrations: Southeast Asia as a Case Study, Springer-Verlag, New York, in press.
Jenny, H. (1980) The Soil Resource: Origin and Behavior, Springer-Verlag, New York.
Lugo, A.E. and Brown, S. (1986) 'Steady state ecosystems and the global carbon cycle', Vegetatio, 68, 83-90.
Lugo, A. E. and Brown, S. (1992) 'Tropical forests as sinks of atmospheric carbon', Forest Ecology and Management, in press.
Lugo, A.E., Brown, S., and Chapman, J. (1988) 'An analytical review of production rates and stemwood biomass of tropical forest plantations', Forest Ecology and Management, 23, 179-200.
Lugo, A.E., Cuevas, E., and Sanchez, M.J. (1990) 'Nutrient and mass in litter and top soil of ten tropical tree plantations', Plant and Soil, 125, 263-280.
Lugo, A.E., Sanchez, M.J., and Brown, S. (1986) 'Land use and organic carbon content of some subtropical soils', Plant and Soil, 96, 185-196.
Ramakrishnan, P. S. and Toky, O. P. (1981) 'Soil nutrient status of hill agro-ecosystems and recovery patterns after slash and burn agriculture (jhum) in north-eastern India', Plant and Soil, 60, 41-64.
Saldarriaga, F. G., West, D. C., and Thorp, M. L. (1986) 'Forest succession in the Upper Rio Negro of Colombia and Venezuela', Environmental Sciences Division Publication No. 2694 (ORNL/TM9712), Oak Ridge National Laboratory, Oak Ridge, TN.
Schlesinger, W.H. (1990) 'Evidence from chronosequence studies for low carbon-storage potential of soils', Nature, 348, 232-234.
Smith, R. F. (1970) 'The vegetation structure of a Puerto Rican rain forest before and after short-term gamma irradiation', in H. T. Odum and R. F. Pigeon (eds.), A Tropical Rain Forest, National Technical Information Service, Springfield, VA, Chapter D-3.
Uhl, C., Buschbacher, R., and Serrao, E. A. S. (1988) 'Abandoned pastures in eastern Amazonia. I. Patterns of plant succession', Journal of Ecology, 76, 663-681.

Uhl, C. and Kauffman, J. B. (1990) 'Deforestation, fire susceptibility, and potential tree response to fire in eastern Amazon', Ecology, 71, 437-449.

Weaver, P. L. (1987) 'Structure and dynamics in the colorado forest of the Luquillo Mountains of Puerto Rico', PhD Dissertation, Michigan State University, East Lansing.

Whitmore, T. C. and Silva, J. N. M. (1990) 'Brazil rain forest timbers are mostly very dense', Commonwealth Forestry Review, 69, 87-90.

FORESTRY OPPORTUNITIES IN THE UNITED STATES TO MITIGATE THE EFFECTS OF GLOBAL WARMING

R. NEIL SAMPSON
American Forestry Association
P.O. Box 2000
Washington, DC 20013 USA

ABSTRACT. There are a variety of opportunities in the United States to expand the area of trees and forests, and to improve their growth, that could have significant impact upon the annual uptake of atmospheric CO_2. Work coordinated by the American Forestry Association has attempted to quantify those opportunities, and demonstrate what kinds of costs and benefits might result from an attempt to begin implementing them. The first section of the work, reported in this paper, has focused on the opportunities that are seldom thought of as regular forestry—planting trees on marginal crop and pasture lands, increasing windbreaks and shelterbelts, growing trees as a biomass energy source, and improving urban tree canopies and placements as an energy-conserving measure. The benefits from such work include the C sequestered in the biomass and soils involved, as well as the carbon emission reductions achieved through energy conservation. These opportunities could add up to a total C impact per year in the range of 141 to 382 x 10^6t—somewhere between 10 and 30% of the current net C emission from fossil fuel in the United States. Additional work is underway to quantify the opportunities inherent in improving the management of existing forestlands, through more traditional forestry. The results of that work will be available in late 1992.

Biomass and Global Warming

Although the ability to predict future climate change is controversial, the fact is that it will, if it occurs, be extremely disruptive. The concerns are significant enough to cause policymakers to look for opportunities to encourage mitigative measures. Forestry is one of those opportunity areas that can provide a degree of protection to help buffer too-rapid changes in climate.

Many of the possible investments in forests, in addition to helping offset atmospheric carbon buildup are, in fact, economically advantageous. Others, particularly those that build long-lived ecosystems, create new resource strength that represents a sound investment in the future. Some replace the need to burn fossil fuels, leaving those fuels available for another day. Some, like the shade trees around homes, make each day more pleasant while saving air-conditioning dollars that will never need to be spent, and reducing fossil-fuel burning in power plants.

Thus, there are several global warming preventative strategies based on forestry that, far from being economically burdensome, represent immediate and tangible

Water, Air, and Soil Pollution **64**: 157–180, 1992.
© 1992 *Kluwer Academic Publishers.*

benefits. This is not to suggest that forestry can "take care of" the threat of global warming. Forestry is just one aspect of an initial agenda. While taking action on that agenda, policymakers must continue to develop the research and information that will help everyone better understand the continuing threat, and the further action agenda that is warranted.

Atmospheric CO_2, the most important of the greenhouse gases, is inextricably tied to growing plants and thus is directly linked to the growth of the trees and forests of the world. The global CO_2 level has fluctuated in the range of 200 to 280 ppm over the past 160,000 years, according to studies of historic records contained in deep icecaps and ocean sediments. Since the beginning of the industrial revolution, however, atmospheric levels of CO_2 have shot up rapidly, from about 280 ppm in 1750 to 350 in 1989. With CO_2 levels continuing to increase at an estimated rate of 0.5% annually, a growing number of scientists are concerned with the potential for significant global warming within the coming century (Hansen et al., 1989, Schneider, 1989b, Trexler, 1991).

An overwhelming proportion—some 80% of the net annual CO_2 addition—comes from fossil-fuel burning, and the United States, with its highly energy-intensive economy, contributes about 22% of that total (Boden et al. 1990). Thus, the reduction of fossil-fuel burning, particularly where it can be done without loss of productive output (through higher efficiencies, for example) becomes a first-order goal in the attempt to reduce atmospheric CO_2 accumulation. Wood used in place of fossil fuel for energy represents a recently formed source of C and one that can readily be replaced by new growth in forests. This means that a recycling C source can replace a nonrecycling fossil source, and the net addition of C to the atmosphere over time can be reduced accordingly.

While paying attention to lessening atmospheric C uptake, it seems prudent to try to increase the rate at which C is taken up and stored in Earth's biophysical systems. The major "sinks" for C are oceans (estimated to contain some 43 x 10^{12}t), soils (estimated to hold 1.5 to 2 x 10^{12}t), and forests (estimated to hold .5 x 10^{12}t) (Trexler, 1991).

Trees and forests are tremendously important as one terrestrial sink of CO_2. About half the dry weight of wood is C, so as trees add mass to trunks, limbs, and roots, C is stored in relatively long-lived structures instead of being released to the atmosphere. Obviously, all plants do this same conversion, but grasses and crops that are eaten by herbivores and digested, or that die and decompose within a season, are not as effective in providing long-term storage as are the woody structures of trees and the wood products made from them.

Significant ways to increase terrestrial C storage include enlarging the area of trees and forests, and increasing the rate of tree growth on those that already exist. Each of these strategies has the advantage of being easier than affecting ocean storage, and each would have the effect of "locking up" additional C in the Earth's biosphere. Eventually, of course, an upper limit on living forest biomass is reached, and additional tree growth is largely offset by mortality, plant respiration, and use so that the net additions to the terrestrial C store become marginal.

There are, however, at least three ways to extend the contribution of trees in storing C and make the expansion or improvement of forest ecosystems even more valuable in addressing the global climate change issue:

(1) Forests can be harvested and converted into long-lived wood, paper, or other products. So long as these products are in use or storage, the C remains out of atmospheric circulation. In addition, where wood is used in place of concrete or steel, significant energy is saved in the production process and lower CO_2 emissions result, making wood an attractive environmental alternative (CORRIM 1976).

(2) Forest biomass can be converted to energy, either directly through burning or indirectly through biological and chemical conversion. If the resulting energy is used to replace fossil energy, the effect is to retain fossil C in storage, a net gain in mitigation efforts.

(3) Properly located and managed trees can reduce the demand for fossil energy through effects such as shading buildings to reduce air conditioning demand, breaking winter winds to lower space-heating needs, preventing soil erosion to lower the amount of fertilizer and fossil fuel that would be needed to produce necessary food and fiber from higher areas of degraded soils, manipulating snow drifts to reduce winter snow-plowing costs and automobile accidents, and sheltering livestock in both winter and summer to reduce stress and improve feed efficiency.

A serious consideration of forest biomass expansion as a global warming mitigation strategy requires that we consider:

** The amount of land biologically suited to expanded or accelerated forest production;

** The new technologies needed to create opportunities for expanded use of wood and other biomass in long-lived applications, or as a replacement for fossil energy sources;

** The market incentives needed to encourage private landowners to expand forests, and the amount of additional public incentive that might be required to meet forest goals in the event that market forces remain inadequate;

** The potential for using trees in ways that curb energy usage in transportation, agriculture, and everyday living;

** The effect of each of these elements on the net fluxes of C between Earth and its atmosphere; and,

** The likely effect of C flux changes in causing a significant global warming trend.

This paper addresses only a portion of that broad agenda, leaving to others the questions of new technology for improving the use of biomass in energy-generating facilities, the economic incentives needed to spur forest expansion, and the question of global climate change in response to atmospheric phenomena. Here we focus mainly on the land, water, and forest resources of the United States, trying to outline:

** The opportunity to expand the land area of the United States that is occupied by forests;
** The impact of those changes on C storage in biomass, and on annual production of usable biomass;
** The potential to utilize a higher percentage of forest biomass for energy, including the growing of biomass specifically as an energy crop; and,
** The C impact possible through expanded use of trees in energy-conserving ways.

The U.S. Land Base, its Use and Potential

The United States has about 916.6 x 10^6 surface ha of land that are currently used in four general categories (Table 1). These land uses are dynamic, with millions of hectares changing from one use to another each year. However, the overall totals don't move a great deal over time. For example, there have been about 160 x 10^6 ha of cropland in the United States since 1920 (Fedkiw, 1989). In that time, however, millions of ha of cropland have been developed for urban and other built-up uses. Those losses have been offset by conversion of pasture, range, and forest to crops.

Table 1. Major uses of land in the United States at selected years

	1950	1969	1987
		10^6 Ha	
Cropland	166	155	171
Pasture and Range	284	280	238
Forest	292	293	296
Other	179	188	211
Total Land Area	920	916	916

Source: Frey and Hexem, 1982; U.S. Department of Agriculture, 1989.

THE IMPACT OF LAND-USE CHANGE ON C DYNAMICS

Before evaluating the opportunity to shift land use through policy mechanisms and incentives, it is useful to review the impact that such changes may contribute to the nation's efforts to address greenhouse gas concentrations. As described above, trees and forests can increase the amount of C held in the terrestrial C sink when compared to other land uses such as cropland, pasture, or range.

In measuring the actual amount of C stored in forests, there is considerable data based on the measurements of trees made by foresters over the years. These measurements, and the resulting tree growth models that have been developed, focus almost exclusively upon the merchantable portion of the tree—the straight trunk portion (or bole) that can be converted into traditional wood products. It is possible, however, to utilize this vast array of scientific information about trees and forests for the purpose of estimating C reserves by making some reasonably straightforward conversions.

Converting merchantable timber volumes to tree biomass estimates can be done in two parts. For only the aboveground portion of the forest, it is estimated that tree biomass averages 1.67 times the computed amount of bolewood (Cost et al., 1990). This calculation would be appropriate if used to convert the estimated annual growth of merchantable timber into an estimate of the total-tree biomass that might become available for an operation such as whole-tree chipping to provide fuel for a biomass energy generator.

Trees also have significant below-ground root structures, and including them in the calculation leads to a conversion ratio that is about 1.91 for softwoods and 2.44 for hardwoods, or about 2.18 for all species (Birdsey, 1992). These factors provide an estimate of the total biomass, both above- and below-ground, for a given volume of merchantable timber.

This figure, calculated as cubic feet of biomass, must be converted to weight and to C content if we are to estimate the effect of forests upon the conversion of atmospheric C into C stored in trees. In converting the wood volume to weight, Birdsey has calculated an average of 490 kg of dry weight per m^3 of wood. The C content is estimated to be 50 percent of the dry weight. Thus, if the estimated annual growth of merchantable timber for a hectare of forest, in cubic meters of wood per year, is known, it can be multiplied by 534 [490 x 2.18/2] to get an estimate of the annual C storage in kg/ha/yr. If the tree species involved is known, Birdsey's work can be used to get a more precise estimate based on the differences between softwood and hardwood species.

For major timber types, Birdsey also calculated the biomass contained in the litter on the forest floor and the soil C, as they are related to the stage of forest growth from seedlings to maturity. Two of his resulting tables are shown graphically as Figure 1. The tables are shown together to illustrate two points: (1) the importance of soil C in the total reservoir; and, (2) the greater impact of soil C in the cooler, more northern forest.

Perhaps the least understood, and least appreciated, C that is stored by trees and forests as they grow is that which is stored within the soil profile itself. This below-ground storage occurs largely as the result of the continuous growth and death of small roots, but decayed leaf and other litter, as well as the large woody debris that falls to the forest floor, contributes significantly as well. Soil C provides the essential food base for the largely unseen, but widely diverse, populations of soil organisms that are so critical to the healthy functioning of ecosystems. Current research indicates that the soil organic C under forest cover is significantly greater than under either cropland or grassland (Birdsey, 1992).

Using this information, it is easy to calculate the potential impact of planting trees upon land that is physically and biologically suited for forest growth, but where forests do not now exist. Much of that land was in natural forest cover in the past but was cleared for a variety of human purposes, such as cropland or grazing. Since people continue to need crops and animals, no one would propose that all these lands be returned to forest for the sole benefit of helping mitigate the buildup of greenhouse gases. There are, however, many areas where the current use is unsustainable—where crops and grazing are ill suited, and where both environmental damage and, in many cases, unprofitable economic performance are realized. It is on those lands where society can look for opportunities that are immediately beneficial and should be pursued.

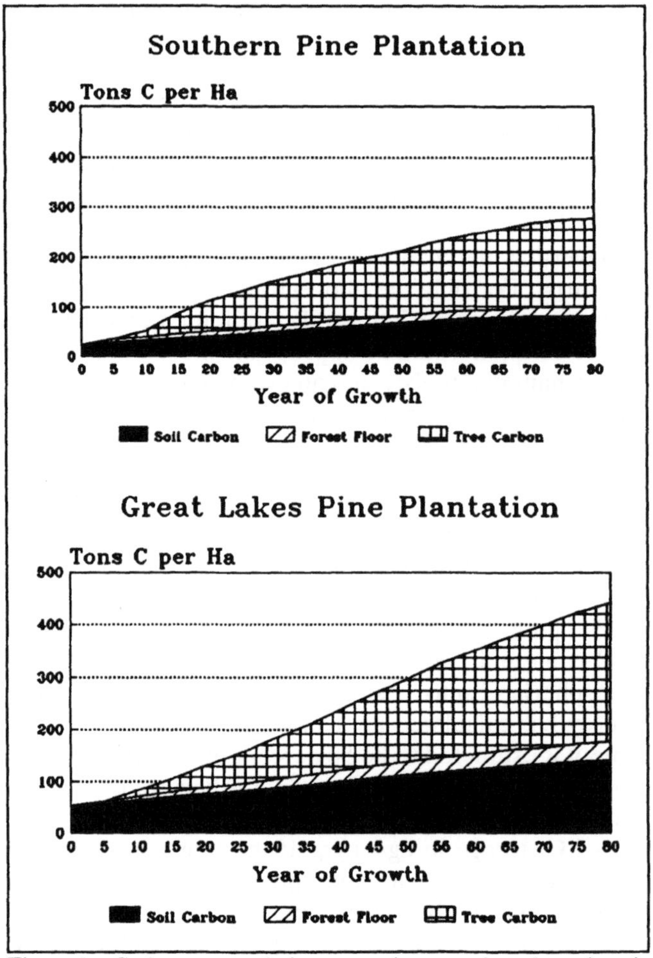

Figure 1. Carbon accumulations over time on pine plantations in two locations, following cropland.

Converting Marginal Crop and Pasture Lands to Forests

There are 47 x 10^6 ha of privately owned crop and pasture land that are marginal for crop and pasture use, using the criteria of the Department of Agriculture, and that are biologically capable of supporting tree growth (Parks, 1992). This land pool, three times as large as the target of the U.S. Conservation Reserve Program contained in the 1985 farm bill, presents a major opportunity for expanding the area of forest land in the United States.

These convertible lands are roughly divided between cropland and pasture in regard to their current use. They are also roughly equally divided between land where softwoods or hardwoods would appear to be most well adapted (Fig. 2).

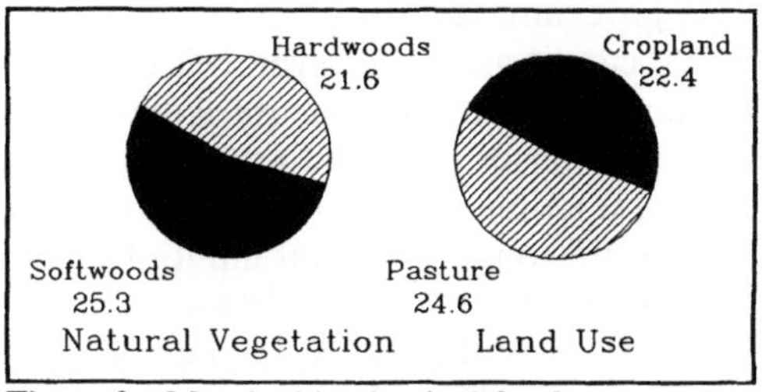

Figure 2. Marginal land suited for forest production, by timber type and current land use (10^6 ha).

On some of these lands, conversion to timber production under current economic situations would be profitable, as well as possible. Of the 25.3 x 10^6 ha adapted primarily to softwoods, Parks estimates that 9.5 x 10^6 ha would be economically more profitable in trees than in their current marginal crop or pasture use. On the 21.6 x 10^6 ha best suited to hardwoods, the lack of forest economic data prevents such analysis, but it is logical to assume that part of those lands would be economically competitive in forest, as well.

For the softwood regions where economic opportunities exist, annual net timber growth on the newly converted crop and pasture lands will average 7 m^3/ha/yr under well-managed plantations. Using the basic conversion factors developed for southern pines by Birdsey, that could mean C storage of 3–4.5 t/ha/yr during the growth cycle of these new forests. This is based on the assumption that about half of the converted land will be cut for pulpwood at age 25 and replanted, and that the rest will be thinned and grown to sawtimber size. Thus the C stored includes some in wood-in-use and in landfills.

In evaluating the opportunity for expanding forests, the 47 x 10^6 ha of marginal crop and pasture land is indicative of the highest amount of forest conversion that could reasonably be expected. Obviously, there are other crop and pasture lands that might be converted, particularly in the case of small areas of better land located in the midst of a surrounding conversion. However, such possibilities have been ignored. For a low estimate of the land that might be converted, the 9.5 x 10^6 ha with an economic advantage to be gained from planting softwoods has been used. While this leaves out land that could perhaps be economically converted to hardwoods, no attempt was made to estimate those areas because economic data were not available.

Lest the conversion of the economically feasible land sounds easy, it is instructive to remember that, while the Conservation Reserve Program was encouraging American farmers to convert 12.3 x 10^6 ha of marginal cropland to permanent cover between 1986 and 1989, only about 1.2 x 10^6 ha of that land was planted to trees (Esseks, et al., 1992). Farmers are reluctant to plant trees for a variety of reasons, including the fact that trees limit their flexibility in using the land in future years, and the evidence that the U.S. Department of Agriculture personnel advising them did not always explain the opportunities for tree planting and commercial harvest after the expiration of the 10-year Reserve contract. Thus, converting 9.5 x 10^6 ha over 10 yrs will not come easily, or without cost and program effort on the part of federal and state governments.

That caveat aside, however, it is estimated that the planting of trees on 9.5 x 10^6 ha of marginal crop and pasture land could in time result in an annual production of 42.5 x 10^6 m^3 of softwood timber per year. That would translate into 30 x 10^6 tons of total C storage per year (Table 2). C emissions from fossil fuels would be reduced by 10 x 10^6 tons. The total input of planting the economically convertible softwood lands would thus be 40 x 10^6 tons a year. The total C impact potential of planting trees on the entire 47 x 10^6 ha that are biologically suited for forests would be approximately 162 x 10^6 tons per year.

Table 2. Potential for mitigating U.S. carbon emissions through converting marginal crop and pasture lands to forests

Current Land Use	Potential	Annual Carbon Storage[1]	Fossil Energy Conservation	Total Impact
	10^6 Ha		*10^6tC/yr*	
Low Estimate of Potential				
Cropland	2.4	3.6	0.9	4.5
Pasture	18.6	26.1	9.0	35.1
Totals	21.1	29.7	9.9	39.6
High Estimate of Potential				
Cropland	49.9	62.1	16.2	78.3
Pasture	54.6	66.6	17.1	83.7
Totals	104.5	128.7	33.3	162.0

Woody Crop Production

The potential for increasing the production of woody crops on what is now cropland can be considered in two contexts: the production of short-rotation, intensive culture wood crops for energy usage and the increased production of food crops from trees. Neither is common today, but that is not to say that increased research, development of appropriate commercial markets and financial incentives, and continued concern for the need to alter land use for improved impact upon environmental factors, including global warming, might not result in significant changes within the coming decades.

Estimating the land base available for energy production is tricky, because the technology is more akin to commercial crop production than to forestry, requiring attention to such details as soil preparation, fertilization, weed control, and mechanical harvesting (Wright, et al., 1992). Thus, it is doubtful that this technology would be economically competitive on land that was marginal for crops or pastures. An exception may be land that is marginal by reason of erosion hazard due to slope, where cultivation of short-rotation woody crops may be feasible as long as slopes are not so steep (over 8%) as to interfere with mechanical harvesting and other operations. In general, however, it is estimated that the land most suitable for wood energy production is not the same land identified by Parks as being marginal for crop or pasture production, so there is little or no duplication when the opportunities for each are calculated separately.

Efficient wood energy production depends on the production of very rapid juvenile woody growth rates, usually in a rotation of 4–8 yrs. Species favored are those that show fast juvenile growth, good stress tolerance and disease resistance, and good coppice growth. These stands are harvested in the winter after leaf fall, and many of the preferred hardwood species then resprout (coppice) from the stump, making recultivation and replanting of the land unnecessary for several rotations. Current yields run 9–16 harvestable dry t/ha/yr in production research trials, although observed maximums range from 16–45, and goals of 16–30 dry t/ha/yr are thought to be reasonable by the year 2010 on moderate-to-good cropland. Costs currently run $37–56 per dry ton, where relatively good cropland can be rented for $100–150 per ha. Of that cost, half to two-thirds is in harvesting and handling—an area where additional research and development appears to offer significant benefits.

Land availability for energy production seems more a function of economic opportunity and technological development than of biological capacity. Of the 171 x 10^6 ha of U.S. cropland, about 124 x 10^6 have the combination of fertility, rainfall, and slope that would make them suitable for short-rotation woody crops. In addition, there are over 34 x 10^6 ha of pasture and forest land that are estimated to have good to moderate potential for conversion to cropland. Most of the suitable lands are in the North Central States (Fig. 3). Of these lands, about 91 x 10^6 ha have the capability to yield 5 or more t/ha of standing dry biomass per year under current technology.

Competition for cropland would not appear to be intense at this time, as national farm policy still encourages the conversion of crop and pasture land to permanent cover. A program such as the Conservation Reserve Program, aimed at encouraging farmers to switch to production of short-rotation woody crops, would need only modest financial incentives if farmers were assured that their "base crop" area would be protected in the event that future national farm policies were once again based upon their crop production history.

Wright, et. al. (1992) established a low estimate of potential land availability for wood energy production at 14 x 10^6 ha and a high estimate of 28 x 10^6. While the high estimate sounds ambitious, it must be noted that soybeans now occupy about 24 x 10^6 ha—up from next to nothing prior to World War II (New Farm and Forests Task Force, 1987). New technologies and economic market inducements can cause significant shifts in U.S. agriculture. While the conversion of 14–28 x 10^6 ha seems feasible, given the estimate in the recent Resources Conservation Act appraisal that the United States could meet crop needs until 2030 while idling as much as 52 x 10^6 ha of current cropland, it is also clear that such a massive conversion will not occur unless biomass markets that provide economic incentives for such a massive shift are developed (U.S. Department of Agriculture, 1989).

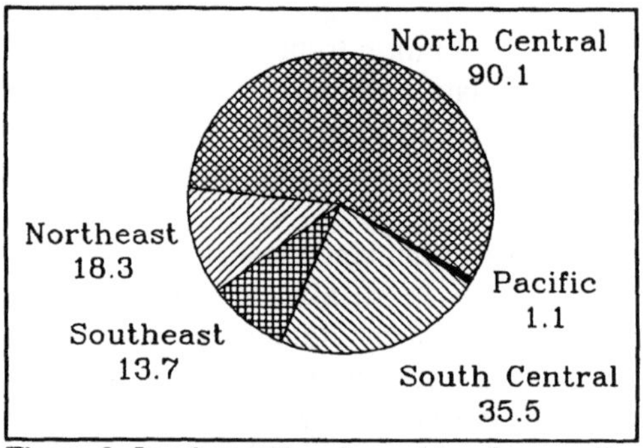

Figure 3. Land considered capable of supporting short-rotation woody crops without irrigation, by major land use regions (10^6 ha).

Another variable, of course, is the conversion efficiency with which dry biomass can be converted to energy. Again, this is an area where research and development can make significant changes. Currently, it is estimated that, with a net dry biomass yield of 11 t/ha/yr and present technology, using wood energy for producing electricity will result in a net C offset of 5 t/ha/yr. If that same biomass is used to produce ethanol, efficiencies are less than half at this time, and the result is estimated to be 1.80 t/ha/yr. Increasing net yields to 22 t/ha/yr and raising conversion efficiencies through research, could, it is proposed, raise those C offset estimates to 10 t/ha/yr for electrical production, and 6 t/ha/yr for ethanol.

The range of total national C impacts from wood energy opportunities are shown in Table 3. The major C storage in wood energy plantations will take place in the root systems and soil C levels because the top growth will be harvested on a regular basis. Each year's root growth and leaf fall will, however, build soil C. On most soils, an increase in soil organic matter of between 0.5 and 1.5% is expected as a result of the conversion to tree crops. In addition, the reduction in soil erosion and the reduced rates of fertilizer and pesticide required could also add to the net beneficial effect.

Energy savings will depend on two factors: whether or not the biomass is utilized to produce ethanol or electricity, and how aggressively improved technology is pursued to improve conversion efficiencies. In Table 3, the "future" column assumes aggressive technology development. Several potentials emerge. A 14-million ha program using current technology to produce ethanol from biomass would have a 90 x 10^6tC/yr impact, representing the low estimate from the table. At the high end, planting 28 x 10^6 ha, using all the biomass for electricity production, and developing future technology could have an annual C impact of 287 x 10^6 tons.

Table 3. Potential for mitigating U.S. carbon emissions through growing short-rotation woody crops to replace fossil fuels in energy production.

Land Utilization	Technology: Current	Technology: Future
	10^6tC/yr	
Low Estimate (14 x 10^6 ha)		
Soil and Plant Carbon Storage	26	26
Energy Conservation—Ethanol	25	85
Total Opportunity	51	111
Soil and Plant Carbon Storage	26	26
Energy Conservation—Electricity	64	144
Total Opportunity	90	170
High Estimate (28 x 10^6 ha)		
Soil and Plant Carbon Storage	52	52
Energy Conservation—Ethanol	50	169
Total Opportunity	103	221
Soil and Vegetation Carbon Storage	52	52
Energy Conservation—Electricity	127	287
Total Opportunity	179	339

Another interesting idea is that of replacing annual crops such as corn and soybeans with specially managed woody crops domesticated and grown for the production of protein, carbohydrates, and oils (Rutter, 1989). Far more revolutionary than many of the other forestry-related options for affecting the emission of greenhouse gases, this proposal foresees research and development leading to a gradual change in the plant cover on significant areas of cropland, both in the United States and around the world.

In asking rhetorically why one would consider such a revolutionary approach, Rutter argues that a switch from annual crops to perennial woody plants captures the natural advantages that woody plants have in sequestering CO_2. Because their leaves develop early in the spring and have more layers of leaf surface, a more complex canopy, and a longer growing season than annual crop plants, woody perennials are able to capture far more sunlight and build far more cellulose per unit area. For example, in May and June, when daily solar energy levels are peaking in North America, woody perennials are fully functioning while corn is still trying to get "knee high by the Fourth of July." In addition, their deep rooting structures often enable woody plants to sustain photosynthesis during dry periods when annual crops struggle to maintain growth rates.

In citing research that calculates the harvestable biomass from a hybrid poplar plantation as representing roughly three times the amount of C content in a total corn crop, Rutter also points out that the C sequestered in the woody plants is more likely to be fixed in long-lived woody structures, including roots, than in the annual crop. In addition, some of the wood production might be used for energy, replacing fossil fuels.

While more revolutionary than other forestry strategies, the systems that Rutter proposes are fertile ground for research and development. In addition to posing enormous potential for increasing the annual uptake of CO_2 on cropland, they offer the potential for increasing the long-range storage of C, both through woody structures and through an increase in soil organic matter, which should be significant in comparison to annual cropping methods. There is also the vision of energy savings due to greatly reduced cultivation on a significant amount of cropland. The proposals are too preliminary at this time to be used as the basis for a realistic estimate of potential impact in the near future, but they are also too interesting to cast aside without further investigation.

Potential To Increase Energy-Saving Trees and Forests

Windbreaks and shelterbelts are associated mainly with the farm and ranch lands of the United States, and their effect upon the root causes of global warming lie in the biomass that they create (within both woody plants and soils and the energy savings their employment facilitates). Brandle, et al. (1992) have looked at this aspect of forestry closely, and drawn estimates leading to the following conclusions:

** The potential for new shelter belt plantings is enormous. It would take 2.25 x 10^6 km of field windbreaks, occupying 1.4 x 10^6 ha, to provide protection to the 27.3

x 10^6 ha of U.S. croplands that were found to be suffering from accelerated rates of wind erosion in 1982 (Trexler 1991). Total tree planting needs are 920 x 10^6.

** An estimated 14.3 x 10^6 ha of additional cropland would benefit from wind protection, needing over 782,000 km of trees that would occupy .5 x 10^6 ha. Total tree planting needs are about 320 x 10^6.

** There are over 1.6 x 10^6 unprotected farmsteads that could utilize 200,000 new farmstead windbreaks involving almost 121,000 ha of trees, adding up to over 55 x 10^6 trees to be planted.

** Facilities housing beef cattle on feed—and dairy cattle in regions where winter conditions would justify wind and/or snow protection—could require over 900 km of windbreaks covering 1,200–2,000 ha, just to protect 10 percent of the exposed livestock. Total tree planting needs are about 1–1.4 x 10^6 trees and shrubs.

** There are significant opportunities for wildlife habitat plantings, shelterbelts, and streamside plantings, but it is impossible to estimate their magnitude.

** The total 1–1.5 x 10^9 trees that could be planted in a complete windbreak and shelterbelt program could contain in the range of 90 x 10^6 tons of C in vegetation and soil by the time they reached 20–25 years of age.

** Indirect C savings are also realized from fuel and fertilizer reductions due to removing land from cultivation (both because the windbreaks take up land and because they raise yields on the remaining land enough to compensate for the land devoted to trees).

There are over 1 x 10^6 km of rural roads in the northern half of the United States that would benefit from the planting of a living snow fence. Living snow fences have only recently been tested in states such as Colorado, Wyoming, and Nebraska. It is estimated that there are only about 182 km in existence today.

The potential benefits, however, are significant. Where road departments now erect temporary slatted snow fences, prices are reported to run as high as $115 per unit of snow storage per km per year, and the structures last an average of 7 years. Large permanent structures, such as one sees along the highways of Wyoming, for example, are cheaper, costing some $12 per unit of snow storage per km per year and having an average service life of 35 years. A 3-row living snow fence, however, drops unit costs for snow storage to about $2 per km per year, with a service life of 75 years. In addition, the trees provide wildlife habitat and increased biodiversity in a region where tree habitats are fairly limited.

Research in Wyoming has demonstrated that snow fence protection can result in a 70 percent reduction in accidents related to winter weather while reducing the cost of snow and ice removal by one-third (Brandle, et al., 1992). The resulting fuel savings add up to emissions reductions of about .5t C/km of snow fence per year.

Assuming that a typical living snow fence will require 2–3 rows of conifers spaced at 3 m, about 1–2 x 10^6 trees on 1,200–2,000 ha will be needed for each 1,600 km of living snow fence. A goal of protecting 1% of the rural roads would add up to 6.4–12.7 x 10^6 trees on 7,500–13,000 ha. Direct C sequestration would be on the order of 27,000 tons per year, and indirect C reductions could run 4,500–9,000 tons per year.

In Table 4, the "low" estimates indicate a "feasible goal" given current information and landowner interest. The "high" estimate shows what the Brandle team calculated as the highest potential opportunity. It is clear that not all these opportunities will be undertaken, but Table 4 shows what the range in opportunity seems to be.

Table 4. Potential for mitigating U.S. carbon emissions through planting trees for soil conservation, windbreaks, and snow control.

Type of Planting	Trees	Hectares	Annual Carbon Impact
Low Estimate of Potential	*(10^6)*	*(10^6)*	*(10^6 t)*
Field Windbreaks	920	1.4	1.4
Crop Protection	320	0.5	0.5
Farmstead Windbreaks	60	0.1	0.4
Other Plantings	30	0.0	0.1
Totals	1,330	2.0	2.4
High Estimate of Potential			
Field Windbreaks	920	1.4	1.4
Crop Protection	1,250	1.9	2.2
Farmstead Windbreaks	110	0.2	0.7
Other Plantings	650	0.9	2.3
Totals	2,930	4.5	6.7

Urban and Community Trees

In addition to sequestering C directly, properly placed urban trees can have a significant impact on atmospheric C buildup through energy conservation. Studies in Florida found that the daily electrical usage for air-conditioning could be reduced by as much as 50% by properly located trees and shrubs (Parker, 1981). Other research has demonstrated that three properly placed trees around homes and small business buildings can cut air-conditioning demand by as much as 50% (Akbari et al., 1988).

Models constructed for Sacramento, CA; Phoenix, AZ; and Lake Charles, LA, indicated that the effect of an additional three trees per house (raising canopy cover by about 30% per dwelling) would result in approximately 30% savings in annual cooling energy used (Huang et al., 1987). McPherson and Woodward (1990) found annual savings of 1,351 kwh for a 137 m^2 house in Tucson, AZ and 1,665 kwh for a similar house in Phoenix.

The impact of large-scale tree plantings in cities can also be very important in reducing the well-documented urban heat-island phenomenon. Researchers have demonstrated that tree planting could save as much as 34, 18, and 44% of residential cooling demand on a hot summer day in Sacramento, CA; Phoenix, AZ; and Los Angeles, CA, respectively (Akbari et al., 1988). If that heat-island impact, which often runs 3–5°C, can be eliminated, or an "oasis" effect created, the cost savings will be enormous.

On the other side of the energy conservation equation, properly placed trees can also reduce winter heating costs by 4–22% (DeWalle, 1978). Heisler (1990) estimates that a good windbreak which does not shade the house in winter will save about 15% of the energy used to heat the home. In residential neighborhoods, windbreaks—as such—are not common, but all the trees act as one huge windbreak. Where tree density is high (canopy cover 77%), Heisler measured wind reductions of 66% in winter and 75% in summer, as compared to open land.

With energy conservation as a major goal, tree planting programs should focus first upon those trees (including species selection and location) that provide direct energy benefits to buildings. The next priority for planting is trees that would provide maximum shade for parking lots, streets, and other dark-surfaced areas. The lowest priority is to plant those open areas where trees could "fill in" the open spaces that, while not directly shading or protecting buildings, would help reduce the urban heat-island effect by modifying albedo and wind patterns as part of the total urban forest.

There have been several attempts to estimate what the C impact of a major urban and community tree program would mean to the nation's efforts to mitigate C emissions. Akbari, et. al. (1988) have estimated that the total impact of planting 100 x 10^6 trees around homes and small businesses, coupled with an effort to reduce albedo through a program of converting dark-colored surfaces such as streets, parking lots, and buildings to light colors, could save about 15 x 10^6 tons of C emissions each year. Their calculations indicated that about half that effect would be due to the trees, for an annual national savings in the range of 5–9 x 10^6 tons of C per year.

Mark Trexler (1991) estimated that the total urban forestry planting opportunities could result in a reduction as high as 15 x 10^6 t/yr, a gain he labelled "modest." He then went on to discount the likelihood of achieving anywhere near that reduction, estimating that even an aggressive policy effort would likely yield a reduction of only 3–5 x 10^6 tC/yr.

Trexler's conclusions were based upon the air-conditioning savings associated with direct shading of small buildings. There are other areas of benefit, however, and they may be more significant than prior work has recognized. Additional trees throughout the urbanized area can reduce the heat-island effect for an entire city. To the extent that peak temperatures can be drawn down on a summer afternoon, everyone's air conditioning works more efficiently, as do most electrical motors and appliances. This citywide improvement in efficiency means a far broader range of savings than will result from directly shading small buildings.

Another area of benefits can come from wintertime savings for space heating that are associated with well-placed windbreaks, as well as with the total windbreak effect of the urban forest itself. Because space heating produces four times the C emissions of air conditioning, such savings could be enormous.

There are about 300,000 ha of new land developed each year, and over half of it comes from crop, pasture, or vacant land uses (U.S. Department of Agriculture, Soil Conservation Service, 1990 unpubl.). Planting trees on that land as part of environmentally sensitive development could result in almost as large a net forest gain as converting marginal farmland to trees, and the potential annual energy savings realized by 2010, if new areas are all planted properly from now on, are significant.

In attempting to factor in all aspects of the urban forest and its effect upon C emissions, we constructed a model that starts with what is known or can be reasonably assumed about the extent of the urban forest, its component land-use portions, and the annual C emissions thought to occur from each land use component. We estimate the current tree canopy and tree population, based on the research data available, and propose a potential goal for a 10-year program aimed at increasing the canopy cover by 10% on residential lands, and 5–20% on other urban lands (Sampson, et al., 1992).

It is then possible to calculate what the attainment of that goal might do in terms of C results in the entire urban forest, not just around small buildings. These estimates are broken down into the additional biomass stored by trees and soils and the C savings due to energy conservation as a direct result of improved shading, evapotranspiration, and reduction of the urban heat island, along with an estimate of wintertime heat savings (Table 5).

The "low estimate" consists of only the annual impact that might be attributable to a good urban forest and tree planting program on the newly developed lands. If urban development continues for 20 yrs at the rate of 300,000 ha/yr, as it did between 1982 and 1987, such a program would require an annual tree planting of almost 17 x 10^6 trees to properly plant the new developments as they occur. For the low estimate, it was assumed that the existing lands would receive only enough attention to maintain current conditions (this would represent a significant urban tree planting program in itself, and a major increase in today's activity in many communities). While this would be better than letting the urban forest continue to decline, it would not result in any significant new energy savings or C sequestration in those areas.

Table 5. Potential for mitigating U.S. carbon emissions through improving urban and community trees and forests.

	Carbon Storage	Energy Conservation	Annual Carbon Impact
		(10^6 tC/yr)	
Low Estimate of Potential	2	7	9
High Estimate of Potential	5	29	34

The "high estimate" calculates the impact that would arise from both planting trees on the newly urbanized lands as they are converted and achieving the goal of increasing canopy cover in existing urban areas by an average of around 10%. About 85% of the total benefits, in terms of C impact, would be realized as a result of energy conservation.

The high estimate represents a huge tree-planting goal. It would require an annual program planting about 42 x 10^6 trees to in-fill the existing urban forest over a 10-yr effort. In addition, if we assume an average 50-yr lifespan for urban trees (which will require a significant improvement in the tree care program of many communities), about 30 x 10^6 trees per year will be needed to provide replacement plants for those that die or are removed. In total, the annual tree planting demand for urban forests appears to be in the range of 89 x 10^6 trees.

The costs of such a planting program are fairly high because urban trees must be specimens of better size and quality than the standard forest planting demands. Site preparation and special protection add significantly to the cost. In total, planting 89 x 10^6 urban trees a year could cost from \$9–20 x 10^9 annually. But much of that cost need not be borne by government. Individual citizens, developers, and businesses can plant and care for the vast majority of these urban trees, if local programs provide the proper education and inducements. Those programs, while not cost free, will consume only a fraction of the total investment, public and private, that the urban forest represents.

The benefits of an improved urban forest are equally large. The C impact, for example, of the high estimate opportunity looks to be about 27 x 10^6 tons of C mitigation per year, either sequestered in the plants and soils, or in reduced emissions due to energy savings. With about 80% of that benefit tied directly to savings in energy usage, the cost savings to consumers alone may be adequate to encourage the adoption of significantly higher levels of planting and care in the urban forest.

Bringing Many Options Together

Row (pers. comm.) evaluated the potential for U.S. military bases to improve their forest management programs as part of a national strategy for mitigating global warming. His results are of interest because military bases are often self-contained communities in their own right, and may provide insight into opportunities elsewhere. Row found that, in general, the forest lands of America's military reservations are managed about as well as can be expected under the conditions imposed by their primary purpose—military training grounds and security. Forests are hard to manage when they are used as training sites, and impossible after they are littered with unexploded ordnance. Even where forest management is not dangerous for work crews and equipment, damage from fires and shell fragments often limits the wood's commercial application. In general, base foresters can do little more than they are already doing (Row, in review).

Where intensive management of the manageable forest land (and occasional salvage from the forests impacted by military training) becomes most feasible is when it is incorporated in a basewide scheme of energy conservation and self-sufficiency. A military base is a well-defined community, with residential areas, office and production areas, land dedicated for transportation, recreation, training and security, and open land that is available for resource management. As such, it is an interesting study area, providing both research and management opportunities that could become models for other communities.

When a base develops an integrated plan for energy conservation and self-sufficiency, that plan normally includes a biomass burner that can utilize nonrecyclable solid waste, landscape trimmings, forest thinnings, nonmerchantable wood, and salvage from military-damaged trees. Once such a plan is operating, timber thinnings, for example, become a positive producer of biomass for energy rather than a nuisance product that must be burned or allowed to rot. Timber stand improvement costs can be partially offset by effective utilization of the thinned wood, and more-intensive management begins to look much more attractive. The burner also offers a way to reduce landfilling of scrap lumber and other solid wastes, as well as lawn and shrub trimmings. It makes wood energy production as part of base operations a feasible option.

The lesson from this look at military installations may be less about the military than it is about the need to consider modest-scale community approaches if there is to be much hope of significantly broadening the use of biomass as an energy resource. Biomass is a high-weight and low-value resource that does not remain economical if it must be moved far from its source to its end use. Burning forest wastes for heat and electricity has long been common at sawmills and paper plants, where such waste abounds, and even has reached commercial application at a few locations around the country where a combination of a large facility needing low-cost energy could be linked to a nearby source of forest products. The Central State Hospital, in Milledgeville, GA, for example, saved $\$1 \times 10^6$ in fuel costs during the first year of operating two wood-fired boilers with wood chips generated from the waste products from nearby state forest lands (Jarrett, 1989). But not very many of those situations exist as yet. If a concerted attempt is to be made to increase the opportunities for biomass to substitute for fossil-fuel energy, research and development in both institutional arrangements and engineering technologies must proceed apace on production and management techniques.

Adding Up the Total Potential

Estimates of the C impact that could be achieved through land conversion and increased use of trees and forests for energy production and conservation are shown in Table 6. Obviously, this estimate must be couched in broad terms to portray the uncertainties at this point in the study of these questions. Science is a long ways from fully exploring all the many opportunities that exist on both public and private lands.

What can be concluded from these calculations, and the studies so far undertaken, is that there are major opportunities through the expansion of trees and forests in the United States to make a significant impact on net C emissions. Those opportunities provide no panacea, nor do they negate the need to consider many other forms of action, including aggressive energy conservation, alternate fuel, and other programs. They are, however, clearly a part of the total approach to the greenhouse gas challenge.

A solid economic case can be constructed for the "low opportunity" levels outlined in Table 6. In regard to conversion of marginal crop and pasture land, for example, it appears that trees can be planted for an average cost of around $150–200 per ha. On a southern pine plantation, over a 40-yr life, that would be a cost of about $1–1.50/t/C sequestered. If, as the National Academy of Sciences has estimated, an opportunity is "low cost" when it can sequester CO_2 for a cost of between $1 and $9 per ton, these costs are low (1991). Since these are the lands identified by Parks as being economically more profitable in trees than in their current usage, there should be no question of whether or not the subsequent land management costs can be recovered. Economic feasibility is less certain for the high opportunity levels shown in Table 6, and many of them would not be feasible without some sort of subsidy to reflect their value as C sequestration practices.

With the total net C emission from fossil fuels for the United States in the range of 1.3×10^9 t/yr (Boden et al. 1990), the forestry opportunities described and quantified in here, if implemented, could offset somewhere between 10 and 30% of that total annual emission level. The major difference between the low and high estimates appears to be the proportion of the impact that is directly related to energy conservation, as opposed to C storage in trees and soils. As more and more of the available opportunities are addressed, a higher percentage of those opportunities is directly connected to the production of biomass energy or the use of trees in energy conservation programs.

It should also be kept in mind that the estimates shown in Table 6 are based on current technology. As illustrated by the data in Table 3, improvements in technology could greatly increase the potential for mitigating C emissions through forestry programs. In addition, these estimates exclude the opportunities to affect C emissions by different management of the nation's timberlands.

Regardless of which estimate is used to portray the forestry opportunities to offset C emissions, several significant factors need to be taken into account. The forestry opportunities described could have many positive impacts on the environment and economy in addition to the effects upon C. Tree planting on marginal crop and pasture lands can reduce soil erosion, prevent water pollution, and improve wildlife habitat. Additional trees in urban areas, as well as shelterbelts and their many uses, means more moderate summer and winter temperatures, less energy used to heat and cool homes and other buildings, and less fuel, fertilizer, and livestock feed needed in production agriculture. There are significant esthetic and positive social impacts as well.

In addition, it should be noted that these opportunities would add to an already-growing forest resource in the United States. In 1987, the USDA Forest Service estimated that net softwood growth (after all mortality and harvests) was 42 $m^3 \times 10^6$ and net hardwood growth was 129 $m^3 \times 10^6$ (AFA, 1989). Converted to net C additions, this adds up to a net increase of about .125 Gt C per year, just in tree biomass on the nation's commercial timberlands. This does not include estimates of surface litter or soil carbon, nor does it estimate the net growth in parks, wilderness areas, non-commercial forests or urban areas. Forests in the U.S. are growing today, and these opportunities would enhance and increase that growth.

Economically, planting trees and more intensively managing the forests that result mean jobs and more wood products to hold down costs to consumers. Where new forms of products emerge, new processing and other facilities may also provide additional economic stimulus. Improving the conservation usage of trees means less money spent on energy, fuel, fertilizer, and feed, and more profit in agriculture.

But, there are cautions with the use of these findings, as well. The data upon which the estimates have been developed are highly variable in quality, as are scientists' understanding of the relationships between tree and forest growth and biomass impact. These are new questions, and a great deal of scientific effort is needed to improve the reliability of these, or any similar, estimates. Much of what is needed can be gleaned from data and research of the past, but there is also the need for a considerable amount of new empirical research.

Increasing the use of biomass as a replacement for fossil fuels can lower net C emissions, but it is not without pollution problems of its own. In addition to the necessity for using new technologies that reduce smoke and particulate release into the atmosphere, scientists must also be wary of the potential for other pollutants, such

Table 6. Opportunities to mitigate U.S. carbon emissions through land conversion, and fossil energy conservation using trees and fore sts

	Carbon Storage	Fossil Energy Conser-vation	Total Im-pact
		10^6 t/C/yr	
Low Estimate			
Converting marginal crop and pasture land to forests	30[1]	10	40
Growing short-rotation, high-intensity wood energy crops	26	64[2]	90
Windbreaks, shelterbelts, and other conservation trees	2	1	3
Urban and community tree planting and forest improvement	2	7	9
Total low opportunity	59	82	141
High Estimate			
Converting marginal crop and pastureland to forests	129	33	162
Growing short-rotation, high-inten-sity wood energy crops	52	127[2]	179
Windbreaks, shelterbelts, and other conservation trees	3	4	6
Urban and community tree planting and forest improvement	5	29	34
Total high opportunity	189	193	382

[1] Includes storage in forests (vegetation and soils) and in products.

[2] The carbon conserved from using woody crops for the production of electricity with current technology.

as heavy metals that may be concentrated in biomass grown in polluted regions or on polluted soils.

The total potential is large, however, and there are many economic and environmental benefits to be gained by pursuing these opportunities, in addition to their impact on the global C cycle. Perhaps continued attention to this subject will result in the kind of public attention and public policy support that can pave the way for these opportunities to be realized.

References Cited

Akbari, H.; Huang, J.; Martien, P.; Rainer, L.; Rosenfeld, A.; Taha, H. 1988. The impact of summer heat islands on cooling energy consumption and global CO_2 concentration. In: *Proceedings of ACEEE 1988 summer study on energy efficiency in buildings*; August 1988; Asilomar, CA. Washington, DC: American Council For An Energy Efficient Economy: 11–23.

American Forestry Association. 1989. *Forests & Forestry in the U.S.A.* Washington, DC: AFA. 32p.

Birdsey, Richard A. 1992. Changes in forest carbon storage from increasing forest area and timber growth. In R. Neil Sampson and Dwight Hair, eds., *Forests and Global Warming.* Washington, DC:American Forestry Association.

Boden, Thomas A.; Kanciruk, P.; Farrell, M. 1990. *Trends '90: a compendium of data on global change.* ORNL/CDIAC-36. Oak Ridge, TN: Oak Ridge National Laboratory. 257 p.

Brandle, James R., Thomas D. Wardle, and Gerald F. Bratton. 1992. Opportunities to increase tree planting in shelterbelts and the potential impacts on carbon storage and conservation. In R. Neil Sampson and Dwight Hair, eds., *Forests and Global Warming.* Washington, DC:American Forestry Association.

CORRIM. 1976. *Renewable resources for industrial materials: Report of the Committee on Renewable Resources for Industrial Materials*, Board of Agriculture and Renewable Resources, Commission on Natural Resources, National Research Council. Washington, DC: National Academy of Sciences. 266 p.

Cost, Noel D.; Howard, J.; Mead, B.; McWilliams, W.; Smith, W.; Van Hooser, D.; Wharton, E. 1990. *The biomass resource of the United States (WO-57).* Washington, DC: U.S. Department of Agriculture, Forest Service.

DeWalle, D. R. 1978. Manipulating urban vegetation for residential energy conservation. In: *Proceedings of the 1st national urban forestry conference*; November 13–16, 1978; Washington, DC. Washington, DC: U.S.D.A. Forest Service: 267-283.

Esseks, J. Dixon, Steven E. Kraft and Robert J. Moulton. 1992. Landowner responses to the forestry option in the conservation reserve program. In R. Neil Sampson and Dwight Hair, eds., *Forests and Global Warming.* Washington, DC: AFA

Fedkiw, John. 1989. *The evolving use and management of the Nation's forests, grasslands, croplands, and related resources.* Gen. Tech. Rep. RM-175. Fort Collins,

CO: U.S. Department of Agriculture, Forest Service, Rocky Mountain Forest and Range Experiment Station. 68 p.

Frey, H. Thomas; Hexem, Roger W. 1982. *Major uses of land in the United States.* AER-535. Washington, DC: U.S. Government Printing Office. 30 p.

Hansen, J.; Rind, D.; DelGenio, A.; Lacis, A.; Lebedeff, S.; Prather, M.; Ruedy, R. 1989. Regional greenhouse climate effects. In: *Proceedings: 2d North American conference on preparing for climate change*; 6–8 December 1988; Washington, DC. Washington, DC: The Climate Institute: 68-81.

Heisler, Gordon M. 1990. Tree plantings that save energy. In: *Make our cities safe for trees: proceedings of the 4th urban forestry conference*; October 15–19, 1989; Washington, DC. Washington, DC: America Forestry Association: 58–62.

Huang, Y. J.; Akbari, H.; Taha, H.; Rosenfeld, A. 1987. The potential of vegetation in reducing summer cooling loads in residential buildings. *Journal of Climate and Applied Meteorology* 26(9): 1103–1116.

Jarrett, Branyon O., Jr. 1989. Central State Hospital: a case study. *Biologue* September/October 1989: 6–9.

McPherson, E. Gregory; Woodward, Gary C. 1990. Cooling the urban heat island with water- and energy-efficient landscapes. *Arizona Review* Spring 1990: 1–8.

National Academy of Sciences. 1991. *Policy Implications of Greenhouse Warming.* Washington, DC: National Academy Press. 128 p.

New Farm and Forests Task Force. 1987. *New farm and forest products: response to the challenges and opportunities facing American agriculture.* Rep. 721-254-1302/60-251. Washington, DC: U.S. Government Printing Office. 55 p.

Parker, John. 1981. *Uses of landscaping for energy conservation.* Miami, FL: Florida International University, Department of Physical Sciences. 102 p.

Parks, Peter J., Susan R. Chadwick and James E. Mitchell. 1992. Opportunities to increase forest area and timber growth on marginal crop and pasture land. In R. Neil Sampson and Dwight Hair, eds., *Forests and Global Warming.* Washington, DC: American Forestry Association.

Rutter, Philip A. 1989. Reducing Earth's greenhouse CO_2 through shifting staples production to woody plants. In: *Proceedings: 2d North American conference on preparing for climate change*; 6–8 December 1988; Washington, DC. Washington, DC: The Climate Institute: 208–213.

Sampson, R. Neil, Gary A. Moll and J. James Kielbaso. 1992. Opportunities to increase urban forests and the potential impacts on carbon storage and conservation. In R. Neil Sampson and Dwight Hair, eds., *Forests and Global Warming.* Washington, DC:American Forestry Association.

Schneider, Stephen H. 1989a. The greenhouse effect: reality or media event. In: *Proceedings: 2d North American conference on preparing for climate change*; 6–8 December 1988; Washington, DC. Washington, DC: The Climate Institute: 57–67.

Schneider, Stephen H. 1989b. *Global warming: are we entering the greenhouse century.* San Francisco: Sierra Club Books. 316 p.

Trexler, Mark C. 1991. *Minding the carbon store: weighing U.S. forestry strategies to slow global warming*. Washington, DC: World Resources Institute. 81 p.
U.S. Department of Agriculture. 1989. *The 2d RCA appraisal: soil, water, and related resources on nonfederal land in the United States: analysis of conditions and trends.* Rep. 242-141/03004. Washington, DC: U.S. Department of Agriculture. 280 p.
U.S. Environmental Protection Agency. 1989. *The potential effects of global climate change on the United States* (with 10 supporting appendices). PM-221. Washington, DC: U.S. Environmental Protection Agency, Policy, Planning and Evaluation. 413 p.
Wright, Lynn L., Robin L. Graham, Anthony F. Turhollow, and Burton C. English. 1992. Opportunities for short-rotation woody crops and the potential impacts on carbon conservation. In R. Neil Sampson and Dwight Hair, eds., *Forests and Global Warming*. Washington, DC: American Forestry Association.

REFERENCES CITED—UNPUBLISHED

Row, Clark. [In review.] Increasing biomass productivity on Department of Defense lands to mitigate global climate change. Report to the Environmental Protection Agency, December 1990.

U.S. Department of Agriculture, Soil Conservation Service. 1990. Unpublished data from the 1987 National Resources Inventory. Provided to the author by Jeff Goebel, Soil Conservation Service, through personal correspondence.

SHOULD WE STORE CARBON IN TREES?

GREGG MARLAND and SCOTT MARLAND†
Environmental Sciences Division
Oak Ridge National Laboratory
Oak Ridge, TN 37831-6335, USA

ABSTRACT. In order to explore for the most effective strategy for using forests to mitigate global climate change, we have constructed a simple model of C uptake during forest growth and the fate of this C when forests are harvested and used as fuel to replace fossil fuels. We suggest that trees are equally effective in preventing the accumulation of CO_2 in the atmosphere if they remove a unit of C from the atmosphere or if they supply a sustainable source of energy that substitutes for a unit of C discharged by burning fossil fuels. The model shows that the most effective strategy for using forest land to minimize increases in atmospheric CO_2 will depend on the current status of the land, the productivity that can be expected, the efficiency with which the forest harvest is used to substitute for fossil fuels, and the time perspective of the analysis. For forests with large standing biomass and low productivity the most effective strategy is to protect the existing forest. For land with little standing biomass and low productivity, the most effective strategy is to reforest or otherwise manage the land for forest growth and C storage. Where high productivity can be expected, the most effective strategy is to manage the forest for a harvestable crop and to use the harvest with maximum efficiency either for long-lived products or to substitute for fossil fuels. The longer the time perspective, the more likely that harvesting and replanting will result in net C benefits.

1. Introduction

The concept of planting trees as part of a strategy to confront the possibility of global climate change is now widely accepted. As trees grow they remove CO_2 from the atmosphere and thus slow the atmospheric build-up of CO_2, an important greenhouse gas. To date, tree planting is the only active component of the official U.S. strategy for responding to increasing anthropogenic emissions of CO_2. The questions being approached now concern the rate at which trees can sequester carbon and the land area which might be available, both in the U.S. and elsewhere, for tree planting. Calculations by Marland (1988), Moulton, and Richards (1990), the U.S. Congress OTA (1991), and many others, show both that it is in theory possible to counteract some of the emissions from fossil-fuel burning and that counteracting a large fraction of current U.S. or world emissions with trees would require very large land areas. Although generally acknowledged to be an optimistic upper limit, the

† Department of Engineering Science and Mechanics, University of Tennesse, Knoxville, TN, 37916, USA.

Water, Air, and Soil Pollution **64**: 181–195, 1992.
© 1992 *Kluwer Academic Publishers*.

Moulton and Richards calculations suggest that the U.S. could offset 56% of its current CO_2 emissions from fossil-fuel burning with trees planted on 140 x 10^6 hectares of economically marginal and environmentally sensitive crop and pasture lands and non-federal forest lands. Planting trees can help but it cannot provide the sole solution.

We also recognize that tree planting is an interim measure. Whereas young, fast-growing trees can take up relatively large quantities of C, we know that on a time scale of decades the trees will approach maturity and their growth rate, and hence rate of net C uptake, will inevitably decline. In their initial paper on the topic, Dyson and Marland (1979) proposed tree planting as an emergency measure to halt or reverse the growth of atmospheric CO_2 while we developed or implemented an energy system not based on fossil fuels.

2. Net Emissions of CO_2 to the Atmosphere

If the longer term contribution of tree plantations is to be limited by the ultimate maturity and senescence of the trees, we ask to what extent it is possible to maintain a young, rapidly-growing stand, vigorously removing CO_2 from the atmosphere, by harvesting the trees and sequestering some of the C in long-lived products such as construction lumber. This is the strategy examined by Harmon et al. (1990) when they inquired to what extent the harvesting and sequestering of lumber from an old-growth forest would result in a net sequestering of C as the forest regrew. The Harmon et al. conclusion was that, for the circumstances they examined, the net amount of C stored in the living forest plus long-lived wood products did not recover for at least 250 yr to the pre-harvest level because of the slow growth of the forest toward its pre-harvest C content (see Table 1).

Table 1. Disposition of carbon on harvest of a 450 year-old Douglas fir forest (Harmon et al., 1990)

	Mg C/ha
Carbon in old-growth boles	325
Carbon in long-lived products	138
Carbon in short-lived products, fuel, paper, and residue	187
Carbon sequestered	138

In a global climate change context, however, the Harmon et al. paper does not quite ask the critical question. If we adopt a single criterion of relevance to global climate change, the issue is not how much C is sequestered in the biosphere and forest products, but how much C is discharged, in net, to the atmosphere. The data required to fully restructure the Harmon et al. accounts are not all in their paper but we can make a start with a couple of rational assumptions. The difference we suggest is in the treatment of the fraction of the wood from this harvest that was used to produce energy for lumber or pulp processing. If we chose to try to sequester that portion of the wood, or to include it in the long-lived products, we would have required additional purchased energy (probably from fossil fuels) to power the lumber and pulping processes. In a very real sense, then, the CO_2 discharged from burning the wood for energy does not, to the extent that the wood burning replaces fossil-fuel burning, represent net emissions of CO_2 to the atmosphere. This assumes, of course, that we insist on making paper, pulp etc. and will use some source of process energy. In our current industrial system, however, the biomass energy probably is not used very efficiently (for the sake of illustration we assume that the thermal efficiency of conversion is 60% of that which would have been achieved with fossil fuels, see below) but to the extent that it avoids use of fossil fuels, its impact on net CO_2 emissions to the atmosphere is the same as for that fraction of the wood that ends up in long-lived products. In effect the C is "sequestered" in unburned fossil fuels. In rough terms, we would recast the Harmon et al. C balance as in Table 2. We assume that of the pulping raw materials, 45% of the C will become paper and 55% will be used to supply process energy. We assume, also, that 50% of what Harmon et al. describe as "defects and breakage" and all of "bark fuel and mulch" will be used for energy generation (Sherman, 1990). Although it is likely that some of the paper will be either put in long-term storage in landfills or used for energy, we assume that it is all oxidized to CO_2 within a relatively short time. (As a second order consideration we should inquire whether these long-lived wood products require more or less fossil energy, and CO_2 emissions, for their production than do the products they replace)

Table 2. Net effect on CO_2 emissions to the atmosphere from harvest of a 450 year-old Douglas fir forest

	Mg C/ha
Carbon in old-growth boles	325
Carbon in long-lived products	138
Carbon in wood-based fuel	113
Fossil-fuel carbon displaced	(113)(0.6)=68
Carbon released to the atmosphere via paper, residues, etc.	74
Net carbon released to the atmosphere	74+(113-68)=119
Effective carbon sequestered	325-119=206

There are thus two potential strategies for using forests to confront the increasing levels of atmospheric CO_2. If we have a choice of whether to plant and protect trees to take up and sequester C over long time periods or to harvest and replant trees for fuel to recycle C and avoid net emissions to the atmosphere from fossil fuels, which is the better choice? Because, with one caveat explored below, it does not affect the carbon balance whether we use the wood to replace fossil fuel or for long-lived products, and because the calculations presented below are simpler if we follow a single path, we will assume that all harvested biomass is burned to displace fossil fuels. Wood for lumber does have greater economic value than wood for fuel but this is more a matter of capital than carbon.

There are at least four factors that will weigh heavily on whether we should grow trees to store C or to recycle C: the amount of biomass which is standing on the site now, the expected productivity of the site, the efficiency with which wood can replace fossil fuels (our caveat from the paragraph above), and the time interval over which we would choose to integrate. In what follows we have constructed a simple model to illustrate the impact of these factors and what they suggest about a forest strategy for confronting global climate change. The model is based solely on C flows and has no implications with respect to economics. We do not presume to suggest that C balance should now be a primary criterion for forest management but illustrate the implications if it were, or how C balance might interact with other forest management objectives. We emphasize that our simulations involve only above-ground, living biomass. By ignoring the implications for litter and soil C, we may exaggerate somewhat the benefits of scenarios involving wood harvest. We also exaggerate slightly the benefits of harvest by not considering the energy-use implications of factors like nutrient replacement. The model does not consider that changing climate and atmospheric CO_2 content could affect the growth rates of trees. Our underlying premise is that land is a limiting resource for forest strategies and that if we want to use forests to address global climate change we need to inquire into the most effective way to use each hectare of forest land.

The reader will be aware that the numeric results displayed below are model dependent. As such, they are illustrative rather that demonstrative. They demonstrate patterns and sensitivities. They are intended to provide perspective and a point of departure for discussion and analysis of realistic, site-dependent values.

3. The Model

An essential component of these model runs is the growth function assumed. Standing stock of total above-ground living biomass (S) is taken to be a function of its original value (S_0), a growth rate (G), and an assumed limiting value for total standing stock which can be supported on the site (S_{max}). Growth is described by the step function

$$S_{n+1} = S_n + G_n$$

where

$$G_n = G \text{ when } S_n \leq S_{max}/2$$

and

$$G_n = (G/S_{max}/2)\,(S_{max} - S_n) \text{ when } S_n > S_{max}/2.$$

With this function the standing stock grows at rate G until it reaches half of the maximum value and growth in subsequent intervals is a constant fraction of the remaining distance to the maximum value. Figure 1 illustrates the function for G = 3 Mg C $ha^{-1}yr^{-1}$. We have assumed throughout that the maximum supportable biomass is 150 Mg C ha^{-1}, a value that might be appropriate for a good site in the southeastern U.S.

In the scenarios involving wood harvest, we assume that anytime the standing stock reaches or passes 40 Mg C ha^{-1} it is harvested. In practice, the time to harvest will depend on the growth curve but also on the harvest method and the purpose for which the harvest is intended. Turhollow and Perlack (1991) report that the energy required for plantation management is dominated by energy use during harvest and haul and is a function of the mass handled. In their analysis of the energy balance for short rotation energy crops, poplar trees

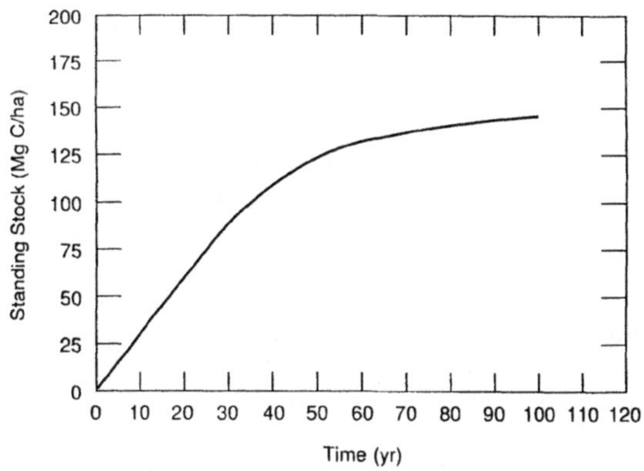

Figure 1. The increase in standing stock with time. This illustration of the functional form of the change in standing stock with time is for Growth Rate (G) equal to 3 Mg C ha^{-1} yr^{-1}. It illustrates that in our model the annual growth rate is constant until the standing stock reaches half of the maximum supportable standing stock. The growth rate then declines so that standing stock approaches the limiting value asymptotically.

are harvested at 42 Mg C ha^{-1} and we use that approximate value so that we can use their value for the energy requirement as well. As a consequence of this early harvest, all growth in our harvest scenarios is linear at rate G.

If we harvest biomass to replace fossil fuels, it is important to know the efficiency with which this substitution takes place. One unit of standing C in a tree cannot substitute directly for one unit of C in coal. We illustrate the sensitivity to this parameter by using two efficiency values based on estimates in Turhollow and Perlack (1991) and Lynd et al. (1991).

Turhollow and Perlack suggest that CO_2 emissions from fossil fuels used for plantation establishment, management, and harvest contribute emissions of CO_2 equivalent to about 6% of the C embodied in the harvest. They estimate further that 15 to 20% of the standing biomass is lost either on site or during harvest and haul. There is also some coal loss and fossil fuel use during coal mining and transport (perhaps 3%, Marland, 1983). Thus 80 to 85% of the C in plantation trees and 97% of the C in coal in the ground arrive at the boiler.

The final step in comparing net C emissions per unit of energy service has to do with the efficiency of conversion of C at the boiler to, say, electricity. As an average, coal has a C to energy ratio of about 24.12 kg C Gj^{-1} while wood (with lower H to C ratio) has 25.32 kg C Gj^{-1}. If both fuels could be converted to electricity with the same thermal efficiency, wood burning would discharge 25.32/24.12 more C per unit of electricity than would coal burning. We assume for this calculation that wood could be used as a boiler fuel with the same thermal efficiency as coal while acknowledging that this is a task for engineering development and that in current practice typical values for efficiency of conversion to electricity are near 33% for coal and at or below 25% for wood. With all of these considerations, the extent to which a unit of wood C can substitute for a unit of coal C in electric power generation is taken to be about:

$$C_s = -0.06\ (C_h) + ((0.80 \text{ to } 0.85)/0.97)\ (24.12/25.32)\ (C_h)$$

or

$$C_s \approx 0.75\ (C_h)$$

where C_s is the net decrease in C emitted to the atmosphere from fossil fuels and C_h is the C in standing stock of trees at harvest time. In other words, a unit of C harvested and oxidized from a forest is able to substitute for about 3/4 unit of coal in terms of CO_2 emissions to the atmosphere.

Wood fuels are being evaluated and researched both as a boiler fuel for electric power generation and as a cellulosic feedstock for ethanol production (Lynd et al., 1991). In this latter processing scheme we still require fossil fuels for plantation operation and have wood lost in harvest and haul. Lynd et al. estimate that (with processes under development) we might envision 45 to 70% thermal efficiency of conversion from wood to ethanol plus waste materials that can be used for excess boiler fuel. For refined petroleum products, roughly 12% of the energy value of crude oil is consumed in production and refining. We take the lower end of the efficiency range suggested by Lynd et al. and, considering again that the wood-based fuels contain less energy per unit of C, use, for illustrative purposes, half of the conversion efficiency above,

$$C_s = 0.375\ (C_h)$$

to represent the C substitution efficiency of liquid fuels derived from wood for those derived from crude oil. Thus a unit of C in standing biomass can substitute for 0.375 unit of C in liquid transportation fuel. We use this value to illustrate the consequences of substituting wood fuel for fossil fuel with a low efficiency of replacement.

There are a number of important considerations that we have not attempted to include in this simple model. Plantation management costs, i.e., fossil fuel supplements, are likely to increase in succeeding harvest cycles because of the need to replace nutrients removed with the harvest. For harvesting the initial biomass in mature forest stands, the wood loss and energy investment are likely to be proportionally smaller per unit mass of harvest than the constant ratio for plantation harvesting assumed here. Stands with high growth rates are likely to have higher limiting values of maximum supportable standing stock (S_{max}). The potential productivity of a site may be significantly changed by an initial harvest and change in management strategy. These factors would affect the quantitative outcome of any detailed analysis but not the qualitative conclusions of this exercise.

4. Model Results

Using the model elaborated above, we have asked how much C can be taken up by a standing forest and how this will compare with the net impact on CO_2 emissions to the atmosphere of harvesting wood and using it to replace fossil fuels (or, by implication, to construct long-lived products).

In all of the figures that follow, the vertical axis shows the amount of C in the standing stock of trees at the time C management is initiated (S_0 above). Values near zero would suggest pasture or agricultural land while values near 150 Mg C ha^{-1} would suggest mature forest near the maximum standing stock which could be supported on the site. Intermediate values represent perhaps open woodland, agricultural land returning to forest, or second growth forest at various levels of stand maturity. The horizontal axis on these plots shows the annual rate of net C assimilation in above-ground biomass that could be expected for young, vigorous forest (G above). Values near zero would represent sites with low rates of growth, perhaps because of nutrient or water deficiency, cold climate, steep slope, etc. Values near 10 Mg C ha^{-1} yr^{-1} have been achieved on managed stands in the tropics and values between 1 and 3 are typical for young commercial forest stands in the continental U.S. The U.S. Department of Energy's short rotation woody crops program has achieved growth rates of 8 Mg C ha^{-1} yr^{-1} in the southeastern U.S. (Wright et al. 1992). Focus should always be on the left half of these figures because these represent the more common circumstances.

Figure 2 shows contours of cumulative net uptake of C by growing forests after 50 yr of growth. For any point on the 50 Mg C contour, for example, the model suggests a net sequestering of 50 Mg C at the end of 50 yr for a forest with the initial values of standing stock and growth rate shown by the coordinates of the point. Figure 2 shows clearly that net

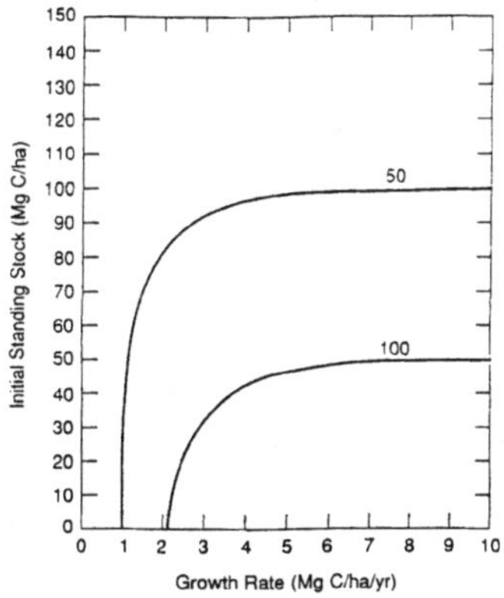

Figure 2. Contours of the increase in C Storage in standing stock after 50 yr of growth. If, for example, the initial biomass on a site has 50 Mg C ha^{-1}, and the growth rate is greater than about 5 Mg C ha^{-1} yr^{-1}, the figure shows that the net C uptake in the biomass after 50 yr will be near 100 Mg C ha^{-1}, the uptake being limited because we have defined the maximum supportable standing stock of C on the site to be 150 Mg C ha^{-1}.

C uptake is minimal for forests with large initial standing biomass. This is because convergence on the maximum value of supportable biomass permits little net C uptake. For forests with high growth rates C saturation also occurs quickly. Figure 3 superimposes the contours for the same forests after 30, 50, and 100 yr of growth and shows that with increasing time, saturation occurs in forests with progressively lower growth rates. In all cases the net cumulative C uptake is limited to the difference between the initial standing biomass and the maximum supportable biomass, and the growth rates simply determine how rapidly saturation (i.e. no further C sequestering capacity) occurs. When envisioning trees as a strategy for offsetting C emissions from fossil fuels we should thus ask how fast they can sequester C, how the growth rate will change with time, and what the maximum supportable biomass will be.

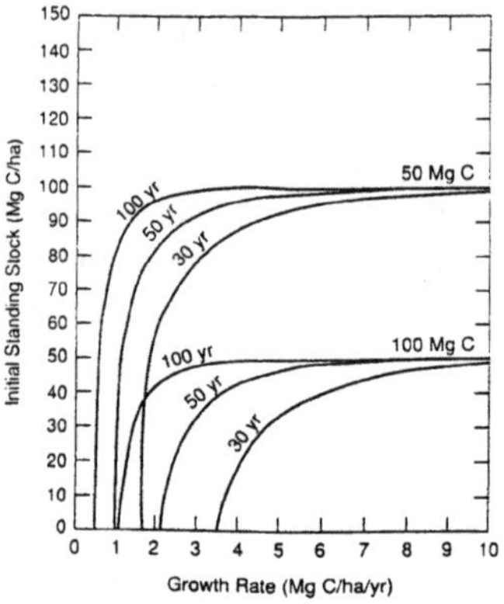

Figure 3. Saturation of C uptake with time. With maximum supportable standing stock defined at 150 Mg C ha^{-1}, forests with high growth rates will saturate, that is, lose their ability to take up additional C, in a short period of time. Forests with progressively lower growth rates will gradually saturate with increasing time. Contours for 50 and 100 Mg C ha^{-1} net uptake are shown for 30, 50 and 100 yr of growth. At the end of 30 yr even the most productive forests are still growing slowly, converging on C saturation. By the end of 50 yr, forests with growth rates greater than about 8 Mg C ha^{-1} yr^{-1} have become saturated and by the end of 100 yr net C uptake has virtually ceased in all forests with growth rates greater than about 3 Mg C ha^{-1} yr^{-1} but continues in forests with lower growth rates.

Figures 4 and 5 address the circumstance where every forest is harvested when it reaches a C stock of 40 Mg C ha^{-1}. Functionally the model harvests any forest with 40 Mg C ha^{-1} or more, replaces it with new forest with 0 starting biomass, and uses the harvested material as a fuel to replace fossil fuels. Forests with less than 40 Mg C ha^{-1} initially are allowed to grow to that point and then are harvested and replanted. Figure 4 shows contours of the cumulative net amount of C sequestered (i.e., either C removed from the atmosphere or fossil-fuel emissions avoided) after 50 yr if the harvested wood substitutes for fossil fuel with .75 C efficiency. In the upper left portion of the figure the net of C sequestered is negative, that is, there are net C emissions to the atmosphere. For these combinations of high initial standing biomass and low growth rate, the net of emissions from the initial harvest is never compensated because of the slow subsequent growth. This is, in essence, the area discussed by Harmon et al. (1990) where net emissions from harvest of a mature forest are not compensated by new growth even over long times. The problem is that 25% of the C in the initial harvest is released to the atmosphere without offsetting comparable emissions from fossil-fuel burning.

The importance of using harvested biomass efficiently, to produce either long-lived products or fossil fuel substitutes, is demonstrated by comparing Figures 4 and 5. Model parameters for Figure 5 were the same as for Figure 4 except that for Figure 5 the efficiency of substitution of biomass C for fossil-fuel C was only .375, suggestive of perhaps the internal process-energy demands for converting wood into a liquid transportation fuel. With low efficiency of substitution for fossil fuels, the portion of the field with net CO_2 emissions after

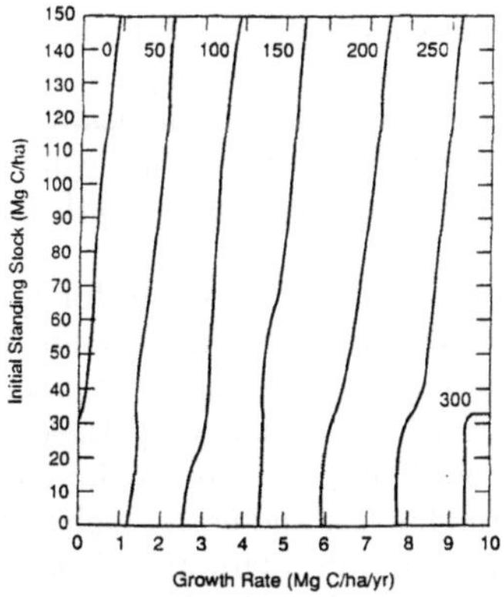

Figure 4. The net of C effectively sequestered after 50 yr when forests are harvested at 40 Mg C ha^{-1} and the harvest is used to offset fossil-fuel use with .75 efficiency of C substitution. Contours are in Mg C ha^{-1}. The contours show the sum of the increase of C in standing vegetation and the fossil-fuel C displaced by renewable biomass fuels. Because of the material lost in handling, the energy cost of managing and harvesting plantation forests, and the differences in fuel chemistry, the equivalent of 25% of the C in harvested biomass is released to the atmosphere without replacing fossil-fuel C releases.

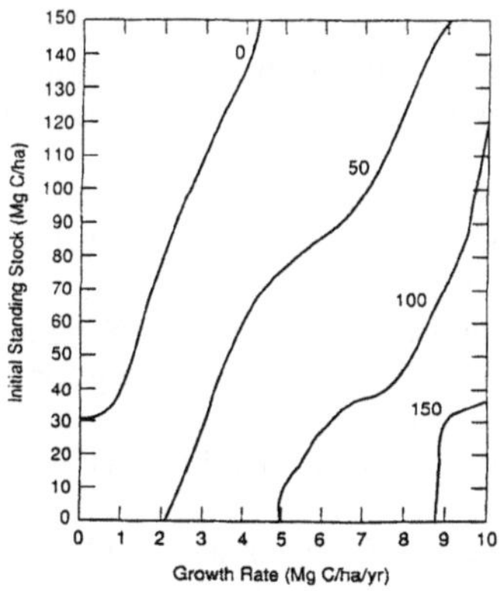

Figure 5. The net of C effectively sequestered after 50 yr when forests are harvested at 40 Mg C ha^{-1} and the harvest is used to offset fossil-fuel use with .375 efficiency of C substitution. Contours are in Mg C ha^{-1}. The low efficiency of C substitution might be because of the internal energy requirements to make liquid transportation fuel from cellulose or because of high costs or inefficient operation of plantations or combustion facilities. The low efficiency of fossil fuel substitution results in a larger area of net CO_2 emissions (to the left of the 0 contour) and lower rates of net carbon sequestering generally than shown in Figure 4.

50 yr is much larger and in other portions of the plot the magnitude of net emissions reduction is much smaller, even for forests with very high growth rates. The shape of the contour lines in Figure 5 emphasizes that our model is indicative only. Odd contour shapes, particularly around 40 Mg C ha^{-1}, are an artifact of contouring discrete values generated from a model with simple step functions for annual biomass growth and harvest. We have done some smoothing of contours.

Having illustrated that net sequestering of C occurs for most combinations of initial standing biomass and growth rate, whether trees are left standing to store C or are harvested to substitute for fossil fuels, we can use our simple model to explore the relative cumulative C offsets of sequestering versus recycling C. Under comparable conditions, which results in the largest offset of net CO_2 emissions? Figure 6 shows the differences in cumulative net carbon sequestered after 50 yr for tree harvest and use at high (.75) C substitution efficiency as opposed to letting the trees grow and store carbon with no harvest. As for Figures 7 through 10, Figure 6 shows four discrete fields. In the field at lower left (stippled in the figures), with low initial biomass and slow growth rate, 40 Mg C ha^{-1} is never reached so harvesting never occurs. In the field at upper left (hachured in the figures), with high initial biomass and low growth rate, there will be net CO_2 emissions if harvest occurs because the growth rate is too slow to compensate (over 50 yr) for net emissions at the time of the initial harvest. For the remainder of the field to the left of the zero contour, there will be net C sequestering in the harvest scenario but the total C sequestered at the end of 50 yr will be less than if the forest were simply allowed to stand and accumulate C. For the remainder of the field, the contour lines show the excess of net effective C sequestering for the harvest scenario as opposed to the no-harvest scenario. That is, for all combinations of initial standing biomass and growth rate to the right of the zero contour line, there would be more C effectively sequestered if the trees were harvested and used efficiently than if they were simply allowed to grow and store C.

In the extreme case illustrated in Figure 6, with high initial biomass and high growth rate, 50 yr of plantation management with harvest and efficient substitution for fossil fuels results in a net effective displacement of over 250 Mg C ha^{-1} more than in the no-harvest scenario, where there is essentially no net C uptake. Coincidentally, Wright et al. (1992) estimate that a productivity of 5.6 Mg C ha^{-1} yr^{-1} is the minimum necessary for economic consideration of short rotation woody crops and this value falls safely into the region of our plot where establishment of energy plantations on non-forest land results in a larger C offset than would reforestation.

The relative effectiveness of recycling C versus storing C in wood is strongly affected by the efficiency with which the harvest is used. Figure 7 shows the difference in net C offset after 50 yr for a case with C recycle through fuels and a case with C storage in standing forest. However, where the harvest in Figure 6 was used to substitute for fossil fuels with a .75 efficiency, in Figure 7 the efficiency of C substitution is only .375. Very high growth rates are required before harvesting with low efficiency use of wood fuels will sequester more carbon than simply letting the trees grow. Also, the portion of Figure 7 which represents net emissions of CO_2 (hachured) is much larger than in Figure 6. Given the parameters of Figure 7, most of the area of the plot corresponding to growth rates which one might expect for new forest established on non-forested land in the U.S. would fall in the region where storing C is more effective than recycling C inefficiently.

Time perspective is another important factor. With the 50 yr perspective of Figure 7 it makes sense, on carbon-sequestering grounds, to harvest and replant forests where high

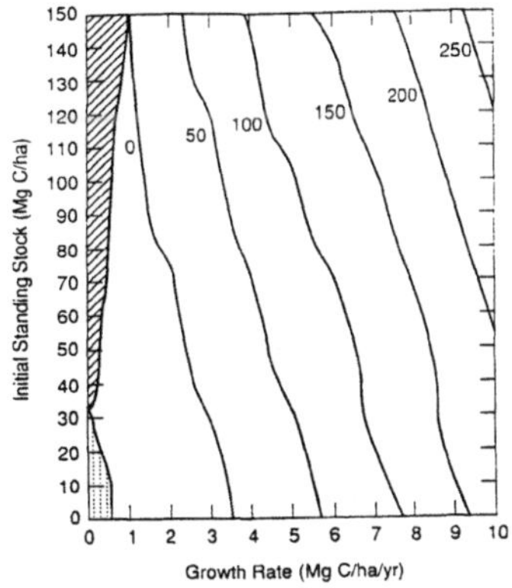

Figure 6. The excess of net effective C sequestering after 50 yr for forest harvest and efficient (.75) use of the product as opposed to taking up and storing C in the standing stock of forests. Contours are in Mg C ha^{-1}. Four regions are represented. In the stippled region the standing stock never reaches 40 Mg C ha^{-1} and hence harvest never occurs. In the hachured region, net biomass growth is so slow that C uptake in regrowth never compensates for the CO_2 released during the initial harvest and there is a net release of C even after 50 years. In the remaining area to the left of the zero contour line, there is net C sequestering for both the harvest and no-harvest cases but the net effective C sequestering is greater in the no-harvest case. To the right of the zero contour line, the contours show the excess of net C sequestered for the harvest-and-substitute-for-fossil-fuels case as contrasted with the no-harvest case. The contours show the mass of additional C that will be withheld from the atmosphere if forests are harvested to offset fossil-fuel burning as opposed to the case where they are allowed to continue to grow and store C.

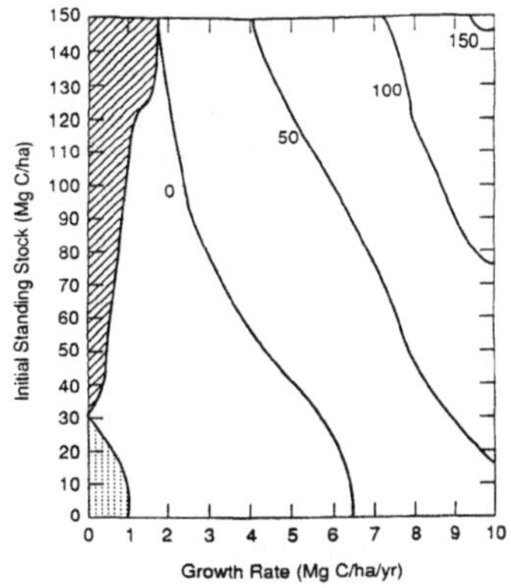

Figure 7. The excess of net effective C sequestering after 50 yr for forest harvest and inefficient (.375) use of the product as opposed to taking up and storing C in the standing stock of forests. See Figure 6 for explanation.

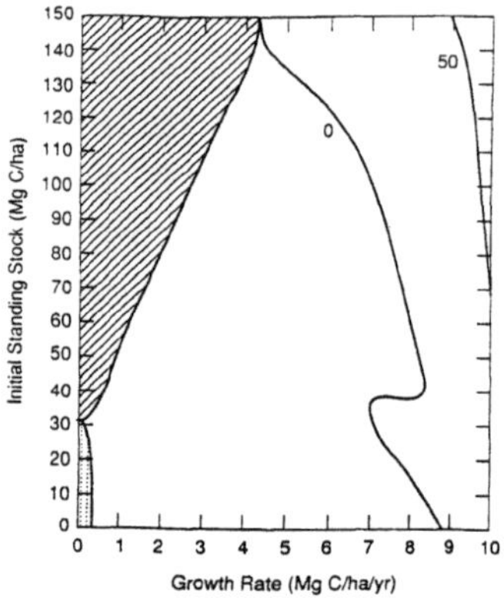

Figure 8. The excess of net effective C sequestering after 30 yr for forest harvest and efficient use to replace fossil fuels as opposed to taking up and storing C in the standing stock of forests. See Figure 6 for explanation.

growth rates can be expected and to establish energy plantations on very productive sites. If the perspective is only 30 years, however, harvesting for fuel is suggested only for very high productivity sites when the wood is to be used efficiently (Figure 8) and under virtually no conditions if the wood will not be used efficiently (Figure 9). Given a sufficiently long perspective, 100 years in Figure 10, several harvesting rotations will insure a C profit for energy plantations with respect to standing forests even if the harvest is used with low efficiency.

As a final sensitivity test with our model, we explored the effect of different values for S_{max}, the maximum supportable standing stock. The boundary in our graphs between the region of net emissions of CO_2 (hachured) and the region of net effective sequestering has a positive slope which depends only on the time perspective and the efficiency of harvest use. Consequently, if the maximum supportable standing stock were increased above the 150 Mg C ha^{-1} shown, perhaps approaching the 325 Mg C ha^{-1} described by Harmon et al. (1990), the hachured region in our figures would continue to expand toward higher growth rates. The higher the initial standing stock, the higher the growth rate required to insure that regrowth will in time compensate for the net CO_2 emissions during the initial harvest.

5. Discussion

Given the choice of using forests to take up and store C or to recycle C by producing an energy crop to substitute for fossil fuels, our simple model of C flows suggests than the most effective strategy for minimizing net CO_2 emissions to the atmosphere depends on the current status of the available land, the expected growth rate of the forest, the efficiency with which the harvest will be used, and the time perspective. The model lacks the refinement to provide quantitative detail but it does confirm that for forest with high standing biomass and low expected growth rates, the best choice is simply to let the existing forest stand. For areas with low standing crop and modest expected growth rates, the best choice is to provide

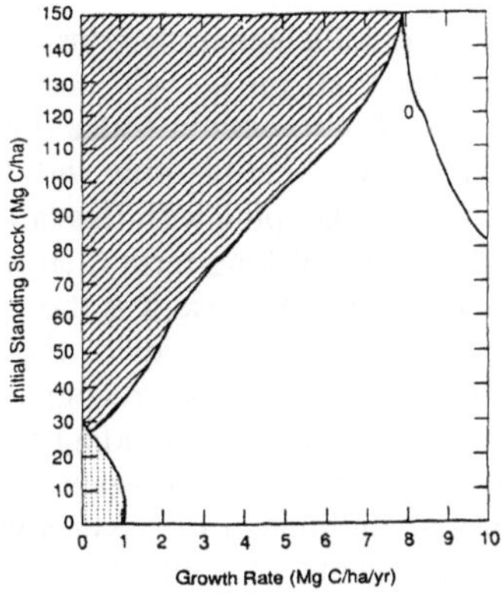

Figure 9. The excess of net effective C sequestering after 30 yr for forest harvest and inefficient use to replace fossil fuels as opposed to taking up and storing C in the standing stock of Forests. See Figure 6 for explanation.

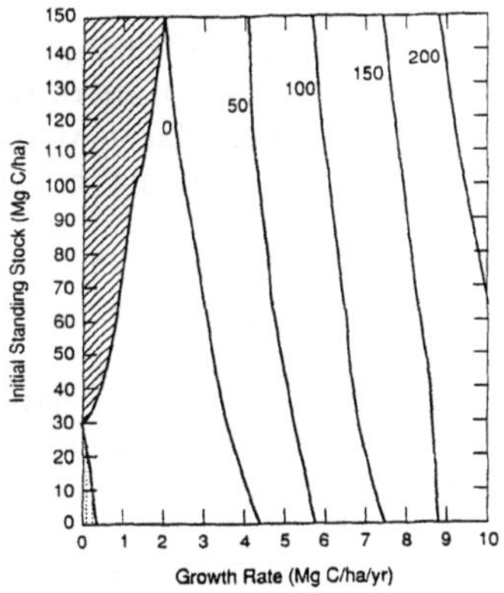

Figure 10. The excess of net effective C sequestering after 100 yr for forest harvest and inefficient use to replace fossil fuels as opposed to taking up and storing C in the standing stock of forests. See Figure 6 for explanation.

management or new planting of forest with the intent of letting the trees stand and accumulate C. However, where higher productivity can be anticipated, net C emissions to the atmosphere can be minimized by planting and subsequently harvesting trees for use either in long-lived products or to supply energy to replace fossil-fuel burning. Efficient utilization of the harvest increases the gain and makes it attractive to harvest trees from less productive sites. In all cases, time is very important because given a sufficiently long planning perspective and enough harvest cycles, any small gain per cycle will eventually total to a net gain. For low productivity forests with low efficiency use of the harvest, this can be a very long time.

The numeric output of this simple model is generally consistent with our intuition. It does confirm that there are some forests that, on C-cycle considerations, should not be harvested and replaced with young stands. It also demonstrates that simply planting trees to store carbon is indeed a short term solution and that planting with the intent of efficient use is the most effective choice in many areas. In keeping with this later observation, the model also suggests that planting trees with the intent of long-term C storage will have very limited impact as an offset for fossil-fuel CO_2 emissions because it is the most effective choice only where productivity will be low or where the circumstances of the forest make harvest and utilization very expensive or inefficient. The model does recognize that even when forests are harvested regularly, they will maintain some average standing stock of C where there may have been little or none previously. Finally, we emphasize again that this discussion is based entirely on C flows and does not consider the many other issues which must be considered to maximize the benefits of forest management. Our intent is only to focus on the factors to be evaluated in managing forests to minimize the net emissions of CO_2 to the atmosphere.

6. Acknowledgements:

This Research was sponsored by the U.S. Department of Energy, Carbon Dioxide Research Program, Atmospheric and Climatic Research Division, Office of Health and Environmental Research, under contract DE-ACO5-84OR21400 with Martin Marietta Energy Systems, Inc. We are grateful for assistance from W. M. Post in preparing the figures for this paper.

7. References

Dyson, F.J., and G. Marland (1979) "Technical Fixes for the Climatic Effects of CO_2", p. 111-118 *in* W.P. Elliott and L. Machta (eds.), Workshop on the Global Effects of Carbon Dioxide from Fossil Fuels, Miami Beach, Florida, March 7-11, 1977, U.S. Department of Energy CONF-770385.

Harmon, M.E., W.K. Ferrell, and J.F. Franklin (1990) "Effects on Carbon Storage of Conversion of Old-Growth Forests to Young Forests", *Science*, 247, 699-701.

Lynd, L.R., J.H. Cushman, R.J. Nichols, and C.E. Wyman (1991) "Fuel Ethanol from Cellulosic Biomass", *Science*, 251, 1318-1323.

Marland, G. (1983) "Carbon Dioxide Emission Rates for Conventional and Synthetic Fuels" *Energy* 8, 981-992.

Marland, G. (1988) The Prospect of Solving the CO_2 Problem Through Global Reforestation, DOE/NBB-0082, U.S. Department of Energy, February, 66p.

Moulton, R.J., and K.R. Richards (1990) Costs of Sequestering Carbon Through Tree Planting and Forest Management in the United States, GTR WO-58, U.S. Department of Agriculture, Forest Service.

Sherman, W.R. (1990) Personal Communication.

Turhollow, A.F. and R.D. Perlack (1991) "Emissions of CO_2 from Energy Crop Production", *Biomass and Bioenergy*, 1, 129-135.

U.S. Congress Office of Technology Assessment (1991) Changing by Degrees: Steps to Reduce Greenhouse Gases, OTA-O-482, U.S. Government Printing Office.

Wright, L.L., R.L. Graham. A.F. Turhollow, and B.C. English (1992) "Opportunities to Mitigate Carbon Dioxide Build-up Using Short Rotation Woody Crops", *in* R.N. Sampson and D. Hair (eds.) Forests and Global Warming, Vol. 1, American Forestry Association, Washington, D.C., in press.

The potential role of temperate forests as sinks for CO_2 - Examples from the german environmental policy against global warming

H.D.Gregor
Federal Environmental Agency (Umweltbundesamt)
Bismarckplatz 1
D-1000 Berlin 33
Germany

ABSTRACT. Among industrialized nations Germany ranks fourth in CO_2 emissions. Most of these originate from the use of fossil fuels. Based on reports of a parliamentary study commission, established in 1987, and other expert groups in Germany this article adresses possible environmental effects of increasing atmospheric CO_2, the sink potential of temperate forests and the influence of forest damage on this potential. A strategy for a 25 to 30% or 250 to 300 x 10^6 t yr^{-1} CO_2 emissions reduction by 2005 (which Germany has itself committed to) is described in which measures to enhance C sequestration by forests play an important role. Expansion of forest area, a further increase of C storage by appropriate management and the restoration and protection of forest health impaired by air pollution would result in an additional storage of 17 to 20 x 10^6 t yr^{-1} of CO_2, equaling 6 to 8% of the reduction target.

1. INTRODUCTION

Despite major efforts to reduce air pollution new environmental problems continue to arise. The history of environmental effects started with problems individual people had with local sources of pollution, sources they could see and often smell. Then larger sources began to affect nearby forest edges, sometimes even killing whole stands, like in Germany around the turn of the century. The first institute of "smoke damage" research was established 150 yr ago in Tharandt, Saxony, in the middle of hard hit forests. In the 1950s and, becoming more widespread in the 1980s, covering approximately 15 x 10^6 ha in Europe today, we experienced for the first time multifactoral problems at a continental and maybe hemispherical scale: forest damage, a concern for many countries, and acidification, a transboundary problem. Today, with the enhanced greenhouse effect, new dimensions of complexity, speed and

Water, Air, and Soil Pollution **64**: 197–212, 1992.
© 1992 *Kluwer Academic Publishers.*

heterogeneity of change and responses appear and we face a global problem.

Many anthropogenic pollutants are radiatively important trace gases (Anon., 1988): N_2O (4%), tropospheric O_3 (8%), CFCs (17%), CH_4 (19%), and CO_2 (50%). This list explains why the main interest is focussed on CO_2.

Atmospheric CO_2 has increased from approximately 280 ppm in 1850 to 350 ppm today. The increase continues at a rate of 0.4% yr^{-1} (Anon., 1988). There are many sources for CO_2. They range from burning fossil fuel as the major source, to the buffering of acid deposition, which contributes only a small fraction (700,000 t in Germany). A large quantity of CO_2 is emitted from burning tropical forests. Before 1950 agricultural sources of CO_2 were more relevant (Anon., 1988). The enormous increase in the utilization of fossil fuels in the last 30 yr has changed that picture completely: today 75% of the anthropogenic CO_2 (5.3×10^9 t C) originates from burning fossil fuel in the industrialized countries. EEC estimates link 19% of all CO_2 to automobiles. Emissions of greenhouse gases (Anon., 1988, Gröbl, 1990) vary between countries, but in countries with similar economic and industrial conditions they reach similar dimensions. In Germany 92% of all anthropogenic CO_2, 88% of the CO, 96 % of SO_2 and 99% of NO_x are emitted by energy-related processes.

The worldwide total emission of CO_2 from fossil fuel use is: 20.5 to 22×10^9 t yr^{-1} (Anon., 1988, Gröbl, 1990). In the northern hemisphere the USA (22.3%), Eastern European countries (21.6%, former USSR alone 17.3%), western Europe (15.4%), Japan (4.3%) and China (12.1%) contribute 78% of the total CO_2 emissions.

Re-unified Germany accounts for 5.3% and ranks fourth in the world. Before reunification, the Federal Republic of Germany emitted 3.7% of the world total which was a per capita emission of 11.7 t yr^{-1}, compared to a per capita emission of 22.4 t yr^{-1} in the former GDR, which was the highest world wide. The US is next with approximately 20 t yr^{-1}. In contrast, developing countries have much lower values: e.g. China 2.4, Brazil 1.3 t/capita yr^{-1} (see table 1).

The goals of this paper are to describe the political approach to the problem in Germany, to review some elements of the present understanding of effects connected with increasing CO_2 in the atmosphere, to rank the sink potential of temperate forests, to address the influence of forest damage on this potential, and to describe the possible use of the forest sink in CO_2 mitigation policies.

TABLE 1. CO_2 emissions and reduction commitments in some selected countries (Anon., 1989).

Country	Target	percent of global emissions (1986)	1986 per capita emissions (tonnes)	Remarks
Germany	25-30% CO_2 reduction by 2005 (based on 1987 figures)	5.3	13.7	
Australia	stabilization of *greenhouse gas* emissions by 2000 20% reduction by 2005 (based on 1988)	1.2	15.2	comprehensive approach; programme applied, if other countries follow
Netherlands	CO_2 stabiliz.by 1995, 3-5%reduction by 2000 (based on average 1989/90)	1.0	13.9	
Denmark	20% CO_2 reduction by 2000, 50% by 2030 (based on 1988)	0.3	13.0	concept approved by parliament
New Zealand	20% CO_2 reduction by 2000 (based on 1990)	0.1	6.2	comprehensive approach; aimed for, not binding
Austria	20% reduction by 2000 (based on 1988)	0.3	7.2	not yet approved by parliament

2. THE POLITICAL APPROACH

Environmental policy in Germany (Töpfer, 1991, Schmidbauer, 1991) as a first priority aims at convincing countries to put a worldwide cap on CO_2 emissions by the yr 2000 followed by stepwise reductions by 2005 and 2010. The final goal should be a 50% reduction by the middle of the next century. Nationally, by a resolution of the Cabinet Ministers in June and November 1990, Germany already committed itself to a 25 to 30% emission reduction for CO_2 (by 2005 on the basis of 1987 emissions). Increasing the efficiency of energy utilization and energy-conscious management and behavior alone provide a potential for about 20% reduction. The German proposal for the Commission of the EC in Brussels for example calls for a stepwise reduction of the average fuel consumption of cars to 5 L/100 km (or 50 mpg) by 2005. This and a number of other national measures will allow Germany to reduce its CO_2-emissions from 1 065 x 10^6 t to about 750-800 x 10^6 t. While projections show that this will be possible, the International Energy Agency (IEA) predicts an increase by 2005 in emissions by western industrialized countries of 20% and 50% in the rest of the world!

The past 20 yr show (Gröbl, 1990) that while the Gross National Product in the former Federal Republic of Germany grew by more than 30%, the use of primary energy did not increase and CO_2 emissions even dropped by almost 9%. It is intended to continue and strengthen this trend of uncoupling economic growth from energy consumption and environmental demands. There still is a large potential for reducing CO_2 emissions if you consider all possible strategies. Even nuclear power may play a role here as we see that by its use in Germany under certain conditions the emission of 110×10^6 t yr^{-1} of CO_2 could be avoided. The role of forests in Germany for the storage of C becomes of considerable significance. A capacity of binding only 1 to 2% of the total excess C becomes a possible contribution of 6 to 8% if compared to the reduction target of 250 to 300×10^6 t yr^{-1}.

If all countries do their share or help others in accomplishing theirs a worldwide reduction of 25% by 2005 to 2010 is not impossible. Germany, in addition to its national measures at home, has doubled its financial commitment to the protection of forest areas, reforestation and combating erosion in tropical forests by contributing 500 million DM in 1989 and 1990.

Finally, the idea behind these plans makes it well possible that along with the international discussion about commitments against global climate changes, arguments will be brought up to balance national per capita emissions of CO_2 by per capita mitigation.

In order to study the options and receive further advice for combating global warming, in 1987 the German Parliament appointed a study commission (Enquete Comission) on Preventive Measures for the Protection of the Earth's Atmosphere. Its first report, "Protecting the Earth's Atmosphere - An International Challenge" (Anon., 1989), was an important precedent in the establishment of the Intergovernmental Panel on Climate Change (IPCC). Its second report "Protection of Tropical Forests - A High Priority International Task" (Anon., 1990), focussed on research and recommendations for political measures. The third Commission report called "Protecting the Earth - A Status Report with Recommendations for a New Energy Policy", (Anon., 1991), gives a concise overview of the whole problem of global change and recommends a number of national as well as international measures for the control of greenhouse gases.

In 1990 the Enquete Commission was reorganized and renamed "Protection of the Earth's Atmosphere". Its advice is now receiving even higher political priority. The present work of the Commission also has shifted to interactions between climate change and agriculture and forestry. In this complex system both can serve as sources and sinks for greenhouse gases and the effects they trigger as emitters are counteracted through cycling or enhanced by feedback phenomena. The activities of the new Commission so far have dealt with agriculture. They will next turn to management requirements in order to increase potential in agriculture and forestry for C sequestration and the reduction of greenhouse gas emissions.

Recent Parliamentary Study Commission Reports are not available yet, as work is still in progress, but the input of the Federal Environmental Agency can be summarized in a short overview, addressing the potential effects of global change on temperate forest ecosystems, the sink potential of these forests, and the influence of forest damage on the sink potential. Finally the role C sequestration in forests may play in reducing greenhouse gases in the atmosphere will be discussed.

3. DISCUSSION

3.1 Effects of Global Change on Temperate Forests.

As we know from basic plant physiology, CO_2, water availability and temperature are, in principle, rate limiting for photosynthesis. Therefore, warming and an increase in atmospheric CO_2 should increase productivity and therefore also increase the sink potential of vegetation, if nutrient supply is adequate and enough moisture is available. There are numerous data available to show that the responses of plant species to rising levels of CO_2 differ considerably and that indirect effects on the ecosystem level have to be assessed separately. Many plants seem to respond 'positively' to higher CO_2 levels (Drake, 1992, Strain and Thomas, 1992). In some cases plants react to rising CO_2 by an almost linear response in dry matter production (Overdieck, 1986). Other investigations show that as long as C-incorporation is growing, the C/N ratio in the plant shifts to lower N and the nutritional value as well as ecological parameters (reduced decomposition rate etc.) of the plant or litter are changed (Overdieck and Forstreuter, 1986, Overdieck et al., 1988). It is suspected that food plants of ruminants with such a decreased N-content and higher lignin will increase their CH_4 exhalation (Burschel et al., 1989). CH_4 has a global warming potential (GWP) 32 times higher than CO_2, therefore such potential CH_4 losses, how ever small they may be, have to be carefully considered.

It was assumed from laboratory experiments and field observations that even when increased CO_2 in an experimental atmosphere increased the rate of photosynthesis, this enhancement was almost completely reversed at the leaf level within a few days, at the level of a whole plant within one or a few yr (Körner, 1989) and at the ecosystem level probably within a decade.

Figure 1: Photosynthetic rate at elevated CO_2 (after Körner, 1989)

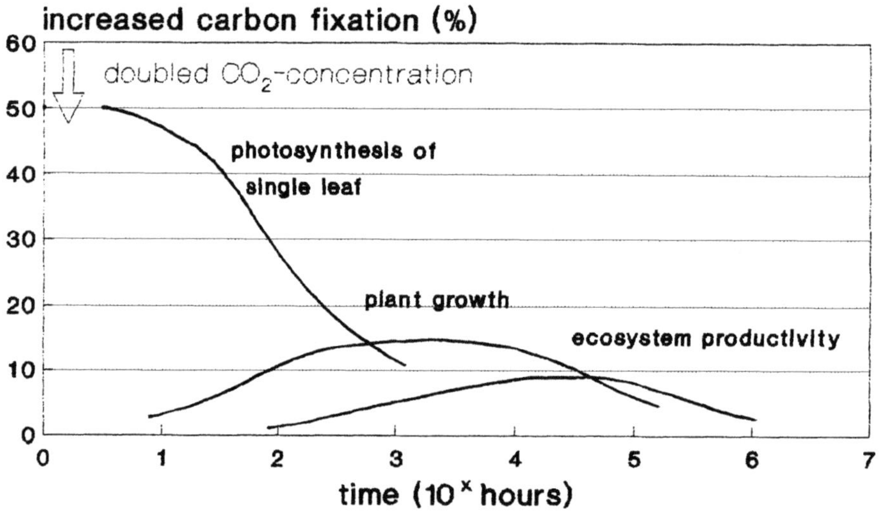

But, while laboratory data already show a wide range of differential reactions in single plants, responses in the field and especially in forest ecosystems are even harder to predict because of potential nutrient and water limitations. In Europe many forests are found growing where nutrient supply is low. These stands have often already reached the peak of their productivity, many of them because today's increasing anthropogenic N deposition accelerated growth beyond what their nutrient supply could support, causing nutrient imbalances.

Increased productivity of some species as a result of higher CO_2 levels will change species distribution within ecosystems. Changes in the C/N ratio in the plant, as described above, will change not only food quality but also plant attractiveness for pathogenic organisms (Overdieck et al., 1988). Further ecological consequences lie in the fact that in some plant species seed production is enhanced (Overdieck, 1986) when CO_2 increases and their propagation could therefore eventually benefit from global changes.

While increased temperature and CO_2 cause reactions in vegetation that at least in some species and for some time suggest a small, maybe temporary, increase in the aboveground storage of C, soils may lose C in the form of CO_2 to the atmosphere, due to higher decomposition rates (Ocio et al., 1991), when the temperature in the upper soil layer increases (Burschel et al., 1989).

Forest ecosystems will be affected by changing temperature and precipitation. Some, but not all, species will benefit from the changing conditions. Shifts in species distribution may be a result (Smith et al., 1992). Adaptation processes require time. Central European forests had appr. 250 tree generations to adapt to the climate change after the last ice age. Predictions now suggest that not even one generation is available to achieve a similar level of adaptation to trace gas-induced climate change.

Assumed climatic consequences of the greenhouse effect in the temperate zones where Germany is located are predicted to strongly affect forests, because in addition to the air pollution damage, expected drier and warmer summers will compensate any possible positive effect of longer vegetation periods. Frequent droughts and heat periods as well as frequent extreme wind storm events are likely to occur and will add to the problems. It should be mentioned here that such stresses will increase instability in the forest ecosystems and increase the danger of further air pollution damage (Addison, 1991).

A concise review of potential effects has recently been published (IPCC, 1991). The effects on temperate forests were also of special concern of the above-mentioned Enquete Commission. A hearing on "Protection of Forests in Temperate and Boreal Zones" took place in August 1989 and its results are also a still important base for conclusions in this paper (Burschel et al., 1989).

3.2 The Sink Potential of Temperate Forests.

Appr. 33% of the earth's soil surface is covered with forests, 40% of which are situated in the temperate and polar latitudes (Körner, 1989, Deutscher Bundestag, 1990). Forests from temperate zones generally do not receive enough attention as far as their potential as sinks for CO_2 is concerned. These forests cover 920 x 10^6 ha in Europe and the former USSR. While today's burning of tropical forests causes worldwide concern, the forest area in Europe was drastically decreased 1000 to 500 yr ago, the period of European forest devastation. The region never fully recovered from this loss of forest area and today the remaining area faces a multitude of anthropogenic stresses that in some areas endanger forest function and survival.

In the second half of this century the European forest area has increased, partially as a result of reforestation and management changes, but also the result of reducing farm production ("set aside plans"). The rate of increase was approximately 2.5% per decade from 1950 to 1990 according to FAO assumptions (Burschel et al., 1989), FAO, 1990, Deutscher Bundestag, 1990) and countries expressed their certainty that this would continue at a similar rate into the next millenium. Timber data also suggest that, for some time, less timber is cut than is produced. This increase in accumulated biomass is highest in Central Europe (Burschel et al., 1989): in Austria, S-Germany, and Switzerland an increased accumulation of 1.1 to 1.5 t ha^{-1} a^{-1} of biomass is reported (for comparison: a spruce forest in this region would have a total of 100 t C ha^{-1} stored in its biomass). Increases in forest biomass are very high in Eastern Europe, whereas in the nordic countries and southern Europe the rate is 0.3 to 0.6 t ha^{-1} a^{-1}. Lack of systematic surveys make these figures somewhat uncertain.

The size of this sink is too small to compensate for tropical forest burning. Its size is also small compared to the C fixing potential of the marine phytoplankton. Data on the storage of C in soils rank its capacity three times higher than that of vegetation. Total C stored in terrestrial vegetation worldwide is suggested to be 560 x 10^9 t while total C stored in soils is given as 1,500 x 10^9 t (Körner, 1989).

The role of agriculture in this respect is not restricted to increasing storage of organic matter in soil, but includes improving the use of regenerative energy/raw materials. If the organic content of soils is raised by only a few percent, this results in sequestration of many million t of CO_2. And replacing fossil fuel consumption with biomass consumption would produce a net decrease in the accumulation rate of C in the atmosphere (Marland, 1992).

3.3 Forest Damage

Forest ecosystems in temperate zones have become subject to impacts that considerably affect their vitality (ECE, 1990, 1991, BML, 1991, EEC, 1991, Gregor, 1991b). In large areas of Europe, and to some extent in North America, forest decline has been observed at considerable rates during the past decade. In the surveyed ECE region during 1989, 15.5 of 101 x 10^6 ha were already in the higher damage classes (Table 2., Gregor, 1991b, ECE, 1990). Air pollutant concentrations in many of these areas are above critical levels and acid deposition to forest soils often exceeds their buffering capacity (Critical levels and loads are shown in Table 3 and Table 4). Systematic forest damage surveys throughout the ECE region suggest that action is required to restore forest health and maintain forest ecosystem stability. This is only possible by a new type of forest policy that combines strict reduction of air pollution with more ecologically-oriented forest management. Other changes are needed in the areas of economic, traffic and industrial policies, which are mainly responsible for the threat to forest ecosystems (Gregor, 1991b).

TABLE 2. Forest damage (defoliation) in European countries in 1990, all tree species, based on national and regional surveys (ECE, 1991).*)

Country	defoliation**) (percent trees affected) none class 0	slight class 1	moderate to severe classes 2 to 4
Spain	79.2	17.0	3.8
Ukrainian SSR	64.1	29.5	6.4
Italy-Bolzano	76.1	16.7	7.2
France	76.0	16.7	7.3
Belgium-Flanders	45.1	46.6	8.3
Austria	50.8	40.1	9.1
Italy (national)	61.4	23.8	14.8
Germany, Fed.Rep.(W)***)	47.1	37.0	15.9
Sweden	50.1	33.7	16.2
Switzerland	39.0	44.0	17.0
Norway	53.8	29.0	17.2
Finland	62.2	20.5	17.3
Greece	39.5	43.0	17.5
Netherlands	53.2	29.0	17.8
Yugoslavia-Slovenia	64.3	17.5	18.2
Belgium-Wallonia	54.6	26.3	19.1
USSR-Lithuania	31.5	48.1	20.4
Denmark	45.3	33.5	21.2
Hungary	49.6	28.7	21.7
USSR-Kaliningrad	29.1	48.4	22.5
Italy-Tuscany	38.4	34.1	27.5
Bulgaria	49.7	21.2	29.1
Portugal	53.1	16.2	30.7
Germany, Fed.Rep.(E)***)	34.3	29.8	35.9
USSR-Latvia	31.0	33.0	36.0
Poland	14.3	47.3	38.4
United Kingdom	26.0	35.0	39.0
Czechoslovakia	23.7	29.7	46.6
Byelorussian SSR	16.0	30.0	54.0

*) Forest damage/defoliation may not only be caused by air pollution
**) defoliation: class 0: up to 10%, class 1: 11 to 25%, classes 2 to 4: 26 to 100% needle/leaf loss observed.
***) the first combined estimate of East and West Germany in 1991 showed the following result (BML, 1991):

	36	39	25

TABLE 3. Critical levels (μg m^{-3}) for the adverse effects of air pollutants on plants, plant communities and ecosystems for short and long-term exposures (UBA, 1988a, deVries and Gregor, 1990, Gregor, 1991a)

Pollutant	Exposure duration: Short-term	Long-term (annual average)
SO_2	70 (24-hr mean)	20
O_3	150 (1-hr mean)	50*
NO_2	95 (95th percentile of 4-hr mean)	30

*based on the growing period mean value which is composed of 7-hr daily means.

C loss to the atmosphere from damaged forests (contribution of damaged forests to the greenhouse effect) could be best avoided by a rapid restoration of forest vigor and C-fixing potential. This in turn is most effectively achieved by drastic measures for the control of N and acid deposition and O_3 formation (Gregor, 1991a, b).

TABLE 4. Examples for critical N deposition loads (mol_c ha^{-1} yr^{-1}) for forest ecosystems (Nilsson and Grennfelt, 1988, deVriess and Gregor, 1990).

Effects considered	Criteria	Critical load
"Eutrophication"	Vegetation changes	700 - 1400
Forest damage	N content in leaves < 1.8%	1000 - 2000
Nutrient imbalances	NH_4/Mg < 5	1000 - 1500
Nitrate leaching to groundwater	NO_3 < 50 mg L^{-1}	1200 - 1400

There are various opinions concerning the impact of forest damage on the C sink potential of forests in Europe. One assumption is that there is no such influence, but instead there are signs of compensation or even overcompensation by the regionally-observed increased productivity. The other opinion points out that not only are there losses of C from forest soils but also losses of C stored in the biomass. Also, more than 100,000 ha of forest have been lost completely due to air pollution in the Erzgebirge and Riesengebirge mountains in the Czech and Slovac Federal Republic. The stored C has been released and no more can be fixed. According to these views even optimistic scenarios of the future development of forests in Germany, for example, suggest losses in average growing stock up to 8% (Burschel et al., 1989).

Forest damage research in Germany (UBA 1988b, FBW, 1990) has yielded an increased understanding of forests under anthropogenic stress. The main fields of research adressed by the German Forest Damage Research Program are shown in Fig. 2.

Figure 2. German Forest Damage Research Programme, 1982 - 1990 (Gregor, 1990).

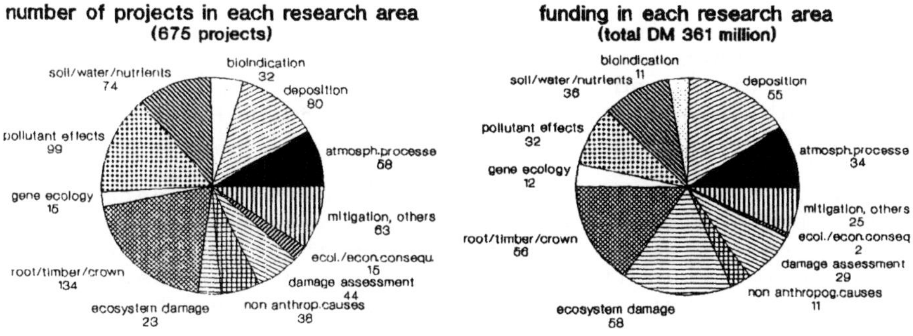

Where there are sufficient data available, especially for spruce, there is evidence for differential reactions (Burschel et al., 1989): severely damaged trees exhibit increment losses, while mildly damaged or undamaged trees or stands show increment increases. This latter feature is observed over a considerable area and may be intensive enough to compensate for the C losses from damaged stands. But as this may be a result of anthropogenic N deposition (reported as 15-50 kg N ha^{-1} a^{-1} (wet deposition) in German forests (Deutscher Bundestag, 1988), mangement changes (Kenk et al., 1990), CO_2 increase or rising temperature, there is every reason to be alarmed about possible ecological consequences.

There are three strategies that could by themselves, or better, if applied together increase C storage by forests. The first is the protection of existing forests and the restoration of the vigor of damaged forests. The second is raising C storage in existing stands by appropriate management. The third is an expansion of forested areas. In this complex of measures the sink potential would be most efficiently and economically increased.

3.4 The Role of Forests in German CO_2 Reduction Plans.

Expert opinion in Germany (Burschel et al., 1989, 1992, IMA, 1990) proposes to maintain and extend CO_2 storage in forests or forest products in the following ways:

- optimum use of increment growth
- prolonged rotation periods
- afforestation, and
- substitution of wood for less energy-consuming materials.

The first of these options still has some significance in Germany, although standing biomass is -according to recent survey figures- already very high. Forests often grow close to what nutrient supply can support. Without further endangering nutrient cycles in forest ecosystems by the input of fertilizers (which in turn may mean increasing energy consumption for their production and application), we would not expect a very high potential. The mean total increment, averaged over all tree species in forests in Germany is 6 m^3 ha^{-1} yr^{-1}. This corresponds to an annual storage of 5.5 t of CO_2 ha^{-1}. The option of prolonging rotation periods up to the time of maximum total increment in the opinion of our agency offers the potential of binding an additional 1.4 t of CO_2 ha^{-1} yr^{-1}. For the total forest area of 10.1 x 10^6 ha this would result in 14.1 x 10^6 t.

This contribution to the reduction target of 250 to 300 x 10^6 t would be 5-6%.

The CO_2 reduction potential for afforestation is estimated as follows: A program for immediate afforestation could result in a trunkwood production of 50 m^3 ha^{-1} until 2005. Annual growth in that time span would bind up to 4.2 t of CO_2 ha^{-1} yr^{-1}. The potential for a conversion of agricultural land to forests is between 0.4 and 1 x 10^6 ha. The potential size of that sink would therefore be 1.7 to 4.2 x 10^6 t of CO_2, meaning another 1 to 2% of a total envisaged 25% reduction.

Utilization of all other kinds of management (e.g., in the form of short rotation woody crops) where possible and a revision of forest management procedures in respect of increasing sink potentials (e.g., by avoiding clearcut harvest and other ecologically riskful management practices) could result in another 1.5 x 10^6 t or roughly 0.5% of the mentioned target.

The total contribution that could result from all measures would lie at least (see also Burschel et al., 1992) in the range of 17.3 to 19.8 x 10^6 t of CO_2, corresponding to 6.5 to 8% of the planned reduction target.

An additional way of storing C is the long term sequestration in long-lived products like furniture, structural timber etc.. German mitigation plans also call for the conservation of C through substitution of fossil fuels as well as energy consuming products and materials like concrete, plastic polymers, steel or aluminum by wooden products. Energy demands or CO_2 release during manufacture of wood compared to aluminum products for example are approximately 1:126 (IMA, 1990).

4. CONCLUSIONS

Increasing CO_2 will, within a certain range, change forest productivity. Other air pollutants serving as nutrients - such as N-compounds - may support this reaction. But noticeable areas of forest ecosystems in Europe under anthropogenic stress have developed symptoms of decline. Declining forests act no longer as sinks but may even become sources themselves. Protection or restoration of their C storage potential through improved air quality is recommended. Healthy or productive temperate forests can sequester C to an extent that makes it possible to include them efficiently in national C reduction plans.

Many uncertainties still exist and additional data are needed, but there is no doubt about the scale of the problem that the earth's atmosphere is beginning to experience. Even though there are still many uncertainties to be resolved by research, it appears that, in the context of all possible measures, the sink potential of temperate forests should not be underestimated. Research requirements though should not be used as pretext for delay in necessary actions.

5. REFERENCES

Addison P.A. (1991) 'Atmospheric Pollution and Forest Effects', 10th World Forestry Congress, Paris, Revue Forestiere Francaise, Hors Serie N° 2, Proceedings 2, 307-315

Anonymous (1988) 'Schutz der Erdatmosphäre: Eine internationale Herausforderung', 1. Bericht der Enquete Kommission des 11. Deutschen Bundestages "Vorsorge zum Schutz der Erdatmosphäre" Bonn, Deutscher Bundestag, Referat Öffentlichkeitsarbeit, Zur Sache 88/5, 69 p.

Anonymous (1989) 'Protecting the Earth's Atmosphere: An Inter-national Challenge'. Interim Report of the Study Commission of the 11th German Bundestag"Preventive Measures to Protect the Earth's Atmosphere", German Bundestag (ed.), Bonn.

Anonymous (1990) 'Protecting the Tropical Forests: A High Priority International Task'. Report of the Study Commission of the 11th German Bundestag"Preventive Measures to Protect the Earth's Atmosphere", German Bundestag (ed.), Bonn.

Anonymous (1991) 'Protecting the Earth: a Status Report with Recommendations for a new Energy Policy' . Report of the Study Commission of the 11th German Bundestag"Preventive Measures to Protect the Earth's Atmosphere", German Bundestag (ed.), Bonn.

BML (1991) 'Waldzustandsbericht des Bundes - Ergebnisse der Waldschadenserhebung', Bundesminister für Ernährung, Landwirtschaft und Forsten, Publ.Sect., Bonn, 89 p.

Burschel P., Gregor D.H., Knabe W., Moosmayer H.U., Prenzel, Schmidbauer B., Schmidt-Vogt, Scholz F., Seiler, Solomon, Steinlin, von Maydell and Wentzel (1989) Enquete Kommission, 'Vorsorge zum Schutz der Erdatmosphäre',Report of 52.Session, Bonn, 192 p.

Burschel P., Weber M. and Kürsten E. (1992) 'Statement for the Study Commission of the 11th German Bundestag "Preventive Measures to Protect the Earth's Atmosphere", January 1992, discussion paper, 22 p.

Deutscher Bundestag (1988) '4. Immissionsschutzbericht der Bundesregierung', Dt. Bundestag, Drucksache 11/2714, BMU (ed.) Bonn, 115p.

Deutscher Bundestag (1990) 'Protecting the Earth's Atmosphäre - An International Challenge', Bonn, 583 p.

de Vries W. and Gregor H.D. (1990) 'Critical Loads and Critical Levels for Environmental Effects of Air Pollutants', in: Acid Depositions in Europe, H.J. Chadwick and M. Hutton (eds.), Stockholm Environment Institute, Sweden, pp. 171-216.
Drake B.G. (1992) 'The Impact of Rising CO_2 and Global Warming on Plant Production in Coastal Wetland Plant Communities', this volume
ECE (1990) 'International Cooperative Programme on the Assessment and Monitoring of Air Pollution Effects on Forests', Report of the 1989 Forest Damage Survey in Europe, Global Environment Monitoring System (ed.), 125 p.
ECE (1991) 'International Cooperative Programme on the Assessment and Monitoring of Air Pollution Effects on Forests', Report of the 1990 Forest Damage Survey in Europe, Global Environment Monitoring System (ed.), 128p.
EEC (1991) 'Waldschadensbericht 1991 - Kurzbericht über die Erhebung 1990' Europäische Gemeinschaften - Kommission (ed.) Luxemburg, 26 p.
FAO (1990) 'Yearbook Forest Products 1977-1988' Rome
FBW (1990) 'Forschungsbeirat Waldschäden/Luftverunreinigungen der Bundesregierung und der Länder', 3.Bericht 1989, Kernforschungszentrum Karlsruhe (ed.), 611 p.
Gregor H.D. (1990) 'Acidification research in the Federal Republic of Germany' in: Acidic Precipitation, Volume 5: International Overview and Assessment,A.H.M. Bresser and W.Salomons (eds.), Springer Verlag, Berlin, pp 139-158
Gregor H.D. (1991a) 'Kritische Belastungswerte für Ozon und deren Kartierung in Europa', Ozon-Symposium, München 1991, TÜV Bayern (ed.) 10 p.
Gregor H.D. (1991b) 'Challenge to Forest Policy - Measures to Control Forest Decline in Temperate Zones' in: 10th World Forestry Congress, Paris 1991, Revue Forestiere Francaise, Hors Serie N° 2, Proceedings 2, 215-225
Gröbl W. (1990) 'Drohende Klimagefahren erfordern weltweites und europäisches Handeln'; BMU-Umwelt - Informationen des Bundesministers für Umwelt, Naturschutz und Reaktorsicherheit, 5/90, 223-224
IMA (1990) 'Report of Working Group V "Agriculture and Forestry", Interministerial Working Group "CO_2-Reduction", discussion paper, BML (ed) Bonn, 24 p.
IPCC (1991) Intergovernmental Panel on Climate Change 'Climate Change - The IPCC Impact Assessment' AGPS, New York, 296 p.
Kenk G., Diener G. and Spiecker H. (1990)'Zum aktuellen und früheren Wachstumsverhalten von Fichten in den Wuchsgebieten Schwarzwald und Schwäbische Alb' Conference Report, KFK-PEF (ed.), Karlsruhe.
Körner C. (1989) 'Bedeutung der Wälder im Naturhaushalt einer vom Menschen veränderten Umwelt' in: Die Bedrohung der Wälder, Schäden, Folgeerscheinungen und Gegenmaßnahmen, Austrian Academy of Science (ed.), Vienna, pp. 7-40

Marland G. and Marland S. (1992) 'Should we Store Carbon in Trees?', this volume
Nilsson J.and Grennfelt P. (1988) 'Critical Loads for Sulphur and Nitrogen', Report from Skokloster Workshop, Sweden, Miljorapport 1988 : 15
Ocio, J. A., Brookes, P.C. and Jenkinson, D.S. (1991) 'Field Incorporation of Straw and its Effects on Soil Microbial Biomass and Soil Inorganic Nitrogen', Soil Biol. and Biochem. 23 (2), 171-176
Overdieck D. (1986) 'Long-term Effects of an Increased CO_2-Concentration on Terrestrial Plants in Model Ecosystems' Int. J. Biometeor, 30, 323-332
Overdieck D.and Forstreuter M. (1986) 'Langzeit-Effekte eines erhöhten CO_2 - Angebotes bei Rotklee-Wiesenschwingelgemeinschaften', Verhandlungen Gesellschaft für Ökologie, Vol.XVI, 197-206
Overdieck D., Reid Ch.and Strain B.R. (1988) 'The Effects of Preindustrial and Future CO_2 Concentrations on Growth, Dry Matter Production and C/N Relationship in Plants at Low Nutrient Supply', Angew. Botanik 62, 119 - 134
Schmidbauer B. (1991) 'Die Rolle des Waldes in einer zukunftsorientierten Umweltpolitik' BMU-Umwelt - Informationen des Bundesministers für Umwelt, Naturschutz und Reaktorsicherheit, 12/91, 548-552
Smith T.M., Bonan G.B., Weishampel J. and Shugart H.H. (1992) 'The Response of Terrestrial Carbon Storage to Climate Change: Modeling Carbon Dynamics at Different Temporal and Spatial Scales', this volume
Strain B.R. and Thomas R.B. (1992) 'Field Measurements of CO_2 Enhancement and Climate Change on Forest and Natural Vegetation', this volume
Töpfer K. (1991) 'Der Schutz unserer Erde' BMU-Umwelt - Informationen des Bundesministers für Umwelt, Naturschutz und Reaktorsicherheit, 12/91, 546-548
UBA (1988a) 'Critical Levels Workshop, Bad Harzburg, Germany', Final Draft Report, Umweltbundesamt (ed.), Berlin, 146 p.
UBA (1988b) 'Gesamtdarstellung der öffentlich geförderten Waldschädenforschung' Umweltbundesamt (ed.), Berlin, Texte 27/88, 352 p.

ESTIMATING THE GLOBAL POTENTIAL OF FOREST AND AGROFOREST MANAGEMENT PRACTICES TO SEQUESTER CARBON

JACK K. WINJUM
National Council for Air and Stream Improvement
USEPA Environmental Research Laboratory
Corvallis, OR 97333

ROBERT K. DIXON
US Environmental Protection Agency
USEPA Environmental Research Laboratory
Corvallis, OR 97333

PAUL E. SCHROEDER
ManTech Environmental Technology, Inc.
USEPA Environmental Research Laboratory
Corvallis, OR 97333

ABSTRACT. Forests play a prominent role in the global C cycle. Occupying one-third of the earth's land area, forest vegetation and soils contain about 60% of the total terrestrial C. Forest biomass productivity can be enhanced by management practices, which suggests that, by this means, forests could store more C globally and thereby slow the increase in atmospheric CO_2. The question is how much C can be sequestered by forest and agroforest management practices. To address the question, a global database of information was compiled to assess quantitatively the potential of forestry practices to sequester C. The database presently has information for 94 forested nations that represent the boreal, temperate and tropical latitudes. Results indicate that the most promising management practices are reforestation in the temperate and tropical latitudes, afforestation in the temperate regions, and agroforestry and natural reforestation in the tropics. Across all practices, the median of the mean C storage values for the boreal latitudes is 16 $tCha^{-1}$ (n=46) while in the temperate and tropical latitudes the median values are 71 $tCha^{-1}$ (n=401) and 66 $tCha^{-1}$ (n=170), respectively. Preliminary projections are that if these practices were implemented on 0.6 to 1.2 x 10^9 ha of available land over a 50-yr period, approximately 50 to 100 GtC could be sequestered.

1. Introduction

Greenhouse gases (e.g., CO_2, CH_4) from anthropogenic and biogenic sources are accumulating in the atmosphere. The concentration of atmospheric CO_2 alone is projected to double from pre-

Water, Air, and Soil Pollution **64**: 213–227, 1992.
© 1992 *Kluwer Academic Publishers.*

industrial levels by the mid to late 21st century. According to general circulation models of climate change, infrared radiation trapped by the additional greenhouse gases in the troposphere could cause the earth's surface temperature to increase 1.5 to 4.5°C and influence frequency and distribution of regional precipitation (Schneider, 1989a, b).

To evaluate a possible means for offsetting part of the increase of atmospheric CO_2, an assessment began in late 1990 to evaluate the potential to enhance C sequestration in global terrestrial ecosystems through forest and agroforest management. This paper presents some of the findings of the assessment.

1.1 THE GLOBAL CARBON CYCLE

The accumulation of CO_2 in the atmosphere has generated increased interest in the global carbon cycle in recent years (Schneider, 1989a; Tans et al., 1990). This accumulation is a result of emissions of CO_2 from natural and anthropogenic sources exceeding removal by the terrestrial biosphere and marine systems (Figure 1). Two sources of CO_2 are especially significant: the combustion of fossil fuels and global deforestation.

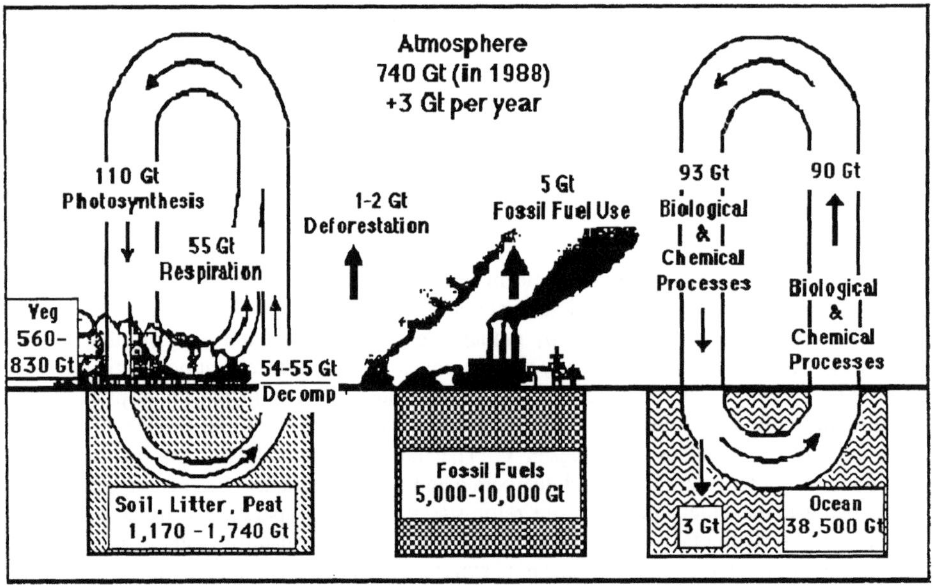

Figure 1. The global carbon cycle including pool size and flux for terrestrial and marine systems (adapted from Schneider, 1989a).

1.2 FORESTS AND THE GLOBAL CARBON CYCLE

Terrestrial ecosystems, especially forests, play a major role in the earth's C cycle. The flux of C through the terrestrial biosphere as a result of plant photosynthesis, respiration and decomposition is approximately 100 $Gtyr^{-1}$. Although terrestrial ecosystems comprise a C pool less than a tenth the size of the ocean's, C flux between the earth's land area and the atmosphere is slightly greater than between the oceans and the atmosphere (Figure 1). Forests also have high rates of ecosystem productivity, and therefore of C sequestration, compared to most other terrestrial ecosystems (i.e., the amount of C photosynthesized is greater than that respired). The world's forests hold approximately 90% (about 740 Gt) of all above-ground terrestrial C and 40% (about 570 Gt) of all below-ground terrestrial C (Waring and Schlesinger, 1985).

Deforestation is a major contributing factor to increased atmospheric CO_2. Estimates of total C released to the atmosphere due to forest clearing between 1860 and 1980 range from 135 to 228 Gt (Woodwell et al., 1983). In 1980, between 1.8 and 4.7 net GtC were released from biotic sources (Table 1), of which 80% was due to deforestation (Detwiler and Hall, 1988). Carbon release from forest burning alone in 1980 has been estimated to be 50% of the annual atmospheric increase of 3 Gt, and the clearing of forest land for agriculture, especially within the tropical latitudes, is now the largest source of C released to the atmosphere globally from biota and soils. Actual estimates differ by a factor of 2 to 3 because of differences in estimated rates of clearing tropical forests and variation in C per unit land area (Detwiler and Hall, 1988; Woodwell et al., 1983). If deforestation increases in proportion to population growth, the biotic release of C could reach about 9 $Gtyr^{-1}$ -- two times the current fossil fuel emissions -- before forests are exhausted in the next century (Postel and Heise, 1988).

Table 1. Anthropogenic and biogenic emissions (1980) of carbon to the atmosphere and potential sequestration of carbon by establishment and intensive management of world forests (Schroeder and Ladd, 1991; Tans et al., 1990; Woodwell et al., 1983).

	Estimated $GtCyr^{-1}$
Carbon Source	
Fossil fuel combustion	4.8 - 6.6
Deforestation and other biotic sources	1.8 - 4.7
Carbon Sink	
Forest establishment 500 x 10^6 ha	3.5 - 4.0
Forest management 300 x 10^6 ha	0.5 - 1.5

Detwiler and Hall (1988) imply that of the 5.1 to 7.5 Gt released annually by fossil fuel combustion and deforestation, between 0.3 and 2.8 Gt are assimilated by the terrestrial biosphere. In addition, in the northern hemisphere, greater land mass and forest cover lead to greater variation in annual atmospheric CO_2 concentration as seasonal changes in photosynthetic uptake of CO_2 occur (Tans et al., 1990). These observations demonstrate that the terrestrial biosphere plays a major role in the global carbon cycle.

1.3 FOREST RESPONSE TO GLOBAL CLIMATE CHANGE

Large uncertainties exist regarding forest response to climate change (Sedjo and Solomon, 1989; Jarvis et al., 1989). As atmospheric concentration of CO_2 increases and climate change occurs, forests could become either a net source or a net sink of C. Climate change will almost certainly cause some forest species in selected regions to decline and migrate (Urban and Shugart, 1989; Woodman and Furiness, 1989). Such shifts in species distribution and composition are projected to occur within the next 50 to 80 yr if current projections of global warming prove accurate. For example, the range of loblolly pine in the southern United States is predicted to shift north several hundred miles (Smith and Tirpak, 1989).

Across forest types, long-term reductions in productivity could also occur, and timber production, biotic habitat, yield of water, site quality and recreation opportunities may be altered (Neilson and King, 1991). Indeed, large areas of world forests (especially temperate and boreal) could experience water stress due to warmer and dryer climatic conditions. This could lead to widespread forest decline and conversion of forests to a major source of atmospheric CO_2 (King et al., 1990). Timing of such changes is unclear, but based on past climate-related events (as with extended droughts), changes might occur in the first half of the next century (Smith and Tirpak, 1989; IPCC, 1990).

Alternatively, atmospheric CO_2 enrichment could increase forest productivity and water use efficiency (Mooney et al., 1990). These positive ecophysiological effects could help forests adapt to global warming. For broadleaf and conifer seedlings, the ecophysiological response to CO_2 enrichment is enhanced in short-term studies (Mooney et al., 1990). That is, increased seedling biomass, leaf area, root-shoot ratios, water-use efficiency and nutrient-use efficiency have been ascribed to short-term CO_2 enrichment. Long-term responses to exposures of high CO_2 are largely unknown. In addition, nutrient and water limitations may not preclude plant growth responses to CO_2 enrichment (Norby and O'Neill, 1989).

1.4 MANAGEMENT OF GLOBAL FORESTS

Forest ecosystems can be managed to increase CO_2 assimilation via photosynthesis and thus to temporarily store larger amounts of C per unit land area (Schroeder and Ladd, 1991). Covering an estimated 29% of the world's surface excluding Antarctica (approximately 4.08×10^9 ha; WRI, 1990; Figure 2), forests have historically contributed to human economic and social progress by providing such resources as shelter, fuel, food, water, recreation and commercial products (Perlin, 1989). These goods and services have largely come from natural forests because only about 11% of the world's forests are actively managed for the production of these human needs (Figure 3).

The extent of management differs among latitudinal regions. Although intensity varies, approximately 20% of the world's temperate forests are managed (Allan and Lanly, 1991). But of boreal and tropical systems, which constitute nearly 75% of world forests, less than 9% receive any degree of management (17% and 4%, respectively). For agroforest systems, global estimates are not available on the amount of land currently managed.

With regard to global reforestation, approximately 15×10^6 ha are reforested annually (WRI, 1990). The breakdown by latitudinal zone is 5.7×10^6 ha in boreal regions (39%), 8.0×10^6 ha in temperate regions (54%), and 1×10^6 ha (7%) in tropical regions. For deforestation, the permanent clearing of forests for agriculture or human settlements, the balance is almost the opposite (WRI, 1990). That is, for the temperate and boreal regions, the net change in forest area

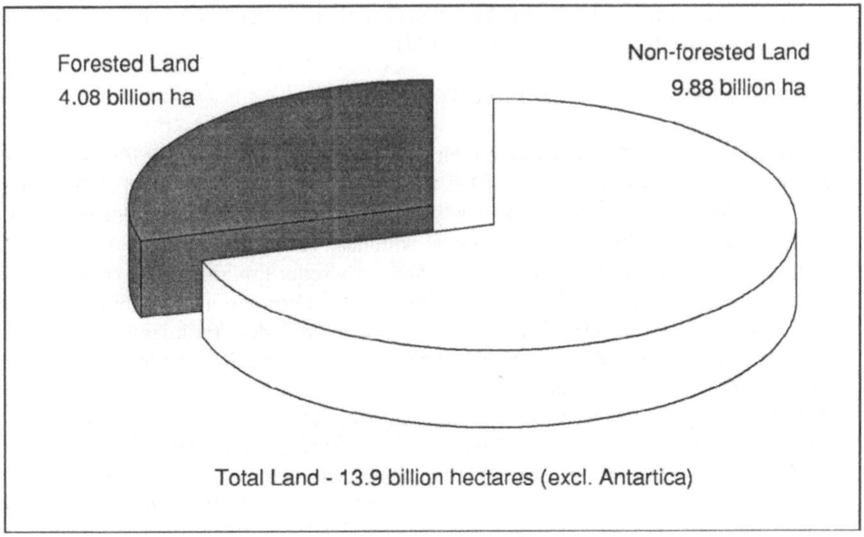

Figure 2. Global forests in relation to total world land area (WRI, 1990).

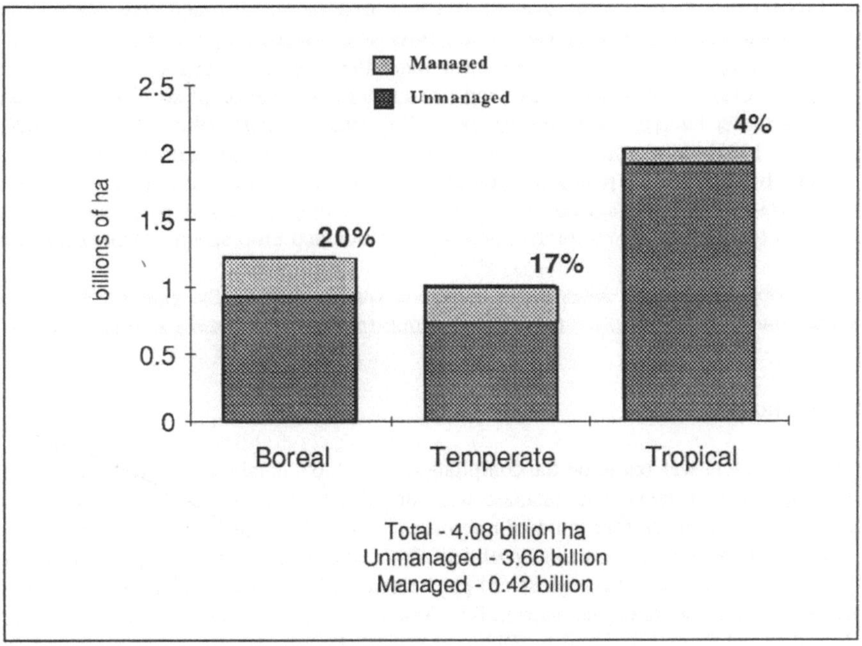

Figure 3. Estimates of managed and unmanaged forests in the world (WRI, 1990; Allan and Lanly, 1991).

is virtually zero at the present, but for tropical regions current estimates of annual deforestation are about 17 x 10^6 ha (Allan and Lanly, 1991).

1.5 ASSESSMENT SCOPE AND OBJECTIVES

The potential role of forest management in C sequestration has recently been evaluated by a number of authors (Andrasko et al., 1991; Grainger, 1991; Houghton et al., 1991; Marland, 1988; Sedjo and Solomon, 1989). These analyses have emphasized the major forest regions on a continental basis, especially within tropical latitudes. Though preliminary, these analyses suggest that forestry and agroforestry could contribute to sequestration and conservation of C while providing a steady flow of goods and services to local communities within many nations.

The scope of the assessment mentioned above included promising forest and agroforest management practices within the forested nations of the world. Representation included the boreal, temperate and tropical regions of latitude (Dixon et al., 1991).

Management practices were considered promising if they were employed widely in the world's forest biomes as evidenced by the literature. For quantitative evaluations, a further requirement was that biomass productivity and implementation cost data were available in the literature. Using these criteria, five categories of management practices became the focus of the assessment's first analyses: reforestation, afforestation, natural regeneration, silviculture and agroforestry.

These practices favor biomass production in forest and agroforest systems. Reforestation -- or artificial regeneration -- is the establishment of tree stands on land that has previously supported forests (Daniel et al., 1979; Young, 1982). Broadcast seeding and tree planting are common reforestation techniques. Afforestation is similar to reforestation but on land that has not supported forests within the last 50 yr (Savill and Evans, 1986). Natural regeneration is the establishment of a new forest by means such as wind or animal dissemination of seed or stump sprouting, i.e., without direct human aid except for protection against catastrophic losses by wildfire or pests (Young, 1982). Silviculture refers to stand treatments designed to increase forest productivity through biological manipulations such as weed control, thinning and fertilization (Smith, 1986). Agroforestry is the retention or introduction of trees or other woody perennials in crop or animal fields to take advantage of interactions between tree and crop species (MacDicken and Vergara, 1990).

The objective of this assessment, therefore, was to evaluate the global potential of these five management practices to contribute to C sequestration by the world's forest systems.

2. Methods

The assessment was based on the compilation of a global database of variables on managed forest and agroforest systems. The database was compiled through an assessment of current published technical information; four hundred references yielded approximately 1,700 data points (Table 2).

For each of the five practices outlined above, the database includes information on two key variables: C that could be sequestered per hectare and land technically suitable and available for the practice throughout the world. Development of the values for these variables is briefly described below and in detail by Dixon et al. (1991).

Table 2. Number of records in global database of promising forest practices by continent and nation.

AFRICA	449
Algeria	8
Angola	8
Benin	8
Botswana	8
Burkina Faso	10
Burundi	9
Cameroon	20
Cape Verde	8
Cent. Af. Rep.	8
Chad	8
Comoros	8
Congo	14
Djibouti	8
Equ. Guinea	8
Egypt	2
Ethiopia	8
Gabon	8
Gambia, The	12
Ghana	17
Guinea	8
Ivory Coast	22
Kenya	5
Lesotho	8
Liberia	9
Madagascar	12
Malawi	10
Mali	9
Mozambique	11
Niger	8
Nigeria	17
Rwanda	11
Senegal	12
Sierra Leone	8
Somalia	8
South Africa	38
Sudan	11
Swaziland	8
Togo	10
Uganda	8
Zaire	16
Zambia	20

N. AMERICA	417
Belize	8
Canada	34
Costa Rica	41
Cuba	3
Guatemala	21
Haiti	5
Honduras	11
Jamaica	1
Mexico	28
Nicaraqua	9
Panama	8
United States	248

S. AMERICA	343
Argentina	135
Bolivia	8
Brazil	70
Chile	27
Colombia	55
Ecuador	21
French Guiana	8
Paraguay	8
Peru	11

OCEANIA	82
Australia	54
Fiji	8
New Zealand	11
Papua N. Guinea	9

ASIA	306
Bangladesh	1
China	57
India	67
Indonesia	37
Israel	1
Malaysia	31
Myanmar	17
Nepal	10
Pakistan	11
Philippines	19
South Korea	2
Sri Lanka	16
Taiwan	3
Thailand	33
Turkey	1

EUROPE	81
Austria	2
Czechoslovakia	3
Finland	3
France	4
Germany	12
Great Britain	13
Greece	1
Ireland	9
Netherlands	1
Poland	7
Portugal	4
Sweden	1

USSR (pre-1991)	21

PRIMARY VARIABLES COLLECTED

1) Biomass productivity
2) Land suitability and availability

The total database includes:		
	Total nations	94
	Total entries	1678

2.1 MEAN CARBON STORAGE

To compare the C sequestration potential among forest and agroforest management practices worldwide, estimates are needed of the amount of C that can be stored *on average* over many decades and rotations. The relevant parameter is the average amount of C on site over an indefinite number of rotations (Schroeder, 1992). If it is assumed that the system is sustainable and there is no yield reduction in later rotations, the result is the same as the average amount of C on site over one full rotation.

Thus, the amount of C potentially sequestered, mean C storage (MCS), was expressed in $tCha^{-1}$ for each promising forest practice identified above (Dixon et al., 1991). This calculation was made by summing the C standing crop for each year in the rotation and dividing by the rotation length, i.e., the calculation of a mean:

$$\text{Mean C Storage} = \frac{\sum_{i=1}^{n} (tCha^{-1})_{\text{standing crop}}}{(\text{rotation length})}$$

Where: n = rotation length.

The MCS values are the average amount of C on site over an indefinite number of rotations as calculated from one rotation's tree stem volume or, where available, the total biomass produced.

2.2 LAND AREA ESTIMATES

Estimates of the amount of land in the world amenable to cropping by forest and agroforest management are essential to a complete assessment, yet these estimates are currently the weakest in terms of reliable data. Land must be both technically suitable -- i.e., ecologically capable of supporting tree crops -- and available from an economic, social and political standpoint. Preliminary estimates compiled in the database indicated that large amounts of land exist that are technically suitable for expanding the area of managed forest and agroforest systems (Dixon et al., 1991).

For the tropical latitudes, Houghton et al. (1991) estimated that more than 1,900 x 10^6 ha of technically suitable land exist upon which to implement forest and agroforest practices. Grainger (1991) estimated that reforestation and afforestation could be implemented on 420 x 10^6 ha in the tropics. Trexler (1991) considered the social, economic and political constraints and concluded that only about one third of the 1,900 x 10^6 ha estimated by Houghton would actually be available for implementing various tree crop systems.

In the boreal and temperate latitudes, estimates of land available for reforestation and afforestation were 100 and 200 x 10^6 ha, respectively (Volz et al., 1991; Moulton and Richards, 1990). Natural regeneration would likely account for additional forest establishment on lower site-quality lands, though estimates of the amount are not available at this time.

2.3 DATA AND STATISTICAL METHODS

The data assembled for this assessment came from a very wide variety of sources and were not selected by random sampling. Plots of the data showed that distributions were not normal but unimodal with a positive skew, i.e., highest frequency for the lowest values of mean C storage and implementation costs (Dixon et al., 1991). Therefore, in the following presentations, results are more properly presented as sample medians with variation indicated by interquartile ranges, the middle 50% of observations (Devore and Peck, 1986). The sample sizes are also shown.

2.4 DATA QUALITY

In a broad review involving large amounts of data from many sources, data quality will vary. Referenced technical data for this assessment were considered the best available. When data on land productivity or initial costs were encountered that were clearly outside reported ranges and were not adequately explained, they were not used in the assessment.

3. Results

The assessment produced productivity data for forest and agroforest management practices in 94 nations. That is, out of a total of 138 nations with forest resources, about two-thirds are represented in the database at present (Table 2). The total forest and woodland area of these 94 nations is about 3.8 billion ha or approximately 93% of the world total.

3.1 CARBON SEQUESTRATION

In terms of MCS, several trends emerge for promising forest practices across latitudinal zones (Table 3). Due to warmer and longer growing seasons toward the tropical latitudes, C sequestration in living biomass generally increases from the boreal to the temperate to the tropical latitudes. Across all practices, the median MCS value for the boreal latitudes is 16 $tCha^{-1}$ (n=46) while in the temperate and tropical latitudes the median values are 71 $tCha^{-1}$ (n=401) and 66 $tCha^{-1}$ (n=170), respectively. Among all the five practices evaluated, the database contains the most information on reforestation. These are the medians and n values for reforestation practices: boreal, 39 $tCha^{-1}$ (n=14); temperate, 56 $tCha^{-1}$ (n=212); and tropical, 65 $tCha^{-1}$ (n=136) (Table 3).

The lowest MCS median values are for silviculture in all three latitudinal zones: boreal, 10 $tCha^{-1}$; temperate, 26 $tCha^{-1}$; and tropical, 34 $tCha^{-1}$. Silviculture treatments, however, will likely play a role in adapting forest stands and plantations to warmer, drier climates that could arise under projections of global warming.

3.1 SUITABLE AND AVAILABLE LAND

Based upon the published estimates noted in the "Methods" section, the following areas were adopted for this analysis as conservative estimates of suitable and available lands: boreal, 100 to 200 x 10^6 ha; temperate, 200 to 400 x 10^6 ha; and tropical, 300 to 600 million ha. That is, these ranges are assumed to be reasonable working numbers for the amount of land suitable and available for expanding world forests through implementation of forest and agroforest management practices over a 50-yr period.

Table 3. Medians and interquartile values of the mean carbon storage (MCS in $tC ha^{-1}$) over a 50-yr period for promising forest and agroforest management practices in the major forested regions of the world. For those practice-region combinations where n equals 5 or less (*), the individual data points are given rather than computed medians and interquartile values. Dashed lines indicate present data limitations.

Practices	Boreal				Temperate				Tropical			
	n	median	Q25	Q75	n	median	Q25	Q75	n	median	Q25	Q75
Reforestation	14	**39**	17	39	212	**56**	31	96	136	**65**	46	105
Afforestation	1*	(15	----	----)	119	**120**	90	178	3*	(29	46	128)
Natural refor.	19	**17**	12	17	6	**44**	23	121	3*	(119	195	195)
Siliviculture	12	**10**	3	10	62	**26**	10	46	12	**34**	14	70
Agroforestry	---	**---**	---	---	2*	(12	57	----)	16	**95**	61	125
All	46	**16**	10	23	401	71	33	116	170	**66**	43	111

3.2 PRELIMINARY PROJECTION

Drawing on the values for C sequestration and the working land estimates, preliminary projections were made for eight promising practice-region combinations. Those combinations were reforestation and natural reforestation in all three latitudinal zones, afforestation in temperate regions and agroforestry in tropical regions. The projections were developed using a simple table format (Table 4).

Table 4. Preliminary projections of the total potential carbon sequestered (GtC over a 50-yr period) for eight practice-region combinations.[1]

(1)	(2)		(3)	(4)	
Region/Practice	Land est. (ha x 10^6)		MCS (tha^{-1})	C sequestered (Gt)	
	a. Lower	b. Upper		a. Lower	b. Upper
Boreal					
Reforestation	50	100	39	2.0	3.9
Natural ref.	50	100	17	0.2	1.7
Temperate					
Reforestation	75	150	56	4.2	8.4
Natural ref.	75	150	44	3.3	6.6
Afforestation	50	100	120	6.0	12.0
Tropical					
Reforestation	100	200	65	6.5	13.0
Natural ref.	100	200	195	19.5	39.0
Agroforestry	100	200	95	9.5	19.0
Total				51.8	103.6

[1]Data sources: Column 2 figures are estimates of land technically suitable and socially available based upon Grainger (1991), Houghton et al. (1991) and Volz et al. (1991). Column 3 is from Table 3. Column 4 is the product of Column 2 (a and b) and Column 3.

While enhancing productivity in existing forest stands and plantations, silviculture does not directly contribute to increasing forest land area. Since increasing forest area was found to be the most important attribute for significant contributions toward increasing global forest C sequestration, silviculture was not included in the preliminary projections in this analysis.

The first step was to estimate how available land areas might be distributed among the eight practice-region combinations identified above and enter them as Columns 2a and 2b in Table 4. Column 3 is the calculated MCS. Column 4 is the product of the values in Columns 2 and 3; it provides an estimated range of C sequestration for each promising practice-region combination.

For example, if reforestation were implemented on 100 x 10^6 ha in the boreal region and the plantations maintained in perpetuity, then 3.9 GtC would be sequestered (100 x 10^6 ha x 39 $tCha^{-1}$ as shown in Table 4).

The sums of the values in Columns 4a and 4b for all practice-region combinations, 51.8 and 103.6 Gt, are important to note. If tree crops were implemented, maintained, harvested at their rotation ages, and regenerated in perpetuity, between 50 and 100 GtC would be sequestered through these eight combinations.

4. Conclusions and Discussion

Forests play an important role in the global C cycle. Although estimates of global C pool size and flux vary, the distribution and productivity of forests influence concentration of greenhouse gases (e.g., CO_2, CH_4) in the atmosphere. Variation in annual atmospheric concentration of CO_2 is greater in the northern latitudes in part due to greater land mass and associated forest cover. Though uncertainties exist regarding projected global climate change, the terrestrial C cycle, and the exact role of forest ecosystems, preliminary evidence suggests that forests could be managed worldwide to aid in reducing the accumulation of CO_2 in the atmosphere.

From the preliminary estimates above, it appears that promising forest and agroforest practices could sequester between 50 and 100 GtC over a 50-yr period. Would this be a significant contribution as a mitigating measure regarding atmospheric CO_2 buildup and projections for global warming? The answer appears affirmative by the following comparisons.

In 1988, global C emissions resulting from human activities ranged from 6 to 8 Gt (Detwiler and Hall, 1988). Assuming that this annual rate continues for the next 50 yr, then 300 to 400 GtC would be emitted to the atmosphere (i.e., 6 and 8 $GtCyr^{-1}$ x 50 yr). In this context, the 50 to 100 Gt range of C sequestered by forest and agroforest practices would appear to be a significant contribution.

If this C sequestration estimate were an order of magnitude larger or smaller, the preliminary conclusion would be quite different. For instance, if the estimated contribution were 10 Gt, it might be considered trivial and not an important global goal. If the estimate were near 1,000 Gt, implementing forest and agroforest practices at the above global scale might be a total solution to the buildup of atmospheric CO_2. However, studies of the whole mitigation issue within the global warming question have failed thus far to identify any single potential mitigating measure. Rather, a collection of smaller contributions is a more likely solution (Schneider, 1989b; Lashof and Tirpak, 1990).

In related work, Winjum et al. (1992) found that the estimated global cost of implementing forest and agroforest practices over a 50-yr period to sequester 50 to 100 GtC would be approximately \$170 to \$340 x 10^9. Dividing these estimated costs by the amount of C sequestered indicates a cost of \$3.40 tC^{-1}. This is well within the "low cost" range for C mitigation measures suggested by the US National Academy of Sciences (NAS, 1991).

As a first approximation, therefore, implementing just the above eight practice-region combinations would seem to contribute significantly to reducing the buildup of atmospheric CO_2. Expanding global forest and agroforest management to make this contribution is a large challenge; it will involve many people in the world's forested nations: government decision makers, land managers, researchers and people at the local level. But these conclusions, though tentative, lend weight to the desirability of developing an international agreement on forests as is currently under discussion (Maini, 1991).

5. Acknowledgments

The authors gratefully acknowledge the effective contributions to this document by the following persons:

P.M. Bradley, *Ability Network*
G.A. Baumgardner, *ManTech Environmental Technology, Inc.*
S.G. McCannell, *Word Design*
J. Van Sickle, *Oregon State University*

This paper was prepared for "Natural Sinks and Sources of CO_2, An International Workshop" held in Puerto Rico February 24-27, 1992, and will be included with all Workshop papers as a summary peer-reviewed document. Portions of the paper will also appear in papers submitted to other peer-reviewed journals.

The research described in this paper has been funded by the US Environmental Protection Agency. The paper has been prepared at the EPA Environmental Research Laboratory in Corvallis, Oregon, USA, through Interagency Agreement No. DW12934530/Grant PNW 91-0051. It has been subjected to the Agency's peer and administrative review process and approved for publication. Mention of trade names or commercial products does not constitute endorsement or recommendation for use.

6. References

Allan, T. and Lanly, J.P. (1991) 'Overview of status and trends of world forests', in D. Howlett and C. Sargent (eds.), Technical Workshop to Explore Options for Global Forestry Management, International Institute Environment and Development, London, pp. 17-39.

Andrasko, K., Heaton, K., and Winnett, S. (1991) 'Evaluating the costs and efficiency of options to manage global forests: A cost curve approach', in D. Howlett and C. Sargent (eds.), Technical Workshop to Explore Options for Global Forestry Management, International Institute Environment and Development, London, pp. 216-233.

Daniel, T.W., Helms, J.A., and Baker, F.S. (1979) Principles of Silviculture, Second Edition, McGraw-Hill, New York.

Detwiler, R.P. and Hall, C.A.S. (1988) 'Tropical forests and the global carbon cycle', Science 239, 42-47.

Devore, J. and Peck, R. (1986) Statistics, The Exploration and Analysis of data, West Publishing Company, St. Paul.

Dixon, R.K., Schroeder P.E., and Winjum, J.K. (eds.) (1991) Assessment of Promising Forest Management Practices and Technologies for Enhancing the Conservation and Sequestration of Atmospheric Carbon and Their Costs at the Site Level (USEPA/600/3-91/067), US Environmental Protection Agency Environmental Research Laboratory, Corvallis, OR.

Grainger, A. (1991) 'Constraints on increasing tropical forest area to combat global climate change', in D. Howlett and C. Sargent (eds.), Technical Workshop to Explore Options for Global Forestry Management, International Institute Environment and Development, London, pp. 196-208.

Houghton, R.A., Unruh, J., and Lefebvre, P.A. (1991) 'Current land use in the tropics and its potential for sequestering carbon', in D. Howlett and C. Sargent (eds.), Technical Workshop to Explore Options for Global Forestry Management, International Institute Environment and Development, London, pp. 211-212.

Intergovernmental Panel on Climate Change (IPCC) (1990) 'Climate Change, The IPCC Scientific Assessment', University Press, Cambridge.

Jarvis, P.G., Monteith, J.L., Shuttleworth, W.J., and Unsworth, M.H. (1989) 'Forests, weather, and climate' (B324), Philosophical Transactions of the Royal Society, London.

King, G.A., Winjum, J.K., Dixon, R.K., and Arnaut, L.Y. (eds.) (1990) Response and Feedbacks of Forest Systems to Global Climate Change (USEPA/600/3-90/080), US Environmental Protection Agency, Washington, D.C.

Lashof, D.A. and Tirpak, D.A. (eds.) (1990) Policy Options for Stabilizing Global Climate: Report to Congress (USEPA-21P-2002.1), US Environmental Protection Agency, Washington, D.C.

MacDicken, K.G. and Vergara, N.T. (1990) Agroforestry: Classification and Management, Wiley, New York.

Maini, J.S. (1991) 'Towards an international instrument on forests', in D. Howlett and C. Sargent (eds.), Technical Workshop to Explore Options for Global Forestry Management, International Institute Environment and Development, London, pp. 278-285.

Marland, G. (1988) The Problem of Solving the CO_2 Problem Through Global Reforestation (DOE/NBB-0082), US Department of Energy, Office of Energy Research, Washington, D.C.

Mooney, H.A., Drake, B.G., Luxmoore, R.J., Oechel, W.C., and Pitelka, L.F. (1990) 'How will terrestrial ecosystems interact with the changing CO_2 concentration of the atmosphere and anticipated climate change?', BioScience 41, 96-104.

Moulton, R.J. and Richards, K.R. (1990) Costs of Sequestering Carbon Through Tree Planting and Forest Management in the United States (GTR WO-58), US Department of Agriculture Forest Service, Washington, D.C.

National Academy of Sciences (NAS) (1991) Policy Implications of Greenhouse Warming, US National Academy Press, Washington, D.C.

Neilson, R.P. and King, G.A. (1991) 'Continental scale biome responses to climatic change', in D. McKenzie, E. Hyatt, and J. MacDonald (eds.), Proceedings of International Symposium: Ecological Indicators. October 16-19, 1990. Ft. Lauderdale, FL, Elsevier Scientific Publications, Ltd., Dordrecht.

Norby, R.J. and O'Neill, E.G. (1989) 'Growth dynamics and water use of seedlings of Quercus alba L. in CO_2 enriched atmospheres', New Phytologist 111, 491-500.

Perlin, J. (1989) A Forest Journey, the Role of Wood in the Development of Civilization, Norton, New York.

Postel, S. and Heise, L. (1988) Reforesting the Earth (Worldwatch Paper 83), Worldwatch Institute, Washington, D.C.

Savill, P.S. and Evans, J. (1986) Plantation Silviculture in Temperate Regions with Special Reference to the British Isles, Clarendon Press, Oxford.

Schneider, S.H. (1989a) 'The changing climate', Scientific American 261, 70-79.

Schneider, S.H. (1989b) 'The greenhouse effect: science and policy', Science 243, 771-781.

Schroeder, P.S. (1992) 'Carbon storage potential of short rotation tropical tree plantations', Forest Ecology and Management. (In press).

Schroeder, P.S. and Ladd, L.B. (1991) 'Slowing the increase of atmospheric carbon dioxide: A biological approach', Climate Change 19, 283-290.

Sedjo, R.A. and Solomon, A.M. (1989) 'Climate and forests', in N.S. Rosenberg, W.E. Easterling, P.R. Crosson, and J. Dormstadter (eds.), Greenhouse Warming: Abatement and Adaptation, Resources for the Future, Washington, D.C.

Smith, D.M. (1986) The Practice of Silviculture, Eighth Edition, Wiley, New York.

Smith, J.B. and Tirpak, D.A. (1989) The Potential Effects of Global Climate Change on the United States (EPA-230-05-89-050), US Environmental Protection Agency, Washington, D.C.

Tans, P.P., Fung I.Y., and Takahashi, T. (1990) 'Observational constraints on the global atmospheric CO_2 budget', Science 247, 1431-1438.

Trexler, M.C. (1991) 'Estimating tropical biomass futures: A tentative scenario', in D. Howlett and C. Sargent (eds.), Technical Workshop to Explore Options for Global Forestry Management, International Institute Environment and Development, London, pp. 311-318.

Urban, D.L. and Shugart, H.H. (1989) 'Forest response to climatic change: a simulation study for southeastern forests', in J.B. Smith and D.A. Tirpak (eds.), The Potential Effects of Global Climate Change on the United States (EPA-230-05-89-054), Appendix D, US Environmental Protection Agency, Washington, D.C.

Volz, H., Kriebitzsch, W.U., and Schneider, T.W. (1991) 'Assessment of potential, feasibility and costs of forestry options in the temperate and boreal zones', in D. Howlett and C. Sargent (eds.), Technical Workshop to Explore Options for Global Forestry Management, International Institute Environment and Development, London, pp. 124-158.

Waring, R.H. and Schlesinger, W.H. (1985) Forest Ecosystems: Concepts and Management, Academic Press, Orlando.

Winjum, J.K., Dixon, R.K., and Schroeder, P.E. (1992) 'Global assessment of promising forest management practices for sequestration of carbon', in G. Wall (ed.), US/Canadian Symposium on the Implications of Climate Change for Pacific Northwest Forest Management, Seattle, October 23-25, 1991, (A peer-reviewed manuscript cleared by ERL-C and submitted for the symposium proceedings scheduled for publication in March, 1992).

Woodman, J.N. and Furiness, C.S. (1989) 'Potential effects of climate change on US forests: case studies of California and the southeast', in J.B. Smith and D.A. Tirpak (eds.), The Potential Effects of Global Climate Change on the United States (EPA230-05-89-054), Appendix D, US Environmental Protection Agency, Washington, D.C.

Woodwell, G.M., Hobbie, J.E., Houghton, R.A., Melillo, J.M., Moore, B., Peterson, B.J., and Shaver, G.R. (1983) 'Global deforestation: contribution to atmospheric carbon dioxide', Science 222, 1081-1086.

World Resources Institute (WRI) (1990) World Resources 1990-91, Oxford University Press, Oxford.

Young, R.A. (1982) Forest Science, Wiley, New York.

Forest fertilization: its potential to increase the CO_2 storage capacity and to alleviate the decline of the global forests

R. F. Huettl
Forestry Department, Kali und Salz AG
P.O. Box 102029, D-3500 Kassel, Germany

H. W. Zoettl
Institute of Soil Science and Forest Nutrition
Albert-Ludwigs-University
Bertoldstr. 17, D-7800 Freiburg, Germany

Abstract

Healthy forests are essential for life on earth. Their multiple benefits include sustainable production of wood and other products, soil protection, air and water purification, wildlife habitat, noise control, various types of recreation, but also climate regulation and CO_2 storage. Owing to man-made influences, forests have been endangered since a long time. In various parts of the world these adverse impacts are now representing a risk for human life. For example, the still increasing destruction of tropical and subtropical forests leads to detrimental soil erosion problems and great losses of soil fertility including soil C. Furthermore, atmospheric pollution brought about forest declines in Europe, North America and elsewhere. In order to overcome these and related problems a variety of efforts have been established. In commercial forests of the temperate/boreal zones forest management practices including fertilization regimes focusing on sustainably increasing yields have been utilized for several centuries. However, new management strategies are under research and/or have been implemented in these forest ecosystems to mitigate the so called new types of forest damage. As these declines are frequently associated with nutritional disturbances fertilizer applications have proven useful when conducted properly. Agro-forestry is one approach to reduce the further deterioration of tropical/subtropical forests. Also rehabilitation of devastated land ist partly concentrated on afforestation. With the ever increasing demands of a fast growing human population and increasing CO_2 concentrations in the atmosphere, proper multi-purpose forestry on a permanent or short rotation basis or in the wide realm of agro-forestry is essential for mankind. In many cases adequate forest management will only be possible when soil fertility is restored and/or maintained through fertilization/manuring considering the demand of a specific forest stand/system under specific site conditions.

Clearly, also the potential to increase the CO_2 storage capacity of forests is often limited by insufficient nutrient availability. However, in general this site specific limitation can be overcome easily by proper nutrient supplementation.

Water, Air, and Soil Pollution **64**: 229–249, 1992.
© 1992 *Kluwer Academic Publishers.*

1. Introduction

Forests and rangelands which together cover about 34 % of the earth's land surface not only supply humans with wood and other products but they also provide many essential benefits for human lives (WRI, 1988). Managing them sustainably is one of our greatest challenges.

The degradation of forests in the tropics is integrally linked to human population growth. More than 90 x 10^6 people are added each year - more than ever before. On its present trajectory the world's population could nearly triple its size reaching 14 x 10^9 before stabilizing. With an heroic effort it could level off at around 9 x 10^9 (WRI, 1990).

From a forestry viewpoint, production of wood and associated products has to increase simultaneously on an ecologically and economically sound basis. In this context the primary aim of forest fertilization is to raise forest productivity. Besides improving site fertility fertilization can also enhance forest vitality by increasing resistance against biotic diseases, frost, drought and immissions. Furthermore, balanced fertilizer applications are practiced in plantation forestry and agro-forestry to replace the nutrients taken from the site by harvesting. Also liming and fertilization have proven useful tools to restabilize declining forest ecosystems impacted by air pollution. With regard to the rising CO_2 concentration in the atmosphere proper nutrient management can help to increase CO_2 storage by increasing forest biomass production, both above and below ground.

2. Forest resources

2.1 Types and extent of the world's forest

The earth is covered by a number of major types of forests and woodlands (cf. WRI, 1988). The upper part of the northern hemisphere is dominated by a vast band of evergreen, coniferous forests stretching over parts of Canada, the United States, Scandinavia, the CIS and China. Further south, as the climate gets warmer, conifers like spruce and pine give way to broad-leaved trees like oak and beech. In upper tropical latitudes the arid or semi-arid lands of northern Africa, Asia and Latin America are covered by savanna woodlands in which trees are widely scattered over grassy plains. Tropical rain forests flourish in the higher temperature and humidity of the low latitudes on either side of the equator. On the fringes of the tropical rain forests the tropical moist deciduous forests also called monsoon forests grow. The tropical rain forests and moist deciduous forests are generally referred to collectively as tropical moist forests.
Closed forests cover 2.8 x 10^9 ha or 21 % of the earth's land surface. Other wooded areas include tropical open woodlands and tropical forest fallow, i.e. the regrowth that occurs after forest land cleared for shifting cultivation has been abandoned. Including these the planet's total woody cover raises to 4.5 x 10^9 ha or 35 %.

Four countries account for more than half of the world's closed forests: the CIS has 792 x 10^6 ha, Brazil 357 x 10^6 ha, Canada 264 x 10^6 ha, the U.S.A. 196 x 10^6 ha. This compares with only 43 x 10^6 ha in the countries of the European Community and 145 x 10^6 ha for all of Europe.

The tropical moist forests occupy just under 1.1 x 10^9 ha and account for 90 % of all tropical closed forests. About half of the area of tropical moist forests is in Latin America. With the other half shared about equally between Africa and the Asia Pacific region. Brazil has the largest area of tropical moist forest, 331 x 10^6 ha, three times the forest area of Indonesia or Zaire. Tropical rain forests account for 66 % of all tropical moist forests.

A large portion of the remaining forests in Europe and other developed nations has either been artificially planted or substantially modified for commercial timber production. Forest plantations in the tropics cover a relatively small area. Out of a total of 11.5 x 10^6 ha 4.4 x 10^6 ha are nonindustrial plantations for fuel wood production or environmental protection and 7.1 x 106 ha are industrial plantations for the production of commercial timber, pulp wood, charcoal and other products.

Trends in world forest cover in recent decades have been influenced by a decline in tropical forests/woodlands and a slow net increase in temperate forest areas because of plantation establishment.

2.2. Tropical deforestation

Concern is rising about the rapid rate of tropical deforestation, the temporary or permanent clearance of forest for agriculture and other purposes (cf. Steinlin, 1989). The main cause of deforestation in tropical closed forests is clearing for agriculture to feed growing numbers of people or - to a lesser degree - to earn foreign exchange from exports for cash crops. Often the only way to increase the agricultural area in densely forested countries is to clear forests or woodlands. In all countries some deforestation is planned or promoted by governments, but a substantial proportion takes place spontaneously and without control.

Logging in the tropical rain forests is not usually a direct cause of deforestation because it is mainly selective and clear cutting is rare. Only a few of the thousands of tropical tree species are commercially acceptable. Logging usually removes 2 to 10 trees ha^{-1} or 3 to 30 % of the total timber volume. However, selective logging does disturb the forest canopy, some of the remaining trees are damaged during felling and log extraction and the composition and functioning of the forest ecosystem are modified to varying degrees. At the very least, logging degrades the forest ecosystem and can also lead to subsequent deforestation. After loggers have left the area, farmers often use logging roads to gain access to clear the forest for growing crops. In dry or montane areas the main causes of deforestations are cutting for fuel wood and heavy lopping of foliage for fodder or direct browsing by goats and other animals.

Every year about 11.1 x 10^6 ha of tropical forests and woodlands are destroyed. About 7.3 x 10^6 ha of tropical closed forests are cleared annually for agriculture. At the moment 0.6 % of the total area of tropical moist forests is cut each year with similar figures for the three tropical regions.

Deforestation and soil fertility

It is now well known that the luxuriant tropical rain forest is more the result of the warm-wet climate than the richness of the soil (cf. Kang and Juo, 1986). Because of intense weathering and leaching soils in the humid tropical regions are often chemically poor. The commonly found oxisols and ultisols are characterized by strong acidity, low cation exchange capacity, low silt content, caolinitic clay mineralogy and low nutrient levels. Moreover, the exchange sites of soil organic matter are usually dominated by Al (Juo and Kamprath, 1979). Thus, in tropical forest ecosystems trees recycle mineral nutrients from decomposed litter in order to meet a major portion of their mineral requirements. Early work by Nye and Greenland (1960) revealed that the biomass of a living tropical forest may contain equal or larger quantities of mineral nutrients such as Ca, Mg and K than the amount present in the upper soil layers (0 to 30 cm). The relative distribution of organic carbon in the plant and in the surface soil also differs greatly between the tropical and temperate forest ecosystem (Kira and Shidei, 1967). As indicated by Figure 1 the surface soil under the temporate forests holds a much larger proportion of total organic carbon than that in tropical forest ecosystems.

On the other hand the less leached alfisols under the secondary forest and savanna vegetation of the sub-humid regions may be chemically better suited for agriculture or forest management provided that proper land clearing methods and appropriate post-clearing soil management practices are used to control soil erosion and degradation.

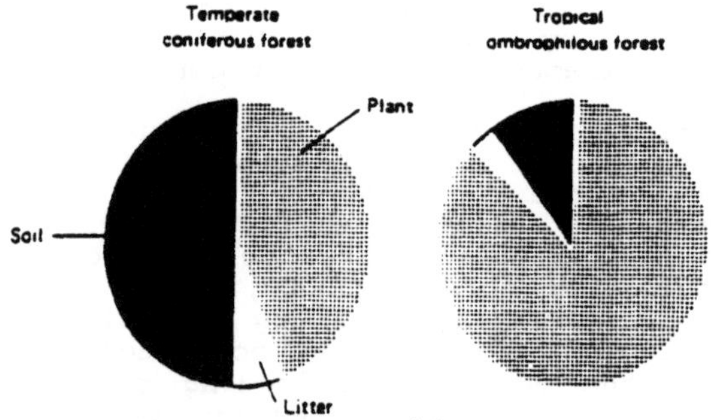

Figure 1: Distribution of organic carbon in two forest ecosystems (from Kira and Shidei, 1967)

Deforestation and soil erosion

This important problem was recently emphasized by Lal (1986). The severity of soil erosion from cleared land depends on method of deforestation, post-clearing soil management and land use. In general, mechanized clearing leads to more erosion than manual clearing and erosion is more severe on a ploughed and harrowed soil than on mulched and/or untilled plots. The data available also indicate that the amount of erosion varies according to the use. The following land uses are listed in descending order of susceptibility to erosion: agriculture > grazing > agroforestry > forest.

Soil erosion also depends on the relationships between soil loss and crop productivity. For some soils in which the effective rooting depth is shallow and in which chemical fertility is mainly concentrated in the surface of soil horizons severe yield reductions can be caused even by low levels of erosion. A loss of fertile surface soil is not always compensated for by additional agricultural inputs. Restoration of the soil through planted fallows may be a useful approach provided that the symptoms of soil degradations are recognized before it has reached the point of no return.

Deforestation without burning

Uexkuell and Mutert (1988) have suggested a method to transform tropical rain forests into plantations and farmland without burning. In order to avoid the adverse effects of burning such as soil erosion, loss of soil fertility and soil C manual clearing in combination with the use of ecologically balanced chemicals for accelerated decomposition of woody biomass should be practiced. Before felling the trees, fast growing ground covering Leguminosae (e.g. Mucuna utilis) should be seeded. The remarkable advantage of this practice is that organic matter is conserved and the soil is continuously covered by living biomass during the cultivation phase.

3. Permanent forestry

Permanent forestry is focused on practices that aim at utilizing land for producing wood and associated products.

3.1. Sustainable forestry

Sustained forest management originated in Germany. It can simply be defined as a forestry practice aiming at constant forest yields. For that purpose the amount of wood (biomass) extracted from a defined forest area within a defined time period must not exceed the growth increment of wood (biomass) within that area and time. Aiming at this approach many practical and theoretical tools have been developed which nowadays provide the basis for the commercial management of many of the northern hemisphere's forest ecosystems. In this respect the basic tool is silviculture (cf. Wenger, 1985).

Silvicultural knowledge and techniques are applied to forest land through prescriptions. A prescription is a record of the examination, diagnosis and treatment regimes recommended for forest stands. A good prescription relates yields and stand conditions to land management objectives. Many silvicultural prescriptions project treatments and their anticipated costs and returns through an entire rotation. In this context forest fertilization may play an important role. However, besides sustained wood production other purposes are provided by this type of forest management such as soil protection, air and water purification, wildlife habitat, noise control and various types of recreation, but also climate regulation and CO_2 storage.

From an historical viewpoint much of the land area of Europe and North America was covered by forest in former times. Due to various impacts of man and its need for wood and farm land large forests were cut down. However, forest land with low site fertility, harsh climate or steep terrain was not useful for agricultural management and remained uncleared . Thus, today most commercial forests of the temperate/boreal zone grow on relatively poor soils and frequently under unfavorable climatic conditions.

It is noteworthy that during times prior to today's highly developed forest management practices, particularly in central Europe and North America manyfold devastations were the rule. For example, woodland grazing, litter-raking as well as uncontrolled biomass output caused partly heavy deteriorations of the nutrient supply at many forest sites.

Due to mainly economical reasons the natural species composition was changed in various commercial forests. Mixed stands were transformed into pure coniferous stands frequently utilizing off site planting stock. Regionally heavy growth reductions resulted from that development. Based on those experiences and a growing demand for wood, forest fertilization was first discussed. As in agriculture, forest fertilization was soon considered a tool to improve site quality, mainly soil fertility, in order to increase wood production.

Hence, forest fertilization is nothing new in sustained forestry (cf. Huettl, 1986). In European forests the first fertilizer trials were initiated about fourteen decades ago. Already in 1847 Chevalier de Valdrome established fertilizer experiments in France and found remarkable growth increases.

In the 1870's Ramann (in Baule and Fricker, 1967) reported about the first German trials where N, P, K and Ca applications had doubled the weight of the fertilized trees. In 1932, Wiedemann found that fertilization revealed good results for Scots pine and even better ones for Norway spruce. Wiedemann suggested that forest fertilization could be a very important silvicultural tool when practiced according to site and stand requirements. He suggested that forest fertilization could enable foresters to overcome silvicultural difficulties in an economically feasable manner. He concluded that forest fertilization could be used to intensify forest management.

Studies carried out by the Swedish Forest Research Institute in the 1950's indicated a need for long-term fertilizer trials where the nutrient status of trees was controlled better than in earlier experiments. New trials were started in 1967 and thereafter in different regions of Sweden on typical forest sites. This program, known as the "optimum nutrition experiments" is focussed on a new way of fertilization, i.e. application of regular small amounts of fertilizers. For example, in middle Sweden growth rates of more than 20 $m^3 \cdot ha^{-1} \cdot a^{-1}$ were achieved in both Norway spruce and Scots pine stands. A general conclusion from this outstanding work was drawn by Tamm (1985): "Regular and carefully adjusted nutrient applications at least on many sites and in the tree species tested produced a much better response than conventional forest fertilizations".

Of course numerous other fertilizer experiments have been carried out between 1847 and the recent Swedish work which at least partially are summarized by Huettl (1991). From these studies as well as from large field applications it can clearly be concluded that forest growth (NPP and NEP) can be significantly improved by proper nutrient supplementation. When utilizing this well established tool on a larger scale it is without question that the CO_2 storage capacity of forests will greatly be enhanced.

Fertilization in declining forests

Under the recent impact of the so called new type of forest damage, forest fertilization and liming are considered as mitigation tools for the restabilization of declining forest ecosystems. In this context forest fertilization and liming are generally seen from an ecological viewpoint as a means to counteract the effect that air pollution and particularly acid rain obviously have on the vitality of forest ecosystems (cf. Huettl, 1992a).

Since the early 1980's the causes of the new type forest damage are investigated internationally. As a result it can be stated that the forest damage is caused by various factors that are related to site and stand conditions. However, it has been accepted that various types of forest decline, particularly in conifers are associated with acute nutrient deficiencies.

Acute nutritional disturbances are not only observed in declining stands of Germany but also in France, the Netherlands, Belgium, Austria, Switzerland, Sweden, Poland, Chechoslovakia, the CIS, the U.S.A., Canada and China.

Based on present findings the nutritional disturbances are at least to some degree caused by adverse impacts of acidic atmospheric deposition leading to increased nutrient losses from the canopy as well as to accelerated soil acidification. The latter process is associated with a loss of cationic nutrients such as Mg, Ca and K from the ecosystem. Therefore the most common nutritional disturbances involve Mg and K deficiencies. Furthermore, increased nitrogen deposition into forest ecosystems clearly enhances this problem by various processes. One such mechanism is through stimulated tree growth on formerly N limited sites - typical of large forest areas of the temperate and boreal life zones - and thus inducing nutrient imbalances in the generally nutrient poor soils. However, in many forests additional N input will increase forest productivity and only after some time growth might cease due to a new limiting factor such as the insufficient availability of another nutrient element. It is therefore not too surprising that forest growth was greatly improved in large areas of Europe that were and are subject to "atmospheric" N fertilization since several decades. For example, in Germany's (former West Germany) commercial forests - based on national inventories - standing timber volume has increased remarkably over the last 30 yr (exact increase numbers will be published in June 1992 by the Federal Ministry of Food, Agriculture and Forestry, Bonn). Only in stands with severe damage symptoms due to e.g. Mg deficiency growth has not increased and indeed has declined during the last 10 to 15 yr. It must be noted here that in addition to better N availability growth in Germany's commercial forests was also stimulated by more advanced and intensified forest management practices (e.g. new thinning regimes, regeneration with more productive species/provenances), changes in age structure (more older stands that carry higher timber volume) and most probably due to the rising atmospheric CO_2 concentration. With regard to the latter impact it should be emphasized that "atmospheric" CO_2 fertilization of forest ecosystems may eventually cause the same growth related problems, i.e. nutrient imbalances, as was observed for "atmospheric" N input. On the other hand, to increase the CO_2 storage capacity of forest ecosystems nutrient supplementation will be necessary on many sites. However, presently no relevant research is conducted.

These observations also demonstrate that negative growth effects coupled with certain forest decline types were by far over-compensated through much enhanced growth in other "damage" types and in healthy forests. Similar results, i.e. overall positive growth trends, despite the so called new type forest damage, have also ben found in France, Switzerland, Austria, Sweden and Finland.

To test whether appropriate nutrient amendments could mitigate or completely remove the damage associated with nutritional disturbances a comprehensive program of diagnostic fertilization experiments was initiated in the early 1980's in Germany (and elsewhere).

A typical example is illustrated in Table 1. In 1983 the Norway spruce stand in the Black Forest was characterized by severe Mg deficiency as is indicated by the insufficient Mg needle content (deficiency threshold: < 0.6 to $0.8\ mg.g^{-1}$ d m). All other nutrients were at an adequate to optimal level. In spring 1984 part of this stand was fertilized with a quick release Mg fertilizer and already in the fall of the same year a remarkable revitalization was observed in the fertilized trees. As can be seen from the 1984 foliar analysis data the fertilized spruce had increased Mg uptake considerably reaching sufficient Mg nutrition. This trend continued for all subsequent years. Furthermore, a balanced nutritional status was correlated with the improved Mg supply as is evident from the element ratios. Similar results as shown for Mg were found with P and K fertilizer experiments on sites where these nutrients were deficient. Also tree growth above <u>and</u> below ground was improved as detailed elsewhere (Huettl 1992b).

Table 1: Diagnostic fertilization trial in 12-yr-old Norway spruce (Picea abies): Elzach, Black Forest, Germany; soil application (150 g per tree): Kieserit ($MgSO_4$ x H_2O; 27 % MgO) in spring 1984; element content in current needles (1st whorl; n=10; fall sampling)

Element	Control Plot						
	1983	1984	1985	1986	1987	1989**	1991
Content, mg x g^{-1} dm							
N	16.0	16.2	17.8	14.6	14.2	17.4	16.4
P	2.6	2.5	2.3	2.4	1.2	3.3	2.0
K	7.5	8.6	7.9	6.4	5.6	8.8	8.9
Ca	2.4	3.0	3.2	1.7	3.2	2.4	2.4
Mg	0.2	0.3	0.3	0.3	0.3	0.3	0.4
Ratio*							
N:P	6.2	6.5	7.7	6.1	12.0	5.3	8.2
N:K	2.1	1.8	2.3	2.3	2.5	2.0	1.8
N:Mg	80.0	51.0	57.0	52.0	57.0	53.0	41.0

Element	Kieserit Plot						
	1983	1984	1985	1986	1987	1989**	1991
Content, mg x g^{-1} dm							
N	16.0	15.1	16.9	14.2	12.8	16.6	17.8
P	2.6	2.6	2.5	2.3	1.9	3.0	3.0
K	7.5	10.1	8.9	6.2	5.9	8.6	9.9
Ca	2.4	2.4	4.0	1.6	2.7	2.2	1.6
Mg	0.2	0.6	0.6	0.6	0.6	0.6	0.6
Ratio*							
N:P	6.2	5.8	6.8	6.2	6.7	5.5	5.9
N:K	2.1	1.5	1.9	2.3	2.2	1.9	1.8
N:Mg	80.0	27.0	30.0	25.0	23.0	29.0	29.0

* In the case of balanced nutrition the following ratios are normal: N:P = 6-12; N:K = 1-3; N:Mg = 8-30 (from Huettl, 1992a)

** no analysis in 1990

The overall positive experiences with fertilization and liming measures in mitigating forest damage have led to a fertilization program in Germany; through this program 80 % of the total fertilization costs (fertilizers, application) are reimbursed to the forest owner by governmental funds.

Since 1983 more than 1 x 10^6 ha of declining forest area has been limed and/or fertilized in Germany. Also within the next 5 to 10 yr this program will be continued with treating 100,000 to 150,000 ha annually. A similar fertilization program is carried out in Quebec, Canada, primarily focused on revitalization of declining sugar maple stands associated with K, P (and Mg) deficiencies.

3.2 Plantation forestry

In the tropics man-made forests for wood production have existed on a small scale for well over a century (Nair, 1984). But it is only during the past two to three decades that plantation forestry has become an economic land use activity in many tropical countries. The techniques developed recently to use mixed tropical hardwoods for the pulp and paper industry have led to large projects based on plantations for pulp mills and a rapid shift from natural forests to man-made softwood forests seems to be the trend in many tropical countries.

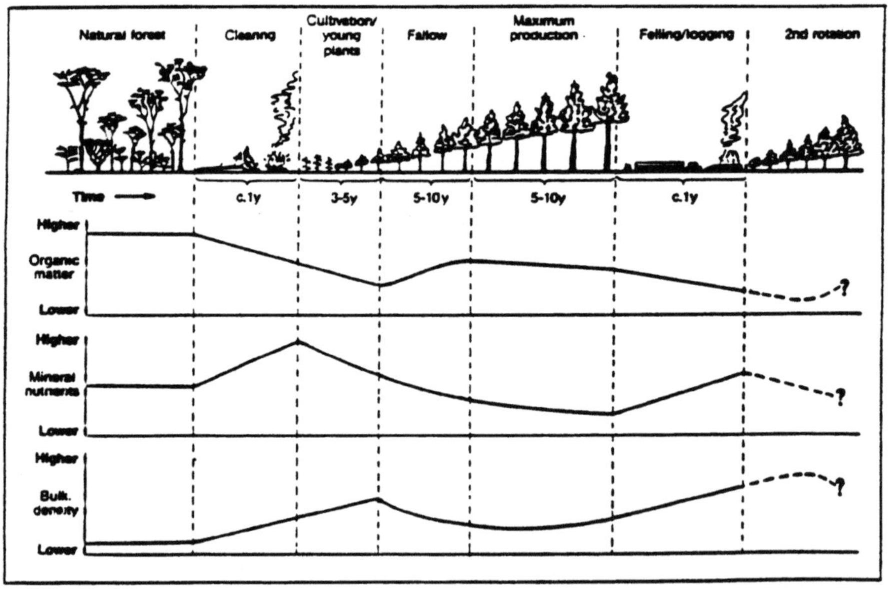

Figure 2: A generalized picture of the likely changes in organic matter status, nutrient content and bulk density of the top-soil during different management phases in the conversion of natural forests to short rotation plantations (from Lundgren, 1978)

Various species of different origin have been tried in tropical plantations, which basically fall under four major groups: Eucalyptus species, tropical hardwoods such as teak and mahagonies and tropical/subtropical conifers mainly Pinus, Cypressus and Araucaria as well as Acacia species.

In this context Lundgren (1978) has presented a generalized view of the likely changes in organic matter status, nutrient content and bulk density of the top-soil during different management phases in the conversion of natural rain forests to short rotation plantations (Figure 2). To strengthen and actualize this picture Huettl (1990) has recently evaluated case studies dealing with nutritional aspects of plantation forestry in tropical (rain) forest areas.

Plantation forestry aiming at maximum wood production for timber, pulp (paper), plywood, panel products, veneer, fuel wood, charcoal and occasionally also for non-wood products such as essential oils, resins, tannins, sterols, waxes, esters, acids, phenols, flavorings, sweeteners, spices, balsams, pesticides, dye-stuffs and last but not least alcohols (e.g. methanol) are practiced in countries such as New Zealand, Australia, South Africa, in parts of Latin America and South East Asia under often poor soil fertility conditions. There fertilization is utilized as an integrative part of forest management. In this context nutrient management starts at the nursery stage, is continued at planting and throughout the often short life span of the plantation trees.

Evans (1982) has presented an overview on nutrition of tree crops in the tropics which is also focused on removing deficiency and/or increasing growth by appropriate fertilization. Just recently, Zech and Drechsel (1990) reviewed mineral deficiencies of tropical trees and their identification by foliar analysis. Nutrient deficiencies and fertilizer use in New Zealand's exotic forests, particularly related to Eucalyptus and pine species, were detailed by Will (1985). Boosma and Hunter (1990) summarized the effects of water and nutrients and their interactions on tree growth and plantation forest management practices in Australia and New Zealand. Fertilization in South African plantation forestry was described by Schoenau (1983). For this area Herbert (1983) provided interesting information on the response of Eucalyptus grandis to fertilization. More recently Herbert and Schoenau (1989) presented a comprehensive review on fertilizing commercial forest tree species in southern Africa. Malvos and Bailly (1979) reported on fertilization of pine plantations in Madagascar. Finally, much research on nutrition in forest plantations was carried out in Latin America. For example, Zoettl (1989) reported about fertilization trials in coniferous plantations of Brazil.

Rapid population growth in Africa and elsewhere has increased the demand for fuel wood. Wood and charcoal are the main sources of energy for more than two billion people. Afforestion with fast growing trees offers a solution to the problem of continued deforestation and C oxidation. Casuarina species originating in the southern hemisphere yield very good fuel wood and have been planted in East, Central and West Africa and the Caribean. Casuarina equisetifolia is the most preferred species because of fast growth, salt tolerance, wind resistance, nitrogen fixation and drought resistance. It is more and more used in the humid and semi-humid Africa. Under beneficial conditions it will yield 15 to 18 $m^3 \cdot ha^{-1}$ with a 7 to 10 year rotation period (Webb et al., 1980).

However, on the coastal sandy soils of West Africa performance is less stimulating. After three rotations reduced growth and crown thinning were observed. According to Zech (1983 and 1986) these effects could be related to nutrient deficiencies.

When investigating healthy and damaged trees in this general area Kaupenjohann and Zech (1988) found that undamaged trees had deeper rooting systems than declining ones, carried more mycorrhizae and were better supplied with N, P and K. In an ongoing project fertilization led to a pronounced revitalization of C. equisetifolia indicating that probably P and K were the main growth limiting factors on this site (Kaupenjohann, pers. com.). Hence, proper fertilization appears to be an important prerequisit for adequate growth under the illustrated conditions.

Concluding the experiences with nutritional research and fertilization in tropical and subtropical plantation forestry it can be stated that proper fertilization has proven an essential and successful practice of forest management to optimize yields and hence to increase the CO_2 storage capacity of these ecosystems, particularly when water supply is adequate. However, much more information is needed in this field, especially for more fragile ecosystems such as the tropical moist forests. In this context CO_2 enrichment studies under field conditions utilizing the ecosystem research approach should be conducted.

3.3 Biomass forestry

The present knowledge and experience on short rotation forest biomass production - another important CO_2 sink - was recently summarized by Weisgerber and Hiege (1989). Like many other countries within the EC, Germany has to contend with surplus production in agriculture and a deficit in wood production. Therefore experts estimate that up to 4 x 10^6 ha might be taken out of food production within the next decade. About 500,000 ha of the surplus area might become available for short rotation biomass production. At present only 52 ha of short rotation plantations exist in Germany. Under certain circumstances biomass forestry may compete economically with agricultural production. In Table 2 biomass yield of short rotation trials in Europe and North America are listed indicating a wide range of productivity.

Cultivation for biomass forestry requires careful site preparation and weeding in order to facilitate rapid juvenile growth and protect cuttings from growth competition of other plants. Owing to the high cultivation cost and low biomass yields on forest sites biomass forestry should be established only on abandoned agricultural land.

Fertilization within the first three years seems to be unnecessary because of the great nutrient supply present at former agricultural sites. But later the intense output of nutrients has to be replaced.

Table 2: Biomass yields in short rotation trials

Country	Tree species	Plants x ha^{-1}	Rotation years	Yield odt x ha^{-1}*
West Germany	Populus	22,000	2	9-29
France	Populus	3,030	7	15
	Eucalyptus	1,250	7	13
Greece	Populus	1,111	6	14-20
	Platanus	10,000	5-6	6-14
U.K.	Salix	20,400	3	14
Ireland	Salix	16,700	3	3.2-15.2
	Populus	16,400	3	12.0-15.5
Yugoslavia	Populus	3,300-8,300	3	13-15
Canada	Populus	35,000	2	16
	Populus	4,000	5	8.35
New Zealand	Eucalyptus	1,250-2,150	10	30
	Pinus	1,775	8	9.5
Sweden	Salix	10,000-20,000	3	15
U.S.A.	Acer	13,900	2	8.0
	Prunus	27,800	2	8.2
	Populus	6,950	2	7.0
	Platanus	6,950	2	6.2
	Salix	27,800	2	8.2

* odt = oven dry t
(from Weisgerber and Hiege, 1989)

With respect to nutrient extraction by biomass harvest, Rehfuess (1986) found that one to two year old shoots of various species withdrew 100 kg N, 10 to 20 kg P, 30 to 50 kg K, 100 kg Ca and 10 kg Mg $ha^{-1} \cdot a^{-1}$ to produce 10 dt $\cdot ha^{-1} \cdot a^{-1}$. The nutrient content removed will rise significantly with an increase in dry matter production. Therefore it seems necessary to preserve the fertility of the soil by fertilization. When burning the biomass for e.g. energy production ash residues should be taken back to the plantation. In this case nitrogen is the major substance which does not remain in the ashes and must be replaced by fertilization.

With further development in this field biomass forestry seems to be a reasonable, cost-effective and practical way to reduce the problem of surplus food production of various industrialized countries provided that markets for these products can be developed.

3.4 Afforestation of degraded land

This important rehabilitation approach and jet another CO_2 sink will be illustrated using the People's Republic of China as a case study. China belongs to the countries with the least forest resources of the world. Owing to an optimistic evaluation about 12 % of the land area are covered by forests and woodlands (cf. Scheifele, 1980).

Because of this situation which also implies an increasing hazard for agricultural production due to e.g. soil erosion, afforestation efforts have been initiated. The government plans to increase the forest area to 20 % by the year 2000 and in the longer term even to 30 %.

To realize this ambitious goal afforestation should best be applied utilizing fast growing tree species. In central and southern China the preferred tree species are Cunninghamia lanceolata, Paulownia fortunei, Robinia pseudoacacia and Eucalyptus as well as Pinus species. In northern China owing to climatic conditions mainly poplar trees are planted.

In this context the Forest Research Institute of the Province Guangxi at Nanning in southeastern China started a comprehensive afforestation project planting various tree species at a variety of sites. Besides C. lanceolata and P. elliottii, also E. citriodora stands were established in the early to mid 1980's.

Eucalyptus citriodora is a fast growing tree (height increment 3 m $\cdot a^{-1}$; up to 45 m) and generally produces an exellent bole form and good timber quality. It can be used as fuel wood and for charcoal production. The leaves of E. citriodora yield a lemon - scented oil rich in citronellal used in the perfume industry. It is also a favorite for bee keepers because of the quantity and the amount of honey it produces. This species is native only at two locations at the coast of Queensland in Australia. However, as the climatic conditions are very similar at various sites in Guangxi, this species was selected for potential afforestation efforts because of its many positive features.

Soon after planting E. citriodora at generally infertile soils degraded by former agricultural practices poor growth was noticed. Because of foliar symptoms indicating K and/or P deficiency, factorial fertilizer trials were established in young plantations. In August 1988 a PK fertilizer was applied at various rates. Only some months later a clear regeneration effect was visible. According to the fertilizer dosis, height and diameter growth was increased remarkably. The same effect could also be noticed with respect to foliar biomass production, i.e. leaf size. Leaf analysis data of November 1989 were well related to the observed growth limitation differences indicating K deficiency as the major nutritional disturbance at the trial sites (Table 3). Similar results were observed for C. lanceolata and P. elliottii when these species were treated with the same fertilizer.

Table 3: Fertilization trial in 2-yr-old Eucalyptus citriodora: Nanning, Guangxi, PR China; soil application of Thomaskali (10 % P_2O_5, 20 % K_2O, 3 % MgO) in August 1988; element contents in sun leaves; sampling November 1989

Plot	N	P	S	K	Ca	Mg
	mg x g^{-1} dm					
Threshold range*	5.8-33.6	0.2-8.5	1.6-4.3	2.7-21.5	2.7-41.0	1.1-6.9
Control	12.5	0.7	2.1	2.0	8.2	1.6
300 kg x ha^{-1}	16.7	0.8	3.3	4.9	8.1	1.5
900 kg x ha^{-1}	17.8	0.8	2.8	12.1	11.5	1.4

* Range of variation of known data for all Eucalyptus spp. (from Huettl, unpublished)

Even though these results are short-term findings they clearly reveal the beneficial effects of improving soils fertility by specific nutrient amendments when afforestation/rehabilitation is aiming at fast biomass production under unfavourable site conditions.

4. Agro-forestry

Agro-forestry is the term given to sustainable land use systems which involve an interacting association of agricultural/horticultural crops and woody perennials (trees, shrubs, palms, wines, bamboos) all on the same unit of land. This form of land use has two main objectives: productivity involving a multiplicity of outputs; and sustainability, which implies the conservation or even improvement of the environmental aspects of the system (Huxley, 1983). In some cases agro-forestry systems may be used to bring about restitution or improvement of soils so as to broaden future land use options. Following two land use systems related to agro-forestry are briefly discussed.

4.1 Shifting cultivation

Shifting cultivation - also known as the bush fallow system, slash and burn agriculture, swidden farming and by a large number of local terms - is still the dominant crop production system practiced in the tropics (cf. Nair, 1984). It is estimated to extend over approximately 30 % of the exploitable soils of the world, 360 x 10^6 ha, and provides 250 x 10^6 people with a means of subsistence (FAO, 1974). The practice can be described in simple terms as a land use system in which forested land is cleared and cropped for a few years and then left for the regeneration of the bush, the duration of the fallow period being much longer than that of the cropping period. Also the system is dominant mainly in the sparsely populated and least developed areas where technological inputs for advanced agriculture such as fertilizers and farm machinery are not available and/or not usually used, it is found all over the tropics in Asia, Africa and Latin America. Even in densely populated South East Asia it is a major agricultural system.

Kyuma et al. (1985) conducted a study related to the effects of shifting cultivation on soil fertility. In this context analysis of the nutrients stored in the biomass and the soil were performed at various sites in the humid tropics before clearing the forest. It was found that the amount of biomass as well as levels of nutrient elements stored in shorter fallows are by a factor of 2 to 5 less than those of mature forests. The level of nutrients in the top-soil under fallow conditions depends more on the parent rock of the soil than on fallow periods. This is particularly true for exchangeable Ca and Mg. The factors that may influence the nutrient dynamics during the cropping period are the losses due to erosion and leaching in addition to crop removal (Tulaphitak et al., 1985).

4.2 Alley cropping/farming

Interest has grown in the development and use of more productive land use technologies involving intercropping of woody species (particularly legumes) with food or forage crops in an agro-forestry system (Steppler and Nair, 1987). Two such technologies are alley cropping for food production (Kang et al., 1981) and alley farming for food and animal production (Okale and Sumberg, 1985). Trees and shrubs feature prominently in traditional farming systems in the tropics because of the many uses and their environmental and socio-economic benefits (Burley, 1987). In the bush fallow slash and burn cultivation system traditional farmers have for generations exploited the potential of trees and shrubs for soil fertility regeneration and weed suppression. When fallow periods are long enough to permit full vegetation regeneration and soil fertility restoration this cultivation system is recognized as ecologically balanced, economically attractive and culturally integrated (Kang et al, 1990). However, under the pressure of increasing population and other competing land use demands long fallow periods are no longer possible in densely populated areas. The possible effects of shortening the fallow cycles on soil productivity are known. In some areas of the tropics traditional farmers respond to this problem by selectively retaining multi-purpose woody species which they consider useful and also more effective in biologically enriching the fallow in a shorter time. Research has recognized he important role of woody legumes for provision of shade in soil conservation and as a source of mulch and green manure and have used them accordingly for some time.

Including woody species in agro-forestry systems is generally considered advantageous for soil fertility maintenance due to their efficient nutrient cycling. Information is available showing the rapid substantial nutrient accumulation in regenerating forest vegetation which is partially recycled through litter fall. However, data on nutrient accumulation and contribution from woody species grown in alley farming are scanty.

The effect of some woody species on micro-site soil enrichment is well known. Certain tree species may change the soil properties considerably so not necessarily permanently under and in the immediate environment of the trunk. This is especially conspicuous in the wooded savanna region.

The long-term effect of addition of prunings from the hedgerows on maintaining the soil carbon and nutrient status under alley farming has been reported by Kang et al. (1984). With continuous addition of Leucaena prunings, higher soil organic matter and nutrient status were maintained than on those receiving no prunings.

Results of a long-term alley farming trial conducted in southern Nigeria on a sandy soil have shown that addition of Leucaena prunings, even without N application, can sustain maize yield at a reasonable level of about 2 $t \cdot ha^{-1}$. To obtain a higher maize yield Leucaena prunings must be supplemented with fertilizer N because of the low direct contribution of the pruning. In addition to N, the hedgerow litter and prunings also contain large amounts of other nutrients, particularly K, Ca and Mg. However, information on the contribution of these nutrients to the associated crops is not available.

Even though there are many variations in agro-forestry practices, some hypotheses about expected soil changes under agroforestry can be made in light of the evidence that has been collected (Nair, 1984).

a) The inclusion of compatible and desirable species of woody perennials on farm lands can result in a marked improvement of soil fertility i.e. carbon and nutrient content, water holding capacity etc. There are several possible mechanisms for this, such as increasing organic matter, more efficient nutrient cycling, biological N fixation, enhanced nutrient economy and shading.

b) The improvement in the organic matter status of the soil can result in increased activity of favourable micro-organisms in the root zones.

c) The role of trees in soil conservation and erosion control is the most important reason in areas that are prone to erosion hazards.

d) The influence of trees on hydrological characteristics can extend from the micro-site level to the farm and regional levels.

The important aspects on the design of ecologically sound agro-forestry systems were discussed by Oldeman (1983). Hence, a concept useful in agro-forestry should consider eco-units that mimic the structure and dynamics of natural eco-units through their successional stages in order to rationalize the design management of man-made agroforests.

5. Counteractions

Much knowledge and experience has been gathered for proper management of the temperate/boreal forests. In these areas more recently research work has been focused on environmental pollution and its impact on commercial as well as uncommercial forests. The findings of these projects indicated that the complex ecological features of forest ecosystems can only be understood adequately when the whole ecosystem is addressed. This is best done through studying all relevant fluxes and pools on the basis of forested watersheds through a number of interacting scientists of different disciplines. As this approach is very costly and as experienced mind-power is limited, careful selection of representative study sites must receive highest priority. Various such studies with a major emphasis on nutrient element turnover have been initiated recently in Europe and North America. However, the carbon cycle has not yet received the attention it deserves when recognizing the fact that the CO_2 concentration in the atmosphere has significantly increased and is still rising.

As indicated above much research is needed related to proper management of tropical/subtropical forests including agroforestry systems as well as afforestation efforts and how to utilize these forests as CO_2 sinks. In this context research must primarily be focused on forest nutrition and forest soil science, forest ecology and silviculture; for in these extremely complex systems carbon and nutrient fluxes, soil fertility and ecological factors play an even much more pronounced role than in temperate forests.

Even from this incomplete list illustrating the lack of knowledge and necessary future activities it becomes evident that the third world countries need financial and technical help in order to achieve these goals that are equally important for the people in these as well as in the more developed countries - a fact that became most evident through the continual enhancement of the CO_2 content of the world's atmosphere and the presumed impact on global climate.

Acknowledgement

The authors wish to thank Mr. B. v. Schenck and Dr. P. Ende for the comprehensive literature review and Ms. I. Mantel, I. Wascher and M. Thies for the excellent clerical effort.

References

BAULE, H. and FRICKER, C. (1967): Die Duengung von Waldbaeumen. 259 p.

BOOMSMA, D.W. and HUNTER, I.R. (1990), Forest Ecology and Management, 30: 455-476.

BURLEY, J. (1987): A decade of development. In: Stepler, H.A. and Nair, P.K.R. (eds.), Agroforestry. ICRAF, Nairobi: 273-287.

EVANS, I. (1982): Plantation forestry in the tropics. Oxford.

FAO (1974): Shifting cultivation and soil conservation in Africa. Soils Bulletin 24, Rome.

FOELSTER, H. (1986): Nutrient loss during forest clearing. In: Lal, R., Sanchez, P.A. and Cummings, R.W., Land clearing and development in the tropics. 450 p.

HERBERT, M.A. (1983), S. Afr. For. J., 124: 4-12.

HERBERT, M.A. and SCHOENAU, A.P.G. (1989): Fertilizing commercial forest species in Southern Africa: Research progress and problems. Symposium "Mineralstoffversorgung tropischer Waldbaeume", July 13, 1989, Bayreuth, FRG.

HUXLEY, P.A. (1983): The role of trees in agroforestry: some comments. In: Huxley, P.A., Plant research and agroforestry. 617 p.

HUETTL, R.F. (1986), The Fertil. Soc., 250, 1-40.

HUETTL, R.F. (1990): Fertilization in multi-purpose forestry and its role in mitigating the decline of the worlds's forest resources. 10th World Fertilizer Congress of CIEC, 21-27 October 1990, Nicosia, Cyprus (plenary paper; proceedings in press).

HUETTL, R.F., ed., (1991): Forest fertilization and tree growth. Fertilizer Research (special issue), 140 p.

HUETTL, R.F. (1992b): Mg deficiency - a' new' phenomenon in declining forests - symptoms and effects, causes, recouperation. Springer Verlag (in press).

JUO, A.S.R. (1989), Agric., 25: 145-163.

JUO, A.S.R. and KAMPRATH, E.J. (1979), Soil Soc. Am. J., 43: 35-38.

KANG, B.T., WILSON, G.F. and SIPKENS, L. (1981): Plant and Soil, 63: 165-179.

KANG, B.T., WILSON, G.F. and LAWSON, T.L. (1984): Alley croping: A stable alternative to shifting cultivation. Ibadan, Nigeria, International Institute of Tropical Agriculture.

KANG, B.T. and JUO, A.S.R. (1986): Effect of forest clearing on soil chemical properties and crop performance. In: Lal, R., Sanchez, P.A. and Cummings, R.W. (eds.), Land clearing and development in the tropics. 450 p.

KANG, B.T., REYNOLDS, L. and ATTA-KRAH, A.N. (1990): Alley farming. In: Brady, N.C. (ed.), Advances in Agronomy.

KAUPENJOHANN, M. and ZECH, W. (1988): Potash Review, 11, 5: 37-41.

KIRA, T. and SHIDEI, T. (1967), Jap. J. Ecol., 17: 70-87.

KYUMA, K., TULAPHITAK, T. and PAIRINTRA, C. (1985), Soil Sci. Plant Nutr., 31: 227-238.

LAL, R. (1986): Deforestation and soil erosion. In: Lal, R., Sanchez, P.A. and Cummings, R.W. (eds.), Land clearing and development in the tropics. 450 p.

LUNDGREN, B. (1978), Reports in forest ecology and forest sils No. 31, Swedish Univ. Agric. Sci., Uppsala, Sweden. 426 p.

MALVOS, C. and BAILLY, C. (1979): Mimeograph translation, Centre Technique Forestier Tropical. Nogent-sur-Marne, France. 43 p.

NAIR, P.K.R. (1984): Soil productivity and aspects of agroforestry. 85 p.

NYE, P.H. and GREENLAND, P.J. (1960): The soil under shifting cultivation. Techn. Comm. No. 51. Harpenden: Commonwealth Bureau of Soils. 156 p.

OLDEMAN, R.A.A. (1983): The design of ecologically sound agroforests. In: Huxley, P.A. (ed.), Plant research and agroforestry. 617 p.

OKALI, C. and SUMBERG, J.E. (1985): Agric. Syst., 18: 39-59.

REHFUESS, K.E. (1986): Standorts- und ernaehrungskundliche Untersuchungen zur Erzeugung von Biomasse mit schnellwachsenden Baumarten. Unveroeffentlichter Exkursionsfuehrer, Lehrstuhl fuer Bodenkunde der Universitaet Muenchen.

SCHEIFELE, M. (1983), Allg. Forstz., 35: 1461-1463.

SCHOENAU, A.P.G. (1983), S. Afr. For. J., 125: 385-397.

STEINLIN, H. (1989): Universitaetsblaetter, 105: 1-62.

STEPPLER, H.A. and NAIR, P.K.R. (1987): Agroforestry - a decade of development. ICRAF, Nairobi.

TAMM, C.O. (1985), K. Skogs- o. Lantbr. akad. tidskr. Suppl., 17: 9-29.

TULAPHITIAK, T., PAIRINTRA, C. and KYUMA, C. (1985), Soil Sci. Plant Nutr., 31: 239-249.

UEXKUELL, H.R. and MUTERT, E. (1988), Kali-Briefe (Buentehof), 19: 185-198.

WEBB, L. J. (1977): Ecological considerations and safeguards in the modern use of tropical lowland rainforests as a source of pulpwood: example in the Modang Area Papua New Guinea. Office of Environment and Conservation, Papua. New Guinea. 36 p.

WEISGERBER, H. and HIEGE, W. (1989): Short rotation forest biomass plantations in the Federal Republic of Germany. In: Hummel, F.C., Palz, W. and Grassi, G. (eds.), Biomass forestry in Europe: a strategy for the future: 325-372.

WENGLER, K.F. (1985): Forestry handbook. 1335 p.

WILL, G. (1985), FRI (Forest Research Institute, Rotorua, New Zealand), Bulletin No. 97: 47-50.

WRI (WORLD RESOURCES INSTITUTE) (1988): World resources 1988-89. 372 p.

WRI (1990): Preserving the global environment: 23 p.

ZECH, W. (1983), Série GTZ No. 154. 384 p.

ZECH, W. (1986): Mineral nutrition of forest plantations near Téné. Sanjoue. Seguie and Mopsi in Ivory Coast. Sodefor., Internat. Report. 35 p.

ZECH, W. and DRECHSEL, P. (1990): Mineral deficiencies of tropical trees and their identification by foliar analysis: a review. Plant and Soil (in press).

ZOETTL, H.W. (1989): Wachstum und Ernaehrungszustand von Koniferenplantagen in Suedbrasilien. Symposium "Mineralstoffversorgung tropischer Waldbaeume", July 13, 1989, Bayreuth, FRG.

THE USE OF HALOPHYTES TO SEQUESTER CARBON

EDWARD P. GLENN -- Environmental Research Lab,
2601 E. Airport Dr., Tucson, AZ 85706, USA

LOUIS F. PITELKA -- Electric Power Research Institute, P.O. Box 10412, Palo Alto, CA 94304, USA

MARY W. OLSEN -- Environmental Research Lab,
2601 E. Airport Dr., Tucson, AZ 85706

Abstract

Wild salt tolerant plants, halophytes, could be grown on 130 x 10^6 ha of potentially arable land in the world's coastal deserts, inland salt deserts and areas of secondary salinization in irrigation districts. Halophytes have potential as biomass crops to directly sequester up to 0.7 Gt C, similar to tree plantations, or they can play an indirect role in absorbing C from the atmosphere by providing food, fodder and energy crops from a new land base. To the extent new cropland can be developed from unused saline land, further land clearing and loss of C storage in forest fallow and old growth forests can be spared, thereby adding to the carbon sequestering potential of all usable ecosystems.

1. Introduction

It is now well established that the concentration of various greenhouse gasses is increasing in the atmosphere and that human activities are largely responsible (IPCC, 1990). Many scientists believe that further increases in these gasses will result in potentially significant changes in global climate. Since a significant change in climate could have major ecological and economic effects, considerable attention is being devoted to finding ways to slow or halt the buildup of greenhouse gasses.

One option that has received attention is reforestation. Various investigators (Marland, 1988; Moulton and Richards, 1990; Sedjo, 1989; Vitousek, 1990) have attempted to calculate how much C might be sequestered per yr by planting new forests. For instance, Vitousek has determined that a massive effort involving the reforestation of 25% of all the land globally that has been deforested over time could result in the sequestration, for several decades, of about 25% of current global C emissions. While such an effort could slow the buildup of CO_2 in the atmosphere and any climate change due to the greenhouse effect, it would not alone solve the problem. Moreover, it seems doubtful that such a massive reforestation effort could realistically be achieved because of the economic and

Water, Air, and Soil Pollution **64**: 251–263, 1992.
© 1992 *Kluwer Academic Publishers.*

social costs of reclaiming so much land that is currently used for other purposes.

It has also been suggested that the ocean ecosystems might be managed in a way so as to increase the uptake and storage of C (e.g. Martin et al., 1990). While it is conceivable that a means to store large quantities of C in the ocean might ultimately be found, it is unlikely that a feasible method having no unacceptable environmental consequences will be found in the near future. Furthermore, it is unlikely that an ocean disposal option will be sufficient to entirely halt the accumulation of CO_2 in the atmosphere (Peng and Broecker, 1991).

The above considerations suggest that a successful strategy for reducing the rate at which greenhouse gasses accumulate in the atmosphere will require multiple approaches, including those that involve sequestration of C, as well as reductions in emissions. One approach to C sequestration that has received relatively little attention is the use of terrestrial plants other than trees. Since reforestation cannot solve the entire CO_2 problem and since forests would need to be planted on land that is useful for other purposes, it would be desirable to identify plants that can grow on less valuable land. We are in the process of evaluating the feasibility of using salt tolerant plants, halophytes, to sequester CO_2 (Glenn et al., 1991a; Glenn et al., 1992b). In this paper we describe the results of an assessment of the C sequestration potential of large scale halophyte plantations and present some preliminary results from field trials.

2. Land Areas Available for Halophyte Plantations

A central dilemma of the reforestation option is that many of the land areas that could support reforestation are either already productive, as farmland or forest fallow, or are already storing significant amounts of C, as old growth forest (Marland, 1988). If agricultural land is converted into forest plantation, total cropland is thereby reduced. However, the world apparently will need more rather than less total cropland in the future. It has been estimated that some 200×10^6 ha of new cropland will be needed to support the projected doubling of population in the tropics in the next 75 yrs, and the most attractive lands for new crops are the same ones that would be used for storing C - the approximately 200×10^6 ha of forest fallow (Lanly, 1982) or the remaining old growth rain forests (Sanchez et al., 1982).

It is apparent that creating tree plantations on a global scale may have a zero net effect on the global C balance if forest fallow and old growth forest are all the more quickly converted to cropland as a result (Figure 1, Top) (see also Harmon et al., 1990, for a discussion on the conversion of old growth forest to tree plantations). On the other hand, if we could increase the total pool of available land through the innovation of cultivating halophytes in the deserts, the balance between old growth forests, managed forests, forest fallow and cropland would no longer be zero sum (Figure 1, Bottom).

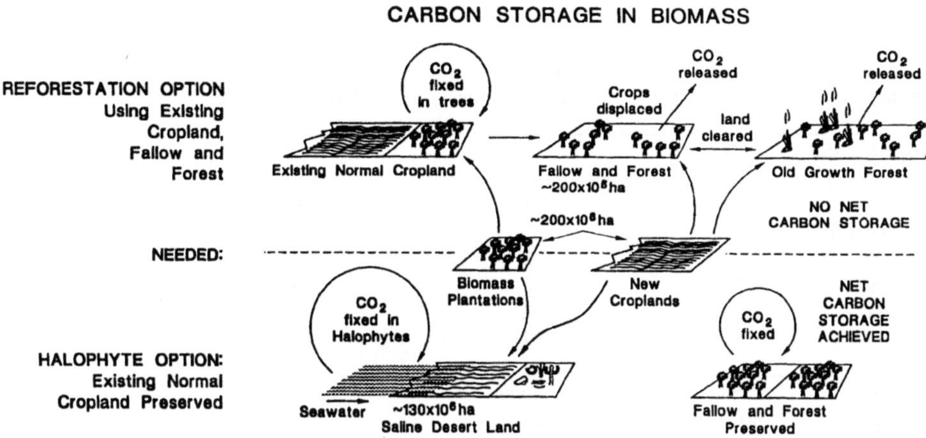

Figure 1. **Two options for storing C in biomass. The needs are for additional land for biomass production plus new cropland. In the reforestation option, tree plantations displace cropland resulting in increased land clearing elsewhere for crop production. In the halophyte option, saline desert land actually adds to the total arable land base.**

Three types of land and water are potentially available for halophyte plantations: coastal deserts, using seawater for irrigation; inland saline basins, using brackish aquifers too saline for conventional agriculture; and unused land in or adjacent to conventional desert irrigation districts, using saline drain water that otherwise goes to disposal. The total land base for halophyte plantations is large. There are approximately 720 x 10^6 ha of coastal desert and inland salt desert in the world. Much of this land is unused by man and unproductive or even barren of natural vegetation, due to extreme aridity and salt accumulation at the soil surface. We attempted to estimate the percentage of this land that might actually be used for halophyte plantations by applying typical agricultural constraints to 36 coastal deserts, 19 inland salt deserts and 9 desert-irrigation regions (Glenn et al., 1991a; Glenn et al., 1992b). Constraints included elevation above sea level, size of available aquifer for irrigation, soil type and slope of land, existing land use and natural vegetation cover. In general only barren, level land within easy reach of the sea or internal saline water was considered usable, and almost all of the regions are classified as extreme desert areas.

We estimated that 17% of coastal desert and 15% of inland salt desert was usable and that an additional areas equal to 6% of existing irrigated land could be irrigated with saline drain water to grow halophytes (Table 1). The total usable land area is approximately 130 x 10^6 ha distributed globally as shown in Figure 2.

Table 1. Total areas and estimated usable areas of coastal desert, inland salinized soil and irrigated agriculture - the resource base for halophyte plantations (from Glenn et al., 1991a).

	——— 10^6 ha ———		
Region	Coastal Desert	Inland Saline Soils	Irrigated Land
Europe (-USSR)	0	3	13
Asia (-USSR)	31	119	165
Africa	230	81	9
N. & Central America	7	2	27
S. America	5	13	7
Australia	22	36	2
USSR	0	171	12
% Usable:	17%	15%	6%
Amount:	50	63	13

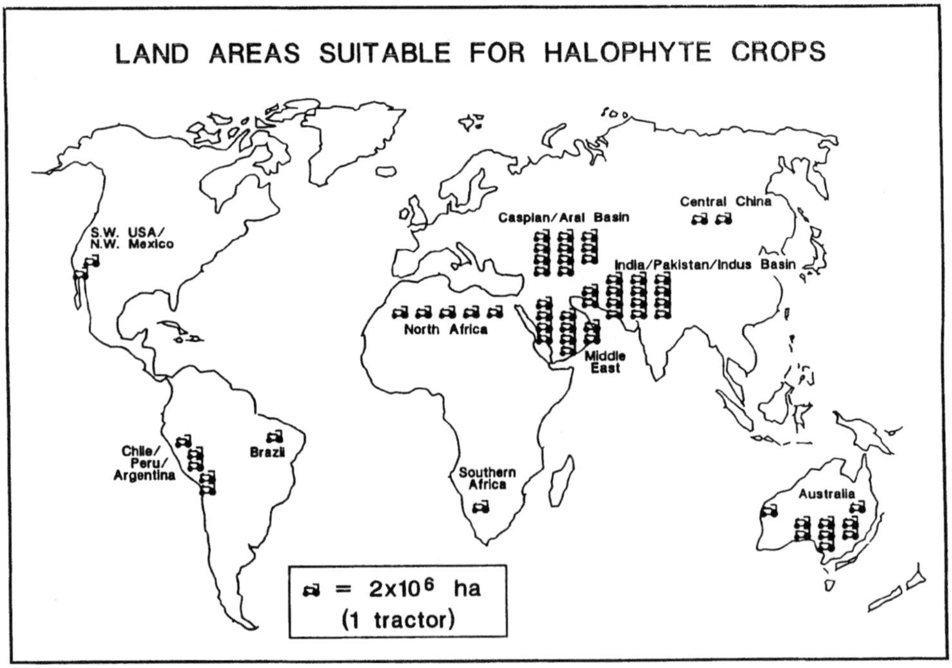

Figure 2.

Although the accuracy of this estimate cannot be known without detailed site work, it appears to be conservative by comparison to other estimates of the amount of saline land that could be used for halophytes (e.g. Barrett-Lennard et al., 1986). There are 70 x 10^6 ha of desert land already under fresh water irrigation (Fukuda, 1976), representing about 2% of total desert land. However, it has been estimated that 22.5% of the world's deserts have no inherent soil constraints for growing crops save lack of water (Hammond and Paden, 1990). Therefore, it appears conservative to conclude that the irrigable desert area can be increased to 6% if the constraint of water availability is relaxed over vast areas of coastal and inland salt desert by the innovation of growing halophytes.

The discovery of 130 x 10^6 ha of potential new cropland could have a major impact on the global C balance. Halophyte plantations could contribute to retiring C from the atmosphere either directly or indirectly: directly, if biomass halophytes are grown specifically for C storage; or indirectly, if they are grown as food, fodder or energy crops, to spare further development of forest fallow and old growth forest for these purposes. Those forest lands could then be managed for optimal C storage.

Questions of concern, reviewed in the following sections, are:

- How productive are halophytes crops?

- How much seawater or other salty water must be supplied for their growth?

- What are the costs of growing halophytes compared to other biomass crops?

- How will the halophyte C be stored?

- What value do halophytes have as food, fodder or energy crops?

3. Productivity of Halophyte Plantations

Most conventional crops suffer growth reductions when the salinity of the irrigation water exceeds more than a few percent of the salt content of seawater. There are few irrigation districts in the world that can utilize water of greater than 1,000 ppm total salts (3% seawater) without damaging crops. This need for high quality water is a major constraint to irrigation, for many of the water sources available in deserts are saline.

There are many different types of halophytes and they exhibit a broad range of salt tolerance (Glenn, 1987; Glenn and O'Leary, 1984) but the most tolerant species have proven to be as productive of biomass and seeds on seawater as conventional crops on fresh water, even in extreme coastal desert environments, in multi-year field trials (Glenn and O'Leary, 1985; Glenn et al., 1991b). Many of the productive species are in the family Chenopodiaceae - salt bushes, succulent salt marsh plants, and other highly tolerant plants native to the salt marshes and salt deserts of the world. Biomass yields under agronomic conditions are in the range of 10 - 20 t ha^{-1}, containing 3 - 7 t ha^{-1} of C.

As an example of halophyte productivity, biomass yields of the annual halophyte, Salicornia bigelovii Torr., during the first yr (1991) of the current seawater field trials in Puerto Penasco, Mexico, were 22.4 mt ha^{-1} (n = 65 square meter quadrants distributed in 22 plots, S.E. = 0.7 mt ha^{-1}). The biomass contained 24.7% C, for a mean C fixation rate of 5.53 mt ha^{-1} in above ground biomass. This compares to yields of 6.2 mt ha^{-1} of C in combined above and below ground biomass obtained in managed tree plantations under favorable soil, rainfall and temperature conditions (Sedjo, 1989). Under marginal conditions tree plantations yield as little as 20% of this value (Marland, 1988).

If the available halophyte lands were fully planted, and fixed an average of 5.5 mt ha^{-1} of C, they would remove 715 x 10^6 mt of C, approximately 12% of the annual C emissions from fossil fuel burning and 24% of the fraction that remains in the atmosphere. Hence, in terms of productivity and area availability, halophyte plantations could make a significant contribution to enhancing biomass storage of C but like trees they could not do the whole job alone.

4. Irrigation requirements and other costs associated with growing halophyte biomass

In seawater field trials we have typically irrigated halophytes more frequently than conventional crops are irrigated in order to leach excess salts below the root zone and insure that the soil is permanently moist. On sandy soils the irrigation amounts have been high (Glenn & O'Leary, 1985). Since irrigation is the single greatest expense in growing halophytes, we are conducting experiments to determine the minimum amounts of water of different salinities that can be used to grow the crops in coastal desert climates. These experiments measure the actual consumptive use of water by halophytes, by using plants grown outdoors in pots that are periodically weighed to determine water usage. Preliminary results with four different species have shown that water use efficiency by halophytes is similar to conventional crops, with the C_4 halophytes (Suaeda, Distichlis and Atriplex) requiring less than the C_3 halophyte (Batis maritima), as expected (Figure 3).

An unexpected finding was the apparently beneficial effect that increased salinity of the irrigation water had on the water use efficiency of the plant. At higher salinity, less water was needed to produce a kg of dry biomass (corrected for mineral content of the tissues) than at lower salinity. This effect was more noticeable for the two succulent species than for the shrub or the grass but was present in all species whether C_3 or C_4. Halophyte crops do require an amount of water above their consumptive use to control salt levels in the soil. The experiments in Figure 3 used a 50% leaching fraction, hence the actual water use was 1.5 times higher than shown in the figure. However, this need for a leaching fraction was partially offset by the greater water use efficiency of the plants on seawater. Overall, halophytes appear to have about the same irrigation requirements as conventional crops grown in a similar climate.

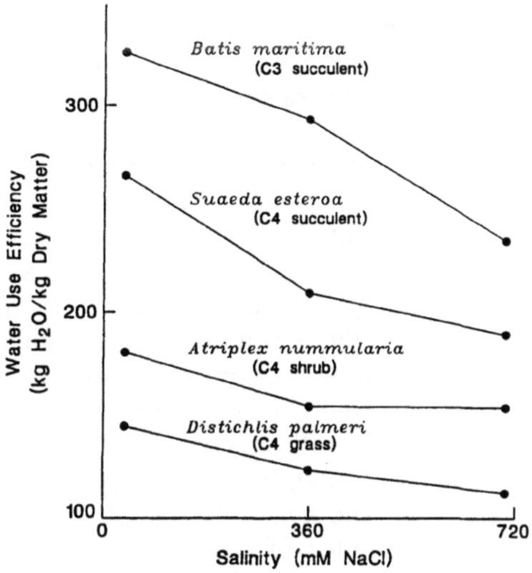

Figure 3. Water use efficiency of halophytes grown at different salinities in outdoor field trials in a coastal desert environment at Puerto Penasco, Sonora, Mexico. Plants were grown for 14 weeks (June - September, 1991) in lysimeter pots, weighed weekly to determine water usage and irrigated with a 50% leaching fraction.

Based on irrigation requirements and other farm costs, we have estimated the dollar and C costs of producing halophyte biomass (Glenn et al., 1992a). The dollar cost estimates, using typical Arizona field costs, are $44 or $53 per dry mt of biomass ($175 or $211 per mt of C) depending on whether brackish water or seawater is used for irrigation. This compares to values of $30 to $45 per dry mt biomass for other biomass crops including tree plantations (Lynd et al., 1991). However, those crops are non-irrigated and therefore can only be grown where there is sufficient rainfall. The C costs of halophyte crops were calculated by assuming that all farm operations, from pumping water to planting, harvesting, baling and hauling the crop, were performed using diesel fuel at 85% C content. The analysis indicated that halophytes require 225 kg to 300 kg of fossil fuel C input for each 1,000 kg of atmospheric C fixed. This compares to estimates of 130 - 180 kg mt^{-1} for non-irrigated cellulosic biomass crops, and 330 - 970 kg mt^{-1} for corn grown as an energy crop, which at the extreme barely breaks even (Lynd et al., 1991).

5. Halophytes for direct sequestering of C

Trees are an attractive C storage crop because of their longevity. Managed tree plantations are productive for 20 to 40 yrs, after which net C fixation slows and eventually reaches a balance with litter decomposition. At this point, the plantation must be harvested and replanted if it is to continue to remove C at an acceptable rate. Since the greenhouse problem will be with us for an estimated

200 yrs, many cycles of harvest and replanting must be envisaged if trees are to be anything more than a short delaying tactic in confronting global warming. This raises the question of what to do with all the harvested lumber, which would accumulate much faster than it could be consumed for housing, furniture or other long-lived wood products.

Halophyte crops cycle much faster than tree crops and present a similar dilemma. The perennial, large, shrubby species such as Atriplex may have a useful life of ten yrs or more and can be treated as trees, but the annual crops such as Salicornia bigelovii are planted and harvested on as short as a 200 day cycle.

A storage option we are testing in the EPRI field trials is to plow halophyte biomass into the soil, where, hopefully, a significant percentage will enter long term storage in the form of soil humus. Desert soils are typically very low in organic matter - less than 0.5% organic C in the case of our halophyte fields at Puerto Penasco. It is already known that the C content of agricultural soils can be increased to 5% or higher in the plow zone, through the plowing under of biomass over time. If the C content of all the available halophyte acreage were to be increased to 5% by the plowing under of halophyte crops, 28 Gt would be eventually be stored, amounting to 4 - 5 yrs of fossil fuel emissions.

The feasibility of this storage strategy depends on the decomposition rate of buried halophyte biomass. If it breaks down very quickly, soil storage will be ineffective as a means to mitigate atmospheric levels of CO_2 over a period of several centuries. We are looking at two options - wet burial and dry burial. In the wet burial trials, halophyte crops are simply plowed back into the same soil in which they are grown, and a succeeding crop is planted. Our hypothesis is that the high salinity of the irrigation water will inhibit microbial decomposition of the buried halophyte material.

Laboratory experiments have so far supported the hypothesis. We loaded incubation vessels with two types of halophyte biomass mixed with soil, and irrigated them with different dilutions of seawater over 140 days (Table 2). They were measured for C losses via soil respiration and leaching of organic C in the drainage water. In fresh water vessels, C losses were twice as great as in seawater vessels. There was also an effect of biomass type: Salicornia biomass decomposed slower than Atriplex at all salinity levels. We are following up this experiment by measuring decomposition rates in field plantings using litter bag techniques. Further measurements are needed before we can project the half-life of halophyte C in seawater-irrigated fields.

If wet burial proves infeasible, dry burial is an option. In this option, halophyte biomass would be baled and hauled to nearby vacant desert land that is unsuitable for farming but on which the bales could be stored or buried. Due to the low rainfall in the target areas decomposition rates would be expected to be very slow. This option is being field tested concurrently with the wet burial trials.

Table 2. Projected first-yr decomposition rates of two types of halophyte biomass at different salinities. Biomass samples were mixed with desert soil and incubated for 140 days in vessels which were irrigated biweekly with seawater solutions ranging from 0 to 40 ppt total salts. Solution that leached through the soil following each irrigation was collected and analyzed. Respiration was measured biweekly following an irrigation. Decomposition was estimated by the amount of CO_2 respired from incubation vessels plus the amount of organic C lost in leachates. The results are expressed as the % of initial C in the vessel lost to decomposition over 1 yr, assuming the final decomposition rate under each treatment continued for the remainder of the yr.

	Carbon Loss (%)					
	Atriplex Biomass			Salicornia Biomass		
	Respir-ation	Leachate	Total	Respir-ation	Leachate	Total
0 ppt	41.6	18.8	60.4	35.2	10.4	45.6
10 ppt	38.9	10.2	49.1	33.6	5.0	38.6
20 ppt	31.5	9.4	40.9	24.3	4.3	28.6
40 ppt	23.3	7.4	30.7	18.5	4.1	22.6

6. Halophytes as food, fodder and biofuel crops

As mentioned, halophytes can indirectly enhance C storage in biomass by providing alternative new croplands, sparing forest fallow and old growth forests from further clear-cutting which is a major anthropogenic source of C releases. Those land could then be devoted to C storage. The best candidate we have discovered so far as a crop plant is Salicornia bigelovii, which is a valuable new oilseed crop (Glenn et al., 1991b). Six yr mean seed and biomass yields on seawater irrigation equalled or exceeded the yields of such fresh water oilseed crops as soybean and sunflower. The oil is suitable for human consumption: it is high in polyunsaturated fatty acids (74% linoleic acid, similar to safflower oil), has a pleasant, nutty taste and a texture similar to olive oil. The seed meal left after oil extraction is 42% protein and is a valuable protein source for ruminants (Moore et al., 1989) or poultry (Glenn et al., 1991b) after treatment to wash out or deactivate saponins. The straw left after seed removal is high in salts but has proven to be equal to conventional fodder sources in goat diets when it is either mixed with other fodder to dilute the dietary salt intake or used as the sole fodder source after washing it in seawater and pressing it to reduce salt levels by about half (Glenn et al., 1992a).

It is noteworthy that just this one halophtye crop can supply vegetable oil, protein meal and fodder. These are among the products in greatest need in tropical and subtropical areas. Brazilian forest clearing, for example, is principally for the purpose of creating pasture land or planting soybeans. Based on comparative yield values of seed and fodder, one hectare of Salicornia bigelovii can replace one hectare of soybean and two or more hectares of non-intensively managed pasture. Other halophyte crops such as Atriplex have are already used as fodder crops in many parts of the world (O'Leary, 1986).

Halophyte biomass can also be used as a renewable fuel for electric power generation (Glenn et al., 1991a). Halophyte or other renewable biomass used for power generation does contribute to retiring C from the atmosphere to the extent it replaces fossil fuel burning. The heating value of halophyte biomass varies with the species, the chief variable being the ash content, but in general halophyte biomass falls in the range of Lignite A or B coals. We considered the case of a 500 MW power plant fueled with a 2:1 ratio of coal to halophyte biomass, which could be burned with minimal plant modification. The fuel mix would produce about 10% less heat per unit weight than coal alone but would result in 25% lower consumption of fossil fuel C per unit of electric power generation.

Finally, we have examined the use of Salicornia bigelovii oil as a diesel fuel (Glenn, Watson and O'Leary, 1982). The refined oil is similar to other vegetable oils in fuel characteristics. Like other vegetable oils, refined halophyte oil can be used in a blend with diesel fuel to power vehicles. Esterification of other vegetable oils enhances their value as diesel fuels (Caruana, 1992) and should be effective with Salicornia oil as well.

7. Conclusions

Any strategy to sequester C in biomass for the purpose of mitigating the effects of fossil fuel burning must take into account the vast new areas of land or sea that must be made productive, and the many ecological, social and economic effects the measures will have aside from the intended effect of reducing C buildup. Previous studies of the action steps that can be taken in the face of an uncertain global warming have concluded that those mitigation measures which would be desirable even in the absence of global warming, should be taken now. Such steps include preserving old growth rain forests to protect biodiversity and practicing conservation to avoid depleting petroleum reserves too rapidly. On the other hand, planting of vast new areas of trees or fertilizing the seas to encourage plankton blooms, strictly for the purpose of storing C, have not been recommended as immediate action steps, but as actions that could be taken at some point in the future if the need materializes.

In our view, the development of halophyte cultivation falls into the first category of action step. There is already a need for additional cropland to meet population demands. Most of the additional cropland is now coming from the world's forests and forest fallows, releasing additional C into the air through land clearing. Forests are presently being cleared at a rate of about 10×10^6 ha per yr (Lanly, 1982). We propose to add to the inventory of arable lands a significant new resource, not presently used, through the use of seawater or other saline sources of water to irrigate halophytes. Such lands and crops can be part of an immediate strategy to sequester C as well as providing additional space to grow food and fodder for the future population. The research now in progress at several institutions (Glenn et al., 1992b) will help clarify the exact roles that halophytes will play in the biomass strategy for C assimilation and storage.

Acknowledgments

This work is supported by the Electric Power Research Institute and the Salt River Project under RP8011-3. This work was inspired and encouraged by the late Walter Orr Roberts of the National Center for Atmospheric Research.

References

Barrett-Lennard, E. G., Malcolm, C. V. Stern, W. R., Wilkins, S. M. (eds.): 1986, Forage and Fuel Production from Salt Affected Wasteland, Plenum Press, New York, 459 pp.

Caruana, C. M.: 1992, "Pollution control drives new interest in biodiesel", Chemical Engineering Progress (February): 14-18.

Fukuda, H.: 1976, Irrigation in the World: Comparative Developments, University of Tokyo Press, Tokyo, 329 pp.

Glenn, E. P.: 1987, "Relationship between cation accumulation and water content of salt-tolerant grasses and a sedge", Plant, Cell & Environment 10: 205-212.

Glenn, E. P., Coates, W. E., Riley, J. J., Kuehl, R. O., and Swingle, R. S.: 1992a (submitted), "Salicornia bigelovii Torr.: a seawater-irrigated forage for goats", Animal Feed Science and Technology.

Glenn, E. P., Hodges, C. N., Lieth, H., Pielke, R., and Pitelka, L.: 1992b (in press), "Halophytes to remove carbon from the atmosphere", Environment.

Glenn, E. P., Kent, K. J., Thompson, T. L. and Frye, R. J.: 1991a, Seaweeds and Halophytes to Remove Carbon from the Atmosphere, Electric Power Research Institute ER/EN-7177, Palo Alto, California, various pagings.

Glenn, E. P. and O'Leary, J. W.: 1984, "Relationship between salt accumulation and water content of dicotyledenous halophytes", Plant, Cell & Environment 3: 309-318.

Glenn, E. P. and O'Leary, J. W.: 1985, "Productivity and irrigation requirements of halophytes irrigated with hypersaline seawater in a coastal desert environment", J. Arid Environments 9: 81-91.

Glenn, E. P., O'Leary, J. W., Watson, M. C., Thompson, T. L., and Kuehl, R. O.: 1991b, "Salicornia bigelovii Torr.: an oilseed halophyte for seawater irrigation", Science 251: 1065-1067.

Glenn, E. P., Watson, M. C., and O'Leary, J. W.: 1982, Evaluation of Oil-Producing Halophytes as Fuel Crops, ASERC 302-81 Final Report, State of Arizona, Phoenix, 22 pp.

Hammond, A. L. and Paden, M. E.: 1990, World Resources 1990-91, Oxford University Press, New York, pp. 287-290.

Harmon, M. E., Ferrell, W. K., and Franklin, J. F.: 1990, "Effects on carbon storage of conversion of old-growth forests to young forests", Science 247: 699-702.

Intergovernmental Panel on Climate Change (IPPC): 1990, Climate Change: the IPPC Scientific Assessment, Cambridge University Press, Cambridge, 363 pp.

Lanly, J. P.: 1982, Tropical Forestry Resources, FAO Paper no. 30, FAO, Rome, 116 pp.

Lynd, L. R., Cushman, J. H., Nichols, R. J., and Wyman, C. E., 1991: "Fuel ethanol from cellulosic biomass", Science 251: 1318-1323.

Marland, G.: 1988, The Prospect of Solving the CO_2 Problem through Global Reforestation, US. D.O.E. Report NBB-0082, U. S. Dept. of Energy, 66 pp.

Martin, J. H., Gordon, R. M., and Fitzwater, S. E.: 1990: "Iron in Antarctic waters", Nature 345: 156-158.

Moore, J. A., Stull, B. E., and Swingle, R. S.: 1989, Evaluation of Salicornia Forage and Seed Meal as Potential Feedstuffs for Ruminants, ERL 89-2, Environmental Research Laboratory, Tucson, Arizona.

Moulton, R. J., and Richards, K. R.: 1990: Costs of Sequestering Carbon through Tree Planting and Forest Management in the United States, U.S.D.A. Forest Service General Technical Paper WO-58, 47 pp.

O'Leary, J. W.: 1986, "A critical analysis of the use of Atriplex species as crop plants for irrigation with highly saline water", in R. Ahmad and A. San Pietro (eds.), Prospects for Biosaline Research, Dept. of Botany, University of Karachi, Pakistan, pp. 415-432.

Peng, T. H., and Broecker, W. S.: 1991, "Dynamical limitations on the Antarctic iron fertilization strategy", Nature 349: 227-229.

Sanchez, P. A., Bundy, D. E., Villachica, J. H., and Nicholaides, J. J.: 1982, "Amazon Basin soils: management for continuous crop production", Science 216: 821-827.

Sedjo, R. A., 1989: "Forests: a tool to moderate global warming?", Environment 31: 14-21.

Vitousek, P., 1990: Reforestation to Offset Carbon Emissions, Electric Power Research Institute EN-6910, Palo Alto, California, various pagings.

CARBON SINKS IN MANGROVES AND THEIR IMPLICATIONS TO CARBON BUDGET OF TROPICAL COASTAL ECOSYSTEMS.

R. R. TWILLEY
R. H. CHEN
T. HARGIS
Department of Biology
University of Southwestern Louisiana
PO Box 42451
Lafayette, LA, USA

ABSTRACT. Nearly 50% of terrigenous materials delivered to the world's oceans are delivered through just twenty-one major river systems. These river-dominated coastal margins (including estuarine and shelf ecosystems) are thus important both to the regional enhancement of productivity and to the global flux of C that is observed in land-margin ecosystems. The tropical regions of the biosphere are the most biogeochemically active coastal regions and represent potentially important sinks of C in the biosphere. Rates of net primary productivity and biomass accumulation depend on a combination of global factors such as latitude and local factors such as hydrology. The global storage of C in mangrove biomass is estimated at 4.03 Pg C; and 70% of this C occurs in coastal margins from 0° to 10° latitude. The average rate of wood production is 12.08 Mg ha^{-1} yr^{-1}, which is equivalent to a global estimate of 0.16 Pg C/yr stored in mangrove biomass. Together with carbon accumulation in mangrove sediments (0.02 Pg C/yr), the net ecosystem production in mangroves is about 0.18 Pg C/yr. Global estimates of export from coastal wetlands is about 0.08 Pg C/yr compared to input of 0.36 Pg C/yr from rivers to coastal ecosystems. Total allochthonous input of 0.44 Pg C/yr is lower than in situ production of 6.65 Pg C/yr. The trophic condition of coastal ecosystems depends on the fate of this total supply of 7.09 Pg C/yr as either contributing to system respiration, or becoming permanently stored in sediments. Accumulation of carbon in coastal sediments is only 0.41 Pg C/yr; about 6% of the total input. The NEP of coastal wetlands also contribute to the C sink of coastal margins, but the source of this C is part of the terrestrial C exchange with the atmosphere. Accumulation of C in wood and sediments of coastal wetlands is 0.205 Pg C/yr, half the estimate for sequestering of C in coastal sediments. Burial of C in shelf sediments is probably underestimated, particularly in tropical river-dominated coastal margins. Better estimates of these two C sinks in the tropics, coastal wetlands and shelf sediments, is needed to better understand the contribution of coastal ecosystems to the global carbon budget.

1. Introduction

1.1 COASTAL MARGIN ECOSYSTEMS

The global flux of C is usually partitioned between terrestrial and oceanic compartments, assuming that terrigenous materials contribute to the budget of C in the oceans. However, there is a shallow transition region between terrestrial and deeper oceanic environments that represent less than 5 % of the earths surface, but is one of the most geochemically and biologically active areas of the biosphere (Figure 1). This coastal region includes

Water, Air, and Soil Pollution **64**: 265–288, 1992.
© 1992 *Kluwer Academic Publishers.*

ecosystems from the upper intertidal regions (including tidal freshwater ecosystems) to the outer edges of the continental shelves. These areas are sites where terrigenous materials are introduced to the ocean, where productive wetlands have developed, and biogenic zones in nearshore waters are the most productive regions of the oceans. The elevated phytoplankton production of neritic waters are due in part to the fertile input of rivers and coastal boundary currents that promote primary productivity. Nearly 50% of terrigenous materials transported to the world's oceans are delivered through just twenty-one major river systems. These river-dominated coastal margins are thus important both to the regional enhancement of productivity and to the global flux of C that is observed in coastal ecosystems. River-dominated coastal margins may be key sites where geophysical and biogeochemical processes represent C sinks in the biosphere.

SINKS OF CARBON IN COASTAL ECOSYSTEMS

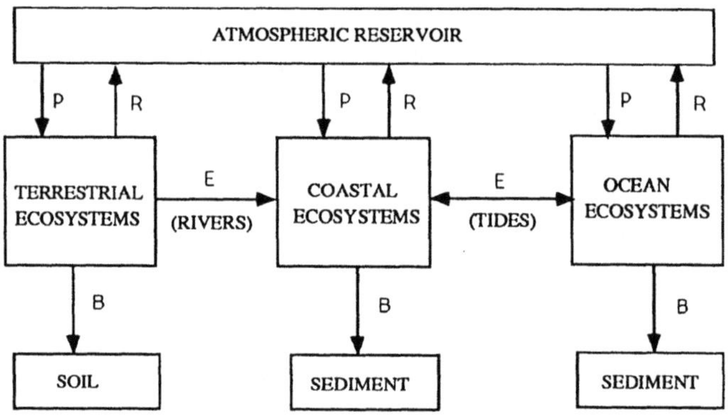

Figure 1. Conceptual diagram of the three major compartments of the biosphere that influence the global cycle of C. P = production, R = respiration, E = exchange, B = burial.

This paper will focus on the factors that control the fate of C in coastal ecosystems of the tropics. Tropical coastal ecosystems are dominated by forested wetlands in the intertidal zone known as mangroves, shallow submersed coral reefs and seagrasses (macrophyte systems), and phytoplankton communities in the neritic waters of the tropical coastal oceans. The ecological processes of these three types of subsystems, together with the fate of riverine input, determine the role of tropical coastal ecosystems as C sinks in the biosphere. The fate of C as a sink in each of these subsystems can be evaluated using a mass balance approach to ecosystem dynamics (Hopkinson, 1988; Twilley, 1988). Net ecosystem production (NEP) of each of the three subsystems is a measure of the mass balance of in situ processes such that:

$$NEP = GPP - (Ra + Rh) - E \qquad (1)$$

where GPP is gross primary productivity, Ra is autotrophic respiration, Rh is heterotrophic respiration, and E is net export from the system. Ecosystem steady state (ESS) dynamics of C flux can be evaluated by a mass balance of gross C fixation (P) and allochthonous C inputs (I) against losses associated with total system respiration (R) and export (E) as follows:

$$ESS = (P + I) - (R + E) \quad (2)$$

If C gains (P +I) are greater than losses (R + E) then the system is non steady state and there is net accumulation and a C sink. If we consider river input to the coastal system as a whole, then ESS for each of the subsystems is equal to NEP. NEP may occur as increase in biomass annually or the accumulation of C in the sediment reservoirs of each respective subsystem. In phytoplankton ecosystems, there is little NEP (or permanent storage), most of the NPP is either respired or deposited to the seabed. However there are estimates that macrophyte ecosystems (including seagrasses and coral reefs) may be an important C sink in the tropics (Smith, 1981).

Mangroves may function as a sink of C in coastal margin ecosystems depending on the balance of production and decomposition within the forest, and net flux of C across the wetland boundary. Many mangrove ecosystems accumulate peat and this storage of C has not been evaluated on a global scale. These forested wetlands are also different from other subsystems of the coastal zone because of the potential for accumulation of C in wood that may be an important C sink. In addition, the export of organic C from mangroves may contribute to sedimentation or storage of C in coastal sediments. This review will synthesize existing information on biomass, wood production, sedimentation, and export of mangroves to make some preliminary estimates of the global scale of these ecological processes. These fluxes of C will be compared to the transport of terrigenous C to tropical waters as comparison to fate of C in these tropical coastal ecosystems. Particular attention will be focussed on the geographical distribution of these processes in the tropics, which should improve our estimates of these C fluxes. Thus we will start with a description of the forcing functions that influence the function of mangrove ecosystems.

1.2. GEOGRAPHICAL PROPERTIES

1.2.1 *Regional classification*: Thom (1982) proposed that the combination of geophysical energies with the geomorphology of the coastal zone is important to establishing the ecological characteristics of mangroves (Twilley, in press). According to Thom, the landform characteristics of a coastal region together with environmental processes control the basic patterns in the structure of coastal forests. He identified five basic types or classes of environmental settings where mangroves occur based on the relative influence of rainfall, river discharge, tidal amplitude, turbidity, and wave power. These geophysical energies are the dominate forcing functions of mangroves and collectively represent the energy signature of mangroves (Twilley, in press). The regional scale description of coastal environments using geomorphology and geophysical processes can be further separated into ecological classification systems. The ecological classification of mangroves as either fringe, basin, or dwarf forests describe the microtopographic effects of hydrology on the formation of forest types (Lugo and Snedaker, 1974). Within the regional boundaries of an environmental setting there may exist all three ecological types of mangrove forest depending on the local effects of tides, waves, and river flow.

The combination of geophysical energies, together with the regional geomorphology, explain the function of mangrove ecosystems. Coastal geomorphology and hydrology have been used to classify the physiognomy and zonation of mangrove forests (Lugo and Snedaker, 1974; Watson, 1928). Recently, the ecological processes or function of wetland ecosystems have also been related to coastal processes such as tides, rivers, and waves (Gosselink and Turner, 1978; Twilley, 1988). Thus the productivity, sedimentation, and coupling of these ecological processes with coastal waters may be specific according to the geomorphology and geophysical characteristics of coastal ecosystems (Twilley, 1988). The geomorphological type of coastal environments can constrain the function, as well as structure, of mangrove ecosystems. Linking ecological function with specific types of environmental conditions of coastal ecosystems should develop greater generality in understanding the role of mangroves as C sinks.

1.2.3 *Global distribution of mangroves*: Information on the distribution of mangroves was gathered from World Resources 1986, a report by the World Resources Institute. This report claims there are 24.00 x 10^6 ha of mangroves in the tropics that dominate the river-dominated delta, lagoon and estuarine coastal environments. Other estimates of mangrove area range from 15.47 to 30 x 10^6 ha, with an average of 21.8 x 10^6 ha (Lugo et al. 1990, Table 4.1). The amount of mangrove area for each country was categorized into four latitudinal belts (0° to 10°, 10° to 20°, 20° to 30°, and 30° to 40°) combining north and south latitudes within each zone. The largest area of mangroves occurs in the 0° to 10° zone with 10.07 x 10^6 ha compared to only 0.25 x 10^6 ha in the 30° to 40° latitudes (Figure 2). The sum of mangrove area based on statistics for each country was short of the estimated total mangrove area of 24.00 x 10^6 ha by 5.81 x 10^6 ha. We distributed this mangrove area among the four latitudinal zones using the relative distribution of the areas given for each country. These distributions on mangrove area were used to develop global estimates of ecological processes within mangrove ecosystems.

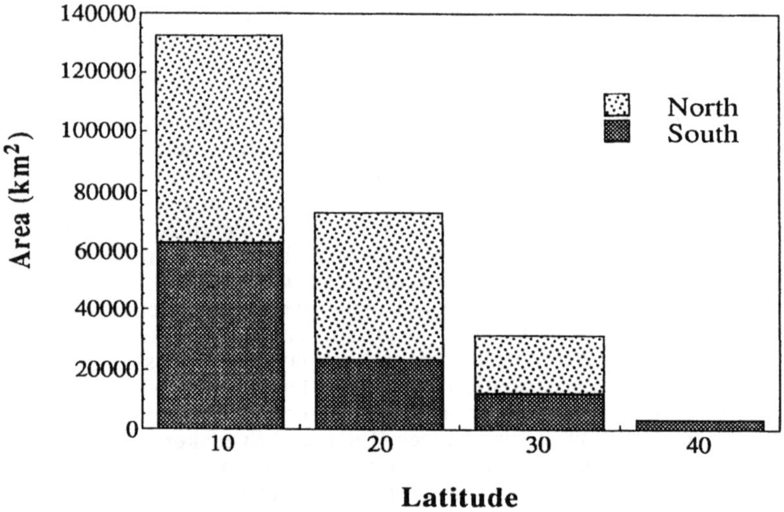

Figure 2. Global distribution of mangroves with latitude.

2. Carbon Sinks in Mangroves

2.1 BIOMASS

Estimates of mangrove biomass are usually restricted to aboveground structures and in many cases, only include the timber that can be harvested from the forest. The distribution of biomass throughout the tropics indicate that higher values occur at lower latitudes (Figure 3). The model for biomass distribution with latitude in Figure 3 is Y = -7.291(X) + 298.5 (R = 0.75), where X is latitude and Y is biomass (Mg/ha). Cintrón and Schaeffer-Novelli (1984) found that mangrove tree height, which is a good indicator of forest biomass, increased at lower latitudes in the neotropics according to the model: Y = 45.8 - 1.28(X) ($r^2 = 0.75$); where Y is tree height (m) and X is latitude north or south. The sensitive response of mangrove biomass and tree height to relatively small changes in latitude is related to the influence of temperature and occurrence of frost on the structure of these ecosystems (Lugo and Patterson-Zucca, 1977). Thus, solar energy represents a major constraint on the distribution and maximum biomass of mangrove ecosystems.

There is much variation of mangrove biomass at any one latitude, particularly in the warmer tropics. Maximum potential biomass at 10° and 35° is about 400 and 100 Mg/ha, respectively, while values less than 50 Mg/ha may occur at either latitude (Figure 3). Thus while the upper limits of mangrove biomass may occur at lower latitudes, there are local effects that may limit the potential for forest development at all latitudes. These local effects include topography and hydrology, including the effects of river and tides on soil characteristics. Cintrón et al. (1978) found that with increasing salinity the values of a number of structural and functional parameters decreased. These included litterfall, tree density, basal area (total cross-sectional area of trunks), and tree height. For example, tree height (Y, m) of mangroves in Puerto Rico is inversely related to soil salinity by the equation Y= -0.20(X) + 16.58 ($r^2 = 0.72$), where X is soil salinity in ppt (Cintrón et al. 1978). Very few mangroves survive above a soil salinity of about 70 - 80 ppt.

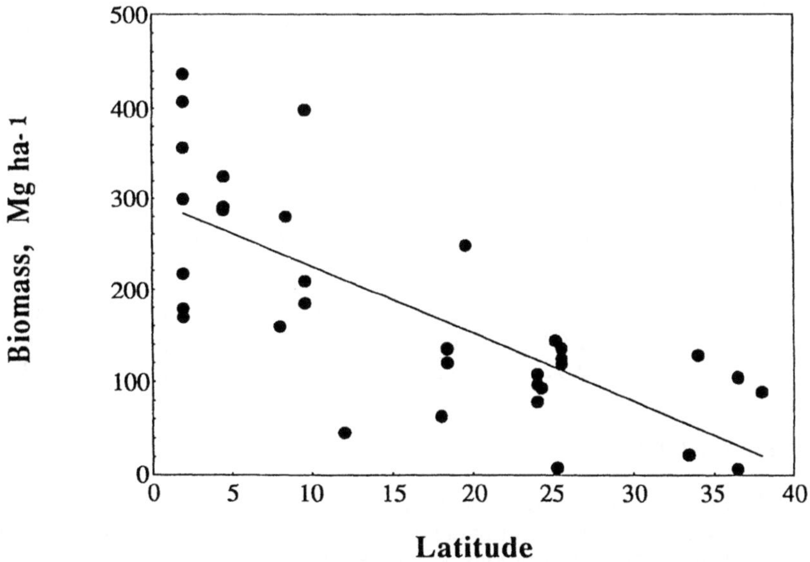

Figure 3. Distribution of aboveground biomass of mangroves with latitude.

Mean values of aboveground and belowground biomass for each of the latitudinal regions together with mangrove area were used to estimate mangrove biomass (Table 1). The average aboveground biomass ranged from 283.6 Mg/ha near the equator to 104.2 Mg/ha in the 30° to 40° zone. Belowground biomass was less and ranged from 171 Mg/ha near the equator to 96 Mg/ha at 30° to 40° latitude. There were fewer estimates of belowground biomass. The average estimate of aboveground (AG) and belowground (BG) biomass for all sites was 178.2 and 146.3 Mg/ha for a BG:AG ratio of about 0.82. Together with a decrease in mangrove area away from the central tropics, 70% of the total mangrove biomass occurs in the 0° to 10° zone (Table 1). Using a conversion of 45% C (per dry mass), the C storage in mangrove biomass at the equator was 2.72 Pg C compared to about 1 Pg C at the 10° to 20° zone. Total mangrove biomass was estimated at 4.03 Pg C.

Table 1. Global estimate of carbon stored in mangrove biomass based on 0.45 gC/g dry mass.

Component	Latitude				
	0° -10°	10° - 20°	20° - 30°	30° - 40°	Total
Area of Mangrove ($x 10^6$ ha)	13.28	7.25	3.14	0.33	24.00
Biomass (Mg/ha)					
Aboveground	283.6	141.6	120.6	104.2	178.2
(Standard deviation)	(90.5)	(77.8)	(16.5)	(64.2)	(112.2)
Belowground	171.2	171.8	69.2	95.5	146.3
(Standard deviation)	(123.6)			(76.3)	(110.2)
Biomass (Pg dry mass)					
Aboveground	3.78	1.03	0.38	0.03	4.98
Belowground	2.27	1.25	0.22	0.03	3.71
Total	6.04	2.27	0.60	0.06	8.69
Biomass (Pg C)					
Aboveground	1.70	0.46	0.17	0.01	2.34
Belowground	1.02	0.56	0.10	0.01	1.69
Total	2.72	1.02	0.27	0.02	4.03

2.2 WOOD PRODUCTION

A variety of factors influences the productivity of coastal wetlands. Most factors are associated with changes in the physical or chemical environment, including solar radiation, temperature, tides, nutrient concentrations, soil type, drainage, oxygen concentration, and pH. The individual plant species present in the intertidal zone can also affect patterns of productivity, because some plants have growth rates that are intrinsically higher than others. Some of these factors cause a difference in productivity over a latitudinal range within wetlands, while others operate at the local level.

Turnover rate of mangrove biomass ranged from 0.041 to 0.126 (Table 2). There was no apparent relationship of turnover rate with latitude. Shifts in ration of litterfall to biomass indicates less allocation of net productivity to wood production in higher latitudes (Warner 1990). Average wood production based on eleven estimates is 12.08 Mg ha^{-1} yr^{-1} (Table 2). Wood production decreased with latitude and a curve was fit to the data ($Y = -0.449 X + 19.88$, $R = 0.60$) . This curve was used to apply a weighted rate of wood production for each of the four latitudinal zones. Most of the wood production occurs in the 0° to 10° zone at about 0.1 Pg C/yr (Figure 4). Total wood production for mangroves is estimated at 0.16 Pg C/yr.

Table 2. Wood production (growth) of mangroves relative to biomass and latitude and the turnover rate of mangrove biomass at each location.

Site	Latitude	Biomass (B) (Mg/ha)	Growth (G) (Mg ha^{-1} yr^{-1})	G:B (yr^{-1})	Reference
Malaysia	4° 50' N	286.8	11.8	0.041	Ong et al. 1979
Malaysia		257.4	24.1	0.094	Ong et al. 1979
Phuket, Thailand	8° N	159	20	0.126	Christensen 1978
Puerto Rico	18° N	62.8	3.07	0.049	Golley et al. 1962
Estero Pargo, Mexico	18° 35' N	120	7.72	0.064	Day et al. 1987
Boca Chica, Mexico	18° 35' N	135	12.06	0.089	Day et al. 1987
Florida, USA			7.31		Sell 1977
Florida, USA			13.33		Sell 1977
Hainan, China	19° 53' N	248.5	11.5	0.046	Lin et al. 1990
Fujian, China	24° 24' N	93.4	8.69	0.093	Lin et al. 1985
Hong Kong	23° N	129.1	13.3	0.103	Lee 1990

2.3 BURIAL

Trapping of particulate materials by wetland plants is an important sedimentary and biogeochemical process in accretionary wetlands (Frey and Basan, 1978). Sedimentation rates in wetland sediments have previously been measured by various methods, including the use of natural and man-made radionuclides, such as ^{137}Cs and ^{210}Pb (Armentano and Woodwell, 1975; Hatton et al., 1983; Sharma et al. ,1987). Lynch et al. 1989 used these techniques to estimate the accretion rates in mangrove forests in southwest Florida and Terminos Lagoon, Mexico. Recent accretion rates based on these techniques ranged from 1.6 to 2.4 mm/yr (Table 3). Higher rates were observed for riverine mangroves in Mexico compared to basin mangroves in Rookery Bay, Florida. Estimates of accretion in mangroves based on aging deep peat cores (6,000 yr time scales) range from 0.3 to 1.88 mm/yr (Table 3). These rates are nearly all for reef mangroves, with little terrigenous inputs. It is surprising that the two techniques give similar estimates for the accretion of material in mangrove sediments. It is possible to suggest from this pattern, that accretion in reef mangroves is less than 1 mm/yr, basin mangroves 1 to 2 mm/yr, and riverine mangroves greater than 2 mm/yr.

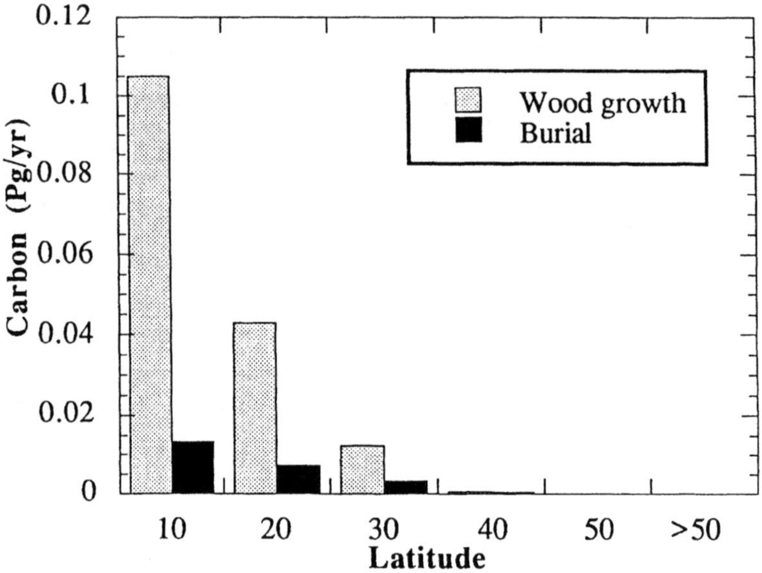

Figure 4. Estimates of annual carbon storage in mangrove wood and sediments.

Sediments suspended in the water column are deposited in mangroves during flooding and this material enriches mangrove soils. The extensive root system of mangroves enhances this trapping process and retards the forces of erosion along the shoreline (Scoffin, 1970). Although this function has been overstated to the extent of calling mangroves "walking trees", roots do contribute to sedimentation in estuaries (Lynch et al., 1989). The accumulation of organic matter in the five sites investigated in Florida and Mexico by Lynch et al. (1989) ranged from 130 to 409 g m^{-2} yr^{-1} (Figure 5). Levels of organic matter accumulation were not correlated with amount of litter production. The higher amount of organic matter accumulation relative to litterfall occurred in basin forests, possibly due to less export at these sites (Twilley et al., 1986). Another source of organic matter is belowground productivity, which has seldom been included in C budgets of mangrove ecosystems. The contribution of inorganic material varied greatly among the five sites and ranged from 133 to 1404 g m^{-2} yr^{-1} (Figure 5). The higher values occurred in the riverine mangrove forest in Terminos Lagoon in Mexico where the Candelaria, Chumpan, and Palizada rivers discharge more than 190 m^3/s of freshwater. The ratio of inorganic to organic material in the sedimentation process was 0.70 and 1.9 for the basin forests in Estero Pargo and Rookery Bay, compared to 4.5 for the riverine forest in Boca Chica. These ratios indicate that the sequestering of C among mangroves in different environmental settings is not as different as the total sedimentation rate. This indicates that the ecological processes associated with C accumulation in mangrove sediments is about 100 gC m^{-2} yr^{-1} among different types of mangrove ecosystems (Figure 5).

Table 3. Critical rates of mangrove peat accretion based on deep cores and ^{14}C ages for low and high islands (from Ellison and Stoddart 1991). Rates for lagoon mangroves from Lynch et al. 1989.

Location	Rate (mm/yr)	Source
	Low Islands	
Taongatapu	0.77	Ellison 1989
Grand Cayman	0.88-0.90	Woodroffe 1981
	High Islands	
Fiji	1.17	Southern 1986
	0.76	Matsushima et al. 1984
	1.31	Latham 1979
Caroline Islands	1.34 - 1.40	Ward 1988
	1.37	Matsumoto et al. 1986
	0.30	Bloom 1970
Samoa	0.99 - 1.05	Matsushima et al. 1984
	1.88	Bloom 1980
South Florida	0.46	Scholl et al. 1969
	Lagoons	
Rookery Bay	1.6	Lynch et al. 1989
Terminos Lagoon, Mexico	2.4	Lynch et al. 1989

Based on an average C burial rate of 100 gC m^{-2} yr^{-1} for all four latitudinal zones, the global estimate of C accumulation in mangrove sediments was much less than in wood (Figure 4). Total C sequestered in mangrove peat is about 0.02 Pg/yr. Thus the global NEP of mangroves based on wood production and burial is estimated at about 0.18 Pg C/yr.

2.4 LITTER DYNAMICS

Litter produced in the canopy of mangrove forests represents a major source of organic matter and nutrients for outwelling to adjacent coastal waters (Odum and Heald, 1972; Twilley, 1988). Thus the dynamics of mangrove litter including productivity, decomposition, and export influence the coupling of mangroves to coastal ecosystems (Twilley, 1988). Litter productivity values for mangrove forests worldwide range from about 2 to 16 Mg ha^{-1} yr^{-1} and also decrease with latitude (Figure 6). Based on the mangrove sites represented in Figure 6, the maximum level of litterfall is about 14 Mg ha^{-1} yr^{-1} from 0° to 20° latitude. Above this latitude, litterfall decreases to less than 10 Mg ha^{-1} yr^{-1}. The lower limit of productivity is 8 Mg ha^{-1} yr^{-1} in the lower latitudes, compared to less than 2 Mg ha^{-1} yr^{-1} in the subtropical zones. This trend indicates that litter productivity is less sensitive to changes in latitude than observed for biomass.

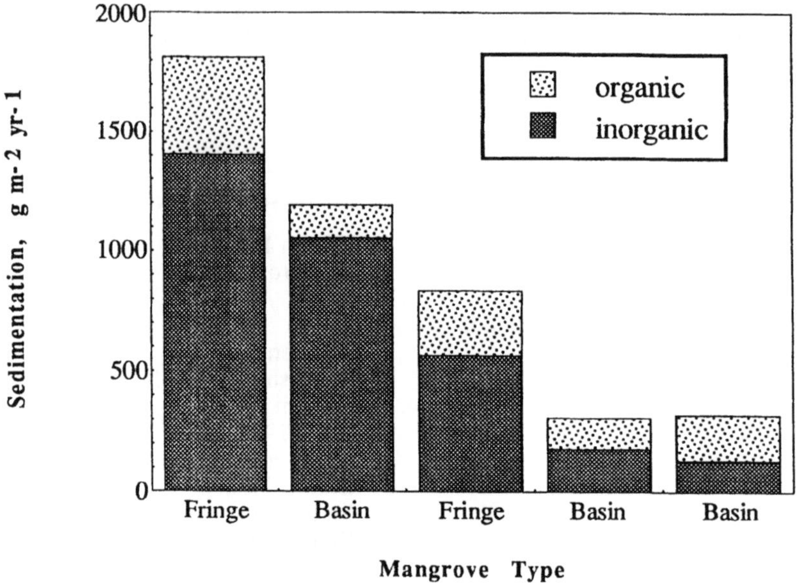

Figure 5. Estimates of sedimentation (organic and inorganic) in mangrove sites in southwest Florida and Mexico (from Lynch 1989). Sites are arranged in order of river discharge.

Export of leaf litter and detritus from mangroves is linked to the hydrology of mangrove forests. Rates of organic C export from basin mangroves are dependent on the volume of tidal water inundating the forest each month, and accordingly export rates are seasonal in response to the seasonal fluctuation in sea level. Rainfall events may also increase organic C export from mangroves (Twilley, 1985), especially dissolved organic C (DOC). Total organic C (TOC) export from infrequently flooded basin mangroves in southwest Florida is 64 gC m^{-2} yr^{-1} , and nearly 75% of this material is DOC (Twilley, 1985). Particulate detritus export from fringe mangroves in south Florida was estimated at 186 gC m^{-2} yr^{-1} (Heald, 1971), compared to 420 gC m^{-2} yr^{-1} for a riverine mangrove forest in Australia (Boto and Bunt, 1981). Estimates of average tidal amplitude in these three forests types are 0.08 m, 0.5 m and 3 m, respectively. Accordingly, as tidal amplitude increased, the magnitude of organic material exchanged at the boundary of the forests increased (Twilley, 1985). Trends for litter productivity and export suggest that as geophysical energy increase, the exchange of organic matter between mangroves and adjacent estuarine waters also increase.

Litter productivity in a riverine forest in Ecuador is similar to a riverine forest in south Florida at about 10 Mg ha^{-1} yr^{-1}. However, the riverine forest in Ecuador has a 3 m tidal amplitude, while the tides in the riverine forests in south Florida are 0.5 m. Leaf litter on the forest floor in Ecuador is absent except for 3 mo of the year (Twilley et al., 1990).

This may be expected to be associated with greater export owing to the effect of tides on the transport of leaf litter from the forest . Yet observations in the mangroves in Ecuador suggest that most of the leaf litter on the forest floor is harvested by the mangrove crab, *Ucides occidentalis*, and transported to sediment burrows (Twilley et al., 1990). During September and October, when the crab aestivates, the standing crop of leaf litter increases on the forest floor. The levels of leaf litter during these 2 mo are still much lower than expected based on daily rates of leaf fall suggesting that leaf export is significant. The influence of mangrove crabs on litter dynamics has been described in other mangrove ecosystems with high geophysical energies and rates of litter turnover above 5 yr^{-1} (Malley, 1978; Leh and Sasekumar, 1985; Robertson and Daniel, 1989). Thus, the use of geophysical forcing functions such as tides to predict export of leaf litter is limited by consideration of ecological factors within the ecosystem such as crabs. In these examples, high rates of litter turnover do not reflect the coupling of mangrove to coastal waters, but the conservation of organic matter within the forest.

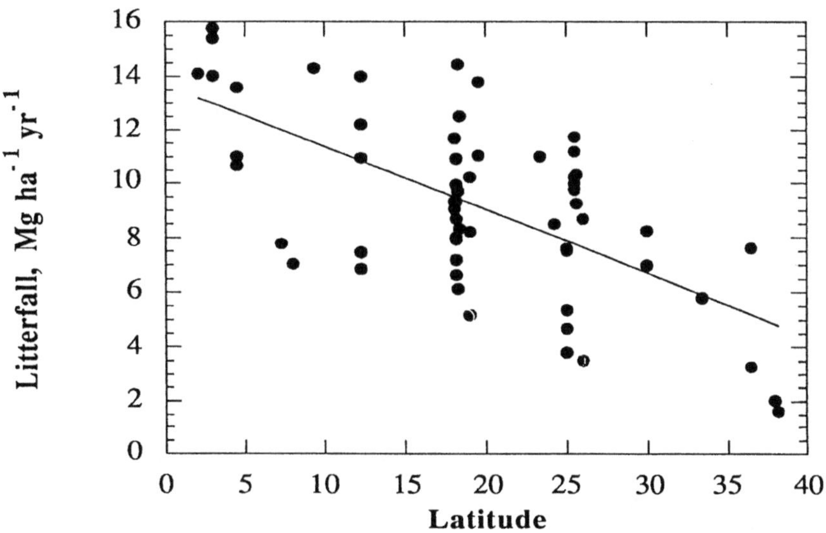

Figure 6. Survey of litterfall of mangrove forests with latitude.

There are now ten estimates of C flux from mangrove ecosystems that range from 1.86 to 401 gC m^{-2} yr^{-1} (Table 4). An average export of C for mangrove ecosystems is about 200 gC m^{-2} yr^{-1}. Nixon (1980) estimated that the average C net export from salt marshes is about 100 gC m^{-2} yr^{-1}. There have been several suggestions to explain why the transport of organic C from mangroves is greater than for coastal marshes, including the buoyancy of mangrove leaf litter and the hydrology of these tropical coastal wetlands (Twilley, 1988). This average export value can be applied to the areal distribution of mangroves to estimate global contribution of mangroves to tropical coastal waters (Figure 7).

Table 4. Export of organic carbon from mangrove forests (gC m^{-2} yr^{-1}).

Site	Export	Reference
Florida, USA	91.2	Lugo and Snedaker 1974
Florida, USA	292	Odum and Heald 1972
Florida, USA	186	Heald 1969
Florida, USA	64	Twilley 1985
Puerto Rico	401	Golley et al. 1962
Hichinbrook, Australia	340	Robertson et al. 1986
Hichinbrook, Australia	420	Boto and Bunt 1981
Matang, Malaysia	193.5	Gong et al. 1990
New Zealand	109.5	Woodroffe 1985
Hong Kong	1.9	Lee 1989
Average	210	

The export of organic C from mangroves ranges from 10 to 50% of litterfall (Twilley 1985). Using an average of 25% and the latitudinal distribution of litterfall in Figure 6, a second estimate of C export from mangrove ecosystems is given in Figure 7. This value of C export is consistently lower than the areal average of 200 gC m^{-2} yr^{-1}. Using the higher estimate, the contribution of organic C from mangroves to coastal waters is 0.05 Pg C/yr.

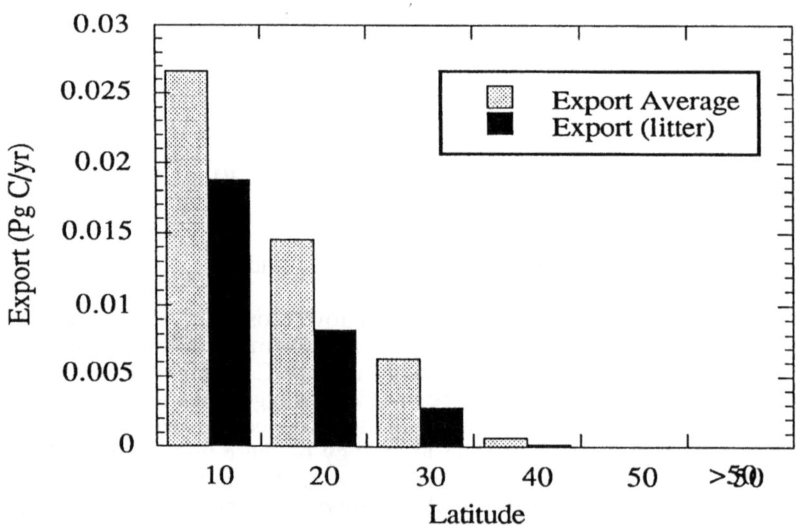

Figure 7. Global estimates of carbon export from mangrove ecosystems with latitude.

3. Carbon Budget Of Coastal Ecosystems

Rivers discharge about 13.5 x 10^9 tons of particulate material annually to the coastal oceans, over half of which is supplied by the world's 21 largest rivers (Milliman and Meade, 1983). Besides their POC load, rivers carry an approximately equal load of DOC - about 200 x 10^{12} g per year. In addition, the nutrient load of rivers may enhance the primary productivity of neritic waters, and some of this C may accumulate in the sediments of continental margins. Present estimates suggest that there is much more organic matter coming down the rivers than can be accounted for in annual accumulation of sediment C in the ocean. And few estimates have synoptically accounted for the other sources of in situ C, such as mangrove, seagrasses, and coral reefs, that may accumulate in coastal ecosystems. Where does all this C go? Or have we grossly underestimated the global C sinks in these ecosystems? The following section is a summary of some mass balances of coastal ecosystems, with emphasis of the tropical regions of the biosphere.

3.1 CARBON SOURCES IN COASTAL ECOSYSTEMS

The drainage area and discharge of the major rivers of the world in Figure 8 are distributed along ten degree latitudinal zones, based on the geographic location of the river mouth. Most of the freshwater input to coastal ecosystems occur in the tropical latitudes; in fact seven of the top ten rivers in discharge empty in the 0° to 30° latitudes. The greater proportion of discharge from watersheds in this region is related to more abundant rainfall. This analysis is important in the consideration of how C export from rivers to coastal margins is estimated. Estimates based on sediment discharge using a ratio of 0.45% of the sediment is particulate organic C are low at only 0.1 Pg C/yr (Walsh, 1983). Rates of organic C transport based on an average export of 6.5 gC m^{-2} (of watershed) yr^{-1} are about 0.2 Pg C/yr (Meybeck, 1982). We feel a more accurate estimate of C export is

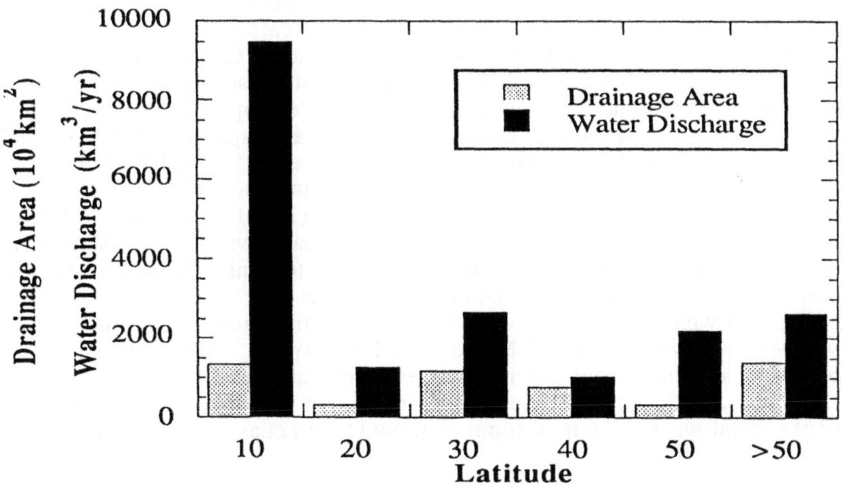

Figure 8. Discharge characteristics of the major river systems in the world with latitude.

based on actual river discharge and an average dissolved plus particulate organic carbon concentration of 20 mg/L (Meybeck, 1982; Milliman and Meade, 1983). This method estimates the delivery of organic C to coastal regions of the world at 0.36 Pg C/yr (Figure 9), similar to the average for recent estimates of river input (Table 5). This value may still underestimate total organic carbon input since the larger particulate pool transported by these rivers is not considered.

Table 5. Estimates of delivery of total organic carbon from the major river systems of the world to coastal ecosystems.

Total Organic Carbon (Pg C/yr)	Reference
0.37	Schlesinger and Melack 1981
0.383	Meybeck 1981
0.302	Meybeck 1988
0.33	Degens et al. 1991
0.35	Average

The contributions of C to coastal ecosystems from terrigenous and mangrove sources are compared in Figure 9. It is interesting to note that nearly 0.26 Pg C/yr of the estimated 0.36 Pg C/yr export from rivers occurs in the 0° to 10° latitude (Figure 9). In the 0° to 10° latitudinal zone, river contribution of TOC is much higher than for mangroves. However, in the 10° to 20° region of the tropics, C input to coastal waters from mangrove is nearly 75% of river contribution.

Global estimates of C sources in coastal ecosystems in Table 6 are identified as either allochthonous or in situ depending on whether the source of CO_2 is from the atmosphere or of ocean origin. River input is terrigenous organic carbon fixed by terrestrial vegetation. TOC export from coastal wetlands also represents C fixed by vegetation directly from the atmosphere, but these are included as part of coastal ecosystems (Figure 10). Coastal wetlands thus should be carefully evaluated in developing carbon budgets of coastal ecosystems since they exchange C directly with the atmosphere and with coastal waters. C export from mangroves is estimated in this study at 0.05 Pg C/yr. Lugo et al. (1990) estimated marsh export at 0.03 Pg C/yr using a mean export of 95 gC m^{-2} yr^{-1}. This is near the average rate of C export suggested by Nixon (1980). Based on this rate of C export for salt marshes, the total contribution of C from coastal wetlands is estimated at 0.08 Pg C/yr. Together with estimate of river input, the total allochthonous input of organic carbon into coastal ecosystems is 0.44 Pg C/yr (Table 6).

Estimates of primary production in coastal environments range from approximately 1.0-7.0 Pg C /yr (Deuser, 1979; Smith and Mackenzie, 1987; Walsh, 1988). Walsh (1984) classified coastal production among macrophyte, estuarine, and shelf systems for total in situ production of 6.65 Pg C/yr (Table 6). Based on these estimates of net primary productivity of coastal waters, the total input of organic C to coastal ecosystems is 7.09 Pg/yr (Table 6).

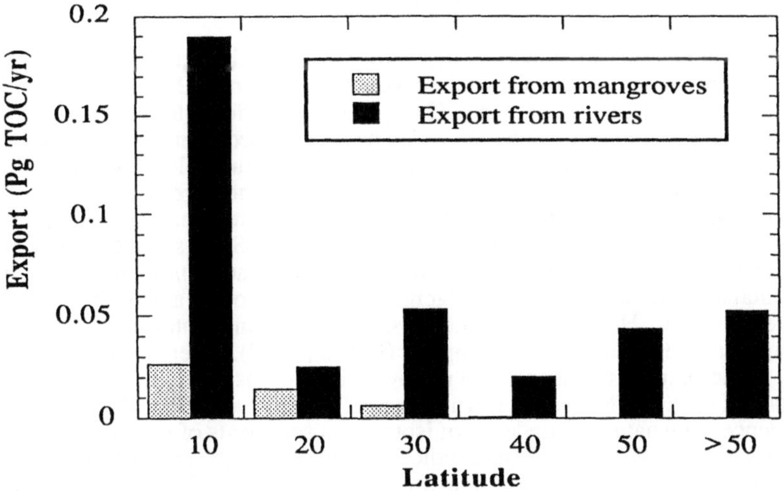

Figure 9. Relative contribution of organic carbon from river transport and mangrove export to tropical continental margins.

Table 6. Estimates of carbon sources and sinks in coastal ecosystems (Pg C/yr)

SOURCES		
Allochthonous	0.44	
Rivers	0.36	This study
Wetlands	0.08	This study
Net Primary Production	6.65	
Shelf Waters	5.40	Walsh 1984
Estuaries/Deltas	0.92	Walsh 1984
Macrophytes	0.33	Walsh 1984
TOTAL	7.09	
SEDIMENT SINKS		
Coral Reefs	0.01	Walsh 1984
Shelf	0.2	Berner 1982
Estuaries	0.20	Walsh 1984
TOTAL	0.41	
WETLAND SINKS		
Mangroves	0.18	This study
Marshes	0.025	Hopkinson 1988
TOTAL	0.205	

3.2 CARBON SINKS IN COASTAL ECOSYSTEMS

Continental margins are important depositories of organic matter for three reasons: (1) the input of terrigenous organic C and the production of marine C result in an abundant supply of organic C to these environments; (2) high sedimentation rates lead to rapid removal of organic material below the oxidized zone near the sediment-water interface causing preferential burial of organic matter (Berner, 1982; Deuser, 1988; Dagg et al. 1991), and (3) coastal wetland sediments and biomass represent large reservoirs of organic C (this study). Most of the 5 to 10 Pg of organic matter ultimately reaching the oceans annually is deposited in estuaries and on continental margins. Shelf sediments may play an important role as a short-term sink for anthropogenic CO_2, accounting for as much as 40% of the total oceanic sink (Smith and MacKenzie, 1987; Wollast and MacKenzie, 1989; Tans et al. 1990). However, estimates of C burial on continental shelves are about 0.2 Pg C /yr; a small fraction of the C supplied (Berner, 1982). There is apparent discrepancy in the estimates of supply and removal of C in coastal environments regarding the magnitude of C export from shelf environments (Walsh et al., 1985; Rowe et al., 1986). Evidence from natural abundance of isotopes in sediments of river-dominated shelf ecosystems such as the Mississippi River indicates that most of the terrigenous C entering coastal margins is deposited in the proximal areas of river plumes (Shultz and Calder, 1976; Gearing et al. 1977; Thayer et al. 1983). In addition, estimates of recent deposition and accumulation of terrigenous materials on the Louisiana shelf in vicinity of the Mississippi River plume demonstrate the strong seasonal nature of these inputs (Dagg et al., 1991). Short term events of extremely high deposition may cause preferential burial of terrigenous C in river-dominated shelf ecosystems. This suggests that in coastal regions dominated by strong rivers, very little of the terrigenous C may be transported to deep water oceanic environments (Figure 10).

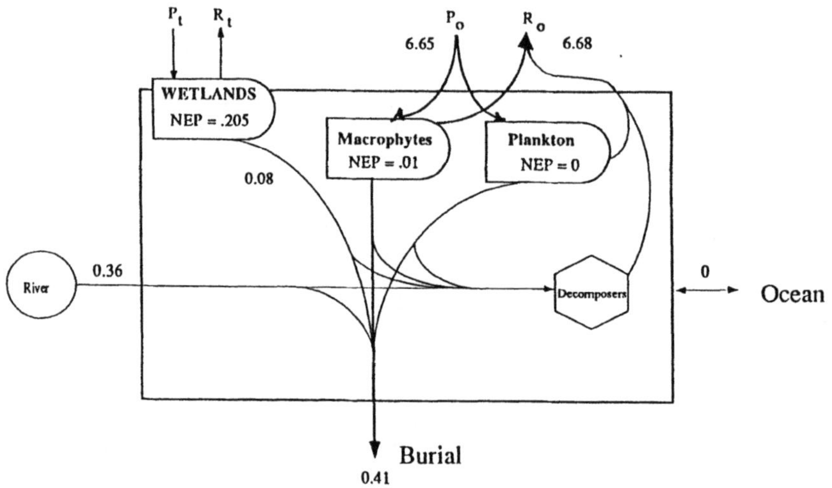

Figure 10. Mass balance of C for coastal ecosystems based on estimates of in situ net production and allochthonous inputs, minus losses associated with burial in coastal sediments. P and R represent net production and heterotrophic respiration, respectively, with exchange of CO_2 directly with atmosphere (t) or coastal waters (o).

In addition, estuaries are considered efficient traps of allochthonous and in situ C with estimates of 0.2 Pg C/yr (Walsh 1984) buried in estuarine sediments (Table 6). Coral reefs may contribute only 0.01 Pg/yr to C sinks in coastal ecosystems. It is unknown what role other macrophyte communities may serve in sequestering C in coastal waters, although biomass and turnover estimates suggest that they may be significant (Smith, 1981). Total sinks in coastal ecosystems is estimated at 0.41 Pg C/yr (Table 6). This is only 6% of the total inputs from allochthonous and in situ sources of C.

Coastal or nearshore environments are considered by some to be net heterotrophic (net respiration exceeds primary production of organic matter), which implies that the high biological activity of coastal ecosystems relies on terrigenous organic C (Smith and MacKenzie, 1987). Smith and Mackenzie (1987) assumed there was no exchange between coastal and oceanic systems and using a mass balance analysis of ecosystem processes (such as equation 2 above), solved for system respiration given inputs and burial. Using the same approach, the respiration of coastal ecosystems in our analysis is about 6.68 Pg C/yr (Figure 10). This gives a P:R ratio (net production to heterotrophic respiration) of about 1.0, symbolic of a steady state system. Smith and Mackenzie assumed that 64% of terrigenous organic carbon is respired and only 1% of the total C input was buried, causing shelf systems to be heterotrophic. Such perspectives of coastal systems rely on accurate estimates of C burial in sediments, which may be underestimated in many shelf environments, particularly in the tropics. Thus the trophic nature of many coastal ecosystems will depend on the relative contribution of respiration and burial to the fate of allochthonous and in situ C sources (Figure 10).

Another sink of C in coastal environments is wetlands. The NEP of mangroves in this study is estimated at 0.18 Pg C/yr, most of which is sequestered in annual increment of wood biomass. Average sedimentation of C in salt marshes along the Atlantic and Gulf coast is 83 gC m^{-2} yr^{-1} (Hopkinson 1988), compared to 100 gC m^{-2} yr^{-1} used in this study for mangrove sedimentation of C. Based on an areal coverage of 0.3 x 10^{12} m^2 (Lugo et al. 1990), the global sink of C in salt marsh sediments is 0.025 Pg/yr. Walsh (1984) estimated C burial in salt marshes at 0.05 Pg/yr. The annual accumulation of C in plant biomass in salt marshes is considered negligible (Hopkinson 1988). Together with mangrove C accumulation in sediments and wood, the total C sequestered in coastal wetlands is 0.205 Pg/yr (Table 6). This is half of the C buried in coastal sediments and represents a major C sink in coastal environments. However, as mentioned above, the source of this C is directly from the atmosphere, and does not influence the balance of C in coastal waters.

3.3. PERTURBATIONS TO CARBON SINKS

This analysis of C sinks in mangrove and coastal ecosystems is based on wide geographic averages of ecological processes. These ecosystems are consistently responding to alterations in the forcing functions that control the ecological processes responsible for the nature of these C sinks. Perturbations, both natural and anthropogenic, can alter the net balance of C in mangrove and coastal ecosystems. These include tropical storms, evolution of landscapes, forest reclamation, and eutrophication.

3.3.1 *Tropical Cyclones.* Another important constraint on the development of biomass in mangroves, particularly in areas such as the Caribbean, is the frequency of hurricanes. It is interesting that of those sites from the Gulf coast and Caribbean islands in Figure 3 (25° latitude), those with the highest biomass are located on west coast of Yucatan peninsula where hurricanes seldom occur (Ruffner, 1978). A basin mangrove forest in Rookery Bay was apparently damaged by a hurricane in 1965, and the regeneration of this forest resulted in high density and low biomass of trees in 1971 compared to 1987. A decline in density and increase in basal area is typical of a developing forest. Biomass was stable by 1987, indicating that a mature forest had developed within 25 yrs following the hurricane disturbance (Warner, 1990). Flores-Verdugo et al. (1986) attributed the poorly developed forest structure in el Verde Lagoon on the Pacific coast of Mexico to frequent hurricanes. Most mangroves in the tropics that are susceptible to damage from high winds and hurricanes are considered young successional forests.

The time scale for a mangrove forest to reach maturity based on steady state levels of biomass is dependant on the regional and local attributes of the environmental setting. Forests in lower latitudes with no natural or human disturbance may reach biomass levels over 250 Mg/ha. For example, biomass of mature forests in the protected areas of Malaysia is 350-400 Mg/ha (Putz and Chan, 1986). Yet forests in higher latitudes do not have the potential for maximum levels of biomass as observed in lower latitude forests (Figure 3). Thus mature levels of biomass in these mangroves may be obtained in less time than in the lower latitudes, given similar rates in wood production. In the Caribbean and Gulf of Mexico, 25 -30 years may be enough time for forests to reach maturity given the maximum biomass that can be supported in these areas. Time scales for recovery and net accumulation of C may occur over longer time scales in warmer tropical regions.

3.3.2 *Mangrove Reclamation.* The destruction of mangroves occurs throughout the tropics for charcoal production, mariculture (shrimp ponds), agriculture, urban development, and dams. In Ecuador there is recently one of the more extreme examples of mangrove reclamation for the production of shrimp ponds. Mangrove loss from 1984 to 1987 ranged from less than 500 to more than 2,500 ha/yr. These are permanent loss of mangrove resources from the coastal zone. The biomass in these forests was largely converted to charcoal, so the C stored was not completely loss. But the wood production, sediment burial and export processes in this coastal environment has diminished in the last 10 yrs. The cumulative impact of mangrove reclamation projects on the global importance of these natural resources has not been evaluated.

3.3.3 *Coastal Geomorphology.* Geomorphology, together with tidal amplitude, will determine the extent of the intertidal zone of a coastal region. Thus geomorphology will to a large extent control the areal distribution of wetlands in any coastal system. The ratio of wetland area to the area of coastal waters has been suggested as important to the ecological function of a coastal wetland (Mann, 1975; Welsh et al., 1982). For example, the Mississippi Deltaic Plain has about 2,500,000 ha of marshes that account for 60% of the coastal wetlands in the lower 48 states (Turner and Gosselink, 1975). However, losses of coastal wetlands in some regions of the delta have been as high as 100 km^2 per year (Gagliano et al., 1981). This has resulted in the conversion of vegetated areas to open water decreasing the wetland:water area ratio. Much of this wetland loss is associated with a high rate of regional subsidence and erosion characteristic of degrading deltas. At the mouth of the Atchafalaya River, however, a wetland system representing the early progradational stages of delta formation is evolving (van Heerden et al., 1983). Here the levels of sediment discharge compensate for the relative increase in water levels due to subsidence and sea level rise. Therefore the Mississippi Deltaic Plain is an environment where the wetland:water ratio is changing due to successional stages in the development of delta ecosystems. A shift in this ratio to less wetland and more water means proportionately less of the organic matter supply will be buried in wetlands. Thus the development of deltas and shifts in the wetland:water ratio in coastal ecosystems will influence the magnitude of C sinks in coastal ecosystems.

3.3.4 *Coastal Eutrophication.* A major change that has occurred, especially during the last 30 to 40 years, is the increase in concentration of nitrogen in river waters, particularly dissolved nitrate (Walsh et al., 1981). Prior to leveeing of the Mississippi River a large portion of this new nitrogen was not transported directly to the shelf system but was processed by coastal bays and marshes before reaching the Gulf. The increase in nitrate concentration has thus occurred simultaneously with changes in how materials from the river are processed in coastal ecosystems along the Louisiana coast. Utilization of nitrate includes assimilation by primary producers in the water column and exchange with wetlands and benthic sediments, where it is either assimilated into biomass or dissimilated to nitrogen gas. A portion of the assimilated nitrate may be buried in wetland and/or benthic sediments, and together with the dissimilated nitrate, represent losses of this nutrient from community metabolism. The direct discharge of nitrate to the shelf is dramatically different from that of an unleveed river, and likely has major impacts on the distribution of biological productivity within the coastal regions of the northern Gulf of Mexico. This region of the Gulf of Mexico has thus been the focus of scientific interests in the coupling of eutrophication of coastal waters with the C balance of the biosphere (Walsh et al., 1981).

4. Conclusions And Recommendations

We propose that a three compartment conceptual model (Figure 1) is needed to focus attention on ecological processes (Figure 10) that influence the global C cycle. This three compartment model includes coastal margin ecosystems in addition to the present terrestrial and oceanic (deep water) ecosystems that are presently being evaluated. These coastal margin ecosystems include the wetland, estuarine, littoral, and shelf environments from the tidal freshwater region to the continental slope. This area is less than 5% of the total land area, but the biogeochemical processes warrant more complete investigations into potential C sinks in the biosphere. For example, the influence of river discharge on coastal margin ecosystems is twofold: 1) the discharge of particulate and dissolved organic C and the fate of this terrigenous C needs more thorough analysis; 2) the nutrients transported to the coastal margins from the ten major river systems in the world (most of which are in the tropics) may enhance the new production of the biogenic regions of these river plumes. In addition, there has not been an integrated approach to the evaluation of these coastal margins to include the synoptic analysis of wetlands (including marshes and mangroves), littoral subsystems (such as seagrasses and coral reefs), together with the fate of new production in phytoplankton communities. This analysis is particularly warranted in the tropical and subtropical regions of the biosphere. This effort should include a hierarchial modelling effort, similar in magnitude to terrestrial and oceanic components, to predict the influence of perturbations to the coastal margins (such as wetland loss, nutrient enrichment, rise in sea level, etc.) on future predictions of global climate change.

5. References

Armentano, T.V. and G.M. Woodwell. (1975) 'Sedimentation rates in a Long Island marsh by ^{210}Pb dating', Limnology Oceanography 20, 452-456.
Berner, R.A. (1982) 'Burial of organic carbon and pyrite in the modern ocean: its geochemical and environmental significance, American Journal of Science 282, 451-

473.
Boto, K.G. and J.S. Bunt. (1981) 'Tidal export of particulate organic matter from a Northern Australian mangrove system', Estuarine, Coastal and Shelf Science 13, 247-255.
Brown, S. and A.E. Lugo. (1982) 'A comparison of structural and functional characteristics of saltwater and freshwater forested wetlands', in B. Gopal, R. Turner, R. Wetzel, and D. Whigham (eds.), Wetlands Ecology and Management. Proceedings of the First International Wetlands Conference, New Delhi. National Institute of Ecology and International Scientific Publications, New Delhi, India. pp. 109 - 130.
Christensen, B. (1978) 'Biomass and primary production of *Rhizophora apiculata* Bl. in a mangrove forest in southern Thailand', Aquatic Botany 4, 43-52.
Cintrón, G., A.E. Lugo, D.J. Pool, and G. Morris. (1978) 'Mangroves of arid environments in Puerto Rico and adjacent islands', Biotropica 10, 110-121.
Cintrón, G. and Y. Schaeffer-Novelli. (1984) 'Caracteristicas y desarrollo estructural de los manglares de Norte y Sur America', Programa Regional de Desarrollo Cientifico y Tecnologico 25, 4-15.
Dagg, M., C. Grimes, S. Lohrenz, B. McKee, R.Twilley, and W. Wiseman, Jr. (1991) 'Continental shelf food chains of the northern Gulf of Mexico', pp. 67-106, in: K. Sherman, L. M. Alexander, and B.D. Gold (eds), Food Chains, Yields, Models, and Management of Large Marine Ecosystems. Westview Press, Boulder.
Day, J., W. Conner, F. Ley-Lou, R. Day, and A. Machado. (1987) 'The productivity and composition of mangrove forests, Laguna de Términos, Mexico', Aquatic Botany 27, 267 - 284.
Degens, E.T., S. Kempe, and J.E. Richey. (1991) 'Summary: Biogeochemistry of major world rivers', pp. 323-347. In: E.T. Degens, S. Kempe and J.E. Richey (eds). Biogeochemistry of Major World Rivers. SCOPE, John Wiley and Sons Ltd.
Deuser, W.G. (1979) 'Marine biota, nearshore sediments, and the global carbon balance', Organic Geochemistry 1, 243-247.
Deuser, W.G. (1988) 'Whither organic carbon?', Nature 332, 396-397.
Ellison, J.C. and D.R. Stoddart. (1991) 'Mangrove ecosystem collapse during predicted sea-level rise: Holocene analogues and implications' Journal of Coastal Research 7, 151-165.
Flores-Verdugo, F.J. Day, and R. Briseño-Dueñas. (1986) 'Structure, litterfall, decomposition, and detritus dynamics of mangroves in a Mexican coastal lagoon with an ephemeral inlet', Marine Ecology Progress Series 35, 83 - 90.
Frey, R.W. and P.B. Basan. (1978) Coastal salt marshes, pp. 101-169. In: R.A. Davis (ed.), Coastal Sedimentary Environments. Springer Verlag, New York.
Gagliano, S.M., K.J. Meyer-Arendt and K.M. Wicker. (1981) 'Land loss in the Mississippi River deltaic plain', Transactions of the Gulf Coast Association of Geological Societies 31, 295-300.
Gearing, P.J., F.E. Plucker, and P.L. Parker. (1977) 'Organic carbon stable isotope ratios of continental margin sediments', Marine Chemistry 5, 251-266.
Golley, F.B., H.T. Odum and R. Wilson. (1962) 'A study of the structure and metabolism of a red mangrove forest in southern Puerto Rico in May', Ecology 43, 9-18.
Gong, W.K. and J.E. Ong. (1990) 'Plant biomass and nutrient flux in a managed mangrove forest in Malaysia', Estuarine, Coastal and Shelf Science 31, 519-530.
Gosselink, J.G. and R.E. Turner. (1978) 'The role of hydrology in freshwater wetland ecosystems', pp, 63-78. In R.E. Good, D.F. Whigham, and R.L. Simpson (eds.), Freshwater Wetlands: Ecological Processes and Management Potential. Academic Press, New York.

Hatton, R.S., R.D. DeLaune, and W.H. Patrick, Jr. (1983) 'Sedimentation, accretion, and subsidence in marshes of Barataria Basin, Louisiana', Limnology and Oceanography 28, 494-502.
Heald, E.J. (1969) 'The production of organic detritus in a south Florida estuary', Ph.D. dissertation, University of Miami, Coral Gables.
Heald, E. (1971) 'The production of organic detritus in a south Florida estuary', University of Miami Sea Grant Technical Bulletin 6. 110 pp.
Hopkinson, C.S.,Jr. (1988) 'Patterns of organic carbon exchange between coastal ecosystems: The mass balance approach in salt marsh ecosystems', pp. 122-154. In: B.O. Jansson (ed.), Coastal-Offshore Ecosystem Interactions. Springer-Verlag, Germany.
Ittekkot, V. (1988) 'Global trends in the nature of organic matter in river suspensions', Nature 332, 436-438
Lee, S.Y. (1989) 'Litter production and turnover of the mangrove *Kandelia candel* (L.) Druce in a Hong Kong tidal shrimp pond', Estuarine, Coastal and Shelf Science 29, 75-87.
Lee, S.Y. (1990) 'Primary productivity and particulate organic matter flow in an estuarine mangrove-wetland in Hong Kong', Marine Biology 106, 453-463.
Leh, C.M.U. and A. Sasekumar. (1985) 'The food of sesarmid crabs in Malaysian mangrove forests', Malay Naturalist Journal 39, 135-145.
Lin, P., C.Y. Lu, G.H. Lin, R.H. Chen and L. Su. (1985) 'The biomass and productivity of *Kandelia candel* community', Journal of Xiamen University 14, 508-514.
Lin, P., C.Y. Lu, G.L. Wang and H.X. Chen. (1990) 'Biomass and productivity of *Bruguiera sexangula* mangrove forest in Hainan Island, China', Journal of Xiamen University 29, 209-213.
Lugo, A.E. (1978) 'Stress and ecosystems', pp. 62-101. In J.H. Thorp and J.W. Gibbons (eds), Energy and Environmental Stress. DOE 771114. Department of Energy, Washington, D.C.
Lugo, A.E., and S.C. Snedaker. (1974) 'The ecology of mangroves', Annual Review of Ecology and Systematics 5, 39-64.
Lugo, A.E. and C. Patterson-Zucca. (1977) 'The impact of low temperature stress on mangrove structure and growth', Tropical Ecology 18, 149-161.
Lugo, A.E., S. Brown, and M.M. Brinson. (1990) 'Concepts in wetland ecology', pp. 53-85. In: A.E. Lugo, M. Brinson, and S. Brown (eds.), Ecosystems of the World 15: Forested Wetlands. Elsevier, Amsterdam.
Lynch, J.C. (1989) 'Sedimentation and nutrient accumulation in mangrove ecosystems of the Gulf of Mexico', M.S. Thesis, University of Southwestern Louisiana, Lafayette, LA. 102 pp.
Lynch, J.C., J.R. Meriwether, B.A. McKee, F. Vera-Herrera, and R.R. Twilley, R.R. (1989) 'Recent accretion in mangrove ecosystems based on ^{137}Cs and ^{210}Pb', Estuaries 12, 284-299.
Malley, D.F. (1978) 'Degradation of mangrove leaf litter by the tropical sesarmid crab *Chiromanthes onychophorum*', Marine Biology 49, 377-386.
Mann, K.H. (1975) 'Relationship between morphometry and biological functioning in three coastal inlets of Nova Scotia', pp. 634-644. In: L.E. Cronin (ed.), Estuarine Research Vol. 1. Academic Press, New York.
Meybeck, M. (1981) 'Flux of organic carbon by rivers to the oceans', pp. 219-269. National Technical Information Service, Springfield, Virginia.
Meybeck, M. (1982) 'Carbon, nitrogen and phosphorus transport by world rivers', American Journal of Science 282, 401-450

Meybeck, M. (1988) 'How to establish and use world budgets of riverine materials', pp. 247-272. In: A. Lerman and M. Meybeck (eds.). Physical and Chemical Weathering in Geochemical Cycles. Kluwer Academic Publishers, Dordrecht.

Milliman, J.D. and R.H. Meade. (1983) 'World-wide delivery of river sediment to the oceans', Journal of Geology 91, 1-21

Nixon, S.W. (1980) 'Between coastal marshes and coastal waters--a review of twenty years of speculation and research on the role of salt marshes in estuarine productivity and water chemistry', pp. 437-525. In P. Hamilton and K.B. MacDonald (eds.), Estuarine and Wetland Processes with Emphasis on Modeling. Plenum Press, N.Y.

Odum, W.E. and E.J. Heald. (1972) 'Trophic analysis of an estuarine mangrove community', Bulletin Marine Science 22, 671-738.

Odum, W.E., C.C. McIvor, and T.J. Smith, III. (1982) The ecology of the mangroves of south Florida: A community profile', Fish and Wildlife Service/Office of Biological Services, Washington, D.C. FWS/OBS-81/24.

Ong, J.E., W.K. Gong, C.H. Wong and Dhanarajan. (1979) 'Productivity of a managed mangrove forest in West Malaysia', Paper presented at International Conference on " Trends in Applied Biology in S.E. Asia", USM Penany, Malaysia.

Pool, D.J., A.E. Lugo, and S.C. Snedaker. (1975) 'Litter production in mangrove forests of southern Florida and Puerto Rico', pp. 213-237. In Proceedings of the International Symposium on Biology and Management of Mangroves (Ed. G.E. Walsh, S.C. Snedaker, and H.J. Teas). Institute of Food and Agricultural Sciences, University of Florida, Gainesville.

Putz, F.E. and Chan, H.T. (1986) 'Tree growth, dynamics and productivity in a mature mangrove forest in Malaysia', Forest Ecology and Management 17, 211-230.

Robertson, A.I. (1986) 'Leaf-buring crabs:their influence on energy flow and export from mixed mangrove forests (*Rhizophora* spp.) in northeastern Australia', Journal of Experimental Marine Biology and Ecology 102, 237-248.

Robertson, A.I. and P.A. Daniel. (1989) 'The influence of crabs on litter processing in high intertidal mangrove forests in tropical Australia', Oecologia 78, 191-198.

Rowe, G. S. Smith, P. Falkowske, T. Whitledge, R. Theroux, W. Phoel and H. Ducklow. (1986) 'Do continental shelves export organic matter?' Nature 324, 55-66

Ruffner, J.A. (1978) 'Climates of the states; with current tables of the National Oceanic and Atmospheric Administration normal 1940 to 1970 and means and extremes to 1975', Vol. I. Gale Research Company, Detroit, MI., USA.

Schlesinger, W.H. and J. M. Melack. (1981) 'Transport of organic carbon in the world's rivers', Tellus 33, 172-187.

Scoffin, T.P. (1970) 'The trapping and binding of subtidal carbonate sediments by marine vegetation in Bimini Lagoon, Bahamas', Journal of Sedimentary Petrology 40, 249-273.

Sell, M.G., Jr. (1977) 'Modelling the response of mangrove ecosystem of herbicide spraying, hurricanes, nutrient enrichment and economic development', Dissertation. University of Florida. Gainesville, Florida, USA.

Sharma, P., L.R. Gardner, W.S. Moore, and M.S. Bollinger. (1987) 'Sedimentation and bioturbation in a salt marsh as revealed by lead-210, cesium-137, and beryllium-7 studies', Limnology and Oceanography 32, 313-326.

Shultz, D.J. and J.A. Calder. (1976) 'Organic carbon $^{13}C/^{12}C$ variations in estuarine sediments', Geochimica et Cosmichimica Acta 40, 381-385.

Smith, S.V. (1981) 'Marine macrophytes as a global carbon sink', Science 211, 838-840.

Smith, S.V. and F.T. Mackenzie. (1987) 'The oceans as a net heterotrophic system: implications for the carbon biogeochemical cycle', Global Biogeochemistry Cycles 1,

187-198

Tans, P.P., I.Y. Fung, and T. Takahashi. (1990) 'Observational constraints on the global atmosphere CO_2 budget', Science 247, 1431-1438.

Thayer, G.W., J.J. Govoni and D. W. Connally. (1983) 'Stable carbon isotope ratios of the planktonic food web in the northern Gulf of Mexico', Bulletin of Marine Science 33, 246-256.

Thom, B.G. (1982) 'Mangrove ecology- a geomorphological perspective', pp 3-17. In: B.F. Clough (ed.), Mangrove Ecosystems in Australia. Australian National University Press, Canberra.

Twilley, R.R. (1982) 'Litter dynamics and organic carbon exchange in black mangrove (*Avicennia germinans*) basin forests in a southwest Florida estuary', Ph.D. dissertation. University of Florida, Gainesville.

Twilley, R.R. (1985) 'The exchange of organic carbon in basin mangrove forests in a southwest Florida estuary', Estuarine, Coastal and Shelf Science 20, 543-557.

Twilley, R.R., A.E. Lugo, and C. Patterson-Zucca. (1986) 'Production, standing crop, and decomposition of litter in basin mangrove forests in southwest Florida', Ecology 67, 670-683.

Twilley, R.R. (1988) 'Coupling of mangroves to the productivity of estuarine and coastal waters', pp 155-180. In: B.O. Jansson (ed.), Coastal-Offshore Ecosystem Interactions. Springer-Verlag, Germany.

Twilley, R.R, R. Zimmerman, L. Solorzano, V.Rivera-Monroy, A. Bodero, R. Zambrano, M. Pozo, V. Garcia, K. Loor, R. Garcia, W. Cardenas, N. Gaibor, J. Espinoza, and J. Lynch. (1990) 'The importance of mangroves in sustaining fisheries and controlling water quality in coastal ecosystems', Interim Report, U.S. Agency for International Development, Program in Science and Technology Cooperation, Washington, D.C.

Twilley, R.R. (in press) 'Energy signature and properties of mangrove ecosystems', In: C. Hall (ed.), Maximum power.

van Heerden, I.L. (1983) 'Deltaic sedimentation in eastern Atchafalaya Bay, Louisiana', Ph.D. thesis, Department of Marine Science. Louisiana State University, Baton Rouge, pp. 116.

Walsh, J.J. (1983) 'Death in the sea: Enigmatic phytoplankton losses', Progress in Oceanography 12, 1-86.

Walsh, J.J. (1984) 'The role of ocean biota in accelerated ecological cycles: A temporal view' BioScience 34, 499-507.

Walsh, J.J. (1988) 'On the nature of continental shelves', Academic Press. 508 pp

Walsh, J.J., G.T. Rowe, R.L. Iverson, and C.P McRoy. (1981) 'Biological export of shelf carbon is a sink of the global CO_2 cycle', Nature 291, 196-201.

Walsh, J.J., E.Premuzic, J.Gaffney, G. Rowe, G. Harbottle, R. Stoenner, W. Balsam, P. Betzer and S.M. (1985) 'Organic stoarage of CO_2 on the continental slope off the mid-Atlantic bight, the southeastern Bering Sea, and the Peru coast', Deep Sea Research 32, 853-883

Warner, J.H. (1990) 'Successional patterns in a mangrove forest in southwestern Florida, USA', M.S. thesis, University of Southwestern Louisiana, Lafayette, LA. 74 pp.

Watson, J. (1928) 'Mangrove forests of the Malay Peninsula', Malayan Forest Records 6. Fraser & Neave, Ltd., Singapore, 275 pp.

Welsh, B.L., R.B. Whitlatch and W.F. Bohlen. (1982) 'Relationship between physical characteristics and organic carbon sources as a basis for comparing estuaries in southern New England', pp. 53-67. In: V.S. Kennedy (ed.). Estuarine Comparisons. Academic Press, N.Y.

Wharton, C.H. and M.M. Brinson. (1979) 'Characteristics of southeastern river systems', pp. 32-40. In: R.R. Johnson and J.F. McCormick (eds), Strategies for Protection and Management of Floodplain Wetlands and Other Riparian Ecosystems. Symposium Proceedings, U.S. Department of Agriculture. Washington, D.C.

Wollast, R.and F.T. MacKenzie. (1989) 'Global biogeochemical cycles and climate', pp. 453-473. In: A. Berger (ed.), Climate and Geo-sciences. Kluver Acad. Publ. Hingham, MA

Woodroffe, C.D. (1985) 'Studies of a mangrove basin, Tuff Crater, New Zealand: II. Comparison of volumetric and velocity-area methods of estimating tidal flux', Estuarine, Coastal and Shelf Science 20, 431-445.

MARINE ALGAE AS A CO_2 SINK

R.L. RITSCHARD
Center for Atmospheric & Biospheric Effects of Technology
Lawrence Berkeley Laboratory, University of California
Berkeley, CA 94720 USA

ABSTRACT. The most effective ways to reduce CO_2 emissions are to improve the energy efficiency of each economic sector and to reduce the cutting of tropical and temperate forests around the world. These options, however, may not fully reach their technical and economic potential due to various political and socioeconomic barriers. Other more innovative and less well developed mitigation measures therefore will be required. The most practical of these is to increase CO_2 sinks through photosynthesis in both standing tree biomass and in ocean primary producers. In this paper, the use of marine algae as CO_2 sinks is reviewed from a technical, engineering/economic, and environmental perspective. Two open ocean options are considered for large-scale CO_2 mitigation: the use of phytoplankton through Fe fertilization and macroalgal (kelp) farms, which can be used for both C sequestering and energy production. It has been estimated that these two approaches can sequester from 0.7 to 3 Gt C yr^{-1} from the atmosphere at an estimated cost of \$5 to 300 t^{-1} C yr^{-1}. Other options currently under study are also mentioned. Numerous questions remain to be answered pertaining to the use of both microalgae and macroalgae for CO_2 assimilation before credible estimates of costs of C removal can be made for either system. In addition, there are several key environmental issues raised by the use of algae. A detailed discussion of these variables, including cost estimates, is presented.

1. INTRODUCTION

The atmosphere is currently absorbing approximately 3.4 $\pm$0.2 Gt of the 7.0 $\pm$1.5 Gt of human-generated CO_2 released each year, which represents 5.4 $\pm$0.5 Gt C yr^{-1} from fossil fuel combustion and 1.6 $\pm$1.0 Gt C yr^{-1} from deforestation (IPCC, 1990). More than 50% of total annual anthropogenic CO_2 production is actually accumulated in the atmosphere, the remainder is found in various terrestrial and oceanic sinks. Today, atmospheric concentrations of CO_2 are about 355 ppm, and are increasing at a rate of about 0.4% yr^{-1}. This atmospheric concentration could easily double within 100 years if fossil fuel consumption increases in both the developed and developing world.

The most effective ways to reduce these CO_2 emissions are to improve the energy efficiency of each economic sector (i.e., buildings, industrial processes, transportation,

Water, Air, and Soil Pollution **64**: 289–303, 1992.
© 1992 *Kluwer Academic Publishers.*

etc.) and to lessen the destruction of tropical and temperate forests around the world. The energy efficiency options although cost-effective may not fully reach their technical and economic potential in the short (or long) term due to various social and political barriers. As in the case of reducing fossil fuel emissions through energy efficiency improvements, reducing or delaying the cutting of forests is also mired with political and socioeconomic constraints. Therefore, other more innovative and less well defined CO_2 mitigation measures are required. The most practical of these innovations is to increase CO_2 sinks through photosynthesis, including increased C storage in standing tree biomass, substitution of fossil fuels with biofuels, increased soil C sequestration, and increased ocean primary productivity. In each case, there still remain many unanswered questions about the technical and economic viability of these schemes as well as about their environmental consequences. For example, are there sufficient inorganic nutrients available to support increased absorption of CO_2 in either terrestrial or oceanic systems (See for example, Orr and Sarmiento, 1992).

In this paper, the use of marine algae (both phytoplankton and macroalgae) as CO_2 sinks will be reviewed from a technical, engineering/economic, and environmental perspective. In the remainder of the paper, we will discuss first the role of the oceans in regulating the uptake of CO_2 and establishing global climate. This discussion is followed by a summary of the Fe limitation hypothesis, including the major technical, engineering and economic, and environmental issues that have been recently identified. Next, we briefly consider macroalgae farms as CO_2 sinks and describe what has been learned about them over the past 18 yr. Finally, we conclude with a comparison of the phytoplankton and macroalgae mitigation systems from the perspective of the potential C that can be sequestered and their estimated costs.

2. ROLE OF THE OCEANS

The oceans play a major role in establishing global climate, both through the transport of heat and supply of water vapor and through the absorption of a large fraction of atmospheric CO_2. Understanding the ocean C cycle, which is ill-defined, is important because the oceans contain about 60 times more C than does the atmosphere. Also, the ocean system operates as a whole: events occurring at the sea surface and in the upper ocean are intimately linked to processes in deeper water and in ocean sediments. Estimates of the net ocean sink for CO_2 range from the traditionally accepted value of 2.0 $\pm$0.8 Gt C yr^{-1}, based on physico-chemical models, to a much lower value of 0.6-1.0 Gt C yr^{-1} recently reported by Tans et al. (1990).

The ocean's role in regulating the uptake capacity of CO_2 involves a combination of inorganic C chemistry, which is generally well understood, ocean mixing, alkalinity, and various biotic processes, including changes in nutrient regimes, biological interactions in the upper ocean, and changes in quantity and quality of C export production. Primary production of phytoplankton (or macroalgae) in the sunlit upper ocean can have several effects on air-sea exchange of CO_2. First, by changing dissolved, inorganic C into organic cell constituents (through photosynthesis), the partial pressure of CO_2 (pCO_2) in surrounding water is reduced, which will promote the drawdown of CO_2 from the atmosphere, with the rate of uptake being a function of the air-sea pCO_2 difference and strength of the wind-driven gas exchange. Second, by taking up the available dissolved nitrate, an important plant nutrient, for "new production", the alkalinity of the surface water will be somewhat increased. This could also further reduce the pCO_2. These two processes can be counterbalanced if inorganic C is precipitated as calcium carbonate (calcite) as part of the phytoplankton cell structure because this will reduce the alkalinity of the surface waters and will increase pCO_2. Although relatively few phytoplankton species produce calcite, those that do can in certain regions of the ocean can be extremely abundant.

Finally, respiration is the process whereby organic compounds are remineralized by the metabolic processes of both autotrophic and heterotrophic organisms releasing CO_2 to the water. This process raises the pCO_2 and, when such water is in contact with the atmosphere, promotes the return of CO_2 to the air. Respiration takes place throughout the water column although mostly below the mixed layer of the ocean. The quantity and quality of particulate export from the upper ocean affect the depth distribution of respiratory activity in the water column, in turn determining the time constant for CO_2 return to the atmosphere. Recent studies suggest that variability in the nature and amount of dissolved organic carbon (DOC) export is probably also important, but its role has yet to be resolved (Sugimura and Suzuki, 1988; Druffel and Williams, 1990)(See Figure 1 for a summary of the ocean cycle).

While the ocean's role in moderating present-day CO_2 levels depends on ocean mixing and alkalinity, the potential role of ocean biota should not be neglected just because it is not generally included in most ocean carbon models. The potential amount of total C that could be utilized by ocean photosynthesis (i.e., gross primary production) has been estimated to be between 30 and 40 Gt yr^{-1} (Moore and Bolin, 1987). This value, however, represents gross C fixation in the ocean and the net effect on the atmosphere will depend on the return flow from decomposition and will eventually reach steady state. Core records from the last Ice Age (Neftel et al., 1982; Barnola et al., 1987) indicate that at that time atmospheric concentrations of CO_2 were from 180 to 200 ppm, while before the Industrial Revolution, atmospheric concentrations of CO_2 were approximately 280 ppm.

In the 1980s, several modeling groups (Broecker, 1982a,b; Knox and McElroy, 1984; Sarmiento and Toggweiler, 1984; Siegenthaler and Wenk, 1984; Toggweiler and Sarmiento, 1985; Keir, 1988) each concluded that surface nutrient concentrations in polar ocean regions were key to understanding the biological uptake of CO_2 from the atmosphere during glacial-interglacial periods. In areas such as the arctic Pacific and the antarctic circumpolar ocean (Southern Ocean), high concentrations of nitrate (NO_3) and phosphate (PO_4^{-3}), the key ingredients for plant growth, are unused. In general, the modeling results appeared to show that changes in Southern Ocean phytoplankton productivity could have resulted in the observed rapid glacial-interglacial changes in atmospheric CO_2. However, why productivity rates changed, if they did, is still unresolved. The model results did suggest that if available nutrients (i.e., N and P) were assimilated, the conversion of CO_2 to organic C could readily account for the ice age "signal". A radical solution to this ice age CO_2 puzzle has been recently proposed by Martin and co-workers for both the arctic Pacific (Martin and Fitzwater, 1988; Martin and Gordon, 1988; Martin et al., 1989) and the Southern Ocean (Martin 1990; Martin et al. 1990a,b). These scientists showed that addition of trace amounts of Fe to natural populations of phytoplankton stimulated photosynthesis in bottle experiments, and they hypothesized that Fe limited the phytoplankton growth in these polar areas (i.e., arctic Pacific and Southern Ocean with high nutrients and low productivity). They also pointed

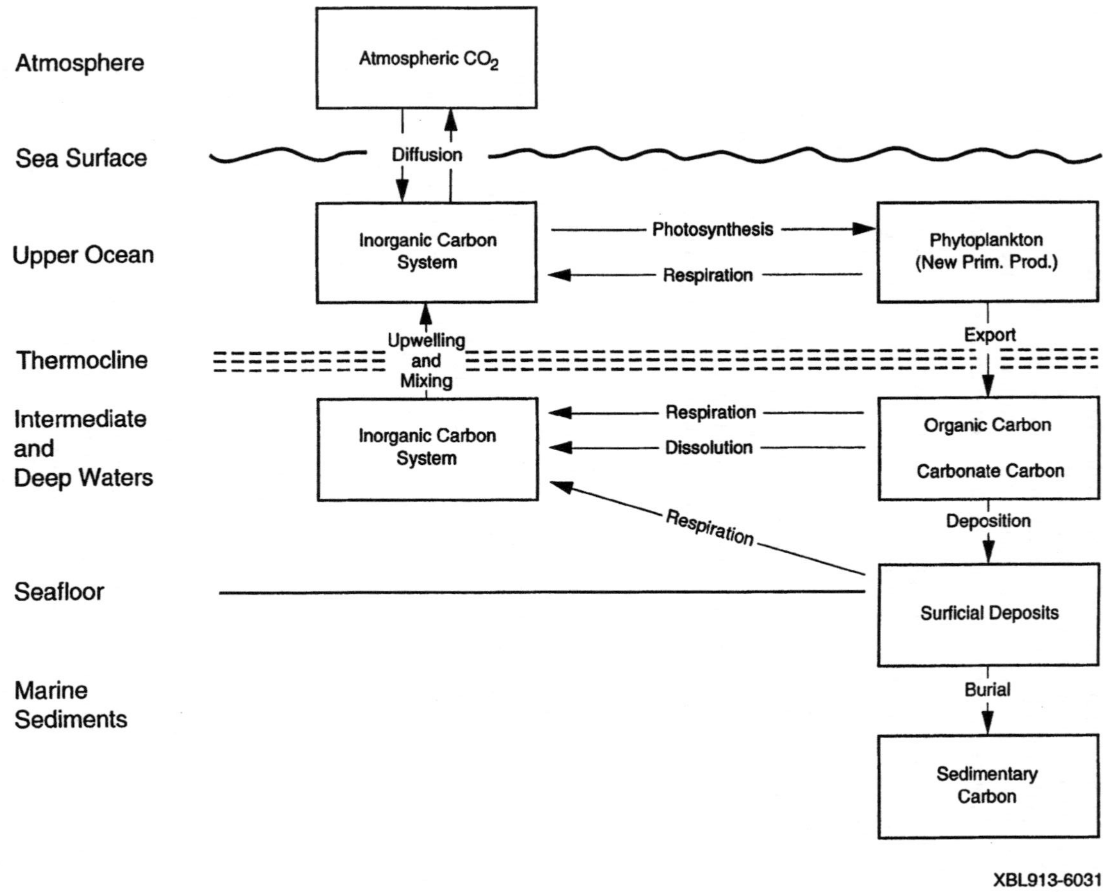

Figure 1. Ocean C cycle showing possible interactions with the atmosphere, through the ocean layers to the deep marine sediments

out that atmospheric dust containing Fe was 50 times higher during the last glacial maximum (DeAngelis et al., 1987). As a result of this Fe enrichment, phytoplankton growth may have been greatly enhanced, larger amounts of upwelled nutrients may have been used, and this stimulation of new productivity may have contributed to a drawdown of atmospheric CO_2 concentrations to less than 200 ppm during the glacial period. This idea, which has been called "Fe limitation", will be described in more detail in the next section.

3. PHYTOPLANKTON AND THE Fe LIMITATION HYPOTHESIS

The "Fe hypothesis" is based on experimental work that suggests the Fe is a limiting factor in the growth of marine algae in some areas of the ocean (arctic Pacific, equatorial Pacific, and Southern Ocean). Iron is essential in plant nutrition, including the synthesis of chlorophyll, the reduction of nitrate (through the activities of nitrate reductase), and atmospheric N fixation. However, unlike other major nutrients such as C, N, and P that have Redfield ratios of 6.6 C:N and 106 C:P, only small quantities of Fe are needed. The molar ratios of C to Fe are of the order of 10,000 to 200,000:1 (Sunda et al., 1991). Also, even though Fe is the fourth most abundant element in the earth's crust, it is scarce in the surface waters of the open ocean, partially because Fe is very insoluble in oxygenated seawater. Ferrous ions rapidly oxidize to ferric ions (insoluble Fe III state). Ferric hydroxide can settle out from the water column and thus be removed from the biotic environment. Without Fe, the productivity in many ocean areas is very low, according to the Fe hypothesis, even when other essential nutrients, such as NO_3^- and PO^{-3}_4, are present in relative abundance.

Martin's Fe hypothesis further states that with such low algal productivity there is relatively little organic C being removed from the surface of these ocean areas to the bottom sediments. Algal photosynthesis in the upper reaches of the oceans, followed by direct (i.e., settling of algal cells) or indirect (conversion through the food chain into fecal matter or animal carcasses that settle into the sediments) removal of the organic matter to the deeper parts of the oceans, is believed to be a major way in which C is removed from the surface of the oceans. This so-called "biological pump" removes the C from the atmosphere and pumps it through the upper layers of the oceans to become trapped in deep ocean layers and sediments. Thus, according to Martin's hypothesis, if sufficient Fe were present, the total amount of additional net primary production that would be removed from the atmosphere through the biological pump would be about 2 to 3 Gt C yr^{-1} in the Antarctic Ocean alone.

3.1 Major Technical Issues of Fe Fertilization

Fe fertilization and the potential use of macroalgae to assimilate large amounts of CO_2 was examined in a 1990 National Research Council (NRC) workshop. Phytoplankton experts at the NRC workshop agreed that it is conceptually feasible to slow the increase in atmospheric CO_2 levels through enhanced new primary production in oceans, resulting in enhanced net transport of CO_2 from the atmosphere to the oceans. From existing models, it was estimated that an additional 1 to 2 Gt C yr^1 (with many arguing for an upper limit of 1 Gt C yr^{-1}) can be removed from the atmosphere if new primary production is enhanced and most unused nutrients (e.g., NO_3^- and PO^{-3}_4) are assimilated.

Two of the major uncertainties with the Fe hypothesis are (1) whether primary production was indeed high during the glacial maximum as proposed by Martin and his co-workers, and (2) how effective the biological pump can operate so that organic C matter will be transferred downward to the bottom layers, including the ocean sediments. On the first major uncertainty, there is conflicting evidence in the literature. Some studies reported evidence that paleoproductivity rates were higher by a factor of three during glacial periods than previous and present interglacial states (Sarnthein et al., 1987; Pedersen et al., 1988), while other paleoceanographic studies (Mortlock et al., 1991), using evidence from tracers of paleoproductivity in Antarctic sediments challenged this view. These more recent data also support widely cited but controversial evidence from $\delta^{13}C$ records of Antarctic planktonic formanifera (Neogloboquadrina pachyderma) for reduced Antarctic productivity during glacials (Labeyrie and Duplessy, 1985; Charles and Fairbanks, 1990; Grobe et al., 1990). The lack of clear evidence for increased Antarctic productivity during the glacial maximum remains the weakest point of Martin's Fe limitation hypothesis.

Two recent modeling studies (Peng and Broecker, 1991; Joos et al., 1991) attempted to address the issue of the effectiveness of the biological pump. Peng and Broecker (1991), for example, argued that on dynamic considerations Fe fertilization is unlikely to succeed as a continuing full-scale pump of CO_2 from the atmosphere to the ocean because vertical transport in the Antarctic Ocean is sufficiently sluggish so that the flow of CO_2 into the ocean would rapidly saturate the surface waters with CO_2. Using a box model, they reported that after 100 yr of totally successful fertilization the CO_2 content of the atmosphere would be lowered by only $10 \pm 5\%$ below that it would have been in the absence of fertilization. The critical issue is thus the rate of vertical mixing in the Southern Ocean. The box models used by Peng and Broecker (1991) and Joos et al. (1991) are based on the assumption that nutrients in surface waters are transferred to intermediate or deep water regions through mixing, diffusion, and advection, which results in the complete removal of surface nutrients (especially PO_4^{3-}) in the Southern Ocean. Using a 3-D ocean model, Sarmiento and Orr (1991) simulated the removal of nutrients from the surface waters of the Southern Ocean resulting in a 0.5 to 1.0 Gt C yr^{-1} reduction in low-latitude new production. Since the biological pump is an implied feature in both the box and 3-D ocean models, whether it can function to rapidly remove organic C from the surface waters to the deep water still needs to be resolved with data on the sedimentation of organic matter in bloom forming areas.

Another hypothesis that might explain the presence of relatively high primary nutrient concentrations in several oceanic areas where chlorophyll (and thus primary production) concentrations are low is grazing control by zooplankton. Dugdale and Wilkerson (1990) found that grazing, which is the source of ammonium (NH_4^+), proceeds in the Southern Ocean at a rate somewhat greater than NO_3^- uptake. A model invoking not only a balance between growth and grazing, but also the preferential uptake of NH_4^+ (as regenerated production), as opposed to NO_3^- (as new production), can explain the observations without direct recourse to Fe limitation (see Figure 2).

The results of Martin's bottle experiments have been interpreted by others in several ways. First, Fe may indeed be limiting. Second, the addition of Fe could have a negative effect on grazers and the consequent relaxation in grazing pressure could yield the increased chlorophyll concentrations reported by Martin and his co-workers (Banse, 1990; 1991; Frost, 1991; Mitchell et al., 1991; Cullen, 1991). Third, Fe could be limiting

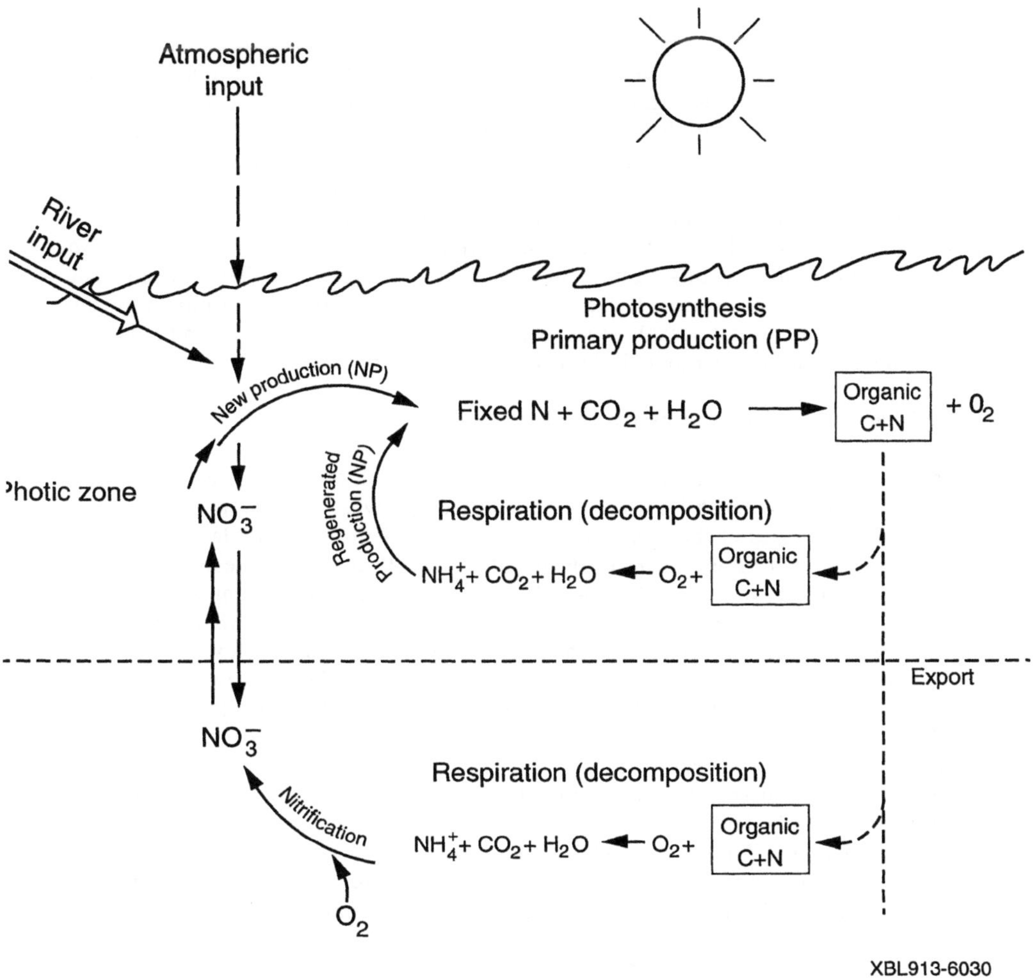

Figure 2. Schematic diagram of C and N transfers in the upper ocean demonstrating the role of NO_3^- and NH_4^+ pathways

only for the larger bloom-forming phytoplankton, i.e., diatoms (Price et al., 1991). Fourth, the addition of Fe to the experimental bottles could reduce the lag time between nutrient uptake and population growth (Sunda et al., 1991).

A meeting was organized by the American Society of Limnology and Oceanography (ASLO) in February 1991 to address further both the scientific issues of limiting factors in primary productivity and specifically, the concept of Fe limitation. At this meeting of about 150 scientists, the consensus was that individual primary producers (usually small cell size) that dominate in low productivity/high nutrient environments may be adapted to the low Fe conditions in which they have evolved, but they may not be limited by the addition of Fe, although Fe may play some regulatory role. It was also concluded that the question of Fe limitation should be investigated further in small-scale experiments in the open ocean. These experiments would help to determine why the phytoplankton of the several oceanic areas (arctic Pacific, equatorial Pacific, and Southern Ocean) grow so poorly, even though these waters are rich in nutrients.

3.2 Engineering and Economic Issues of Phytoplankton

Before any large Fe fertilization experiment is conducted in the Antarctic open ocean, there are issues of how much Fe needs to be delivered, in what form, how often, and in what manner. Estimates of these parameters are required before a credible cost analysis can be completed. Very preliminary estimates of the supply concept and economics were discussed at both the NRC workshop and ASLO meetings and were summarized in the National Academy of Science's Report of the Mitigation Panel (NAS/NAE/IOM, 1991). The major engineering and economic considerations are summarized below.

There are several preliminary bases for the cost of Fe fertilization over the Southern Ocean. One is based on Martin's work, the second comes from the NRC workshop, and the third represents initial estimates by the Electric Power Research Institute (EPRI). Although these estimates are crude because of the uncertainties of the engineering design concept, they do potentially provide upper and lower limits. Martin's studies estimate that Fe fertilization over the entire 16% of Antarctic Ocean surface would require only 0.43 x 10^6 t Fe yr^{-1} and would remove 2 to 3 Gt C yr^{-1} (Martin et al., 1990a). The form of Fe usually absorbed most easily by living organisms is the ferrous form. The cheapest bulk ferrous is ferrous sulphate costing about \$10 to \$15 t^{1} (NAS/NAE/IOM, 1991) for a total of \$12 to \$18 x10^6 yr^{-1} for Fe as fertilizer.

The NRC workshop assumptions about fertilizer costs were somewhat different. A removal of about 1.8 Gt C yr^{-1} over a 100-yr period was assumed with an Fe application of 1 to 5 x10^6 t Fe yr^{-1} in the form of ferrous chloride ($FeCl_2$), covering an ocean surface area of approximately 4 x10^9 ha. Since the estimated cost of $FeCl_2$ was assumed to be much higher, \$220 t^{1} Fe, the required fertilizer would cost from \$0.22 to \$1.1 x10^9 yr^{-1}. Also, $FeCl_2$ is normally available in 15% solution, which is too low. Therefore, the initial $FeCl_2$ solution would need to be concentrated to a higher level (30%) to reduce transportation costs. This conversion, in turn, might require capital costs for the development of concentration plants. Another uncertainty is the availability of such large quantities of ferrous Fe which will affect the overall price. To account for these uncertainties, the NAS/NAE/IOM Mitigation Panel assumed that the price of Fe fertilizer will be less than 100 times the current price of $FeCl_2$ for a total range of cost for fertilizer of \$0.10 to \$100 x10^9 yr^{-1} (NAS/NAE/IOM, 1991). These cost estimates for Fe, however, would be significantly less for the removal of 1 Gt C yr^{-1}, the upper estimate resulting from the consensus of participants at the 1991 ASLO meeting. In the case of sequestering 1 Gt C yr^{-1}, it was estimated that a total of 100,000 to 1 x10^6 t of Fe-containing fertilizer would need to be applied to the ocean.

The major engineering issue is how to physically fertilize the oceans. The large area to be covered and the relatively small amounts of Fe to be delivered, suggest aerial delivery. The question of ship vs airplane delivery has also been considered, but most estimates assume some type of ship delivery system will be used. The most complete cost analysis was performed by the Mitigation Panel in which it was estimated that ship costs and operations would be \$9 x10^9 yr^{-1} plus \$1 x10^9 yr^{-1} for general system operations, for a total operations cost of \$10 x10^9 yr^{-1} (NAS/NAE/IOM, 1991). Adding these operating costs to the cost of Fe resulted in an estimated cost range of \$10 to \$110 x10^9 yr^{-1} to mitigate 1.8 to 3 Gt C, or \$5 t^{-1} C yr^{-1} to \$60 t^{-1} C yr^{-1}. The upper end of this range corresponds reasonably well to the previous estimates of \$50 to \$100 t^{-1} C yr^{-1} for potential carbon sequestering of 1 Gt C yr^{-1}, presented by EPRI at the 1990 NRC workshop.

3.3 Major Environmental Issues of Phytoplankton

In addition to the many technical and economic uncertainties, several environmental issues have been raised about this ocean engineering experiment. It appears generally accepted that one of the effects of Fe fertilization may be to change species dominance of the primary producers and this may have profound, though uncertain, effects on the entire food chain. For example, Hudson and Morel (1990) suggested that cell size as well as cellular requirements may be an important factor relating Fe availability to the nitrogenous nutrition of marine phytoplankton. Goldman (1988) proposed that small phytoplankton grow on rapidly reduced organic N (the microbial loop), while larger species of phytoplankton are primarily responsible for consumption of NO_3 and new production. Size-differentiated limitation of growth rates may explain Martin and his co-workers observations of increased numbers of large diatoms upon addition of Fe to samples from iron-limited waters (Martin and Fitzwater, 1988; Martin et al., 1989). More recently, metal enrichment in the Southern Ocean showed a selection advantage for diatoms upon addition of Fe (Buma, 1990). This change in species dominance is expected to set up a food chain consisting of larger grazers and, in turn, other consumers, including the production of krill, fish, whales, and birds. In general, it would be expected that such a change in the biotic community, which would become similar to near-shore conditions, would result in a marked increase in the settling of this primary production (i.e., organic C) below the surface waters (Mitchell and Holm-Hansen, 1991).

A second and equally important environmental concern, demonstrated in several model studies (Knox and McElroy, 1984; Toggweiler and Sarmiento, 1985; Sarmiento, et al, 1988; Broecker, 1990; Peng and Broecker, 1991; Joos et al, 1991; Sarmiento and Orr, 1991), is that the increased organic matter produced through the enhanced primary production that will end up in the deeper part of the oceans would result in a decrease in deep water O_2 levels, possibly resulting in anoxic conditions. The 3-D model results of Sarmiento and Orr (1991) are the most interesting. These authors found that the loss of O_2 in a 1600-yr Southern Ocean nutrient-depletion scenario was much smaller than that obtained by box model studies (Knox and McElroy, 1984; Toggweiler and Sarmiento,

1985; Sarmiento et al, 1988). The three-box model scenario, for example, drove the ocean nearly to complete anoxia, while the 3-D model results suggested that anoxia would be isolated only to the southwestern Indian Ocean.

Another critical environmental consideration, illustrated in the model results, is the possible recycling of the organic matter back to the surface layers if the Fe fertilization process once started is stopped. Joos et al. (1991) concluded from their model experiments that Fe addition must be continued year after year. If it is stopped, then much of the CO_2 drawn into the Antarctic Ocean would be released again to the atmosphere. Sarmiento and Orr (1991), using a 3-D ocean model, also stopped the Fe fertilization process after 50 yr. In this sensitivity test, the excess CO_2 captured by the ocean during the 50 yr gradually escaped back into the atmosphere during the subsequent years of the simulation. Therefore, the biological pump must maintain surface nutrients and total C at lower levels, or CO_2 will escape back into the atmosphere. If these model results prove to be correct, it seems unlikely that any large-scale test of the Fe fertilization will be forthcoming. On the other hand, smaller-scale experiments to test the feasibility of annually fertilizing the Southern Ocean and to learn more about the basic science of this oceanic system seem to be in the realm of possibility during the next decade.

4.0 MACROALGAE AS CO_2 SINKS

The concept of ocean farms using macroalgae (seaweeds) has been considered since 1974 when a marine biomass energy program was initiated by the American Gas Association (AGA) and the U.S. Energy Research and Development Administration (ERDA), the forerunner of the Department of Energy (DOE). Two years later the prime contract was awarded to the General Electric Company, and in 1977 AGA's research program was transferred to the Gas Research Institute (GRI). Over a period of about 10 yr nearly \$20 $\times 10^6$ was spent under the joint sponsorship of GRI and DOE to utilize the concept of growing macroalgae on an open ocean farm for conversion by anaerobic digestion to pipeline quality substitute gas, methane, and other possible by-products (e.g., fertilizer, animal feed supplement, other chemicals).

One of the major problems with the ocean farm design was that oceanic areas that would support marine algae were also nutrient-limited for most of the year because of a lack of upwelling. Therefore, fertilizing operations were necessary to produce high yields in macroalgal species such as the California giant kelp (*Macrocystis pyrifera*), which were to be grown on man-made structures in the open ocean. The most astounding result from the analysis of this concept was that the cost of energy to upwell nutrient-rich waters to the ocean farm would be 6 to 7 times greater than the energy produced by the farm (i.e., synthetic gas). In the conceptual design of the algal farm, it was assumed that the upwelling system would be a direct displacement wave pump, with an upwelling requirement of 3217 l/min/ha. The hypothetical ocean farm was assumed to be about 3 $\times 10^5$ ha; this size farm would theoretically contribute about 3.2 $\times 10^{14}$ kJ of substitute natural gas.

In addition to the major technical issues mentioned above, several other potential impacts were also identified (Ritschard et al., 1981). These included the impact of kelp farming on ocean O_2 budgets because of the particulate organic matter flux from the surface to the deep water column. Also, the dynamics of upwelled water especially related to nutrient availability and uptake would need to be understood more fully. From an engineering and operations perspective, a critical farm experiment was suggested to test the conceptual design structure, including its associated pumps (wave-driven) and attachment lines. A small test farm was actually established off the coast of southern California, but during a severe winter's storm it was completely washed away. The concept of ocean algal farms as an energy source was abandoned in the 1980s for both technical and economic reasons.

The open ocean macroalgal farm has been re-examined by EPRI's CO_2 mitigation program in recent years as a means to sequester atmospheric CO_2 into the oceans (Spencer, 1991). The prototypical ocean farm would be composed of 100, 405 ha floating algal mats (Marcrocystis kelp) attached by 12m tethers about 9m below the mean ocean water surface. The upwelling requirement for such a farm system would be 1982 m^3 m^{-1}. In order to make the process more cost-effective, each farm's system would also contain a central processing plant to convert the algae (approximately 25,000 t d^{-1}) to synthetic gas and other valuable products. It has been estimated that such open ocean systems stimulated by upwelled nutrients could sequester 1 to 5 Gt C yr^{-1} at an estimated annual cost of \$200 t^{-1} C (Alpert et al., 1992).

As part of the CO_2 mitigation program, EPRI is also conducting fundamental studies in other ocean-related areas, including algal refossilization of CO_2 and algal-induced CO_2 fixation, and CO_2 clathrate formation and sequestration. In the first process, CO_2 would be fixed through photosynthesis by calcareous marine algae grown on large open ocean farms (Neushul, 1991). The idea is that the inorganic C would remain sequestered for long periods if it sank into the deep oceans and organic C would be digested and decomposed by bacterial or grazers. The problems with this scheme are the rates of CO_2 fixation in these species, and the uncertainties about the flux of organic and inorganic C and its ultimate deposition to the deep ocean waters. It should also be noted that when calcium carbonate (calcite) is sequestered, CO_2 can be released to the atmosphere, which defeats the purpose of this process. The formation of CO_2 clathrates is a relatively new scheme that the Japanese have been pursuing recently through a major research program. These clathrates, which are generally on the deep ocean floor, would sequester CO_2 and remove it from the surface layers. A fundamental problem of this option will exist if CO_2 is released to the atmosphere during the injection of liquid CO_2 into the ocean to form the clathrates. EPRI recently has been studying deep injection of CO_2 without clathrate formation. The EPRI studies are still very preliminary, but also critical to all ocean sequestration approaches, since the final repository for the CO_2 will most likely be the oceans.

In sum, the most promising macroalgal system for the open ocean remains the ocean farm using kelp like Macrocystis although there are still many uncertainties, both with the conceptual design and with the economics. Before this process can proceed, it is recommended that there be more quantitative information on the nutrient requirements, the value of the by-products, environmental impacts, and most importantly, studies of where the C sequestered by the ocean farm will eventually reside in the water column. Also, a new conceptual design, including both the structured (seaweed attached) and free-floating systems has been suggested by several scientists familiar with this approach.

5. CONCLUSIONS

Three opportunities exist for large-scale CO_2 mitigation (i.e., removal of at least 1 Gt C

yr^{-1}) in the oceans: the use of phytoplankton through nutrient (i.e., Fe) fertilization, macroalgal farms in the open oceans, and deep sea injection of O_2 to form clathrates. In all cases, these options are plagued by technical, engineering, and environmental uncertainties, which limit our ability to perform sound cost analyses. In Table 1, we summarize the preliminary results of the cost analyses done to date for the macroalgae and phytoplankton options. It should be noted that these estimates assume that the overall systems will operate as designed and will sequester a significant quantity of CO_2 from the atmosphere and transfer it to the bottom layers of the ocean.

Table 1. Potential for C emission reduction and estimated costs of ocean algal mitigation systems

Ocean Algal System	Potential C Sequestered (Gt C yr^{-1})	Estimated Costs (\$ t^{-1} C yr^{-1})
Macroalgae*		
Continental shelf	0.7	300
Open ocean	2-7	200
Phytoplankton+	1	50-100
Phytoplankton#	1.8-3	5-60

* Alpert et al., 1992
\+ Spencer, 1991
\# NAS/NAE/IOM, 1991

In a further comparison of microalgae vs macroalgae it should be noted that the ocean macroalgal farm has a decade or more of previous research and analysis funded by the federal government and by the gas industry, which should provide useful insights to future studies. Also, to sequester an equal amount of CO_2 from the atmosphere an area of about 1 $x10^8$ ha would be needed for ocean farms compared to nearly 40 $x10^9$ ha for the phytoplankton-Fe experiments. Both systems will be in the open ocean to maximize the available space and to take advantage of the nutrient requirements. In addition, both ocean systems will require some application of "fertilizer" whether it is $FeCl_2$ delivered by ships or airplanes, or upwelled nutrients from the lower depths of the ocean. In the case of the macroalgal ocean farm, the issue of how to upwell nutrients is a serious engineering limitation to the overall system. Finally, there are numerous environmental impacts that are possible from either system. Any large ocean engineering experiment would be expected to have unforeseen environmental consequences.

In the mid- to long-term, it seems prudent to pursue options that increase ocean primary productivity. In this regard, a recent workshop, supported by the National Science Foundation through the U.S. JGOFS Planning Office and by the Environmental Protection Agency was held to design a mesoscale experiment (1,000 to 10,000 ha) that would supply Fe to an area of the ocean in sufficient quantities to permit study of plankton response over 3 to 10 days. Workshop participants selected an oceanic region south of the Galapagos Islands for the first Fe enrichment experiment. Estimates of cost for the initial experiment were \$1.5 2 x 10^6, including ship time. Workshop participants agreed that this size venture was large enough to analyze planktonic food web dynamics in response to Fe fertilization, but small enough to avoid any long-term environmental impact. Although the objective of these studies are to test the Fe limitation hypothesis, rather than CO_2 mitigation, the results will be useful in determining if increased levels of Fe in the Southern Ocean will increase the effectiveness of the biological pump, and in turn, sequester more CO_2. The ocean macroalgal farms will also continue to receive some funding from EPRI and others interested in both C sequestration, as well as other by-products, including synthetic gas, chemicals, and fertilizers. The reduction of greenhouse gases will most likely not result from a single technical fix, but rather from a series of mitigation options that can collectively maintain atmospheric levels of CO_2 and other gases at more optimal levels. Basic research, however, will need to be considered over the next decade that addresses the major uncertainties of these ocean-related systems so that they might be considered more seriously, if needed, at some future time.

ACKNOWLEDGMENTS

This work is supported by Lawrence Berkeley Laboratory Exploratory Research & Development project through the U.S. Department of Energy under Contract No. DE-AC03-76SF00098. We would also like to thank James Orr, AOS Program, Princeton University, for his thoughtful review comments.

REFERENCES

Alpert, S.B., Spencer, D.F., and Hidy, G.: 1992, 'Biospheric options for mitigating atmospheric carbon dioxide levels', EPRI Discussion Paper.

Banse, K.: 1990, Limnol. Oceanogr. 35, 772.

Banse, K.: 1991, Limnol. Oceanogr. 36 (in press).

Barnola, J.M., Raynaud, D., Korotkevich, Y.S., and Lorius, D.: 1987, Nature 329, 408.

Broecker, W.S.: 1982a, Prog. Oceanogr. 11, 151.

Broecker, W.S.: 1982b, Geochim. Cosmochim. Acta 46, 1689.

Broecker, W.S.: 1990, Global Biogeochem. Cycles 4, 3.

Buma, A.: 1990, Global Biogeochem. Cycles 4, 5.

Charles, C.D. and Fairbanks, R.C.: 1990, in U. Bleil and J. Thiede (eds.), Geological History of the Polar Oceans: Arctic vs Antarctic, Kluwer Academic Publishers, Dordrecht, pp. 519-539.

Cullen, J.: 1991, Limnol. Oceanogr. 36 (in press).

DeAngelis, M., Barkov, N.I., and Petrov, V.N.: 1987, Nature 325, 318.

Druffel, E.R.M. and Williams, P.M.: 1990, Nature 347, 172.

Dugdale, R.C. and Wilkerson, F.P.: 1990, Global Biogeochem. Cycles 4, 13.

El-Sayed, S.Z.: 1985, 'Plankton of the Antarctic Seas', in W.R. Siegfried, et al. (eds.), Antarctic Nutrient Cycles and Food Webs, Springer-Verlag, New York, pp. 135-153.

Frost, B.: 1991, Limnol. Oceanogr, 36 (in press).

Goldman, J.C.: 1988, 'Spatial and temporal discontinuities of biological processes in pelagic surface waters', in B.J. Rothschild (ed.), Towards a theory on apelagicheory on Biological and Physical Processes in the World Ocean, Kluwer Academic Publishers, Dordrecht, pp. 273-296.

Grobe, H., Mackensen, A., Hubberten, H.-W., Spiess, V., and Fütterer, D.K.: 1990, in U.Bleil and J. Thiede (eds.), Geological History of the Polar Oceans: Arctic vs Antarctic, Kluwer Academic Publishers, Dordrecht, pp. 539-572.
Hudson, R.J.M. and Morel, F.M.M.: 1990, Limnol. Oceanogr. 35, 1002.
Intergovernmental Panel on Climate Change (IPCC): 1990, Climate Change: The IPCC Scientific Assessment, J.T. Houghton, G.J. Jenkins, and J.J. Ephraumus (eds.), Cambridge University Press, New York.
Joos, F., Sarmiento, J.L., and Siegenthaler, U.: 1991, Nature 349, 772.
Keir, R.S.: 1988, Paleoceanography 3, 413.
Knox, F. and McElroy, M.S.: 1984, J. Geophys. Res. 89, 4629.
Labeyrie, L.D. and Duplessy, J.-C.: 1985, Palaeogr. Palaeoclimatol. Palaeoecol. 50, 217.
Legrand, M., Feniet-Saigne, C., Saltzman, E.S., Germain, C., Barkov, N.I., and Petrov, V.N.: 1991, Nature 350, 144.
Martin, J.H.: 1990, Paleoceanography 5, 1.
Martin, J.H. and Fitzwater, S.E.: 1988, Nature 331, 341.
Martin, J.H. and Gordon, R.M.: 1988, Deep-Sea Res. 35, 177.
Martin, J.H., Gordon, R.M., Fitzwater, S.E., and Broenkow, W.W.: 1989, Deep-Sea Res. 36, 649.
Martin, J.H., Fitzwater, S.E., and Gordon, R.M.: 1990a, Global Biogeochem. Cycles 4, 5.
Martin, J.H., Gordon, R.M., and Fitzwater, S.E.: 1990b, Nature 345, 156.
Mitchell, B.G., and Holm-Hansen, O.: 1991, Deep-Sea Research, 38, 981.
Mitchell, B.G., Brody, E.A., Holm-Hansen, O. McClain, C., and Bishop, J.: 1991, Limnol. Oceanogr., 36 (in press).
Moore, B. and Bolin, B.: 1987, Oceans, 29, 9.
Mortlock, R.A., Charles, C.D., Froelich, P.N., Zibello, M.A., Saltzman, J., Hays, J.D., and Burckle, L.H.: 1991, Nature 351, 220.
National Academy of Science, National Academy of Engineering, Institute of Medicine (NAS/NAE/IOM): 1991, National Policy Implications of Greenhouse Warming: Report of the Mitigation Panel, National Academy Press, Washington DC, pp. 10.9-10.10.
Neftel, A., Oeschger, H., Schwander, J., Stauffer, B., and Zumbrunn, R.: 1982, Nature 295, 220.
Neushul, M.: 1991, 'Algae refossilization of atmospheric carbon dioxide,' EPRI Report (EAK7-401) Electric Power Research Institute, Palo Alto, CA.
Orr, J.C. and Sarmiento, J.L: 1992, Water, Air, and Soil Pollution (this volume).
Pedersen, T.F., Pickering, M., Vogel, J.S., Southon, J.N., and Nelson, D.E.: 1988, Paleoceanography 3, 157.
Peng, T-H. and Broecker, W.S.: 1991, Nature 349, 227.
Price, N.M., Anderson, L.F., and Morel, F.M.M.: 1991, Deep-Sea Research 38, 1361.
Ritschard, R., Berg, V., and Killeen, S.: 1981, 'Proceedings of a workshop on environmental impacts of marine biomass', Gas Research Institute (GRI-80.0076), Chicago, IL.
Sarmiento, J.L. and Toggweiler, L.R.: 1984, Nature 308, 621.
Sarmiento, J.L., Herbert, T.D., and Toggweiler, J. R.: 1988, Global Biogeochem. Cycles 2, 115.
Sarmiento, J.L. and Orr, J.C.: 1991, Limnol. Oceanogr. 36 (in press).
Sarnthein, M., Winn, K., and Zahn, R.: 1987, 'Paleoproductivity of oceanic upwelling

and the effect on atmospheric CO_2 and climatic change during glaciation times', in W.H. Berger and L.D. Labeyrie (eds.), Abrupt Climatic Change, D. Reikel, Hingham, MA, pp. 331-337.

Siegenthaler, U. and Wenk, T.: 1984, Nature 308, 624.

Spencer, D.F.: 1991, 'Open ocean macroalgal farms for CO_2 mitigation and energy production', Proceedings of IEA Conference on Technology Responses to Global Environmental Challenges, 6-8 November 1991.

Sugimura, Y. and Suzuki, Y.: 1988, Mar. Chem. 24, 105.

Sunda, W.G., Swift, D.G., and Huntsman, S.A.: 1991, Nature 351, 55.

Tans, P.P., Fung, I.Y., and Takahashi, T.: 1990, Science 24, 1431.

Toggweiler, J.R. and Sarmiento, J.L.: 1985, 'Glacial to interglacial changes in atmospheric carbon dioxide: The critical role of ocean surface water in high latitudes', in E.T. Sundquist and W.S. Broecker (eds.), The Carbon Cycle and Atmospheric CO_2: Natural Variations Archean to Present, Geophys. Monogr. Ser. 32, AGU, Washington, DC, pp. 163-184.

MODELING CARBON FLUXES

PART IV

THE RESPONSE OF TERRESTRIAL C STORAGE TO CLIMATE CHANGE: MODELING C DYNAMICS AT VARYING TEMPORAL AND SPATIAL SCALES

T.M. SMITH, J.F. WEISHAMPEL and H.H. SHUGART
Department of Environmental Sciences
University of Virginia
Charlottesville, VA 22903 USA

G.B. BONAN
National Center for Atmospheric Research
P.O. Box 3000
Boulder, CO 80307 USA

ABSTRACT. The response of terrestrial **C** storage to GCM derived climate change scenarios was investigated over a range of temporal and spatial scales. The potential changes in the global distribution of major ecosystem complexes were examined by combining changes in land cover and corresponding soil type with **C** storage estimates for each of the ecosystem types. All scenarios consistently showed an overall increase in global carbon storage. On a more regional basis, the potential impacts of climate change on the structure, composition and biomass dynamics of major forest types within the North American Boreal zone were investigated using individual based stand models. Biomass fluxes were found to be dependent on the tree species, site and GCM parameters. A method to simulate corresponding changes in intra- and inter-annual patterns of CO_2 flux by combining a gap model with an ecosystem model which incorporates photosynthesis, respiration (both canopy and decomposer) and transpiration is demonstrated.

1. Introduction

Processes that affect terrestrial **C** flux operate at varying spatio-temporal scales. The **C** dynamics from forested ecosystems, representing the largest component of the terrestrial biotic **C** pool, function at multiple levels (Shugart *et al.*, 1986; Urban *et al.*, 1987; King *et al.*, 1990). The cellular to leaf scales respond on the order of seconds to years representing diurnal and seasonal fluctuations. Processes which define these scales, such as photosynthesis, respiration, and leaf abscission, are influenced by conditions such as inter-cellular CO_2 concentrations and leaf surface temperature. The tree to gap scales respond on the order of years to centuries. Defining processes include **C** allocation, demographics, and succession, and are influenced by disturbances and weather patterns. The forest stand to biome scales respond on the order of centuries to millenia. Defining processes include species migration and extinction, and are

Water, Air, and Soil Pollution **64**: 307–326, 1992.
© 1992 *Kluwer Academic Publishers.*

influenced primarily by climate.

Models simulating C flux typically are restricted to one of the scales described above. As a result, voids exist as certain processes which are scale dependent may not be extrapolated to other scales without necessary feedbacks. In this paper, we review multiple scale approaches to modeling terrestrial C dynamics and present a method that combines an existing gap model with a process based model to predict intra-annual C fluxes under changing climate scenarios.

2. Global Scale C Dynamics

2.1. HOLDRIDGE LIFE-ZONE MODEL

The Holdridge Life-Zone Classification method (Holdridge, 1947, 1964) relates vegetation associations to climate indices. The life-zones are depicted by a series of hexagons formed by intersecting axes of climate variables on a triangular coordinate system (Figure 1). Two climate variables, biotemperature and annual precipitation, determine the classification. Average biotemperature is the temperature sum over a year with the unit temperature values (e.g., average daily weekly or monthly temperatures) that are used in computing the index. A temperature value is set to 0°C if it is less than 0°C.

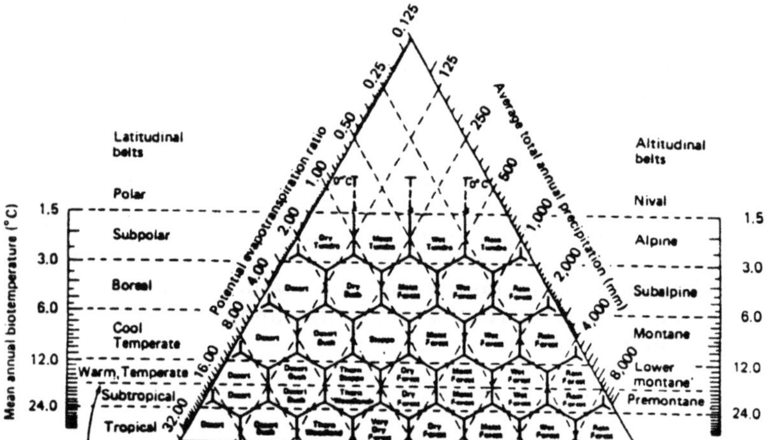

Figure 1. Holdridge Life-Zone classification scheme.

Average annual precipitation forms two sides of an equilateral triangle. The ratio of potential evapo-transpiration (PET) to precipitation forms the third side. Holdridge (1959) assumes, based on the data from several ecosystem types, that PET is proportional to biotemperature (constant of proportionality, 58.93). The axis for biotemperature is oriented perpendicularly to the base of the triangle.

One additional division in the Holdridge classification is based on the presence of a killing frost. This division is along a critical temperature line that divides hexagons between 12 and 24°C into warm temperate and subtropical zones. The complete Holdridge Classification at this level includes 37 life-zones.

2.2. MODELING GLOBAL C FLUXES UNDER CLIMATE CHANGE

In manner similar to Emanuel *et al.* (1985), Smith *et al.* (1992, in press) estimated the effects of CO_2-induced climate change on patterns of global **C** storage using the Holdridge Classification. Each life-zone was assigned a value for **C** in both above-ground biomass and soil. Soil **C** estimates for the Holdridge Life-Zones were taken directly from Post *et al.* (1982). Estimates of **C** in above-ground biomass for each life-zone were calculated using data from Olson *et al.* (1983). Because Olson *et al.* (1983) does not classify ecosystems according to the Holdridge Life-Zone system, it was necessary to estimate **C** values for the life-zones directly from the climate variables used in the Holdridge Life-Zone Classification. By combining the biotemperature and annual precipitation data for each 0.5°x0.5° cell with the **C** estimates from the corresponding cell of the Olson *et al.* database (also 0.5°x0.5°), an average value of **C** storage in aboveground biomass was calculated for the climate space corresponding to each Holdridge Life-Zone.

Expected current distributions of life-zones were mapped using a global climate data base of mean monthly precipitation and temperature at a 0.5°x0.5° (latitude and longitude) resolution (Leemans and Cramer, 1990). Simulations of current ($1xCO_2$) and $2xCO_2$ climates from four General Circulation Models (GCMs) (Table 1) were used to construct change scenarios. Changes in mean monthly precipitation and temperature were calculated for each GCM scenario for each

Table 1. General circulation models used to construct climate change scenarios.

GCM	Resolution (lat/lon)	Change in Mean Global: Temperature (°)	Change in Mean Global: Precipitation (%)
[1]Oregon State University (OSU)	4 x 5°	2.84	7.8
[2]Geophysical Fluid Dynamics Laboratory (GFDL)	4.5 x 7.5°	4.00	8.7
[3]Goddard Institute for Space Studies (GISS)	7.8 x 10°	4.20	11.0
[4]United Kingdom Meteorological Office (UKMO)	5 x 7.5°	5.20	15.0

[1] Schlesinger and Zhao, (1988)
[2] Manabe and Wetherald, (1987)
[3] Hansen *et al.*, (1988)
[4] Mitchell, (1983)

computational grid element by taking the difference between simulated current and $2xCO_2$ climates. Temperatures were expressed as absolute difference ($2xCO_2$-$1xCO_2$) and precipitation as the ratio of $2xCO_2$ to $1xCO_2$. These data from each GCM were interpolated to 0.5°x0.5° using the same technique as applied in the development of the database for current climate (Leemans and Cramer, 1990). The technique used was a triangulation of all datapoints (algorithm developed by Green and Sibson, 1978) followed by a smooth surface fitting (Akima, 1978). Changes in monthly precipitation and temperature were then applied to the global climate data base to provide a change scenario. The altered data bases corresponding to each of the four GCM scenarios were then used to reclassify the grid cells (0.5°x0.5°) using the Holdridge Classification. This approach provides a single estimate of above- and below-ground **C** for each life-zone, and therefore, changes in **C** storage under the $2xCO_2$ scenarios are a function of changes in the extent of the various life-zones.

Global terrestrial **C** increased under all four climate change scenarios (Table 2). The increase ranged from 0.4% (8.5 Gt) for the UKMO scenario to 9.5% (180.5 Gt) for the OSU scenario. Although the UKMO and GFDL derived scenarios show a decrease in soil **C**; above-ground **C** increased under all four scenarios. If the atmospheric **C** is not replenished, these potential increases in terrestrial **C** storage would result in a reduction in atmospheric CO_2 levels from concentrations used in the $2xCO_2$ GCM simulations. These potential reductions range from 4 to 85 ppm. The increase in **C** storage is primarily due to the; 1) poleward shift of the forest zones, with an increase in the extent of tropical forest and a northern movement of the boreal forest zone into areas currently occupied by tundra, and 2) a decrease in the global extent of desert.

Table 2. Changes in **C** storage (Gt) in above-ground biomass and soil for four climate change scenarios. Values in parentheses are percentage change from current.

Scenario	Aboveground biomass	Soil	Total	Change
Current	737.2	1158.5	1895.7	-
OSU	860.4 (16.7)	1215.8 (4.9)	2076.2	180.5 (9.5)
GFDL	782.3 (6.1)	1151.3 (-0.6)	1933.6	37.9 (2.0)
GISS	829.6 (12.5)	1213.0 (4.7)	2042.6	146.9 (7.7)
UKMO	765.2 (3.8)	1139.0 (-1.7)	1904.2	8.5 (0.4)

2.3. CAVEATS OF MODEL ASSUMPTIONS

Despite the qualitative agreement among scenarios, the results discussed above must be viewed in light of the limitation of the methodology. The Holdridge Life-Zone model is a climate classification rather than a classification based on actual vegetation distribution, such as the system developed by Box (1981). As with any classification scheme, the Holdridge model is an abstraction of the actual pattern. The limited number of categories of vegetation or biome types and the fixed **C** values within a type results in a coarse resolution of **C** dynamics. This discrete view of vegetation pattern results in a threshold phenomenon when applying change scenarios. **C** values for a grid cell will not vary with climate until the climate varies beyond the boundary conditions of the zone to which the cell is classified. As such, the dynamics are inherently scaled to the classification scheme, and the more aggregated the scheme, the more pronounced the threshold effect.

This approach has the Clementsian underpinnings that the vegetation unit or biome moves as a fixed unit in time and space to a stable endpoint determined by the macroclimate. This assumption may not hold, especially under conditions where the changed climate has no current analogue. Climate-vegetation classification models are correlative and based on a limited set of variables. Although the Holdridge bioclimatic indices may do a sufficient job of bounding present vegetation patterns, the actual patterns are a function of additional factors not explicitly considered in the model (i.e., soils) which may vary differently (both temporally and spatially) under the changed climate conditions. For example, a lag between climate and soils may also have a major influence on changes in vegetation response, especially in those areas where predicted transitions would be dependent on large scale changes in the current soil conditions (i.e., polar desert tundra-boreal forest transitions).

Perhaps most importantly, the results represent equilibrium solutions for both climate (i.e., $2xCO_2$) and vegetation dynamics. Although changes in the climate pattern as suggested by the GCM simulations may occur on a timescale of decades to a century, the response of vegetation and soils to those changes may occur at different and varying timescales. In areas where biomass values decrease due to moisture stress (i.e., higher PET ratio) the changes may occur quickly as the environmental conditions become such that the present vegetation can no longer be supported (i.e., forest to grassland).

In contrast, increases in biomass may require much longer periods of time. In some cases the present vegetation may show increased growth or recruitment under the more favorable conditions. However, major shifts of forest type (i.e., warm temperate to tropical rainforest, or boreal to cool temperate forest) are dependent on the movement of species across the landscape and the ability of new species to invade existent communities. These changes in forest type would operate on timescales related to the lifecycle or longevity of the component species. In the case of many forest species this may be on the order of centuries.

Changes in soil **C** may require longer periods of time to track changes in climate conditions. This difference in the timescale associated with vegetation and soil response to changing climate could have a major influence on the temporal dynamics of **C** flux between atmospheric and terrestrial pools. If changes in soil **C** are slow in comparison to vegetation, the initial differences between the scenarios may not be as great as expected under the combined **C** storage in soil and above-ground vegetation estimates (Table 2). Although all scenarios predict an increase in above-ground biomass, they differ in their predictions for soil C. The GFDL and UKMO predict a decline in soil **C** while the GISS and OSU scenarios predict an increase. It is this divergence in patterns of soil **C** which account for the large differences between scenarios in total **C**.

3. Gap Scale C Dynamics Across A Region

The equilibrium based approach discussed above implicitly assumes a time-scale sufficient for vegetation migration and eventual equilibrium of vegetation to the new "changed" climate patterns. In contrast, a dynamic view requires the explicit consideration of plant demographic processes to capture the transient nature of vegetation response to changing environmental conditions. There are numerous models of vegetation dynamics which simulate the demographics of plant populations (Shugart and West, 1980), however, we will focus on a single class of models known as gap models.

3.1. INDIVIDUAL-BASED FOREST GAP MODELS

Forest gap models simulate the establishment, growth, and mortality of trees at an annual time-step on a plot of defined size (Botkin *et al.*, 1972; Shugart, 1984). Although the models differ in their inclusion of processes which may be important in the dynamics of the particular site being simulated (i.e., hurricane disturbance, flooding), all forest gap models share a common set of characteristics and demographic processes.

Each individual plant is modeled as a unique entity with respect to the processes of establishment, growth, and mortality. This allows the model to track species- and size-specific demographic behaviors. The model structure includes two features important to a dynamic description of vegetation: 1) the response of the individual plant to the prevailing environmental conditions, and 2) how the individual modifies those environmental conditions (i.e., the feedback between vegetation structure/composition and the environment).

3.1.1. *Response of Individual to Environment.* The growth of an individual is calculated using a function that is species specific and predicts, under optimal conditions, an expected growth increment given an individual's current size (Figure 2). This optimal increment is then modified by a set of environmental response functions, and the realized increment is added to the individual.

Environmental responses are modeled via a "constrained potential" paradigm. In this, an individual has a maximum potential behavior under optimal

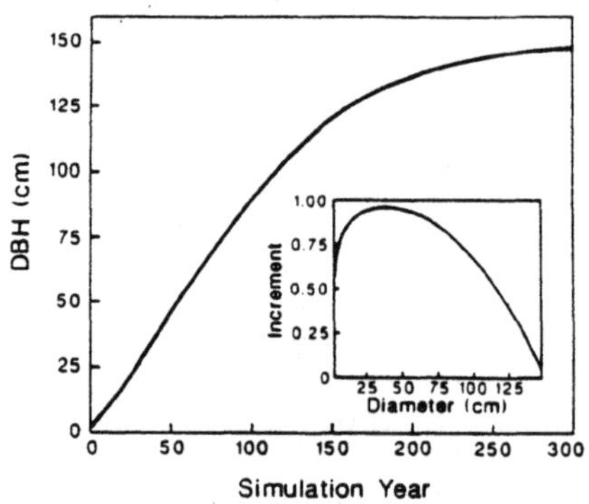

Figure 2. Diameter growth as a function of tree age, and (inset) diameter increment as a function of current diameter, as simulated by gap models.

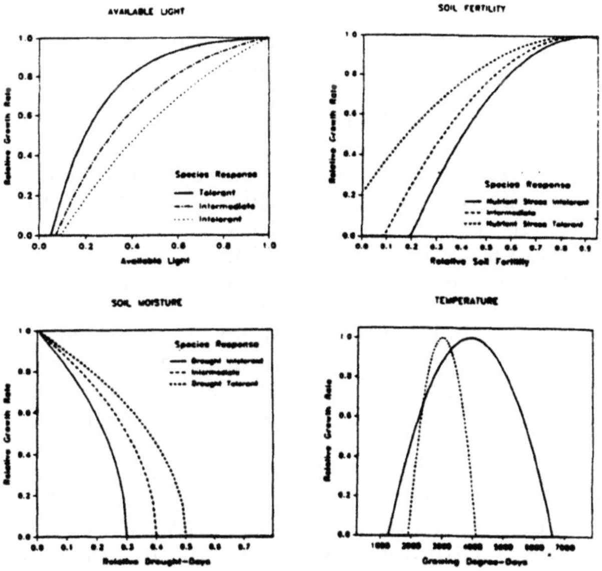

Figure 3. Environmental response functions implemented by traditional gap models. Light, soil fertility, and soil moisture are ranked tolerances, whereas temperature is a function of minimum and maximum growing degree-days.

conditions (i.e., maximum diameter increment, survivorship, or establishment rate). This optimum is then reduced according to the environmental context of the plot (e.g., shading, drought), to yield the realized behavior under ambient conditions. The functions describing species response to environmental resources tend to be generic curves that scale between 0.0 and 1.0, and species are often categorized into a small number of functional types (Figure 3).

The horizontal position of each individual on the plot is not considered. The assumption is made that each individual on the plot is influenced by (and influences) the growth of all other individuals on the plot. Given this assumption of "horizontal homogeneity", the size of the simulated plot is critical (Shugart and West, 1979). The spatial scale at which these models operate is an area corresponding to the zone of influence of a single individual of maximum size. This allows for an individual growing on the plot to achieve maximum size while at the same time allowing for the death of a large individual to significantly influence the environment on the plot (i.e., gap formation).

Competition in the model is indirect and depends on the relative performance of the individuals under the environmental conditions on the model plot. These environmental conditions may be influenced by the trees themselves (i.e., a tree's leaf area influences light available beneath it), or may be modeled as extrinsic (i.e., temperature). Competitive success thus depends on the environmental conditions on the plot, which species are present, and the relative sizes of the individuals; each of these varies through time in the model.

The death of individuals is modeled as a stochastic process. Most gap models have two components of mortality: age-related and stress induced. The age-related component applies equally to all individuals of a species and depends on the maximum expected longevity for the species. Stress is defined with respect to a minimal growth increment; individuals failing to meet this minimal condition are subjected to an elevated mortality rate. Establishment and regeneration are largely stochastic, with maximum potential establishment rates constrained by the same environmental factors that modify growth.

3.1.2. *Effect of Individual on Environment.* Although horizontal homogeneity on the plot is assumed, the vertical structure of the canopy is modeled explicitly. The sizes of individuals (height and leaf area, which are allometric on diameter) are used to construct a vertical leaf-area profile. Using a light extinction equation, the vertical profile of available light is then calculated so that the light environment for each individual can be defined.

All gap models simulate individual tree influence on light availability at height intervals on the plot. Plant influence on other environmental factors are incorporated to varying degrees in different versions of the model; these include soil moisture, fertility, temperature, as well as disturbances such as fire and herbivory.

Plant influence on soil moisture has been incorporated using a variety of soil hydrology/evapotranspiration models (i.e., Priestly-Taylor, Penman-Monteith) which allow for the distribution of roots and leaf area of the plot to influence the soil moisture profile (Pastor and Post, 1986; Bonan, 1989; Martin, 1990). In return, soil moisture integrated over the growing season is used to modify plant growth (e.g., root biomass and leaf area).

Soil nutrient availability on the plot is influenced by the plants in terms of the quantity and quality of litter input into the soil organic layer for decomposition (Aber *et al.*, 1982; Weinstein *et al.*, 1982; Pastor and Post, 1986). Species are classified into functional categories based on leaf characteristics (e.g., C/N, lignin content) which relate to their decomposability. Nitrogen dynamics (and other mineral nutrients) of the plot are simulated as a function of the prevailing environmental conditions at the soil surface and the quantity and quality of litter material.

Soil temperature is influenced by the collective leaf area of individuals on the plot by reducing incoming radiation to the soil surface as well as the insulating effect of litter material on the plot. In turn, soil temperatures are needed as input for calculating decomposition rates and the development of a permafrost layer at northern latitudes. These processes, together with direct temperature effects on growth, provide a feedback to plant processes (Bonan, 1989, 1990a).

In actuality, the modeling framework outlined above has two components or modules; one involving plant demographics, and the other involving the processes, both climatic and biophysical, which define the environment on each plot. An extensive discussion of the biophysical models is beyond the scope of this manuscript, but a complete discussion of these models and their incorporation into the gap model structure can be found in the references provided in the above text.

In general, these biophysical models require the transfer of certain parameters from the demographic module describing the vegetation of the plot and in return define the environmental conditions as a function of the geology and soils, prevailing climate and the modifying influence of the plant structure. The gap model approach has been applied to a wide array of forested systems ranging from boreal (Bonan, 1988; Leemans and Prentice, 1987) to subtropical rainforest (Shugart et. al., 1980; Doyle, 1981) and has recently been extended to grassland dominated ecosystems (Coffin and Lauenroth, 1989).

3.2. MODELING FOREST C DYNAMICS UNDER CLIMATE CHANGE

Gap models have been applied to examine the response of forested systems to climate changes, both in the reconstruction of prehistoric Quaternary forests (Solomon *et al.*, 1980, 1981; Solomon and Shugart, 1984; Solomon and Webb, 1985; Bonan and Hayden, 1990) as well as to project possible consequences of future climate change (Solomon *et al.*, 1984; Solomon, 1986; Pastor and Post, 1988; Bonan *et al.*, 1990; Urban and Shugart, 1989; Overpeck *et al.*, 1990). In contrast to the modeling approach discussed in the previous section, the gap model approach is high resolution in that it can predict species composition, vegetation structure and associated productivity and standing biomass through time, however, it is limited in spatial scale. That is, it is limited in the spatial extent to which the results can be extrapolated. The reason for this limitation is that the information required to parameterize/initialize a model which can address changes in these state variables (i.e., species composition and productivity through time) relate to site specific features such as topographic position, soil characteristics, land-use history, history of disturbance, and present vegetation structure, all of which may vary over short distances. The application of fine resolution models to provide total coverage over broad regions would be virtually impossible due to both computational and data limitations. As an alternative, sampling approaches have been put forward to provide large scale coverage over broad environmental gradients (Solomon, 1986; Bonan, 1990a,b).

The large scale Holdridge model discussed earlier can be used to develop a systematic process of site selection based on the implied shifts in the distribution of major ecosystem complexes. The shift in the global boreal forest zone for the UKMO and OSU based climate change scenarios shown in Table 3. involves an array of transitions between life zones and can be used to locate sites of potential importance for application of stand level models. We will use examples from the North American boreal zone.

Table 3. Major shifts in Holdridge Life-Zones for the boreal forest region (current area is 1503 ha x 10^5).

		Surface Area (ha x 10^5)			
From	To	UKMO	GFDL	GISS	OSU
Boreal Forest	Cool Temperate Steppe	368	674	185	269
Boreal Forest	Cool Temperate Forest	938	685	668	471
Tundra (including Polar Desert)	Boreal Forest	882	818	682	642
	Stable	130	126	597	742

A forest gap model (Bonan, 1988) was applied to a series of sites in the North American boreal zone to explore the implied transitions based on the Holdridge Life Zone System (Smith *et al.*, 1992, in press). The model simulates plant demographics following the approach outlined above for gap models and includes biophysical models for solar radiation, permafrost, decomposition and soil hydrology/ evapotranspiration (Bonan, 1989). Site selection was based on: 1) representation of differing life zone transitions based on the Holdridge analyses, and 2) availability of data for model validation under current climate. Site descriptions, parameterization and validation of the model for the locations are discussed in Bonan (1988, 1989).

For each site, stand structure and composition were simulated under current climate conditions, defined by monthly statistics (mean and standard deviation) for precipitation, temperature and cloudiness. The resulting stand description after 500 yr of simulation was used to initialize simulations for control (continued current climate statistics) and changed climate conditions based on the four GCMs discussed earlier. Changes in monthly temperature and precipitation for each site were gleaned from the global databases for current and changed climate for the 0.5° cell corresponding to the site location. To provide a transient climate scenario based on the temperature and precipitation changes from the equilibrium GCM simulations, the changes were assumed to occur over a 70 yr period (years 0 to 70 of simulation). Annual values for each month were then determined by linearly interpolating between current and $2xCO_2$ values over this period. Climate statistics were then assumed to remain constant after this period for the remainder of the simulation. Only a local seed pool was considered in light of the relatively short duration of the simulations relative to the timescale of observed patterns of species migration (Davis and Botkin, 1985; Davis, 1984, 1989).

The simulations from a site on a south-facing slope in central Alaska are presented in Figure 4. The site is currently dominated by white spruce (*Picea glauca*), birch (*Betula papyrifera*) and aspen (*Populus tremuloides*). Under the warmer, drier conditions predicted under all scenarios, the stand declines as a function of moisture stress. The site would eventually be dominated by herbaceous ground cover. The differences in the timing of the decline among the scenarios reflects the varying degree of warming predicted. This result is in agreement with the transition from boreal forest to cool temperate steppe (grassland) predicted for the area using the Holdridge Life Zone System.

In contrast to the predicted decline in forest cover on the south-facing slope, simulation of a stand in the same location on a north-facing slope (Figure 5a) shows an increase in biomass of the dominant species, black spruce (*Picea mariana*). This difference is due to the link between the stand water balance and solar radiation. The lower input of solar radiation to the northfacing slope (in comparison to the south-facing) reduces the moisture stress associated with increasing temperature.

Simulation of a site in the boreal forest-tundra transition zone of northern Quebec, Canada is shown in Figure 5b. The site is currently an open-canopy woodland of black spruce with relatively low productivity and standing biomass. With the warming for this region predicted under all scenarios, the stand increases in biomass to a closed-canopy forest. This transition would be associated with a decline in the present ground cover of moss and lichen. The occurrence of fire would facilitate this transition in that the establishment of seedlings on the site is currently hindered by ground cover.

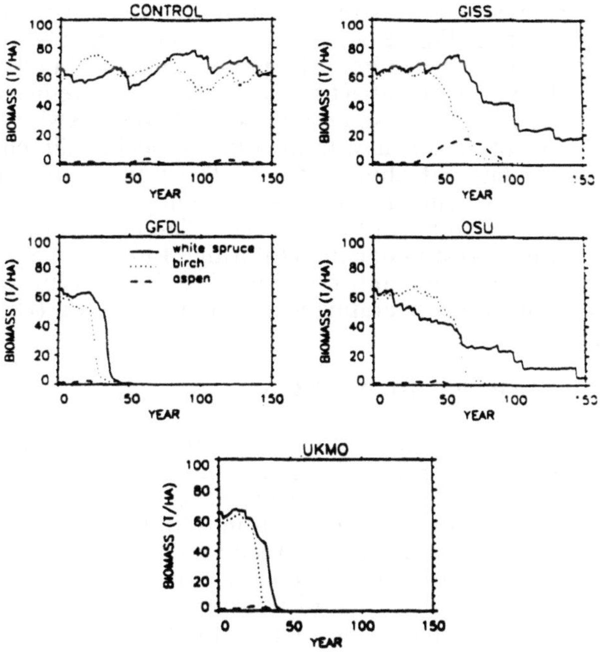

Figure 4. Modeled biomass fluxes on a south-facing slope in central Alaska simulated under CO_2-induced climate change scenarios.

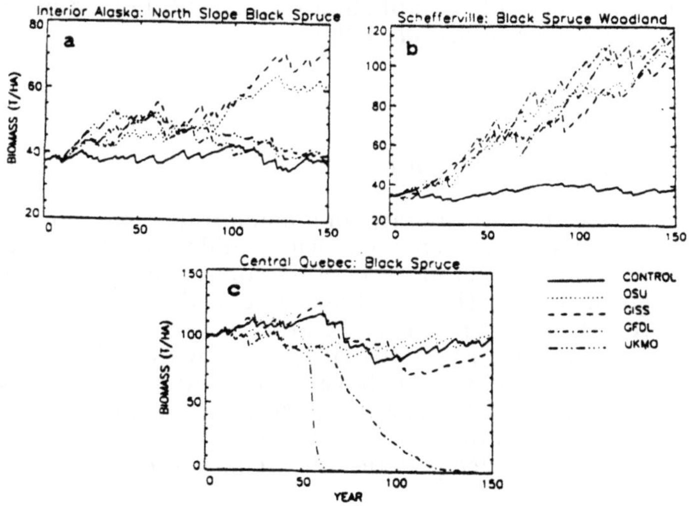

Figure 5. Black spruce biomass fluxes simulated for: a) a north-facing slope in central Alaska; b) a stand in the boreal forest-tundra transition zone in northern Quebec; and c) a stand in central Quebec under different GCM scenarios.

In contrast to the northern Quebec site, the predicted dynamics of a black spruce stand in central Quebec (Figure 5c) vary among the scenarios. The OSU and GISS scenarios show little change from the pattern predicted under current climate (e.g., control). These results reflect the lower warming predicted for the region by these two GCM's. In contrast, both the GFDL and UKMO scenarios predict a complete dieback of the stand, with the rates of decline directly related to the degree of warming. These declines are not a function of moisture stress, but result from temperatures rising above those currently associated with the southern limit to the distribution of black spruce. The area is predicted to change to a cool temperate forest under the Holdridge Life Zone System, and the establishment of species characteristic of that forest type would be dependent on a range of processes operating at the landscape level (i.e., rates of dispersal and establishment).

3.3. LIMITATIONS OF GAP MODEL APPROACH

These results point out a limitation in our current understanding of the distribution of forest zones at a continental scale. The predicted decline in the boreal forest at its southern boundary under increased temperatures has an implied assumption that the southern distributional limits of the component species are physiologically determined (e.g., exceed temperature tolerance for germination, establishment or growth). In contrast, if competition between boreal and northern hardwood (i.e., cool temperate forest) species is a determining factor in the southern distributional limit of boreal forest, then the predicted decline may not occur. In fact, the increased temperatures may result in increased productivity of these stands and the transition to cool temperate forest would involve the eventual invasion of northern hardwood species into established boreal forest. The rate of this transition would be related to the ability of the species to invade established stands; influenced by rates of disturbances such as fire (Overpeck *et al.*, 1990).

4. Within Stand C Dynamics

4.1. FOREST STAND C FLUX MODEL

Models which simulate **C** flux operate on a much finer timescale than the Holdridge or gap models. One such model which addresses the processes of photosynthesis and transpiration at the level of the forest canopy is the model of Running and Coughlan (1988). The model uses a defined leaf area index (LAI) to quantify the forest structure and treats the processes of canopy interception and evaporation, transpiration, photosynthesis, growth and maintenance respiration, **C** allocation to below and above ground, litterfall, decomposition and **N** mineralization. The model has been applied to simulate water and **C** flux across an environmental gradient at a continental scale and to develop a protocol for validation of predicted patterns using remotely sensed data (Nemani and Running 1989).

Bonan (1991a,b) has developed a stand-level model to examine the patterns of **C** flux in the North American boreal forest. The model combines the ecophysiological approach of Running and Coughlan (1988) with the biophysical model developed by Bonan (1991a). Given forest structure and site conditions, the ecophysiological model calculates stomatal resistance, foliage temperature, soil temperature, soil moisture, evapotranspiration and CO_2 fluxes on a daily timestep (Figure 6). The surface energy budget of an n-layer forest canopy is written in terms of n unknown temperatures. These equations are solved simultaneously, and with the ground surface temperature, and evapotranspiration known, soil temperature and moisture in a multilayer soil are updated. A full description of the biophysical and ecophysiological calculations involved in the model are available in Bonan (1989; 1991a,b).

Required stand level parameters for the model are: canopy height, leaf area index, foliage nitrogen, biomass in sapwood and root, and forest floor biomass and nitrogen. Initial tests of the model for a series of stands in Alaska have found patterns of simulated soil temperature, foliage water potential, and evapotranspiration consistent with observed data (Bonan, 1991c). No data are available on CO_2 flux rates, but simulated soil respiration and decomposition were consistent with observed data (Bonan, 1991b).

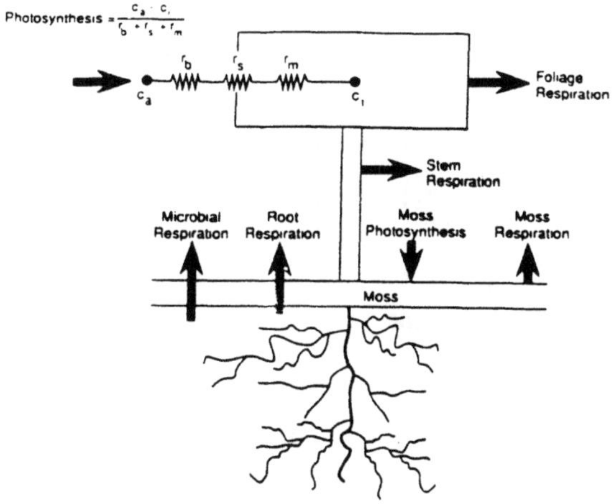

Figure 6. Schematic detailing CO_2 fluxes from the Bonan (1991a,b) physiologically based model.

5. Linking Individual Based Models With CO_2 Flux Models

The models described above each address a set of patterns and processes operating at very different temporal scales. The forest gap models operate at an annual timestep, simulating the demographics of trees over a period of years to centuries. These models can predict changes in forest structure and composition as a function of changing environmental conditions. However, these models do not address the question of C and water flux over the growing season. In contrast, the physiological models of photosynthesis and transpiration at the scale of the forest canopy can address these short term patterns of flux rates, but require that information on the structure of the forest be provided. They do not simulate the changes in forest structure through time.

A methodology has been proposed by King *et al.* (1990) to overcome the temporal scale problems posed by linking the gap models with the physiological models. Additionally, Friend *et al.* (in press) has incorporated physiological processes that simulate within-tree C fluxes directly into the framework of a gap model.

5.1. PREDICTING SEASONAL STAND CO_2 FLUXES UNDER CLIMATE CHANGE

Our approach represents a more indirect method to combine these two model types. To study seasonal and interannual C dynamics under a given set of climate and site conditions, individual-based forest gap models are used initially to simulate forest structure, composition and additional parameters necessary for the C flux model. These data from the forest gap models will then be used as input to initialize the CO_2-flux model described above. Bonan (in review) has used this approach of combining these two models to examine the influence of fire on CO_2 flux rates in the boreal forests of interior Alaska (Figure 7).

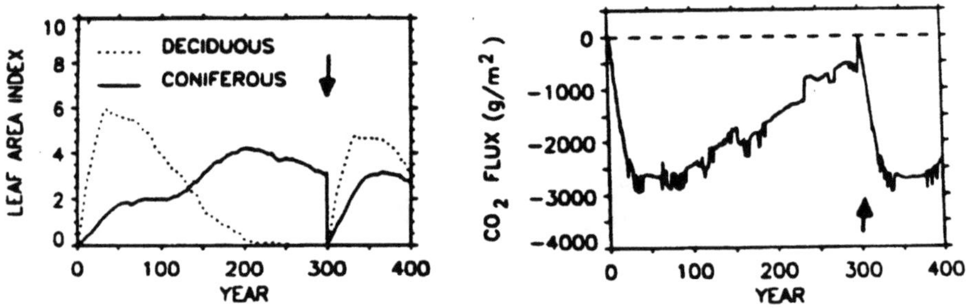

Figure 7. Stand structure and annual CO_2 flux during post-fire succession on a well-drained south-facing slope in central Alaska. The timing of the fire is indicated by the arrow.

This methodology was used to simulate the seasonal CO_2 flux for the site in central Alaska (same as Figure 5a). The gap model was run for a black spruce stand on a north facing slope with permafrost using ramp functions to the projected temperature and precipitation levels predicted by the four GCMs. The C flux model was parameterized with the monthly temperature and precipitation predicted by the four GCMs (Figure 8) for the specified grid cell and the stand characteristics after 150 yr of the gap model run. The seasonal C fluxes for the climate scenarios are shown in Figure 9. The two warmer scenarios (GFDL and UKMO) show a reduction in C fixed over a year, while the less drastic scenarios have fluxes comparable to the current climate. This suggests that microbial and/or moss respiration has increased (and/or moss photosynthesis has decreased) with all four scenarios given the increase in black spruce biomass (Figure 5a) for the OSU and GISS scenarios and the negligible change with the GFDL and UKMO scenarios. Linking the model in this fashion may help to understand the seasonal CO_2 fluctuations which have shown an increase in amplitude (Bacastow *et al.*, 1985) which has been attributed to the metabolism of the biosphere (Houghton, 1987).

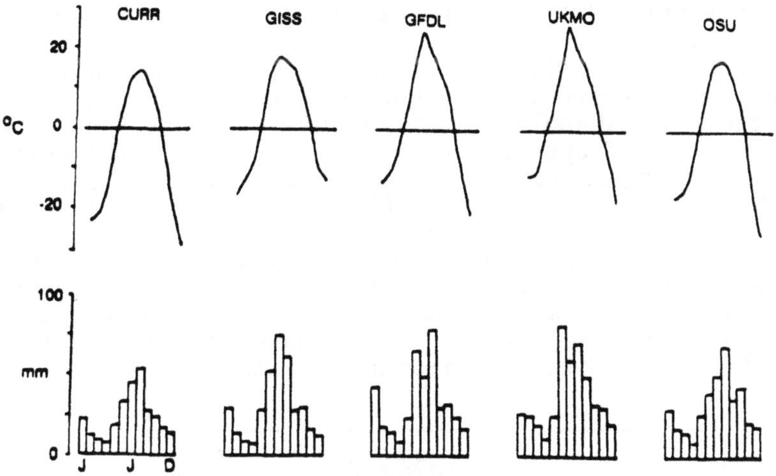

Figure 8. Monthly temperature and precipitation values predicted for an area near Fairbanks, Alaska under CO_2-induced climate change conditions.

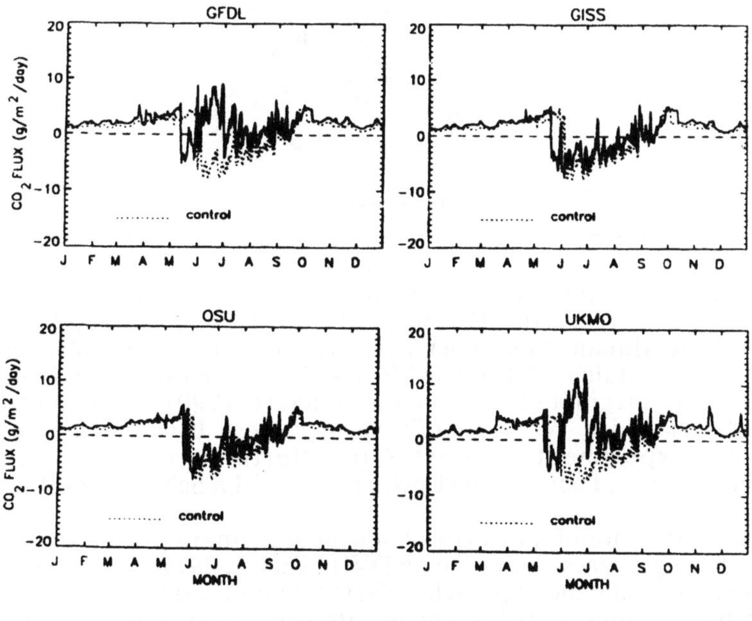

Figure 9. Modeled daily ecosystem CO_2 flux from January 1 to December 31 for a black spruce stand on a north-facing slope with permafrost near Fairbanks, Alaska. The flux is the sum of tree photosynthesis and respiration, moss photosynthesis and respiration, and microbial respiration. A negative flux indicates CO_2 uptake.

5.2. GLOBAL APPLICATIONS OF THE COMBINED MODEL APPROACH

Presently, we are applying the above methodology to a series of forested sites representing a gradient of climatic conditions from boreal to tropical. The sites have been chosen based on the criterion that forest gap models have either already been developed, applied and tested for the site in the past, or model application and testing is already currently underway at the site. The locations of sites and model citations are shown in Figure 10. Climate data bases (>30 yr) are available for each site to construct transient climate scenarios for simulations of current forest patterns. 50 to 100 plots will be simulated for each location to provide a sufficient sample of stand composition and structure for the site conditions. Flux rates will be simulated for each of the forest stands and averaged to provide a timeseries for the site. Calculation of predicted changes in CO_2 flux and C storage will be made to construct new latitudinal profiles.

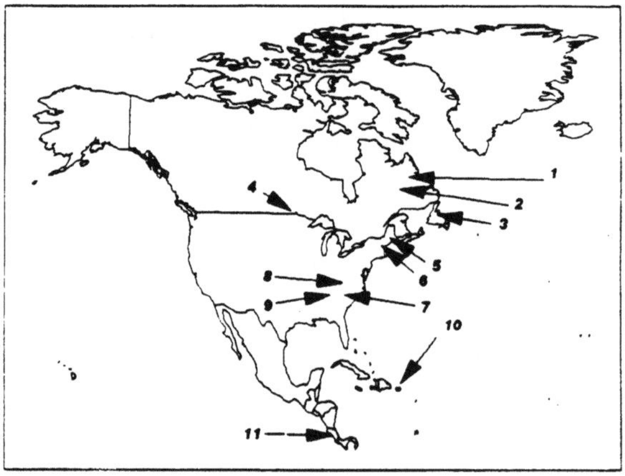

Figure 10. Location of sites for modeling seasonal C flux along a latitudinal gradient: 1) Schefferville, Quebec; 2) Sept Iles, Quebec; 3) Gander, Newfoundland (Bonan, 1990b); 4) Voyaguers National Park, Minnesota; 5) Howland Forest, Maine; 6) Hubbard Brook, New Hampshire; 7) Shenandoah National Park, Virginia (Harrison and Shugart, 1990); 8) Walker Branch Watershed, Oak Ridge, Tennessee (Shugart and West, 1977, Smith et al., 1981); 9) Coweeta Experimental Forest, Otto, North Carolina; 10) Loquillo Experimental Forest, Puerto Rico (Doyle, 1981); 11) La Selva, Costa Rica.

To validate the model, patterns of seasonal primary productivity will be compared to a greenness-index (i.e., NDVI) (Tucker, 1979; Tucker *et al.*, 1985) for the sites following the approach of Nemani and Running (1989). This will allow for both comparisons for a given site over the period monitored (i.e., seasonal or annual patterns), as well as a comparison across sites. Patterns of CO_2 flux will then be plotted to provide portrait of forest fluxes as function of the latitudinal gradient.

The process of linking dynamic models of forest structure and CO_2 flux should provide an important step in the development of a methodology for predicting feedback between terrestrial and atmospheric C pools under changing climate conditions. Although the proposed work focuses on forested ecosystems, the potential exits to expand the approach to other ecosystem types. The individual-based modeling approach (e.g., gap model) has been recently applied to grassland ecosystems (Coffin and Lauenroth, 1989), and work is currently in progress to expand the approach to savanna/woodland ecosystems (Smith *et al.*, 1989).

REFERENCES

Aber, J.D., Hendry, G.R., Francis, A.J., Botkin, D.B. and Melillo, J.M. (1982) 'Potential effects of acid precipitation on soil nitrogen and productivity of forest ecosystems', in F.M. Ditri (ed.), Acid Precipitation: Effects on Ecological Systems, Ann Arbor Science, Ann Arbor, MI, pp. 411-434.

Akima, H. (1978) 'A method of bivariate interpolation and smooth surface fitting for irregular disturbed datapoints', ACM Trans. Math. Software 4, 148-159.

Bacastow, R.B., Keeling, C.D., and Whorf, T.P. (1985) 'Seasonal amplitude increase in atmospheric CO_2 concentration at Mauna Loa, Hawaii, 1959-1982', J. Geophys. Res. 90, 10529-10540.

Bonan, G.B. (1988) Environmental Processes and Vegetation Patterns in Boreal Forests, Ph.D. dissertation, University of Virginia, Charlottesville, VA.

Bonan, G.B. (1989) 'A computer model of the solar radiation, soil moisture, and soil thermal regimes in boreal forests', Ecol. Model. 45:275-306.

Bonan, G.B. (1990a) 'Carbon and nitrogen cycling in North American boreal forests. I. Litter quality and soil thermal effects in interior Alaska', Biogeochemistry 10:1-28.

Bonan, G.B. (1990b) 'Carbon and nitrogen cycling in North American boreal forests. II. Biogeographic patterns', Canadian J. Forest Res. 20:1077-1088.

Bonan, G.B. and Hayden, B.P. (1990) 'Using a forest stand simulation model to examine the ecological and climatic significance of the late-Quaternary pine-spruce pollen zone in eastern Virginia, USA', Quat. Res. 33, 204-218.

Bonan, G.B., H.H. Shugart, and D.L. Urban. (1990) 'The sensitivity of some high-latitude boreal forests to climatic parameters', Clim. Change 16:9-29.

Bonan, G.B. (1991a) 'A biophysical surface energy budget analysis of soil temperature in the boreal forests of interior Alaska', Water Resources Res. 27, 767-781.

Bonan, G.B. (1991b) 'Atmosphere-biosphere exchange of carbon dioxide in boreal forests', J. Geophys. Res. 96, 7301-7312.

Bonan, G.B. (1991c) 'Seasonal and annual carbon fluxes in a boreal forest landscape', J. Geophys. Res. 96, 17329-17338.

Bonan, G.B. (in review) 'Fire, succession and CO_2 flux in the boreal forests of interior Alaska', Canadian J. of Forest Res.

Botkin, D. B., Janak, J. F., and Wallis, J. R. (1972) 'Some ecological consequences of a computer model of forest growth', J. Ecol. 60, 849-873.

Box, E.O. (1981) Macroclimate and Plant Forms: An Introduction to Predictive Modeling in Phytogeography, Junk, Hague.

Coffin, D.P. and W.K. Lauenroth. (1989) 'A gap dynamics simulation model of succession in a semi-arid grassland', Ecol. Model. 49, 229-266.

Davis, M.B. (1984) 'Climatic instability, time lags and community disequilibrium', in J. Diamond and T.J. Case (eds.), Community Ecology, Harper and Row, New York, pp 269-284.

Davis, M.B. (1989) 'Lags in vegetation response to greenhouse warming', Clim. Change 15, 75-82.
Doyle, T.W. (1981) 'The role of disturbance in the gap dynamics of a montane rain forest: An application of a tropical forest succession model', in D.C. West, H.H. Shugart, and D.B. Botkin (eds.), Forest Succession: Concepts and Applications. Springer-Verlag, New York, pp. 56-73.
Emanuel, W.R., Shugart, H.H., and Stevenson, M.P. (1985) 'Climatic change and the broad-scale distribution of terrestrial ecosystem complexes', Clim. Change 7, 29-43.
Friend, A.D., Shugart, H.H. and Running, S.W. (in press) 'A physiology-based gap model of forest dynamics', Ecol. Model.
Green, P.J. and Sibson, R. (1978) 'Computing Dirichlet tessellations in the plane', The Computer J., 21, 168-173.
Hansen, J., Fung, I., Lacis, A., Rind, D., Russell, G., Lebedeff, S., Reudy, R. and Stone, P. (1988) 'Global climate changes as forecast by the GISS-3-D model', J. Geophys. Res. 93, 9341-9364.
Harrison, E.A. and Shugart, H.H. 1990. 'Evaluating performance of an Appalachian oak forest dynamics model', Vegetatio 86, 1-13.
Holdridge, L.R. (1947) 'Determinations of world formulations from simple climatic data', Science 105, 367-368.
Holdridge, L.R. (1959) 'Simple method for determining potential evapotranspiration from temperature data', Science 130, 572.
Holdridge, L.R. (1967) Life Zone Ecology, Tropical Science Center, San Jose.
Houghton, R.A. (1987) 'Biotic changes consistent with the increased seasonal amplitude of atmospheric CO_2 concentrations', J. Geophys. Res. 92, 4223-4230.
King, A.W., Emanuel, W.R. and O'Neill, R.V. (1990) 'Linking mechanistic models of tree physiology with models of forest dynamics: problems of temporal scale', in R.K. Dixon, R.S. Meldahl, G.A. Ruark and W.G. Warren (eds.), Process Modeling of Forest Growth Responses to Environmental Stress, Timber Press, Portland, OR, pp. 241-248.
Leemans, R. and Cramer, W. (1990) The IIASA climate database for land area on a grid of 0.5o resolution, WP-41, International Institute for Applied Systems Analysis, Laxenburg.
Leemans, R. and Prentice, I.C. (1987) 'Description and simulation of tree layer composition and size distribution in a primaeval *Picea-Pinus* forest', Vegetatio 69, 147-156.
Martin, P. (1990) Forest Succession and Climate Change: Coupling Land-Surface Processes and Ecological Dynamics, Ph.D. dissertation, University of California, Berkeley, CA.
Manabe, S. and Wetherald, R.T. (1987) 'Large scale changes in soil wetness induced by an increase in carbon dioxide', J. Atm. Sci. 44, 1211-1235.
Mitchell, J.F.B. (1983) 'The seasonal response of a general circulation model to changes in CO_2 and sea temperatures', Q. J. Roy. Met. Soc. 109, 113-152.
Nemani, R.R. and Running, S.W. (1989) 'Testing a theoretical climate-soil-leaf area hydrologic equilibrium of forests using satellite data and ecosystem simulation', Agr. For. Met. 44, 245-260.
Olson, J.S., Watts, J.A., and Allison, L.J. (1983) Carbon in live vegetation of major world ecosystems, ESD Pub. No. 1997, Oak Ridge National Laboratory, TN.
Overpeck, J.T., Rind, D., and Goldberg, R. (1990) 'Climate-induced changes in

forest disturbance and vegetation. Nature 343, 51-53.
Pastor, J. and Post, W.M. (1986) 'Influences of climate, soil moisture, and succession on forest carbon and nitrogen cycles', Biogeochemistry 2, 3-27.
Pastor, J. and Post, W.M. (1988) 'Response of northern forests to COP_2-induced climate change', Nature 334, 55-58.
Post, W.M., Emanuel, W.R., Zinke, P.J., and Stangenberger, A.G. (1982) 'Soil carbon pools and world life zones', Nature 298, 156-159.
Running, S.W. and J.C. Coughlan. (1988) 'A general model of forest ecosystem processes for regional applications. I. Hydrological balance, canopy gas exchange and primary production processes', Ecol. Model. 42, 125-154.
Schlesinger, M. and Zhao, Z. (1988) Seasonal climatic changes induced by doubled CO_2 as simulated by the OSU atmospheric GCM/mixed layer ocean model, Climate Research Institute, Oregon St. Univ., Corvallis, OR.
Shugart, H.H. (1984) A Theory of Forest Dynamics. Springer-Verlag, New York.
Shugart, H.H. and West, D.C. (1977) 'Development of an Appalachian deciduous forest succession model and its application to assessment of the impact of chestnut blight', J. Environ. Manage. 5, 161-179.
Shugart, H.H. and West, D.C. (1979) 'Size and pattern of simulated forest stands', For. Sci. 25, 120-122.
Shugart, H.H. and West, D.C. (1980) 'Forest succession models', BioScience 30, 308-313.
Shugart, H.H., M.S. Hopkins, I.P. Burgess, and A.T. Mortlock. (1980) 'The development of a succession model for subtropical rainforest and its application to assess the effects of timber harvest at Wiangaree State Forest, New South Wales', J. Environ. Manage. 11, 243-265.
Shugart, H.H., Antonovsky, M.Y, Jarvis, P.G. and Sandford, A.P. (1986) 'CO_2, climatic change and forest ecosystems', in B. Bolin, B.R. Döös, J. Jäger and R.A. Warrick (eds.), The Greenhouse Effect, Climatic Change, and Ecosystems (SCOPE 29), John Wiley and Sons, New York, NY, pp. 475-521.
Smith, T.M., Bonan, G.B., Shugart, H.H., and Smith, J.B. (1992) 'Modeling the potential response of vegetation to global climate change', Adv. Ecol. Res. 22, 93-116.
Smith, T.M., R. Leemans, and H.H. Shugart. (in press) 'Sensitivity of terrestrial carbon storage to CO_2 -induced climate change: Comparison of four scenarios based on general circulation models', Clim. Change.
Smith, T.M., H.H. Shugart, D.L. Urban, W.K. Lauenroth, D.P. Coffin, and T.B. Kirchner. (1989) 'Modeling vegetation across biomes: Forest-grassland transition', in E. Sjögren (ed.), Studies in Plant Ecology Vol. 18: Forests of the World: Diversity and Dynamics. Almquist and Wiksell Int., Stockholm, Sweden.
Solomon, A.M. (1986) 'Transient responses of forests to CO_2-induced climate change: Simulation modeling experiments in eastern North America', Oecol. 68, 567-569.
Solomon, A.M., Delcourt, H.R., West, D.C., and Blasings, T.J. (1980) 'Testing a simulation model for reconstruction of prehistoric forest-stand dynamics', Quat. Res. 14, 275-293.
Solomon, A.M. and Shugart, H.H. (1984) 'Integrating forest-stand simulations with paleoecological records to examine long-term forest dynamics', in G.I. Agren (ed.), State and Change of Forest Ecosystems, Swedish University of Agricultural Sciences, Upsalla, Sweden, pp. 333-357.

Solomon, A.M., Tharp, M.L., West, D.C., Taylor, G.E., Webb, J.M. and Trimble, J.L. (1984) Response of unmanaged forests to CO_2-induced climate change: available information, initial tests and data requirements, US DOE Carbon Dioxide Research Division, Washington, D.C.
Solomon, A.M. and Webb, T. III. (1985) 'Computer-aided reconstruction of late quaternary landscape dynamics', Ann. Rev. Ecol. Syst. 16, 63-84.
Solomon, A.M., West, D.C., and Solomon, J.A. (1981) 'Simulating the role of climate change and species immigration in forest succession', in D.C. West, H.H. Shugart and D.B. Botkin (eds.), Forest Succession, Springer-Verlag, New York, pp. 154-177.
Tucker, C.J. (1979) 'Red and photographic infrared linear combinations for monitoring vegetation', Rem. Sens. Environ. 8, 127-150.
Tucker, C.J., Townshend, J.R.G., and Goff, T.E. (1985) 'African land cover classification using satellite data', Science 227, 369-374.
Urban, D.L., O'Neill, R.V. and Shugart, H.H. (1987) 'Landscape ecology', BioScience 37, 119-127.
Urban, D.L. and Shugart, H.H. (1989) 'Forest response to climate change: a simulation study for Southeastern forests', in J. Smith and D. Tirpak (eds.), The Potential Effects of Climate Change on the United States, US Environmental Protection Agency, Washington, D.C. pp. 3-1 to 3-45.
Weinstein, D.A., Shugart, H.H. and West, D.C. (1982) The Long-term Nutrient Retention Properties of Forest Ecosystems: A Simulation Investigation, ORNL/TM-8472, Oak Ridge National Laboratory, Oak Ridge, TN.

CHANGES IN C STORAGE BY TERRESTRIAL ECOSYSTEMS: HOW C-N INTERACTIONS RESTRICT RESPONSES TO CO_2 AND TEMPERATURE

E. B. RASTETTER, R. B. McKANE, G. R. SHAVER, and J. M. MELILLO
The Ecosystems Center
Marine Biological Laboratory
Woods Hole, Massachusetts 02543
USA

ABSTRACT. A general model of ecosystem biogeochemistry was used to examine the responses of arctic tundra and temperate hardwood forests to a doubling of CO_2 concentration and to a 5°C increase in average growing season temperature. The amount of C stored in both ecosystems increased with both increased CO_2 and temperature. Under increased CO_2, the increase in C storage was due to increases in the C:N ratio of both vegetation and soils. Under increased temperature, the increased C storage in the forest was due to a shift in N from soils (with low C:N ratios) to vegetation (with high C:N ratios). In the tundra, both a shift in N from soils to vegetation and an increase in C:N ratios contributed to increased C storage under higher temperatures. Neither ecosystem sequestered N from external sources because the supply rate was low.

1. Introduction

The potential for C sequestration or loss by terrestrial ecosystems as a result of increased atmospheric CO_2 and climate change has been intensely debated (Kramer 1981, Billings et al. 1982, Strain and Bazzaz 1983, Gates 1985, Strain and Cure 1985, 1986, Idso 1986, 1989, Tissue and Oechel 1987, Drake et al. 1989, Tans et al. 1990, Bazzaz 1990, Melillo et al. 1990, Rastetter et al. 1991). Whether terrestrial ecosystems are sources or sinks of atmospheric CO_2 will depend almost certainly upon the interactions of the C cycle with the cycles of other vital elements like N and P (Chapin et al. 1980, Bolin and Cook 1983, Melillo and Gosz 1983, Billings et al. 1984, Hobbie et al. 1984, Vitousek et al. 1988, Raich et al. 1991, Shaver et al. 1992, Rastetter and Shaver 1992). In this paper, we will use a general model of ecosystem C and N cycles (Rastetter et al. 1991) to examine the importance of C-N interactions on C sequestration in arctic tundra and temperate hardwood forest ecosystems in response to increased CO_2 concentration and temperature.

Water, Air, and Soil Pollution **64**: 327–344, 1992.
© 1992 *Kluwer Academic Publishers.*

Interactions between the C and N cycles are complex and have ramifications on a diversity of ecosystem properties from tissue function, on the fine scale, to whole system C storage, on the coarse scale. Our aim is to generalize from these complex interactions to identify the key properties controlling C storage. We suggest that changes in C storage in terrestrial ecosystems can be explained largely in terms of three key biogeochemical properties (Shaver et al. 1992). The three properties are,

1) **The relative importance of external sources of N and other vital elements versus internally recycled sources of these elements**,

2) **The distribution of C, N and other vital elements between vegetation and soils**, and

3) **The flexibility of the element ratios in both vegetation and soils.**

The importance of these three key biogeochemical properties has also been noted by other researchers (e.g., Fisher et al. 1982, several papers in Bolin and Cook 1983, Vitousek et al. 1988, Aerts and Berendse 1989, Berendse 1990). In this paper, we build upon these earlier studies by developing explicit comparisons of the relative importance of these three properties in regulating changes in C storage in response to changes in CO_2 and temperature.

The first of these properties, the relative importance of external sources of N and other vital elements versus internally recycled sources of these elements, is important because the supply of vital elements ultimately limits ecosystem C-storage capacity (Vitousek and Reiners 1975, Fisher et al. 1982, Bormann and Likens 1979, Anderson 1988, Melillo and Gosz 1983, Peterson and Melillo 1985). An ecosystem with a ready, external supply of vital elements (e.g., N) can increase its C storage without a redistribution of its internal resources or a change in the stoichiometry of its components. To increase C storage without a ready, external supply of vital elements, resources like N would either have to be redistributed from components with low C:N ratios to components with high C:N ratios, or the C:N ratios of the components themselves would have to increase. Overall, the more open the ecosystem is to external sources of vital elements relative to its dependence on internally recycled sources of these elements, the more responsive should be its C-storage capacity to disturbances like long-term changes in CO_2 and temperature.

The second property, the distribution of C, N and other vital elements between vegetation and soils, is important because the stoichiometry of vegetation and soils differs greatly (Melillo and Gosz 1983, Vitousek et al. 1988, Anderson 1991, Parton et al. 1988). For example, the C:N ratio of vegetation ranges from about 30 to 600 (by mass), depending largely on the amount of woody tissue; that of soils ranges from about 8 to 50, depending on many factors including temperature and the degree of water saturation. For N-limited ecosystems, this wide difference between vegetation and soils means that if, in response to a disturbance, N is mineralized from the soil organic matter (with a low C:N ratio) and taken up and incorporated into the vegetation (with a high C:N ratio), then there will be a net increase in the storage of C in the ecosystem, even without sequestering additional N from outside the ecosystem. On the other hand, if the disturbance causes the vegetation to degrade so that N is transferred from vegetation biomass (with a high C:N ratio) to the soil organic matter (with a low C:N ratio), then eventually there will be a net loss of C from the ecosystem, even if there is no net loss of N. Thus, the initial distribution of organic matter between vegetation and soils places limits on the amounts of vital elements that can be redistributed and therefore defines, to some degree, the limits on the C-storage capacity of ecosystems. Overall, the greater the difference in C:N ratios of vegetation and soils, the more important are the controls on N distribution between vegetation and soils in determining the C-storage response of ecosystems to disturbances like increased CO_2 and temperature.

The third property, the flexibility of the element ratios in vegetation and soils, is important because it determines how tightly linked the various element cycles are to one another (Melillo and Gosz 1983, Vitousek et al. 1988, Berendse et al. 1987, 1989). For example, if the C:N ratio of the vegetation is very rigid (i.e., a severely restricted range of vegetation C:N ratios), then the rate at which C is taken up and incorporated into vegetation biomass is very tightly constrained by the rate at which N can be taken up and incorporated into biomass. On the other hand, if the C:N ratio is more flexible and the vegetation can adjust to decreased N availability by decreasing its N concentration (increased N-use efficiency), then the cycling of C through that vegetation will be less severely constrained by the N cycle. An ecosystem with components that have flexible element ratios might be expected to respond more strongly to disturbances like increased CO_2 and temperature than one with components that have more rigid element requirements.

In this paper, we will briefly describe the Marine Biological Laboratory's General Ecosystem Model (MBL-GEM) and derive a means of attributing changes in total-ecosystem C storage to the three ecosystem properties described above. We will then apply the model to tundra and temperate forest ecosystems to examine their responses to increased CO_2 and temperature and compare these responses based on the key properties we have just discussed. Next, because both of these ecosystems cycle N tightly and very little of their N capital is derived from external N sources, we will use the model to artificially "open" the N cycle and examine the consequences on C storage. Finally, we will discuss the consequences of C-N interactions on the potential for increasing C storage in the terrestrial biosphere.

2. Model Description

The structure of MBL-GEM (Figure 1) is described in detail in Rastetter

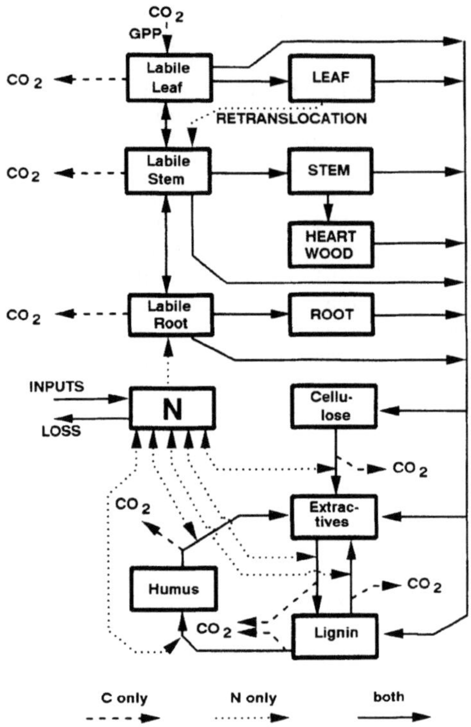

Figure 1: Diagrammatic representation of the Marine Biological Laboratory's General Ecosystem Model (MBL-GEM). Compartments represent storages of both C and N, except the compartment labeled "N", which represents inorganic soil N. Dashed arrows represent fluxes of C, dotted arrows represent fluxes of N, and solid arrows represent fluxes of both. The model is described in more detail in the text and in Rastetter et al. (1991).

et al. (1991). Therefore, only a brief, general description will be given here.

The model simulates changes in the amounts of C and N in 11 organic-matter pools plus inorganic soil N (a total of 23 "state" variables) in response to changes in CO_2 concentration, temperature, top-of-the-canopy irradiance, soil moisture, and inorganic N inputs. The 11 organic-matter pools are 1) labile C and N in foliage, 2) labile C and N in stems, 3) labile C and N in fine roots, 4) structural C and N in foliage, 5) structural C and N in stems, 6) structural C and N in heartwood, 7) structural C and N in fine roots, 8) "extractive" (hot water and dichloromethane extractable) C and N in litter, 9) "cellulose" (acid soluble) C and N in litter, 10) "lignin" (acid insoluble) C and N in litter, and 11) C and N in soil humus (to the bottom of the rooting zone). Changes in these pools are simulated based on uptake of C and N by vegetation, allocation of that C and N among vegetation tissues, production of litter from those tissues, subsequent decomposition of that litter, net mineralization of N during decomposition, and losses of inorganic N from the ecosystem as a whole.

While MBL-GEM provides a detailed, process-level description of ecosystems, it is designed to have broad applicability. That is, the same model structure can be easily applied to many different types of terrestrial ecosystems merely by recalibrating model parameters. Calibrations for the two ecosystems examined here were done by assuming year-to-year equilibrium and specifying the following: 1) the amounts of C and N for each of the vegetation state variables, 2) the amount of inorganic N, 3) the total amount of soil plus litter C and N, 4) the rates of gross primary production (GPP), N uptake by vegetation, and the loss of C from each of the three vegetation tissues, 5) the environmental conditions (i.e., CO_2 concentration, temperature, top-of-the-canopy irradiance, soil moisture, and inorganic N inputs), and 6) the value of all parameters *except* the rate constants for vegetation processes and humus decomposition and the C:N ratio of humus. The rate constants for vegetation processes and humus decomposition and the C:N ratio of humus were then "back calculated" to be consistent with the equilibrium assumption.

3. Partitioning the Change in C Storage Among Three Factors

The underlying premise of our thesis is that changes in C storage in terrestrial ecosystems can be attributed to three factors: 1) changes in the total amount of N in the ecosystem, 2) net shifts in N between vegetation and soils, and 3) changes in the C:N ratio of vegetation and soils. In this section we derive an algebraic partitioning for the change in C storage among these three factors to quantify the importance of each.

First, we define several variables: $C_T(t)$, $C_V(t)$, and $C_S(t)$ are the amounts of C at time t in the whole ecosystem, in vegetation, and in soils, respectively; $N_T(t)$, $N_V(t)$, and $N_S(t)$ are the amounts of N in these same pools at time t; and $\Psi_T(t) = C_T(t)/N_T(t)$, $\Psi_V(t) = C_V(t)/N_V(t)$, and $\Psi_S(t) = C_S(t)/N_S(t)$ are the C:N ratios of these pools. We use Δ to indicate deviations in these variables from their initial values (e.g., $\Delta C_T(t) = C_T(t) - C_T(0)$). For convenience, we will drop the "(t)" except to indicate a specific time (e.g., $t = 0$).

Next, we derive expressions for the change in the total amount of C stored in ecosystems that can be attributed to each of the three factors listed above. For each factor, we will assume that that factor changed by the amount predicted in the simulations but that the other two factors remained constant. For example, how much would the total amount of C have changed if the amount of N in vegetation and soils was unchanged and the C:N ratios of vegetation and soils changed by the amount predicted in the simulations? The answer is simply the sum of the initial amounts of N in vegetation and soils multiplied by the change in their respective C:N ratios. That is,

$$\Delta C_{T,C:N} = N_V(0)\ \Delta\Psi_V + N_S(0)\ \Delta\Psi_S \tag{1}$$

where the "C:N" subscript identifies the factor to which the change has been associated.

The change in the total amount of C in the ecosystem that can be attributed to a net input of N from outside the ecosystem, assuming that the C:N ratios of vegetation and soils remain constant and that the relative distribution of N between vegetation and soils remains constant, is the product of the change in total N in the ecosystem and the initial C:N ratio for the ecosystem. That is,

$$\Delta C_{T,Nin} = \Delta N_T\ \Psi_T(0) \tag{2}$$

where the "Nin" subscript indicates that the change has been associated with the net input of N into the ecosystem.

The change in total C associated with shifts in N between vegetation and soils is more difficult to calculate because the changes in N_V and N_S must be partitioned into that change resulting from a shift in N between vegetation and soils and that change resulting from the net input of N from outside the ecosystem (i.e., associated with the previous factor). To make this calculation, we retain the assumption from the previous derivation that the net N input to the ecosystem as a whole is distributed between vegetation and soils in proportion to the amounts of N initially in vegetation and soils. That is, the changes in the amounts of N in vegetation and soils that can be attributed to the net input of N to the ecosystem are equal to

$$\Delta N_{V,Nin} = \frac{N_V(0)}{N_T(0)}\ \Delta N_T, \tag{3}$$

and

$$\Delta N_{S,Nin} = \frac{N_S(0)}{N_T(0)}\ \Delta N_T. \tag{4}$$

Subtracting equations 3 and 4 from the total change in vegetation and soil N, and noting that $N_T(0) = N_V(0) + N_S(0)$ and $\Delta N_T(0) = \Delta N_V(0) + \Delta N_S(0)$, yields the change that is attributable to the shift of N between vegetation and soils,

$$\begin{aligned} \Delta N_{V,shift} &= \Delta N_V - \Delta N_{V,Nin} \\ &= \Delta N_V - \Delta N_T \frac{N_V(0)}{N_T(0)} \\ &= \frac{\Delta N_V\ N_S(0) - \Delta N_S\ N_V(0)}{N_T(0)} \end{aligned} \tag{5}$$

$$\Delta N_{S,shift} = \Delta N_S - \Delta N_{S,Nin}$$

$$= \Delta N_S - \Delta N_T \frac{N_S(0)}{N_T(0)}$$

$$= \frac{\Delta N_S\ N_V(0) - \Delta N_V\ N_S(0)}{N_T(0)} \tag{6}$$

where the "shift" subscript indicates that the change in N has been attributed to a shift of N between vegetation and soils. The change in C storage associated with this shift in N is equal to the sum of these two N changes times the respective initial C:N ratios for vegetation and soils. The changes in N associated with this shift are, of course, equal in magnitude but opposite in sign for vegetation and soils. Thus the total change in C associated with this factor is

$$\Delta C_{T,shift} = \frac{\Delta N_V\ N_S(0) - \Delta N_S\ N_V(0)}{N_T(0)} \{\Psi_V(0) - \Psi_S(0)\} \tag{7}$$

Finally, there is a term that incorporates the interactions among these three factors. For example, a net gain in N from outside the ecosystem will result in a larger increase in C storage if the C:N ratio of the components of the ecosystem also increases. The interactive term is simply

$$\Delta C_{T,inter} = \Delta N_V\ \Delta\Psi_V + \Delta N_S\ \Delta\Psi_S \tag{8}$$

where the "inter" subscript indicates the interaction of the three factors described above.

A simple algebraic manipulation will reveal that in fact

$$\Delta C_T = \Delta C_{T,C:N} + \Delta C_{T,Nin} + \Delta C_{T,shift} + \Delta C_{T,inter} \tag{9}$$

4. Simulations

We calibrated MBL-GEM to arctic tundra and temperate hardwood forest ecosystems based on data from the Alaskan tundra (Shaver and Chapin 1991, Giblin et al. 1991, Shaver et al. 1990) and Harvard Forest (Aber et al. 1983, K. J. Nadelhoffer unpublished data) Long-Term Ecological Research (LTER) sites. Tundra and temperate hardwood forest were chosen because of our extensive experience with them (McClaugherty et al. 1982, Chapin and Shaver 1985, Shaver et al. 1986, Aber et al. 1983, 1991) and because the two ecosystems differ greatly with respect to the three key properties that we suggest are important determinants of their response to long-term changes in CO_2 concentration and temperature. The inorganic N input to the forest ecosystem is about 10 times higher than that of the tundra ecosystem (1 gN m^{-2} yr^{-1} and 0.06 gN m^{-2} yr^{-1}, respectively). However, both ecosystems are N limited (Mitchel and Chandler 1939, Shaver et al. 1986) and external inputs of inorganic N are small relative to the internal recycling of N. External N inputs account for about 3% of the annual uptake requirements of the tundra vegetation and about 10% of the annual uptake requirements of the forest vegetation. The tundra has a total of about 27,000 gC m^{-2} with about 98% of that in soils and only 2% in vegetation. On the other hand, the forest has a comparable amount of total C (30,000 gC m^{-2}) but only 40% of it is in soils and 60% in vegetation. The C:N ratios of the soils plus litter in both ecosystems is about 20. However, the C:N ratios of the vegetation differ greatly. Tundra vegetation has little woody material and therefore has a low C:N ratio near 40. The woody vegetation of the forest has a C:N ratio that is about 10 times higher.

We used the model to examine the responses of these ecosystems to a doubling of CO_2 concentration (340 to 680 ppmv), a 5°C increase in growing season temperature (7 to 12°C in tundra, 17 to 22°C in temperate forest), and a combined doubling of CO_2 and 5°C temperature increase. Simulations were of 50-year duration, using average growing season conditions and each year being compressed to the duration of the respective growing season (3 mo for tundra, 5 mo for forest). At the beginning of the simulations, the ecosystems were in equilibrium with ambient CO_2 and temperature. At the beginning of the third year of simulation, either the CO_2 concentration was doubled or the temperature was increased by 5°C or both.

5. Results

Model predictions of the responses of the two ecosystems to increased CO_2 concentration and temperature are qualitatively similar (Figure 2). The integrated response of both ecosystems was to increase C storage in all three treatments with little or no increase in the total amount of N. This increased storage of C was accompanied by shifts in N between vegetation and soils and by changes in the C:N ratios of vegetation and soils.

Under increased CO_2 concentration, the major cause of the change in C storage in both ecosystems was an increase in the C:N ratios of vegetation and soils (Figure 3). The C:N ratio of vegetation increased because the increase in CO_2 stimulated photosynthesis. Litter derived from this C-rich vegetation also had a high C:N ratio and immobilized more inorganic N. This additional demand on the inorganic N in the soil placed the vegetation under higher N limitation and amplified the increase in vegetation C:N ratio. Additionally, because of the increase in N immobilized into soil organic matter, there was a net shift of N from vegetation and to soils (Figure 2 upper left). By itself, this shift would decrease C storage (Figure 3). However, the increases in vegetation and soil C:N ratios were more than enough to compensate for the C loss that would have been associated with a shift in N from vegetation (high C:N ratio) to soils (low C:N ratio). The absolute increase in C storage in the forest ecosystem was much higher than in the tundra because of the large C-storage capacity and high C:N ratio of woody tissue. However, the relative increase in C storage in vegetation was 3 times higher in tundra (12% increase after 50 yr) than forest (4%

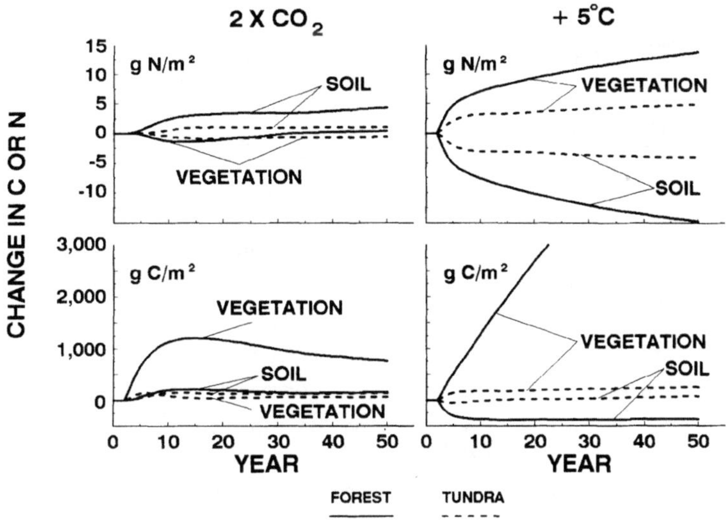

Figure 2: Simulated changes in vegetation and soil C and N as a result of a doubling of CO_2 concentration (2 X CO_2) and a 5°C increase in temperature (+ 5°C) for hardwood forest and tundra ecosystems. Ecosystems were in equilibrium with the CO_2 and temperature at year 0. At the beginning of the third year, either CO_2 was doubled (left) or temperature was increased by 5°C (right). Upper graphs are for absolute changes in the amounts of organic N in vegetation and soils, lower graphs are for absolute changes in C. Solid lines are for hardwood forest and dashed lines are for tundra.

increase after 50 yr). New N from outside the ecosystem contributed very little to the increase in C storage in either ecosystem (Figure 3). The net increase in total organic N was only about 1% in the forest ecosystem and less than 0.1% in the tundra.

Under increased temperature, a major cause of increased C storage in both ecosystems was the internal redistribution of N (Figure 3). At the higher temperatures, the rates of decomposition and N mineralization increased. This increase in mineralization increased N availability to vegetation and stimulated productivity. The net effect was a large shift in N from soils to vegetation (Figure 2 upper right). The increase in mineralization was accompanied by a loss of C from soils, but the uptake of that N by vegetation was accompanied by an even larger gain in C by vegetation. Because the C:N ratio of forest vegetation is about 10 times higher than that of soils, the increase in C storage as a result of this soil to vegetation N shift was very large in the forest ecosystem (Figure 2 lower right). In the tundra the C:N ratio of vegetation is only about 2 times higher than that of soils and the increase in C storage associated with the N shift was therefore

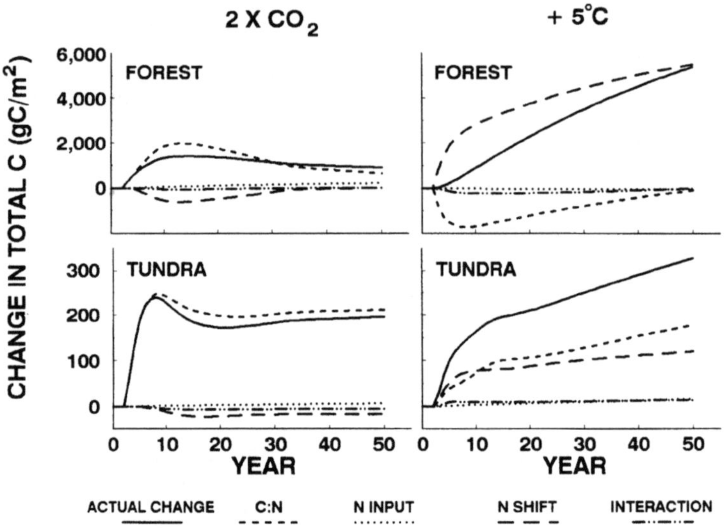

Figure 3: Partitioning the simulated change in total ecosystem C storage among three factors and their interaction. Solid line is the simulated change in total C storage. The short-dashed line is the change in C storage associated with changes in the C:N ratios of vegetation and soils, the dotted line is the change associated with N input from outside the ecosystem, the long-dashed line is the change associated with a shift in N from soils to vegetation, and the chained line is the change associated with the interactions of all three factors. The ecosystems were in equilibrium with CO_2 and temperature at year 0. At the beginning of the third year, either CO_2 was doubled (left) or temperature was increased by 5°C (right). Upper graphs are for hardwood forest, lower graphs are for tundra.

substantially smaller.

The contribution of changes in C:N ratios to changes in C storage under increased temperature differed greatly between the two ecosystems (Figure 3). In the forest, the C:N ratio of soils decreased while that of vegetation remained almost constant, which by itself would result in a decrease in C storage. In the tundra, the C:N ratios of both vegetation and soils increased, which contributed to an increase in C storage. The differences in the responses of the C:N ratios in the two ecosystems are primarily due to the magnitude of the temperature-induced increase in net C fixation relative to the amount of C already in the vegetation. In the forest, the amount of additional C fixed as a result of the temperature increase is only a small proportion of the total vegetation C, so there was not a strong tendency to increase the C:N ratio. In addition, the increase in N availability at higher temperatures helped prevent an already high C:N ratio from getting higher. In the tundra, the additional amount of C fixed is a large proportion of the total vegetation C, so the tendency to increase the C:N ratio was strong. The increase in N availability was not large enough to compensate for this tendency, a condition that was exacerbated by the high N immobilization potential of litter produced by vegetation having a higher C:N ratio.

Changes in the total amount of N in the ecosystems as a result of increased temperature were small and the effects on C storage were consequently small (Figure 3). The forest ecosystem actually lost about 0.1% of its total N capital, mostly from soils. The tundra gained less than 0.1%.

The effects of the combined increase in CO_2 and temperature were virtually equal to the sum of the effects of the two independent treatments. Because the temperature responses were much stronger than the CO_2 responses, they dominated the responses to the combined treatment. Thus, the major cause of increased C storage was a shift in N from soils to vegetation. This increased C storage was enhanced in the tundra due to an increase in the C:N ratios of both soils and vegetation. Again, new N from outside the ecosystem did not contribute significantly to the increase in C storage of either ecosystem. Total N increased by about 1% in the forest and less than 0.1% in the tundra.

6. Opening the N Cycle

In neither the forest nor the tundra was N input from outside the ecosystem an important component of the annual N requirement of the ecosystem. Relatively "closed" N cycles such as these are probably representative of many natural ecosystems. However, more "open" N cycles would be expected, for example, for tidal ecosystems that are regularly flooded and therefore have soil solutions that are regularly flushed or for agricultural ecosystems that have large N inputs from fertilizer and large N losses through leaching and denitrification. To simulate these conditions, we simply increased the N input to each ecosystem to 10 times its original value and adjusted the N loss parameter so that N outputs again equaled N inputs. This change is not a fertilization in the normal sense, it increases the flux of N through the inorganic N pool but does not change the concentration of inorganic N for equilibrium conditions. However, it does buffer any transient change in the inorganic N concentration resulting from a response to disturbance.

When the ecosystem N cycles are opened, increased C storage resulting from N sequestration from outside the ecosystems becomes more important (Figure 4). For the "opened" tundra, responses to both increased CO_2 concentration and temperature were similar to those described in the previous section, except that there was an additional increase in C storage resulting from the N input. This additional increase in storage accounted for about 20% of the total increase in C storage under elevated CO_2 concentration, about 40% of the total increase in C storage under increased temperature, and about 20% in the combined treatment. Again, the effect of each factor under the combined

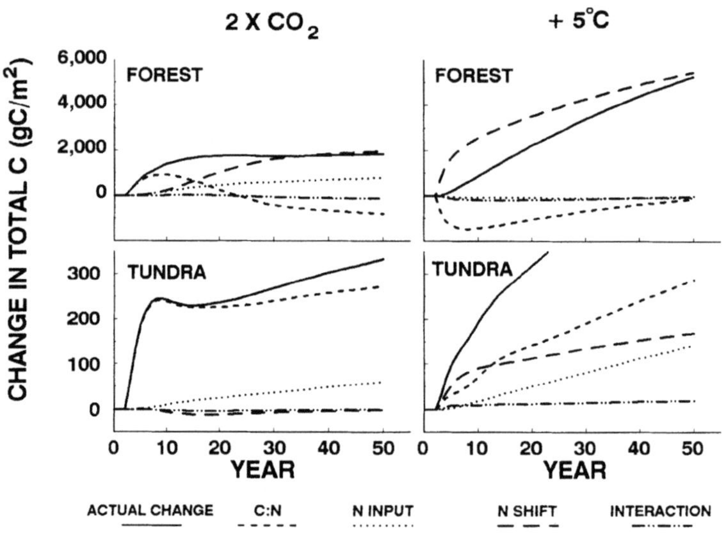

Figure 4: Partitioning the simulated change in total ecosystem C storage among three factors and their interaction after increasing both the inputs and losses of inorganic N by a factor of 10. Simulations are otherwise identical to those portrayed in Figure 3.

treatment was about equal to the sum of the effects of the factors for the two treatments individually, except for the interactive term. The effect of the interactive term was about four times as large as the sum of its effects on the two treatments individually, but still accounted for less than 10% of the total change in C storage.

For the "opened" forest, responses were more complex. Under increased CO_2 concentration, increased C storage associated with new N from outside the ecosystem was about 40% of the total increase in C storage. However, the "opened" N cycle also altered the changes in C storage associated with N shifts between soils and vegetation and those associated with changes in the C:N ratios. In particular, the "opened" N cycle resulted in a shift of N from soils to vegetation; the reverse was observed in the simulations with a more "closed" N cycle. In addition, increases in C:N ratios tended to increase C storage just after the CO_2 increase. However, as the ecosystem acclimated to higher CO_2 concentrations, the C:N ratios decreased and C storage associated with changes in C:N ratios declined. At the end of the simulation, the effects of increased total N in the ecosystem and the changes in C:N ratio almost exactly canceled one another, so that the total change in C storage was equal to that attributable to the shift in N from soils to vegetation.

Under increased temperature, the partitioning for the forest with the "opened" N cycle was almost identical to that for the forest with the more closed N cycle. In neither case was there a substantial change in the total N in the ecosystem; all of the change in storage was due to a shift in N from soils to vegetation. Changes in the C:N ratios decreased the effect of the N shift during the early part of the simulation but the C:N effect was negligible by year 50.

The combined CO_2 and temperature treatment for the "opened" forest was the only case for which the effects of CO_2 and temperature were not virtually additive. Increased C storage associated with the shift of N from soils to vegetation was about 25% higher than the sum of the changes in C storage associated with the same factor for the two treatments independently. Increased C storage associated with new N from outside the ecosystem was nearly 40% higher than that for the two treatments independently. The loss of C associated with the decrease in C:N ratios was about 100% higher than the sum of the changes in C storage for the two treatments independently. Associated with these non-additive responses to the combined treatment was a large interactive term, which decreased the total change in C storage by about 15%. This is the only treatment for which the interactive term was important. However, the additional effects of these factors, beyond the purely additive effects, tended to cancel one another so that the total change in C storage in the combined treatment was almost equal to the sum of the total change in C storage in the independent treatments.

7. Discussion

The factors controlling C storage in terrestrial ecosystems are inextricably linked to the cycles of vital elements like N. We suggest that differences in the C-storage capacities of terrestrial ecosystems can be understood largely from the three key biogeochemical properties we have examined in this paper. The potential to increase C storage in response to increases in CO_2 concentration and temperature should increase with 1) the ability to sequester new N and other vital elements from external sources, 2) the ability to move nutrients from soils (with low C:nutrient ratios) to vegetation (with high C:nutrient ratios), and 3) the ability to increase the C:nutrient ratios of ecosystem components.

The two ecosystems we examined were both mature and initially at equilibrium. Neither had a high potential for sequestering additional N from external sources, at least within the 50 year duration of the simulations. The ability to sequester external nutrient sources is probably low for most unmanaged ecosystems simply because the supply of external nutrients is low relative to internal recycling. Thus, the sequestration of C in response to changes in CO_2 and temperature must depend primarily on the redistribution of nutrients within the ecosystem and on increases in the C:nutrient ratios, at least for the first few decades. The exceptions to this are likely to be wet ecosystems with a high turnover of water (e.g., tidal and riparian ecosystems) and ecosystems with large atmospheric inputs of nutrients (e.g., N in precipitation). Nevertheless, in the long run, most ecosystems should be able to increase C storage by sequestering external nutrients.

However, because the supply rate of these nutrients is so slow, significant nutrient sequestration and the associated increase in C storage may take a very long time.

The redistribution of N between vegetation and soils was important in the responses of both ecosystems to increased CO_2 and temperature. The effects of redistributing N was more pronounced in the forest ecosystem than the tundra because the C:N ratio of the woody forest vegetation was much higher relative to soils than was the C:N ratio of the non-woody, tundra vegetation. The capacity of ecosystems with large woody components to increase C storage by redistributing nutrients from soils to vegetation should be greater than that of non-woody ecosystems. An exception may be some tropical forests growing in highly weathered soils, where the amounts of nutrients in soil organic matter are low relative to those in vegetation. In such ecosystems, the amounts of soil nutrients (especially P) that can be redistributed into vegetation are too small to significantly affect the C-storage capacity of the ecosystem. However, temperature changes in the tropics are expected to be relatively small and thus CO_2 effects may be more important. Our simulations suggest that under these circumstances the redistribution of nutrients will be from vegetation to soils (not from soils to vegetation) due to increased C:nutrient ratios in the vegetation and higher immobilization of nutrients in soils. Moving nutrients out of vegetation and into soils should tend to decrease C storage in these ecosystems. Thus, if the movement of nutrients from vegetation to soils were the only factor involved, tropical forests might tend to exacerbate an increase in atmospheric CO_2. However, our simulations also indicate that there will be a substantial increase in the C:nutrient ratios of the ecosystem components that will more than compensate for this tendency to decrease C storage. In addition, changes in cloudiness and moisture may be particularly important in the tropics, but are not considered in our analysis.

Changes in the C:N ratios of vegetation and soils contributed to changes in C storage in both ecosystems. Under increased CO_2 concentration, the major cause for increased C storage in both ecosystems was an increase in the C:N ratios of both vegetation and soils. On a global basis, increases in the C:nutrient ratios of ecosystem components are likely to be important in decreasing the rate of CO_2 increase in the atmosphere. However, the direct effects of CO_2 on ecosystems are likely to be confounded by changes in climatic variables like temperature. Under increased temperature, C:N ratios of the forest soils decreased, resulting in a reduction of the C-storage capacity. In the tundra, C:N ratios of both vegetation and soils increased, resulting in an increased C-storage capacity. The difference between the two ecosystems is related to the magnitude of the temperature-induced increase in net C fixation relative to the amount of C already in the vegetation. If the temperature-induced increase in C fixation is small relative to the amount of C in the vegetation, then the C:nutrient ratio of the vegetation will not increase substantially. This is likely to be the case for forested ecosystems. On the other hand, if the temperature-induced increase in C fixation is large relative to the amount of C in the vegetation, then the C:nutrient ratio of the vegetation will probably increase. This is likely to be the case for tundra, grasslands, and other non-woody ecosystems. Because of the temperature increases that are predicted to occur with increasing atmospheric CO_2, ecosystems dominated by non-woody species will increase their C storage as a result of a temperature-induced increase in the C:nutrient ratios of vegetation and of the litter derived from that vegetation. Ecosystems dominated by woody species are not likely to increase their C storage as a result of a temperature-induced increase in C:nutrient ratios of vegetation or litter. However, other temperature-induced factors may indeed result in increased C storage in woody ecosystems (e.g., movement of nutrients from soils to vegetation).

Of course these three factors will not act in isolation from each other. All three must be considered collectively and their net effect on C storage evaluated. Predictions of the net effect of all three are more difficult to generalize because the factors sometimes work in opposition and the relative magnitudes of the effects will change depending on the magnitude of the changes in CO_2 and temperature. In addition, other ecosystem properties like soil cation exchange capacity and moisture holding capacity might also be altered in conjunction with the changes we analyzed. In many ecosystems other elements, especially P, may be more important than N in restricting changes in C storage. The effects of changes in these additional properties were not addressed in our analysis. Nevertheless, based on the simulations presented here, the capacity to increase C storage in terrestrial ecosystems is high and most terrestrial ecosystems are likely to be sinks for CO_2 as a result of increases in both CO_2 concentration and temperature. The strength of these sinks will be strongly affected by interactions between the C cycle and the cycles of other vital elements.

Acknowledgements: This research was supported by funds from the National Science Foundation (BSR-8718426, BSR-8702328, and BSR-8811764).

References

Aber, J.D., J.M. Melillo, C.A. McClaugherty, and K.N. Eshleman. 1983. Potential sinks for mineralized nitrogen following disturbance in forest ecosystems. Ecological Bulletins (Stockholm) 35: 179-192.

Aber, J. D., J. M. Melillo, K. J. Nadelhoffer, J. Pastor, and R. D. Boone. 1991. Factors controlling nitrogen cycling and nitrogen saturation in northern temperate forest ecosystems. Ecological Applications 1:303-315.

Aerts, R., and F. Berendse. 1989. An analysis of competition in heathland ecosystems. I. Competition for nutrients. Chapter 9 in R. Aerts (ed.) *Plant Strategies and Nutrient Cycling in Heathland Ecosystems*. Foundation for Fundamental Biological Research, Netherlands Organization for Scientific Research, The Netherlands.

Anderson, D.W. 1988. The effect of parent material and soil development on nutrient cycling in temperate ecosystems. Biogeochemistry 5:71-97.

Anderson, J.M. 1991. The effects of climate change on decomposition processes in grassland and coniferous forests. Ecological Applications 1:326-347.

Bazzaz, F.A. 1990. The response of natural ecosystems to the rising global CO_2 levels. Annual Reviews of Ecology and Systematics 21:167-196.

Berendse, F. 1990. Organic matter accumulation during secondary succession in heathland ecosystems. Journal of Ecology 8:413-427.

Berendse, F., H. Oudhof, and J. Bol. 1987. A comparative study on nutrient cycling in wet heathland ecosystems. Oecologia (Berlin) 74:174-184.

Berendse, F., R. Robbink, and G. Rouwenhorst. 1989. A comparative study on nutrient cycling in wet heathland ecosystems: II. Litter decomposition and nutrient mineralization. Oecologia 78:338-348.

Billings, W. D., J. O. Luken, D. A. Mortensen and K. M. Peterson. 1982. Arctic tundra: A source or sink for atmospheric carbon dioxide in a changing environment? Oecologia 53:7-11.

Billings, W. D., K. M. Peterson, J. O. Luken and D. A. Mortensen. 1984. Interaction of increasing atmospheric carbon dioxide and soil nitrogen on the carbon balance of tundra microcosms. Oecologia 65:26-29.

Bolin, B. and R. B. Cook (eds). 1983. *The Major Biogeochemical Cycles and Their Interactions.* John Wiley and Sons, Chichester, 532 pp.

Bormann, F. H., and G. E. Likens. 1979. *Pattern and Process in a Forested Ecosystem.* Springer-Verlag, New York.

Chapin, F. S. III, P. C. Miller, W. D. Billings, and P. I. Coyne 1980. Carbon and nutrient budgets and their control in coastal tundra. Pages 458-484 in J. Brown, P. C. Miller, L. Tieszen, and F. L Bunnell (eds.) *An Arctic Ecosystem: The Coastal Tundra at Barrow Alaska.* Dowden, Hutchinson, and Ross. Stroudsburg, PA.

Chapin, F. S. III, and G. R. Shaver. 1985. The physiological ecology of arctic plants. Pages 16-40 in B. Chabot and H. A. Mooney (eds.), *Physiological Ecology of North American Plant Communities.* Chapman and Hall, London.

Drake, B. G., P. W. Leadly, W. Arp, P. S. Curtis, and D. Whigham. 1989. The effect of elevated atmospheric CO_2 on C_3 and C_4 vegetation on Chesapeake Bay. Proceedings of the Symposium on the Physiological Ecology of Aquatic Plants. Aarhus, Denmark.

Fisher, S., L. Gray, N. Grimm and D. Busch. 1982. Temporal succession in a desert stream ecosystem following flash flooding. Ecol. Monogr 52:93-110.

Gates, D. M. 1985. Global biospheric response to increased atmospheric carbon dioxide concentration. Pages 174-184 in B. R. Strain and J. D. Cure (eds), *Direct Effects of Increased Carbon Dioxide on Vegetation.* U. S. Dept. Energy, Washington D. C.

Giblin, A.E., K.J. Nadelhoffer, G.R. Shaver, J.A. Laundre, and A.J. McKerrow. 1991. Biogeochemical diversity along a riverside toposequence in Arctic Alaska. Ecological Monographs 61(4):415-435.

Hobbie, J. E., J. Cole, J. Dungan, R. A. Houghton, and B. Peterson. 1984. Role of biota in global CO_2 balance: The controversy. BioScience 34:492-498.

Idso, S. B. 1986. Industrial age leading to the greening of the earth? Nature 320:22.
Idso, S. B. 1989. Carbon dioxide, soil moisture, and future crop production. Soil Science 147:305-307.
Kramer, P. J. 1981. Carbon dioxide concentration, photosynthesis, and dry matter production. BioScience 31:29-33.
McClaugherty, C. A., J. D. Aber, and J. M. Melillo. 1982. The role of fine roots in the organic matter and nitrogen budgets of two forested ecosystems. Ecology 63:1481-1490.
Melillo, J. M., T. V. Callaghan, F. I. Woodward, E. Salati, and S. K. Sinha. 1990. Effects on Ecosystems. Pages 282-310 in J. T. Houghton, G. J. Jenkins, and J. J. Ephraums (eds), *Climate Change: the IPCC Scientific Assessment*. Cambridge University Press, Cambridge.
Melillo, J. M., and J. R. Gosz. 1983. Interactions of the biogeochemical cycles in forest ecosystems. Pages 177-222 in B. Bolin and R. B. Cook (eds) *The Major Biogeochemical Cycles and Their Interactions*. John Wiley and Sons, New York.
Mitchell, H. L. and R. F. Chandler. 1939. The nitrogen nutrition and growth of certain deciduous trees of the northeastern United States. Black Rock Forest Bull. 11.
Parton, W.J., J. W. B. Stewart, and C. V. Cole. 1988. Dynamics of C, N, P, and S in grassland soils: A model. Biogeochemistry 5:109-131.
Peterson, B. J., and J. M. Melillo. 1985. The potential storage of carbon caused by eutrophication of the biosphere. Tellus 37B:117-127.
Raich, J. W., E. B. Rastetter, J. M. Melillo, D. W. Kicklighter, P. A. Steudler, B. J. Peterson, A. L. Grace, B. Moore, and C. J. Vorosmarty. 1991. Potential net primary productivity in South America: Application of a global model. Ecological Applications 1:399-429.
Rastetter, E. B., M. G. Ryan, G. R. Shaver, J. M. Melillo, K. J. Nadelhoffer, J. E. Hobbie, and J. D. Aber. 1991. A general model describing the responses of the C and N cycles in terrestrial ecosystems to changes in CO_2, climate, and N deposition. Tree Physiology 9:101-126.
Rastetter, E. B., and G. R. Shaver. 1992. A model of multiple element limitation for acclimating vegetation. Ecology 73, in press.
Shaver, G. R., W. D. Billings, F. S. Chapin, A. E. Giblin, K. J. Nadelhoffer, W. C. Oechel, and E. B. Rastetter. 1992. Global change and the carbon balance of arctic ecosystems. BioScience, in press.
Shaver, G.R. and F.S. Chapin, III. 1991. Production:biomass relationships and element cycling in contrasting arctic vegetation types. Ecological Monographs 61(1):1-31.
Shaver, G. R., F. S. Chapin, and B. L. Gartner. 1986. Factors limiting seasonal growth and peak biomass accumulation in *Eriophorum vaginatum* in Alaskan tussock tundra. Journal of Ecology 74:257-278.
Shaver, G.R., K.J. Nadelhoffer, and A.E. Giblin. 1990. Biogeochemical diversity and element transport in a heterogeneous landscape, the North Slope of Alaska. Pages 105-125 in M.G. Turner and R.H. Gardner (eds.), *Quantitative Methods in Landscape Ecology*. Springer-Verlag,

New York, New York, USA.
Strain, B. R., and F. A. Bazzaz. 1983. Terrestrial plant communities. Pages 177-222 in E. R. Lemon (ed), CO_2, and Plants: The response of Plants to Rising Levels of Atmospheric Carbon Dioxide. Westview Press, Boulder
Strain, B. R., and J. D. Cure (eds). 1985. *Direct Effects of Increasing Carbon Dioxide on Vegetation*. U. S. Dept. Energy, Washington D.C.
Strain, B. R., and J. D. Cure. 1986. *Direct Effects of Atmospheric CO_2 Enrichment on Plants and Ecosystems: A Bibliography with Abstracts*. U. S. Dept. Energy, Washington D.C.
Tans, P. P., I. Y. Fung, and T. Takahashi. 1990. Latitudinal distribution of sources and sinks of atmospheric carbon dioxide. J. Geophys. Res. 94:5151-5172.
Tissue, D.T., and W.C. Oechel. 1987. Response of *Eriophorum vaginatum* to elevated CO_2 and temperature in the Alaskan tussock tundra. Ecology 68:401-410.
Vitousek, P., T. Fahey, D. Johnson and M. Swift. 1988. Element interactions in forest ecosystems: Succession, allometry and input-output budgets. Biogeochemistry 5:7-34.
Vitousek, P. M. and W. A. Reiners. 1975. Ecosystem succession and nutrient retention: A hypothesis. BioScience 25:376-381.

ASPECTS OF THE INTERACTION BETWEEN VEGETATION AND SOIL UNDER GLOBAL CHANGE†

W. M. POST, J. PASTOR‡
A. W. KING, and W. R. EMANUEL
Oak Ridge National Laboratory
P. O. Box 2008, Bldg. 1000
Oak Ridge, Tennessee 37831–6335
USA

ABSTRACT. Responses of terrestrial ecosystems to a world undergoing a change in atmospheric CO_2 concentration presents a formidable challenge to terrestrial ecosystem scientists. Strong relationships among climate, atmosphere, soils and biota at many different temporal and spatial scales make the understanding and prediction of changes in net ecosystem production (NEP) at a global scale difficult. Global C cycle models have implicitly attempted to account for some of this complexity by adapting lower pool sizes and smaller flux rates representing large regions and long temporal averages than values appropriate for a small area. However, it is becoming increasingly evident that terrestrial ecosystems may be experiencing a strong transient forcing as a result of increasing levels of atmospheric CO_2 that will require a finer temporal and spatial representation of terrestrial systems than the parameters for current global C cycle models allow. To adequately represent terrestrial systems in the global C cycle it is necessary to explicitly model the response of terrestrial systems to primary environmental factors. While considerable progress has been made experimentally and conceptually in aspects of photosynthetic responses, and gross and net primary production, the application of this understanding to NEP at individual sites is not well developed. This is an essential step in determining effects of plant physiological responses on the global C cycle. We use a forest stand succession model to explore the effects of several possible plant responses to elevated atmospheric CO_2 concentration. These simulations show that ecosystem C storage can be increased by increases in individual tree growth rate, reduced transpiration, or increases in fine root production commensurate with experimental observations.

† Research sponsored by U.S. Department of Energy, Carbon Dioxide Research Program, Atmospheric and Climate Research Division, Office of Health and Environmental Research, under contract DE–AC05–84OR21400 with Martin Marietta Energy Systems, Inc.

‡ Natural Resources Research Institute, 5013 Miller Trunk Highway, Duluth, Minnesota 55811, USA

Water, Air, and Soil Pollution **64**: 345–363, 1992.
© 1992 *Kluwer Academic Publishers.*

1. Terrestrial Vegetation and the Global C Cycle

The global C cycle is composed of three major reservoirs: the atmosphere, the oceans, and the terrestrial system. The atmosphere provides a major link between the other two reservoirs. The amount of C stored in global C reservoirs and the exchange rates between them vary to differing degrees depending on the time scale considered. Over the last 160,000 years the atmospheric CO_2 concentration has varied from 200 $\mu L \cdot L^{-1}$ during the height of glaciations to between 260 and 300 $\mu L \cdot L^{-1}$ during interglacial times (Barnola et al. 1987). Over the past 6,000 to 8,000 years the concentration of CO_2 in the atmosphere has remained fairly constant at 280 to 290 $\mu L \cdot L^{-1}$ until the large scale burning of fossil fuels in the mid 1800's. Olson (1985) estimates terrestrial plant C during the mid-Holocene 6,000 years ago to be 823 Pg and modern storage (i.e., in 1980) to be 560 Pg. Similar changes have occurred in C storage in various ocean subreservoirs over long time scales. Over shorter time periods during the next few decades or centuries, the transfers of C between the terrestrial and oceanic systems through the atmosphere as well as the assimilation of inputs from fossil fuel burning will be governed by the same processes operating today. The C flux rate of these processes is influenced by today's inventories of C in various reservoirs and their turnover rates, which vary as functions of environmental factors.

The balance between assimilation by photosynthesis, its partioning among several terrestrial pools of varying turnover times, and releases of C from both living and dead material through respiration and decomposition dictate the magnitude of the net exchange of C between the atmosphere and the world's terrestrial systems. At shorter than decade time scales the balance between C uptake and release certainly shifts at specific locations with nutrient availability, weather changes, and various disturbances including wildfire, violent storms, pathogen outbreaks, and widespread animal activities. At even shorter, subannual time scales, the seasonality of net primary production (NPP) and decomposition are sufficiently temporally separated to contribute to the annual oscillation of atmospheric CO_2 concentration (Tans et al. 1990; Houghton 1987). For the major portion of terrestrial ecosystems, uptake and loss fluxes are frequently assumed to reach a balance over sufficiently large spatial extent and long time periods. This assumption of balance does not imply that each hectare of land surface is at some kind of pristine equilibrium with maximum biomass. For this reason, most global C cycle models that explicitly treat the terrestrial system (Emanuel et al. 1984, Houghton et al. 1987) use lower estimates of pool sizes and flux rates than ones appropriate for mature stands (Whittaker and Likens 1973, Houghton et al. 1983, Esser 1987). This is an expression of the notion that in any suitably long span of time, the amount of C accumulating into aggrading systems is matched by an equivalent amount released by periodic or episodic disturbances and growth declines. This is consistent with the history of CO_2 concentration recorded in the Siple and South Pole ice cores (Post et al. 1990). Frequently used estimates of global terrestrial ecosystem NPP center around the value 60 Pg $C \cdot yr^{-1}$, assumed to be in balance with an equivalent flux of CO_2 from decomposition (Olson et al. 1983, Ajtay et al. 1979). This is about 10 times the current release of CO_2 from fossil fuel burning (Marland and Boden 1989).

Lugo and Brown (1986) present a dissenting view concerning the balance between NPP and decomposition at the global level. Their argument has two components. The first one is that even "equilibrium" ecosystems may be net sinks of C relative to the atmosphere. C involved in ecosystem dynamics may reach a constant amount after time, but a fraction of ecosystem production may be accumulated in inactive pools such as peat in wet ecosystems or soil carbonates in dry ones, or exported dissolved in rivers and groundwater and eventually to the ocean and not the atmosphere. Locally or over geological time scales, such factors may be important, but globally the rates attributed to these C fluxes are small. Accumulation of C in organic soils of the world is estimated at 0.14 Pg $C{\cdot}yr^{-1}$ (Armentano 1980). Soil carbonate formation proceeds at the rate of 0.01 Pg $C{\cdot}yr^{-1}$ (Schlesinger 1985), and somewhere between 0.44 to 0.47 Pg $C{\cdot}yr^{-1}$ reaches the oceans from rivers (Maybeck 1981; Schlesinger and Melack 1981).

The second part of Lugo and Brown (1986)'s argument hinges on the speculation that we cannot assume that there are enough reasonably undisturbed regions of the terrestrial world for an equilibrium assumption to be usefully applied in global C cycle calculations. This argument has considerable merit, but there is not enough information currently to address the complexity that this supposition introduces. It is becoming increasingly clear that the global C cycle is responding to strong transient forcings that can influence rates of terrestrial NPP and decomposition. Our knowledge of the atmosphere and ocean components of the global C cycle indicate that the global terrestrial system may currently be sequestering 1 to 3 $Pg{\cdot}yr^{-1}$ (Post et al. 1990). A change that causes an imbalance between NPP and decomposition on the order of 3% to 5% of NPP could result in a global terrestrial NEP of 2 to 3 $Pg{\cdot}yr^{-1}$. Equilibrium depictions of terrestrial system C dynamics will be increasingly invalid if global changes affect the processes that regulate the exchange rates of CO_2 between terrestrial ecosystems and the atmosphere.

Identifying a 1 to 3 $Pg{\cdot}yr^{-1}$ net terrestrial flux presents a formidable challenge to terrestrial ecosystem scientists who do not have a precise methodology to measure or observe an estimated 60 $Pg{\cdot}yr^{-1}$ global uptake and release as NPP and decomposition. While this may have the appearance of looking for a needle in a haystack, a flux of 1 to 3 $Pg{\cdot}yr^{-1}$ is very significant. This is approximately the rate at which atmospheric CO_2 is increasing. One of the most plausible hypotheses that terrestrial systems may not be presently in equilibrium is that they are responding directly to the increase in atmospheric CO_2 as observed in small scale experiments through increased rates of photosynthesis and perhaps decreased rates of respiration. There are many processes that intervene between photosynthetic rate and NEP that are not nearly as well known as those that occur at the physiological level. Integration of single plant responses to heterogeneous canopies, landscapes, etc., existence of photosynthate sinks, plant allocation, interactions with other limiting resources, effects of plant life history and demography, interactions with multiple species and life forms which modify environmental conditions and much more need to be considered in extrapolating experimental results to the global C cycle. Here we describe an ecosystem model that can extrapolate whole tree responses to environmental conditions on a 1/12 hectare plot and project NEP for that plot. This is an essential step in determining the effects of plant physiological responses on the global C cycle.

2. Sensitivity of Terrestrial NEP to Primary Environmental Factors

Evaluating the potential impacts of atmospheric and climatic changes on terrestrial ecosystems requires a detailed understanding of the effects of environmental factors on ecosystem processes. While considerable experimental and theoretical work is being done on investigating the response of leaf and plant photosynthesis to environmental factors (Oechel and Strain 1985; Eamus and Jarvis 1989), very little work has been done on the responses of NEP to changes such as elevated CO_2 or climate change. Several studies have examined the response of tundra samples to climate variables and CO_2 changes in a phytotron (Billings et al. 1983, 1984) and in the field (Grulke et al. 1990). A fairly complete field experiment on salt marsh vegetation (Drake and Leadley 1991) is also a valuable contribution to our knowledge of the effects of elevated CO_2 on net ecosystem production. However, for forest ecosystems, comprehensive field experiments assessing NEP under elevated CO_2 conditions have yet to be attempted. The stature of the vegetation and the life spans of trees makes such experiments challenging. Until experimental results are available to guide modeling activities, we must make use of what knowledge we currently have available for components of ecosystem processes involved.

Most models that are reported as relevant to terrestrial ecosystem responses to global changes are physiological models describing photosynthesis rates and or respiration rates, perhaps even C allocation to different plant tissues. While these are essential components of forest ecosystems responses to environmental change, the understanding of changes in NEP is not complete until the fate of the photosynthate is followed into living and dead organic matter pools or back to CO_2. There are several forest ecosystems models that do not only model C fixation as a function of environmental factors, but also keep track of C as it is allocated to different plant parts, shed as litter, and decompose into soil organic matter and CO_2. Most notable are Forest-BGC (Running and Coughlan 1988) and BIOMASS (McMurtrie et al. 1990). These models are made manageable by limiting simulations to single species stands or plantations with a horizontally uniform canopy structure. As a result they are not immediately applicable to the majority of forests that contain several to many species of varying sizes and ages.

Species composition and population dynamics play an important role in determining the NEP dynamics of forest ecosystems. Since species composition and tree population dynamics can also be modified by changes in environmental factors it then becomes important to evaluate their role under global change. In the following discussion we will emphasize decomposition and treat production very generally. This emphasis is not meant to take any importance away from primacy of production on NEP, rather it is to examine the interaction between vegetation and soil that may result in long term sequestration of C in the large soil organic matter pool or increases in aboveground storage resulting from indirect effects of environmental factors on nutrient cycling.

2.1 FOREST ECOSYSTEM ORGANIZATION

Forest ecosystem processes of resource competition, nutrient cycling, and primary productivity are significantly influenced by individual trees and therefore operate

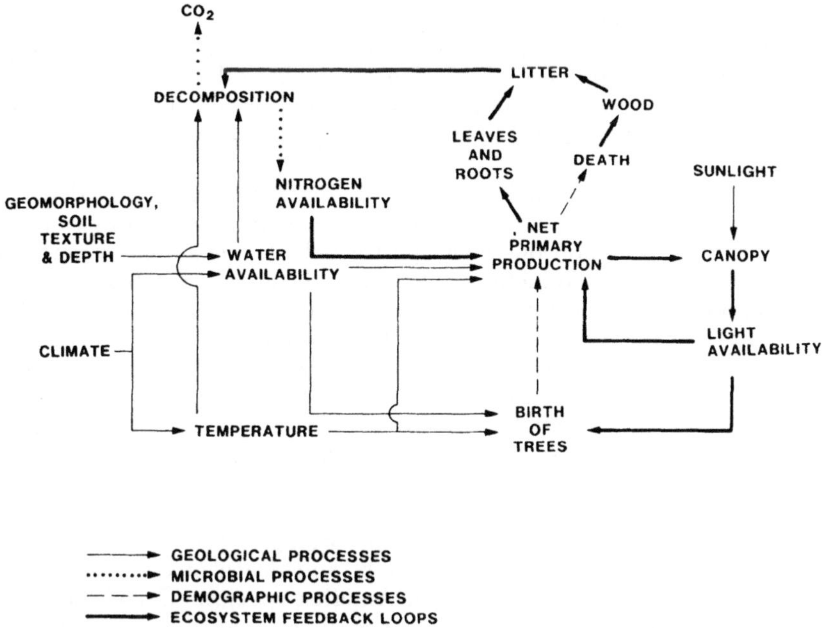

Figure 1. Interactions between geological, microbial, tree demographic processes, and ecosystem feedback loops contained in the individual-based forest gap model LINKAGES (from Pastor and Post 1985).

largely at the temporal and spatial scale of single large trees. Some models keep track of the establishment, growth, and mortality of individual trees by introducing, solving, or deleting an equation for each individual tree when appropriate. These models have been used to examine the successional dynamics of forests at numerous locations spread over at least four continents (Shugart 1984). Aber et al. (1982) and Weinstein et al. (1982) modified this approach to simulate ecosystem nutrient cycles. Solomon et al. (1984) developed a version that could predict species composition of most upland forests of eastern North America from climatic and soil moisture factors. Using Solomon et al. (1984)'s model for aboveground population dynamics we added C and N decomposition dynamics using a different approach from that of Aber et al. (1982) and Weinstein et al. (1982). In our model, the aboveground population dynamics are extended into the soil environment by tracking changes in C and N content of the individual tree's annual production of litter. This model, therefore simulates the individual's influence on ecosystem processes during the decomposition of its shed parts. The details of this model, referred to as LINKAGES, have been published previously (Pastor and Post 1985, 1986). Our objective here is to show the importance of this "individual-based" modeling approach (Huston et al. 1988) in linking the population, community, and ecosystem dynamics of forests in predicting their response to a changing environment.

The LINKAGES model simulates ecosystem C and N storage and cycling by considering the interactions between processes determining individual tree growth, demographic processes determining tree population dynamics, microbial processes determining N availability, and processes determining water availability. Figure 1 is a diagram of these hypothesized relationships. According to this depiction of ecosystem function, soil and climate are constraints within which feedback relationships between vegetation and light availability and between vegetation and N availability operate.

The feedback loops shown in Figure 1 involving light availability depict negative feedbacks. A feedback is called negative when it involves processes that tend to restore steady state conditions when perturbed. For example, a perturbation that reduces net primary production (NPP) will result in a smaller canopy mass, increasing canopy light penetration, enabling the establishment of new individual trees which, in turn, increase NPP. The feedback loop in Figure 1 involving decomposition and resultant N availability introduces a positive feedback into the model of a forest ecosystem. Initial deviations, in NPP for example, are amplified by this feedback loop. Increases in NPP result in more litter production, decomposition, and available N, eventually resulting in further increases in NPP. Similarly, declines in NPP will set into play processes that result in reduced N availability and further declines in NPP.

LINKAGES assembles a forest ecosystem according to the configuration in Figure 1 by modeling the establishment, growth, and death of trees that represent individuals from any of 72 eastern North America species. Correlation between species properties, particularly shade tolerance, litter quality (lignin:N ratio), and drought tolerance, and the effect of these individual's properties on population and ecosystem processes results in surprising ecosystem dynamics that are explored below in two sets of simulations — one that results in internal dynamics that lead to spruce decline, and another that shows a divergence in ecosystem response to climate change at the boreal/northern-hardwood border.

2.2 SHADE TOLERANCE AND LITTER QUALITY

Growth declines of white spruce in mature Canadian stands are well documented (Dendron Resource Surveys 1985). There are two intrinsic properties of black and white spruce (*Picea mariana* and *P. glauca*) and other northern conifer ecosystems pertinent to the problem of spruce decline: (*i*) net primary production in spruce and other boreal forests is limited by N availability (Weetman 1968, Stewart and Swan 1970, Krause 1981, Mahendrappa and Salonius 1982, Van Cleve and Zasada 1976, Van Cleve and Oliver 1982); (*ii*) spruce litter depresses soil N availability because it decays and mineralizes N slowly and has a high immobilization potential as a result of its high lignin and phenolic and low N contents (Davies et al. 1964, Piene and Van Cleve 1978, Melillo et al. 1982, Flanagan and Van Cleve 1983, Moore 1984, Van Cleve and Harrison 1985, Carlyle and Malcolm 1986). This depression of soil N availability by spruce and other conifer litter can cause slow growth (Weetman and Nykvist 1963, Heilman 1966, Viro 1967, Williams 1972, Van Cleve et al. 1983, Matson and Boone 1984) and increased susceptibility to insects or pathogens (Matson and Waring 1984).

Two sets of simulations for a region in central Ontario were completed using the LINKAGES model. In one set, N availability was affected by the litter production

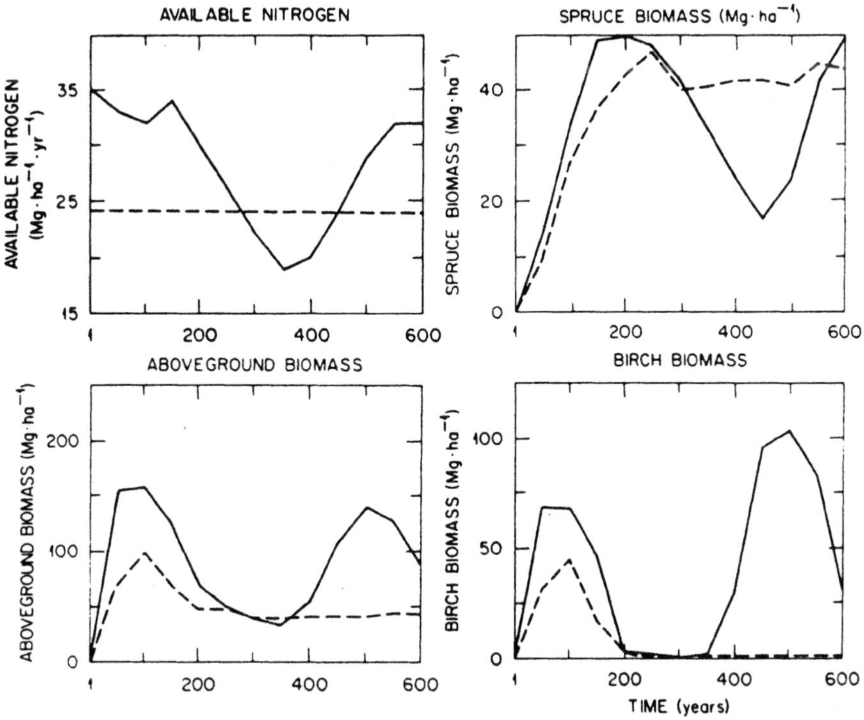

Figure 2. Simulated Ontario aspen, white birch and black spruce forest with N availability as a function of litter chemistry (———), and N availability held constant at 24 kg/ha/yr (- - - -) (after Pastor et al. 1987).

and chemical composition of the tree individuals on the site. In the other set, N availability was held constant at 24 kg·ha^{-1}·yr^{-1}, a reasonable amount for boreal forests (Larson 1980). Climate data from Hare and Thomas (1979) was used as input to the model. The resulting simulated forest stands, in both simulations, consisted initially of aspen and paper birch followed by black and white spruce. After the initial successional dynamics, the simulations differed in the pattern of dominance with these two cover types. The climate is near optimal for these species and simulations were run for a deep silty clay loam with field moisture capacity of 38.8 cm and wilting point of 20.0 cm, so temperature and drought are not limiting factors in the resultant model responses. The remaining factors limiting to individual tree growth, N and light availability, are therefore responsible for differences observed between the two sets of simulations (Figure 2).

The results of this modeling exercise are presented in Figure 2. For the case in which N availability varies as a function of litter chemistry, N is highest in the early successional forest dominated by the shade-intolerant species paper birch with small amounts of aspen. Both of these species have easily decomposable litter with high N content. As the shade- tolerant white and black spruce gain dominance around year 150, the slowly decomposing and low N litter they produce begins to depress N availability. As N availability declines to 20 $kg \cdot ha^{-1} \cdot yr^{-1}$, there is a steady decline over the next 125–200 years in live spruce biomass and density. This decline causes the canopy to open, increasing decomposition and N availability. The open canopy also allows growth of the shade-intolerant, rapid growing birch with higher litter quality. N availability and leaf litter N are further increased. This increased N availability causes aboveground production and biomass to increase.

The modeled alternation between spruce and birch dominance is caused by the feedbacks between vegetation and soils that is set in motion due to an inverse relationship between litter quality and shade-tolerance of the dominant species. When N availability was held constant at 24 $kg \cdot ha^{-1} \cdot yr^{-1}$, thus breaking the influence of vegetation on soil N availability, spruce maintained dominance and the ecosystem reached a steady spruce biomass and aboveground biomass. It may be concluded from this modeling exercise that spruce dominance and decline alternating with dominance by early successional species may be a logical consequence of C-N linkages in boreal forests (Van Cleve et al. 1983, Pastor et al. 1987).

2.3 CLIMATE CHANGE – DIVERGENCE IN POTENTIAL RESPONSES

Previous evaluations of forest response to CO_2-induced climate changes using simulation models (Solomon 1986, Shugart et al. 1986, Davis and Botkin 1985) have not considered concomitant changes in soil water and N availability and their interaction with shifts in vegetation composition. Simulations using LINKAGES have shown that the potential forest response to an altered climate can be very sensitive to soil water holding capacity and the subsequent positive feedbacks that are initiated as the result of changes in the N cycle (Pastor and Post 1988). The greatest changes are predicted to occur at current transition zones. At regions far from the transition zones, climate change resulted in direct effects on stand growth, increasing production where climate change resulted in more favorable conditions for extant species, and decreasing production where the opposite was the case. At transition zones, direct effects of climate change usually resulted in much larger changes in productivity due to the indirect effect of species composition changes and amplification of ecosystem responses caused by the subsequent modifications in nutrient cycling.

Figure 3 shows simulation results for forests at the current boreal/northern-hardwood border in northeastern Minnesota. The LINKAGES model was run for five hundred years at each site under current climate, which was then altered linearly over the next 100 years to reach 2 x CO_2 climate for the region, then run for 200 more years with the new climate. Two contrasting soils, sand and silty clay loam, were selected to give estimates of the range of soil water availability. Available water is assumed to be 18.3 $cm \cdot m^{-1}$ soil depth for the silty clay loam and 10 $cm \cdot m^{-1}$ soil depth for the sand. On the silty clay loam soil there was no decrease in soil water availability with a doubling of CO_2, and the current mixed spruce-fir/northern-hardwood forest was replaced by a more productive northern-hardwood forest. Northern-hardwood forest species have faster intrinsic growth rates and can attain greater biomass than either spruce or fir. In addition,

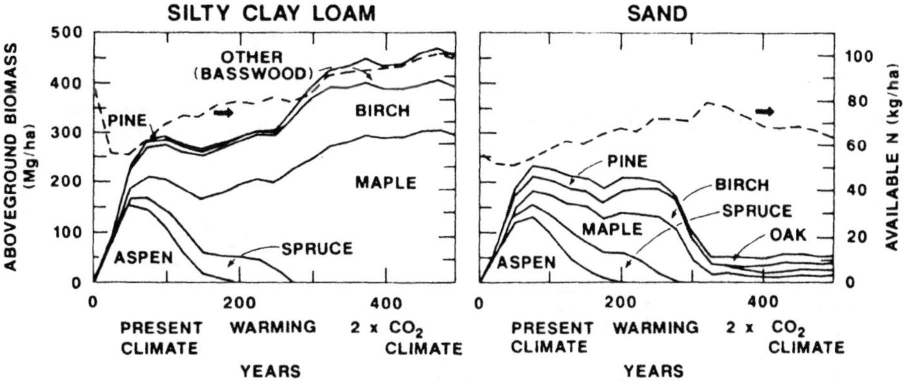

Figure 3. Simulated changes in biomass, species composition, and soil N availability (- - - -) for northeastern Minnesota. Horizontal axes represent model runs for 200 years under current climate, which are then altered linearly over the next 100 years ot reach a 2 x CO_2 climate, then run for 200 more years with the new climate (Modified from Pastor and Post 1988).

the warmer climate as well as the higher N and lower lignin content of northern-hardwood forest species litter enhance soil N availability which amplifies the effect of warming on productivity.

On the sandy soil there were increases in the proportion of the growing season with soil water below wilting point. The simulated mixed spruce-fir/northern-hardwood forests were replaced by a stunted pine-oak forest with a low C storage and productivity characteristic of savannas. Although the warmer temperatures would increase decomposition rate and N mineralization, the higher lignin and lower N contents of oak and particularly pine litter compensated for this and reduced decomposition.

The divergence in projected responses to the same climate changes on different soils results from the positive correlation between drought tolerance and the lignin:N ratio of recently shed leaves. For the 17 species that occur in the northeast Minnesota simulations described above, species leaf litter lignin:N ratio tends to increase (and therefore litter quality decreases) as species drought tolerance increases (Figure 4). This ensures that as the importance of drought increases, the species composition is shifted in such a way that both N and insufficient soil moisture will limit growth.

2.4 POTENTIAL EFFECTS OF CO_2 ON FOREST CARBON STORAGE

Individual based forest succession models that LINKAGES has been derived from have been used in computer experiments to examine the potential effects of CO_2 fertilization on model response (Botkin et al. 1973, Shugart and Emanuel 1985, Luxmoore et al. 1990). The general result of all these simulation experiments was that increasing the growth rate of some or all species in the model had a small impact on the amount of biomass. LINKAGES differs in two important aspects from these earlier models that can make a significant difference in the amount

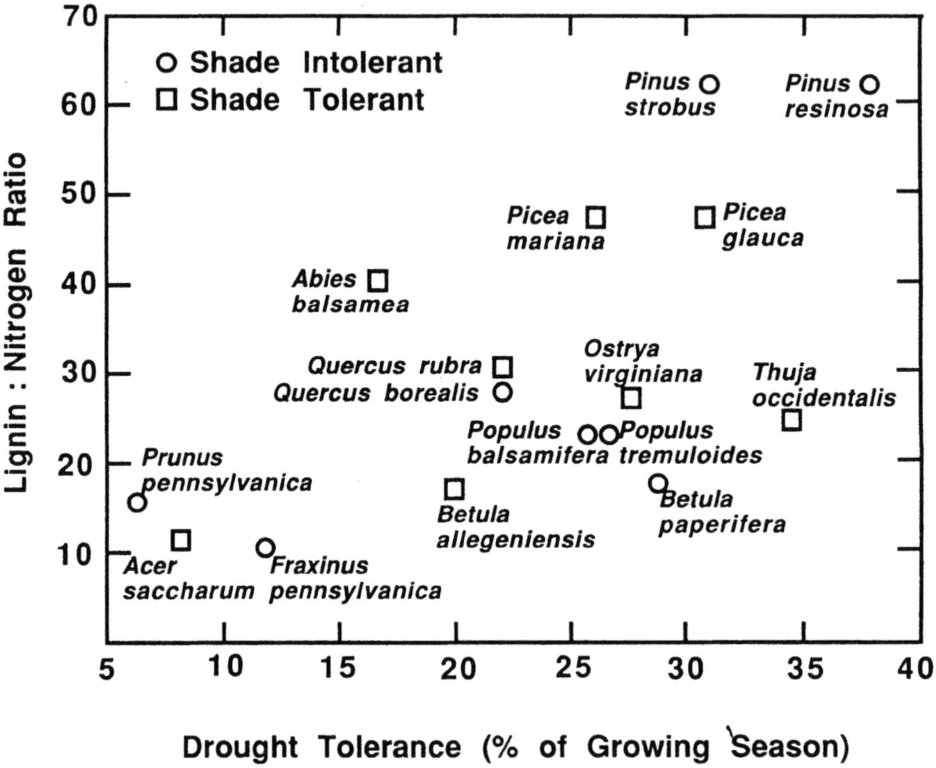

Figure 4. A scatter plot of species parameters — ratio of the mass of lignin to the mass of N in freshly fallen litter on the ordinate and the tolerance of a species to drought on the abscissa. Drought tolerance is expressed as the percentage of the growing season for which there must be adequate moisture for the species to exhibit a growth increment for the year (modified from Post and Pastor 1990).

of organic matter that can accumulate in a CO_2 fertilization model experiment. First, LINKAGES includes litter and soil organic matter dynamics that were not considered in the models mentioned above. Second, these earlier models assumed a maximum basal area per plot. As total basal area of all individuals approached this maximum, growth is reduced, eventually to zero if necessary, in order not to exceed this maximum basal area. In LINKAGES this constraint is removed. Instead, the amount of basal area is completely determined by birth, growth, and mortality processes, which are affected by N availability. In a certain sense, the maximum basal area constraint has been replaced by N limitation that is free to go up or down with production and decomposition. We performed three model experiments using LINKAGES to explore potential effects of increased atmospheric CO_2 on the amount of organic matter in forest ecosystems. The results of these experiments for climate conditions on the Cumberland Plateau, Tennessee, are given in Figures 5 and 6. For both sites simulation were done using two contrasting soil types and two climate regimes. Soil parameters representing a sand soil were 15 cm for field capacity and 5 cm for wilting point. A silty clay loam soil was represented with 38.3 cm for field capacity and 20 cm for wilting point. Climate inputs for present day climate simulations are values from meteorological observations. The 2 x CO_2 climate inputs are from Mitchell (1983) and are derived from climate GCM experiments. All results are averages of 50 simulated plots or individual runs of the LINKAGES model each started from a bare plot with soil organic matter and soil N levels appropriate for the combination of soil texture and climate.

The nominal runs for the four combinations of soil and climate conditions show the results of the standard LINKAGES model. The model runs with sand soil simulate less litter and soil organic matter and less aboveground biomass than the simulation with silty clay loam soil. On the sand soil, soil water is limiting and results in communities of drought tolerant species largely of hickories and oaks. The silty clay loam soil allows the growth of some highly productive beech, maple, and hackberry in addition to hickories and oaks. Drought on the sand soil under 2 x CO_2 climate is so severe that an oak savanna vegetation is predicted. Drought does not restrict growth in the simulations on the silty clay loam soil, either under present climate or under 2 x CO_2 climate conditions. The same patterns are reflected in N availability (Figure 6). The presence of mesic species in the silty clay loam soil runs results in slightly higher N availabilities than on sand soil under present climate. The sand soil under 2 x CO_2 climate has very low N available due to low production in addition to the low litter quality of the oak species present in the model simulations.

To simulate the potential direct effect of higher atmospheric CO_2 concentration on tree growth we increased the potential annual basal area increment of all species by 50%. This amount of growth enhancement is observed in many experiments with plants grown at 600 $\mu L \cdot L^{-1}$. This was accomplished in the simulations by multiplying the maximum ring with increment, G, for each species by 1.22. Through allometric relationships, this increases the amounts of leaves which will affect the competition for light among individual trees. It also increases the amount of leaf, bole, and root litter that will then increase the demand for N, both for growth and decomposition. As light and N constraints increase, the potential increase of 50% in growth rate is diminished but still has a noticeable impact on ecosystem organic matter stocks. This 50% increase resulted in an average increase in aboveground biomass of approximately 150 Mg/ha. At this Cumberland Plateau site this is a 25% increase in biomass under three of the four environmental conditions. On the sand soil under the 2 x CO_2 climate, this is a 100% increase in C storage from the nominal run but still represents considerably less C storage than is possible for a more favorable climate or less droughty soil. While such increase in growth also results in increases in the production of leaf, bole, and root litter inputs, very little change occurred in soil and litter organic matter levels. This increase in material entering the decomposition cycle, however, did increase the amounts of available N (Figure 6). The manner in which N concentration in the biomass and therefore the litter is determined was not modified in these simulations (that is, we did not alter the litter quality of organic material produced as has been hypothesized for plants grown in high CO_2 atmospheres).

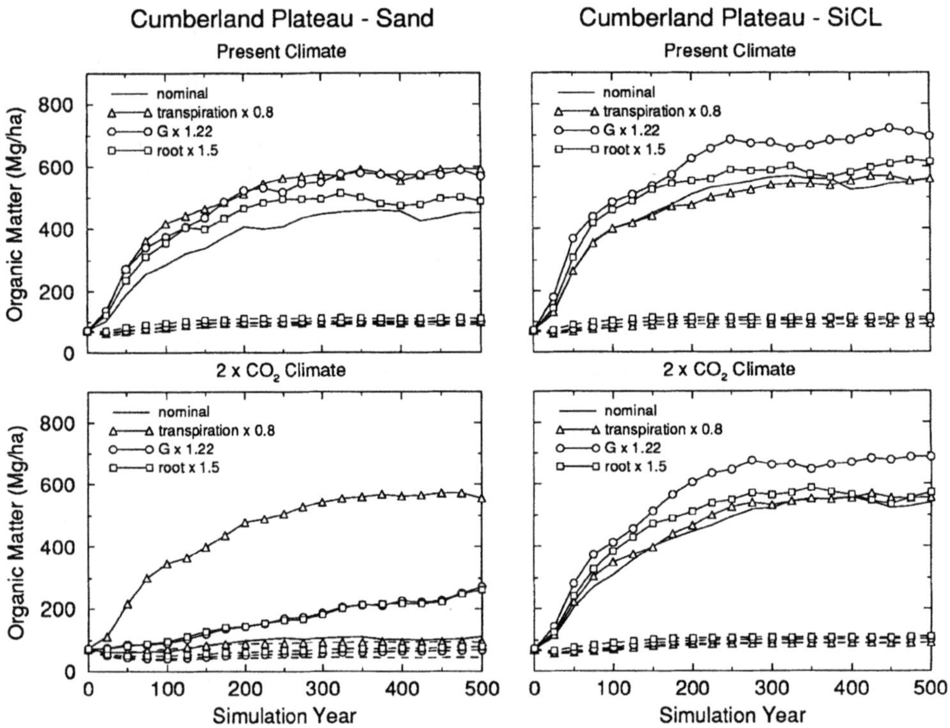

Figure 5. Changes in ecosystem organic matter content during a 500 year simulations with LINKAGES under two climate regimes (present climate and a warmer, drier 2 x CO_2 climate) and two soils with contrasting soil water holding capacities - low (sand) and high (silty clay loam, SiCL). Litter + soil organic matter is indicated by dashed lines. Aboveground biomass, expressed as dry weight, is the distance between the dashed line and solid line which represent the total ecosystem organic matter. Different symbols indicate various model modifications to simulate the potential effects of elevated CO_2 on forest organic matter dynamics. Nominal model runs against which other runs should be compared have no symbols plotted. Lines marked G x 1.22 simulate the effect of a 50% increase in annual basal area increment. Lines marked transpiration x 0.8 have transpiration reduced by 20% during the growing season. Lines marked root x 1.5 have a 50% increase in root production per year.

In the next experiment we explored the possible effects of increased water use efficiency on organic matter dynamics. In this set of simulations we assume that during the growing season evapotranspiration is reduced by 20% (i.e. multiplied by 0.8). We do this in a very unrealistic way in that it is not related to leaf area or CO_2 uptake in any way. We merely reduce water loss by a proportion of the amount that was computed in the nominal case using the method of Thornthwaite and Mather (1957) which is based on empirical relationships with temperature. This experiment

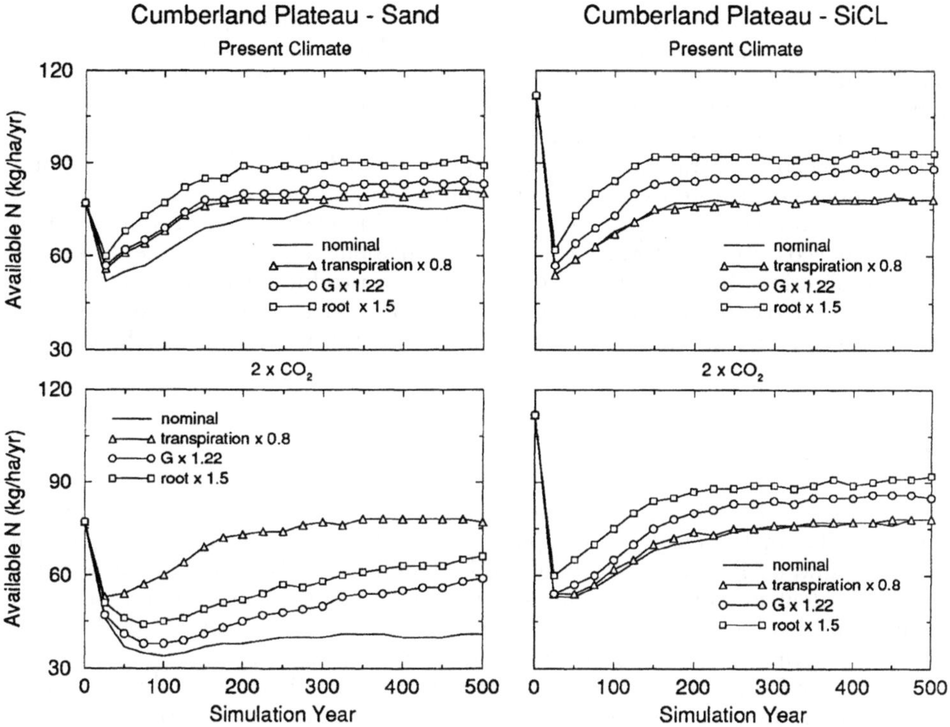

Figure 6. Changes in available N during 500 year simulations of LINKAGES. For fuller description of legends see Figure 5 and text.

simulated a modified water balance that could occur as the result of decreases in plant water use. These simulations show very little change in organic matter dynamics on the silty clay loam soils. This is expected since water availability was not an important factor in limiting growth in the nominal models runs on this soil. On the sand soil, however, reductions in evapotranspiration had dramatic effects, and completely eliminated the differences in aboveground and litter and soil organic matter amounts between the silty clay loam soil and the sand soil. Under current climate conditions, all the changes are in aboveground biomass. Under the warmer 2 x CO_2 climate, some of the increases come from additional soil organic matter over the nominal simulations on the sand soil. Again, we see corresponding changes in available N to support the changes in production. Some of the changes in N availability are the result of the increased growth of mesic species with higher leaf litter qualities.

The final experiment involved increasing the amount of root litter production by a factor of 1.5, without changing basal area increment, leaf production, or transpiration. Such changes in root production have been consistent with experiments using young trees grown under approximately double CO_2 (Norby 1992). Under all conditions this resulted in small increases in soil organic matter. It also has the effect of increasing aboveground organic matter by increasing N availability. In LINKAGES, over several simulated years, accumulation of fixed N through immobilization processes resulting from increased inputs of organic matter eventually, when this material mineralizes N, it leads to increased N availability. Increasing root litter inputs by 50% lead to greater N availability than increasing aboveground production by 50%. A fraction of the aboveground production remains in wood and therefore does not lead to as much N immobilization/mineralization as roots which decompose relatively soon after their production. Increasing root production does not increase ecosystem C storage as much as equivalent relative increases in aboveground production. In addition to the fact that root production (mostly in the form of fine roots) decompose more rapidly than wood, a 50% increase in root production results in a smaller increment in total organic matter production than a 50% increase in aboveground production.

3. Conclusions

The terrestrial biosphere is an important component in determining the concentration of CO_2 in the earth's atmosphere. Net primary production at the global scale of around 60 Pg/yr may or may not be balanced for the past 100 or so years by an equivalent amount of decomposition. Significant evidence has accumulated from experimental manipulations that plant C dynamics for most terrestrial species is altered as atmospheric CO_2 concentrations change. It is possible that changes in atmospheric CO_2 since 1860 have changed the balance between global NPP and decomposition. In order to evaluate this possibility we need to determine whether the experimental observations on plant growth affect C storage in ecosystems. Currently, experimental results for forest ecosystems are not available. We used a forest gap model to evaluate several potential effects of elevated atmospheric CO_2 on ecosystem organic matter storage.

All potential plant responses examined (increased annual basal area increment, decreased water use, increased allocation of C and N to roots) have the potential to increase organic matter in forest ecosystems. The amount of increased organic matter storage depends on species composition, soil water holding capacity, and resultant N dynamics. Most changes in storage occurred in aboveground biomass. The litter and soil organic matter pools changed very little between model experiments. The enhancement of net ecosystem production is relatively short lived. The difference in ecosystem C storage between nominal plant response and potential CO_2 enhanced responses does not increase after 100 or so years. C storage over long periods of time on soils with good water holding capacity can be increased on average by 25% with 50% increases in basal area increment. Increases in C storage can be even relatively greater in water limited forest ecosystems. The indirect effect of reduced transpiration can have even larger effects than the direct CO_2 growth effects on C storage in water limited forest ecosystems.

These analyses focus on some aspects of ecosystem C dynamics, but they ignore many processes and interactions that will be necessary to give accurate estimates of changes in forest ecosystem C storage resulting from global changes of climate and atmospheric CO_2 concentration. Contrary to previous CO_2 fertilization experiments using forest gap models, we show that considerable potential exists for increasing C storage in forest ecosystems resulting from plant responses to elevated CO_2 levels.

References

Aber J. D., Melillo J. M., and Federer C. A. (1982) 'Predicting the effects of rotation length, harvest intensity, and fertilization on fiber yield from northern hardwood forests in New England', Forest Science 28,31–45.

Ajtay, G. L., P. Ketner, and P. Duvigneaud. (1979) 'Terrestrial primary production and phytomass', in B. Bolin, E. T. Degens, S. Kempe, and P. Ketner (eds.), The Global Carbon Cycle, SCOPE 13, John Wiley and Sons, New York, pp. 129–181.

Armentano T. V. (ed.) (1980) The role of organic soils in the world carbon cycle CONF-7905135, United States Department of Energy, Washington, D. C.

Barnola J. M., Raynaud D., Korotkevich Y. S., and Lorius C. (1987) 'Vostok ice core provides 1600,000-year record of atmospheric CO_2', Nature 329,408–414.

Billings W. C., Luken, J. O., Mortensen, D. A., Peterson, K. M. (1983) 'Increasing carbon dioxide: Possible effects on arctic tundra', Oecologia 58, 286–289.

Billings W. D., Peterson, K. M., Luken, J. O., Mortensen, D. A. (1984) 'Interaction of increasing atmospheric carbon dioxide and soil nitrogen on the carbon balance of tundra microcosms', Oecologia, 65, 26–29.

Botkin, D. B., Janak, J. F., and Wallis, J. R. (1973) 'Estimating effects of carbon fertilization on forest composition by ecosystem simulation', in G. M. Woodwell and E. V. Pecan (eds.), Carbon and the Biosphere, U.S. Atomic Energy Commission, Washington D.C., pp. 328–344.

Carlyle J. C., and Malcolm D. C. (1986) 'Nitrogen availability beneath pure spruce and mixed Larch + spruce stands growing on a deep peat. 1. Net N mineralization measured by field and laboratory incubations', Plant and Soil 93,95–113.

Davies R. I., Coulson C. B., and Lewis D. A. (1964) 'Polyphenols in plant, humus, and soil. Factors leading to increase in biosynthesis of polyphenol in leaves and their relationship to mull and mor formation', Journal of Soil Science 15,310–318.

Davis M. B., and Botkin D. B. (1985) 'Sensitivity of cool-temperate forests and their fossil pollen to rapid temperature change', Quaternary Research 23,327–340.

Dendron Resource Surveys, Ltd. (1985) Impact of acid rain on forest growth in Canada. Report completed for the Canadian Forest Service, Ottawa.

Drake B. G., and Leadley P. W. (1991) 'Canopy photosynthesis of crops and native plant communities exposed to long-term elevated CO_2', Plant, Cell and Environment 14,853–860.

Eamus D., and Jarvis P. G. (1989) 'The direct effects of increases in the global atmospheric concentration on natural and commercial temperate trees and forests', Advances in Ecological Research 19,2–57.

Emanuel W. R., Killough G. G., Post W. M., and Shugart H. H.. (1984) 'Modeling terrestrial ecosystems in the global carbon cycle with shifts in carbon storage capacity by land-use change', Ecology 65,970–983.

Esser, G. (1987) 'Sensitivity of global carbon pools and fluxes to human and potential climatic impacts', Tellus 39B,245–260.

Flanagan P.W., and Van Cleve K. (1983) 'Nutrient cycling in relation to decomposition and organic matter quality in tiaga ecosystems', Canadian Journal of Forest Research 13,795–817.

Grulke N. E., Riechers G. H., Oechel W. C., Hjelm U., and Jaeger C. (1990) 'Carbon balance in tussock tundra under ambient and elevated atmospheric CO_2', Oecologia 83,485–494.

Hare F.K., and Thomas M. K. (1979) Climate Canada. John Wiley and Sons, Toronto, Canada.

Heilman P.A. (1966) 'Change in distribution and availability of nitrogen with forest succession on north slopes of interior Alaska', Ecology 47,825–834.

Houghton, R. A. (1987) 'Terrestrial metabolism and the seasonality of atmospheric CO_2 concentrations', Bioscience 37,672–678.

Houghton, R. A., Hobbie J. E., Melillo J. M., B. Moore, Peterson B. J., Shaver G. R., and Woodwell G. M. (1983) 'Changes in the carbon content of terrestrial biota and soils between 1860 and 1980: Net release of CO_2 to the atmosphere', Ecological Monographs 53,235–262.

Houghton, R. A., Boone R. D., Fruci J. R., Hobbie J. E., Melillo J. M., Palm C. A., Peterson B. J., Shaver G., Woodwell G. M., Moore B., Skole D. L., and Myers N. (1987) 'The flux of carbon from terrestrial ecosystems to the atmosphere in 1980 due to changes in land use: Geographic distribution of global flux', Tellus 39B,122–139.

Huston M., DeAngelis D. L., and Post W. M. (1988) 'New computer models unify ecological theory', Bioscience 38,682–691.

Krause H. H. (1981) Factorial experiments with nitrogen, phosphorus, and potassium in spruce and fir stands of New Brunswick: 10- year results, Canadian Forest Service Maritime Forest Research Center Information Report M–X–123.

Larsen J.A. (1980) The Boreal Ecosystem, Academic Press, New York.

Lugo A. E., and Brown S. (1986) 'Steady state terrestrial ecosystems and the global carbon cycle', Vegetatio 68,83–90.

Luxmoore R. J., Tharp M. L., and West D. C. (1990) 'Simulating the physiological basis of tree-ring responses to environmental changes', in R. K. Dixon, R. S. Meldahl, G. A Ruark, and W. G. Warren (eds.), Process Modeling of Forest Growth Responses to Environmental Stress, Timber press, Portland, pp. 393–401.

Mahendrappa M. K., and Salonius P. O. (1982) 'Nutrient dynamics and growth response in a fertilized black spruce stand', Soil Science Society of America, Journal 46,127–133.

Marland G. and Boden T. A. (1991) 'CO_2 Emissions - Modern Record', in T. A. Boden, R. J. Sepanski, and F. W. Stoss (eds.), Trends 91: A Compendium of Data on Global Change ORNL/CDIAC-46, Carbon Dioxide Information Analysis Center, Oak Ridge National Laboratory, Oak Ridge , Tennessee, pp. 386-507.

Matson P. A., and Boone R. D. (1984) 'Natural disturbance and nitrogen mineralization in wave-form dieback of mountain hemlock in the Oregon Cascades', Ecology 65,1511–1516.

Matson P. A., and Waring R. H. (1984) 'Effects of nutrient and light limitations on mountain hemlock: susceptibility to laminated root rot', Ecology 65,1517–1524.
Maybeck M. (1981) 'River transport of organic carbon to the ocean', in G. E. Likens, et al. (eds.), Flux of Organic Carbon by Rivers to the Oceans CONF-8009140, National Technical Information Service, Springfield, pp. 219–269.
McMurtrie R. E., Rook D. A., and Kellher F. M. (1990) 'Modeling the yield of *Pinus*', Ecology and Management 30,381–413.
Melillo J.M., Aber J. D., and Muratore J. F. (1982) 'Nitrogen and lignin control of hardwood leaf litter decomposition dynamics', Ecology 63,621–626.
Mitchell J. F. B. (1983) 'The seasonal response of a general circulation model to changes in CO_2 and sea temperature', Quarterly Journal of the Royal Meteorological Society 109,113–152.
Moore T. R. (1984) 'Litter decomposition in a subarctic spruce-lichen woodland, eastern Canada', Ecology 65,299–308.
Norby R. (1992) Personal Communication, Oak Ridge National Laboratory, Oak Ridge, TN.
Oechel W. C., and Strain B. R. (1985) 'Native species responses to increased carbon dioxide', in B. R. Strain and J. D. Cure (eds.), Direct Effects of Increasing Carbon Dioxide on Vegetation, National Technical Information Service, U.S. Department of Commerce, Springfield, VA, pp. 117–154.
Olson J. S. (1985) 'Cenozoic fluctuations in biotic parts of the global carbon cycle', in E. T. Sundquist and W. S. Broecker (eds.), The Carbon Cycle and Atmospheric CO_2: Natural Variations Archean to Present, Geophysical Monograph 32, American Geophysical Union, Washington D.C., pp. 377–396.
Olson J. S., Watts J. A., and Allison L. J. (1983) Carbon in live vegetation of major world ecosystems ORNL–5862, Oak Ridge National Laboratory, Oak Ridge, Tennessee.
Pastor J., and Post W. M. (1985) Development of a linked forest productivity-soil process model ORNL/TM-9519, Oak Ridge National Laboratory, Oak Ridge, Tennessee.
Pastor J., and Post W. M. (1986) 'Influence of climate, soil moisture, and succession on forest carbon and nitrogen cycles', Biogeochemistry 2,3-27.
Pastor J., and Post W. M. (1988) 'Response of northern forests to CO_2-induced climatic change: Dependence on soil water and nitrogen availabilities', Nature 334,55–58.
Pastor J., Gardner R. H., Dale V. H., Post W. M. (1987) 'Successional changes in nitrogen availability as a potential factor contributing to spruce declines in boreal North America', Canadian Journal of Forest Research 17,1394–1400.
Piene H., and Van Cleve K. (1978) 'Weight loss of litter and cellulose bags in a thinned white spruce forest in interior Alaska', Canadian Journal of Forest Research 8,42–46.
Post W. M., and Pastor, J. (1990) 'An individual-based forest ecosystem model for projecting forest response to nutrient cycling and climate changes', in L. Wensel and G. Biging (eds.), Forest Simulation Systems: Proceedings of the IUFRO Conference, Berkeley, California, November 2–5, 1988, University of California, Division of Agriculture and Natural Resources, Bulletin 1927, Berkeley, pp. 61–74.

Post, W. M., Peng T.-H., Emanuel W. R., King A. W., Dale V. H., and DeAngelis D. L. (1990) 'The Global Carbon Cycle', American Scientist 78,310–326.

Running S. W., and Coughlan J. C. (1988) 'A general model of forest ecosystem processes for regional applications: I. Hydrologic balance, canopy gas exchange and primary production processes', Ecological Modelling 42,125–154.

Schlesinger W. H. (1985) 'The formation of caliche in soils of the Mojave Desert, California', Geochim. Cosmochim. Acta 49,57–66.

Schlesinger W. H., and Melack J. M. (1981) 'Transport of organic carbon in the world's rivers', Tellus 33,172–187.

Shugart H.H. (1984) A theory of forest dynamics. Springer-Verlag, New York.

Shugart H. H., Antonovsky M. Ya., Jarvis P. J., and Sandford A. P. (1986) 'CO_2, climatic change, and forest ecosystems', in B. Bolin, B. R. Döös, J. Jæger, and R. A. Warrick (eds.), The Greenhouse Effect, Climatic Change, and Ecosystems SCOPE 29, John Wiley and Sons, New York. pp. 475–521.

Shugart H. H., and Emanuel W. R. (1985) 'Carbon dioxide increase: the implications at the ecosystem level', Plant, Cell and Environment 8,381–386.

Solomon A. M. (1986) 'Transient response of forests to CO2-induced climate change: Simulation modeling experiments in eastern North America', Oecologia 68,567-579.

Solomon A.M., Tharp M. L., West D. C., Taylor G. E., Webb J. M., and J. M. Trimble. (1984) Responses of unmanaged forests to CO_2- induced climatic change: available data, initial tests and data requirements DOE/NDD-0053-TR009, U.S. Department of Energy, Washington, D.C.

Stewart H., and Swan D. (1970) Relationships Between Nutrient Supply, Growth, and Nutrient Concentrations in the Foliage of Black Spruce and Jack Pine, Pulp and Paper Research Institute Canada Woodlands, Paper Number 19.

Tans P. P., Fung I. Y., and Takahashi T. (1990) 'Observational constraints on the global atmospheric CO_2 budget', Science 247,1431–1438.

Thornthwaite C. W., and Mather J. R. (1957) 'Instructions and Tables for Computing Potential Evapotranspiration and the Water Balance', Publications in Climatology 10,183–311.

Van Cleve K., and Harrison A. F. (1985) 'Bioassay of forest floor phosphorus supply for plant growth', Canadian Journal of Forest Research 15,156–162.

Van Cleve K., and Oliver L. K. (1982) 'Growth response of postfire quaking aspen (it Populus tremuloides Michx.) to N, P, and K fertilization', Canadian Journal of Forest Research 12,160–165.

Van Cleve K, and Zasada J. (1976) 'Response of 70-year old white spruce to thinning and fertilization in interior Alaska', Canadian Journal of Forest Research 6,145–152.

Van Cleve K., Oliver L., Schlenter R., Viereck L. A., and Dryness C. T. (1983) 'Productivity and nutrient cycling in tiaga forest ecosystems', Canadian Journal of Forest Research 13,747–766.

Viro, P. (1967) 'Forest manuring on mineral soils', Messelser det Norske Skogforsøksvesen 23,113–136.

Weetman, G. F. (1968) The nitrogen fertilization of three black spruce stands, Pulp and Paper Research Institute Canada Woodlands, Paper Number 6.

Weetman G. F., and Nykvist N. B. (1963) 'Some more humus, regeneration and nutrition problems and practices in north Sweden', Forest Chronical 39,188–198.
Weinstein D.A., Shugart H. H., and West D. C. (1982) The long-term nutrient retention properties of forest ecosystems: A simulation investigation, ORNL/TM–8472, Oak Ridge National Laboratory, Oak Ridge, Tennessee.
Whittaker R. H., and Likens G. E. (1973) 'Carbon in the biota', in G. M. Woodwell and E. V. Pecan (eds.), Carbon and the Biosphere, U.S. Atomic Energy Commission, Washington D.C, pp. 86–135.
Williams, B. L. (1972) 'Nitrogen mineralization and organic matter decomposition in Scots pine humus', Forestry 45,177–188.

THE TRANSIENT RESPONSE OF VEGETATION TO CLIMATE CHANGE: A POTENTIAL SOURCE OF CO_2 TO THE ATMOSPHERE

GEORGE A. KING
ManTech Environmental Technology Inc.
and
RONALD P. NEILSON
Oregon State University
U.S. EPA Environmental Research Laboratory
200 SW 35th Street
Corvallis, OR 97333, USA

ABSTRACT. Global climate change as currently simulated could result in the broad-scale redistribution of vegetation across the planet. Vegetation change could occur through drought-induced dieback and fire. The direct combustion of vegetation and the decay of dead biomass could result in a release of carbon from the biosphere to the atmosphere over a 50- to 150-year time frame. A simple model that tracks dieback and regrowth of extra-tropical forests is used to estimate the possible magnitude of this carbon pulse to the atmosphere. Depending on the climate scenario and model assumptions, the carbon pulse could range from 0 to 3 Gt of C yr^{-1}. The wide range of pulse estimates is a function of uncertainties in the rate of future vegetation change and in the values of key model parameters.

1. Introduction

The potential for significant changes in global climate caused by increases in radiatively important trace gases has generated a large international research effort to improve our ability to predict the magnitude of future climate change and the associated environmental impacts. A number of studies over the past decade have used state-of-the-art atmospheric, biospheric, and oceanographic models to estimate the potential magnitude of future climate change and related impacts on the biosphere (e.g., Smith and Tirpak et al., 1989; IPCC, 1990; Jenkinson et al., 1991; Smith et al., 1991, in press). These studies have demonstrated that the rate of climate change could be unprecedented in recent geologic history and have large effects on the terrestrial biosphere. They have also demonstrated current limitations in our ability to simulate earth system dynamics, especially feedbacks between the biosphere and the atmosphere. These feedbacks could accelerate or dampen future climate change and are important to quantify in order to improve our predictive power. This paper focuses on the linkage of the terrestrial biosphere and atmospheric dynamics and suggests that the time-dependent (transient) response of terrestrial vegetation to climate change could act as a positive feedback to climate change through the rapid release of stored C.

Water, Air, and Soil Pollution **64**: 365–383, 1992.
© 1992 *Kluwer Academic Publishers.*

1.1. BACKGROUND - EQUILIBRIUM SIMULATIONS OF TERRESTRIAL VEGETATION RESPONSE TO CLIMATE WARMING

The sensitivity of terrestrial vegetation to future global climate change has been demonstrated through a number of steady-state (equilibrium) simulations using a variety of vegetation models and climate scenarios (Emanuel, 1985a,b; Lashof, 1989; K. Prentice and Fung, 1990; Smith et al., 1991, in press). The vegetation redistribution scenarios predicted by these models all assume that the vegetation has had sufficient time to respond to the new set of climatic conditions, a process that may in fact take several hundred yrs (Davis, 1986; I.C. Prentice et al., 1991). As an example, Smith et al. (1991, in press) used the Holdridge life-zone classification system (Holdridge, 1947, 1967) to project the future location of biomes in response to four climate change scenarios (Figure 1). The four double-CO_2 climate scenarios (Table 1) they used were produced with output from the OSU (Schlesinger and Zhao, 1989), GISS (Hansen et al., 1988), GFDL (Manabe and Wetherald, 1987), and UKMO (Mitchell, 1989) general circulation models (GCMs). Tundra and boreal forest area decreases under all climate scenarios, while temperate and tropical forests and grasslands/shrublands increase in areal extent. Extra-tropical forests show a poleward shift in distribution (Figure 1). Depending upon the magnitude of the warming, 16% to 56% of the earth's vegetated surface changes from one biome type to another in their simulations.

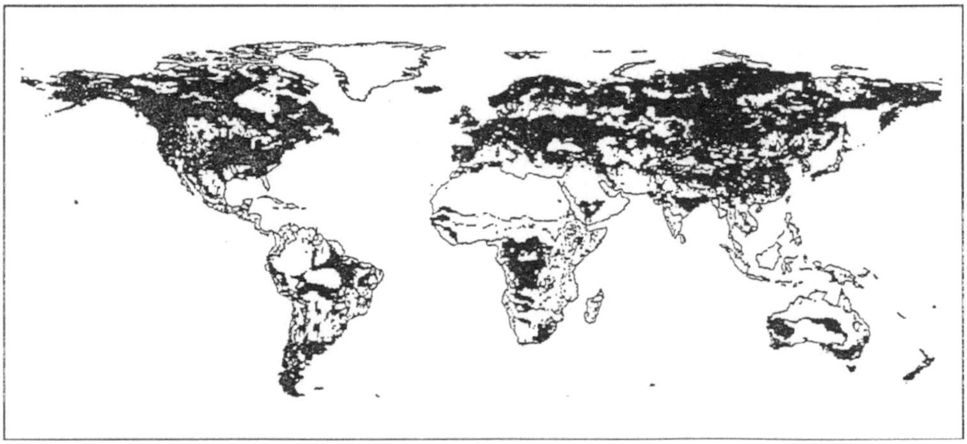

Figure 1. Area of vegetation change (black) between current and double-CO_2 conditions as simulated by Smith et al. (1991, in press) using the Holdridge life zone classification system (Holdridge 1947, 1967) and the UKMO climate change scenario (Mitchell, 1989).

Vegetation changes of this magnitude could have large impacts on terrestrial C storage. Using the Holdridge vegetation change scenarios, Smith et al. (in press) estimated that above-ground C storage would increase 4% to 17% depending on the climate scenario (Table 2). Below-ground storage decreases slightly under the UKMO and GFDL scenarios and increases about 5% under the OSU and GISS scenarios. Overall C storage would increase under the GISS and OSU scenarios, and remain unchanged under the GFDL and UKMO scenarios. K. Prentice and Fung (1990) also simulated an increase in C storage for the GISS scenario using a modified Holdridge model (K. Prentice, 1990), although their estimated increase in terrestrial above-ground C storage is twice as great as the Smith et al. (1991, in press) estimate. These

Table 1. Summary of predicted changes in global mean temperature and mean annual precipitation as simulated by four General Circulation Models. These data were used to generate the climate change scenarios used by Smith et al. (1991, in press) to simulate changes in vegetation and terrestrial C storage.

GCM	Temperature (°C)	Precipitation (%)
Oregon State University[1] (OSU)	+ 2.8	+ 7.8
Geophysical Fluid Dynamics Laboratory[2] (GFDL)	+ 4.0	+ 8.7
Goddard Institute for Space Studies[3] (GISS)	+ 4.2	+ 11.0
United Kingdom Meteorological Office[4] (UKMO)	+ 5.2	+ 15.0

[1] Schlesinger and Zhao, 1988
[2] Manabe and Wetherald, 1987
[3] Hansen et al., 1988
[4] Mitchell, 1989

Table 2. Estimated changes in terrestrial above-ground C and total terrestrial C storage (above-ground and soil) from current levels as simulated by Smith et al. (1991, in press) for equilibrium double-CO_2 climate conditions. The percent change from current conditions are given in parentheses.

Climate Scenario	Above-Ground C (Gt)	Total C (Gt)	Change in Total C (Gt)
Current	737	1896	
OSU	860 (16.7)	2076	180 (9.5)
GFDL	782 (6.1)	1934	38 (2.0)
GISS	830 (12.5)	2043	147 (7.7)
UKMO	765 (3.8)	1904	9 (0.4)

results suggest that under new equilibrium conditions the terrestrial biosphere may store ~0% to 30% more C above ground than it currently does and thus could act as a negative feedback to climate change. Much of the increase in C storage is the result of the simulated increase in tropical forest area; however, this is unlikely to occur because of current land use trends in the tropics (e.g. Houghton, 1990).

1.2. POTENTIAL TRANSIENT ECOSYSTEM DYNAMICS AND THEIR IMPACT ON C FLUXES

The equilibrium simulations of global vegetation change show major shifts in the location of major world biomes, but do not directly reveal anything about the dynamics of the vegetation redistribution. At what rate will vegetation change, and through what process will that change take place? The implications of these considerations on global C dynamics are explored in the remainder of this paper. Vegetation change could proceed along two fundamentally different paths. Changes in forest composition and distribution could proceed slowly through the gradual competitive displacement of overstory tree species with species better suited to the new climatic conditions. Or, changes in composition could occur catastrophically through massive, drought-induced dieback of overstory vegetation concomitant with, or perhaps followed by, large wildfires (Neilson and King, in press). The first scenario of vegetation change would produce little effect on the global C cycle apart from gradual changes in terrestrial C storage in the direction implied by the equilibrium analyses. The accelerated dieback scenario could produce a transient pulse of C to the atmosphere as C stored in rapidly declining forests is released through decomposition and fire and then later resequesterd by vegetation regrowth. If the rate of tree mortality, decomposition, and combustion is fast enough, the resulting release of C could cause an additional increase in atmospheric C concentrations besides that from fossil fuel emissions. This would cause additional climate changes through increased radiative forcing.

Current theory on climate-vegetation interactions combined with the expected magnitude of future climate change suggest that vegetation change may in fact proceed through catastrophic decline (Neilson and King in press). The argument can be stated as follows. Regional water balance is the principal control of biomass, leaf area, and physiognomic growth form for most of the world's biomes (Woodward, 1987; Neilson et al., 1989; Stephenson, 1990; Neilson et al., 1992; Neilson et al., in press). Vegetation within a biome appears to transpire most of the soil moisture available to it during the growing season. Over the long term, if more water were available, more leaf area would be produced. Consequently, most of the additional water would be lost through transpiration. If soil moisture is depleted before the end of the growing season, water stress would cause dieback of vegetation and reduced leaf area. Thus, the amount of water transpired would decrease (Woodward, 1987). The seasonality and amount of water available also appear to control the growth form of vegetation at a particular location, determining for example, whether trees or shrubs predominate (e.g., Daubenmire, 1968; Whittaker, 1975; Neilson et al., 1992; Neilson et al., in press).

Empirical evidence supporting these hypotheses can be obtained from streamflow data (U.S. West 1988). These data show that streamflow throughout the U.S. is minimal or absent at the end of the summer growing season (Neilson et al., 1989; Neilson et al. 1992), suggesting that most of the water across a landscape has been lost through evapotranspiration and runoff during the preceding months. The above arguments can be extended to mid and low latitude vegetation, but high latitude vegetation (e.g. the boreal forest-tundra ecotone) may be more constrained by regional energy balance as opposed to moisture availability (Bonan and Shugart, 1989).

Most double-CO_2 climate scenarios indicate that temperatures will rise over most land areas, resulting in increased potential evapotranspiration (PET). Changes in precipitation are more variable in direction across a continent, but globally are expected to increase 3% to 15% (IPCC, 1990). For the U.S., PET may rise 50 to over 1000 mm yr^1 depending on the climate scenario and method of calculating PET (Neilson et al., 1989; Gucinski et al., 1990). Simple water balance calculations using the PET estimates and estimates of future precipitation suggest that less water will be available for plants across much of the U.S. (Neilson et al., 1989; Dolph et al., in press) and that increased drought stress will result in regional declines in leaf-area.

Similar water balance calculations have been made for Canada, where drought-stress is also expected to increase over much of the central part of the country (J. Lenihan, Oregon State University, personal communication).

Drought stress under double-CO_2 conditions will also be exacerbated by increases in the growing season caused by warmer temperatures. Longer growing seasons will increase transpirational demands by existing vegetation (Smith and Tirpak, 1989). Warmer winters will result in decreased precipitation falling as snow and reduce regional snow packs, resulting in less carry-over of moisture from the winter to the growing season (Gleick, 1987a,b; Lettenmaier and Gan, 1990). If recharge of deep soil layers is dependent on snow melt, less water will be stored in the deep soil layers and thus less water will be available for vegetation in the following growing season. Again, drought-induced dieback will be the likely effect of these changes.

Concomitant with increased drought stress could come increases in disturbance frequency, particularly fires. Analysis of GCM output suggests that meteorological conditions favoring lightning will increase in a warmer world, increasing fire frequencies (Overpeck et al., 1990) If significant dieback occurs, the higher ignition probabilities combined with increases in dry fuel could produce large increases in fire frequencies.

In sum, if climate change proceeds to currently simulated double-CO_2 conditions, drought-induced vegetation dieback could be the principal mode of vegetation change resulting in releases of stored C to the atmosphere. Since estimates of above ground terrestrial C storage range from 56 to 110% of the atmospheric pool of C (Olson et al., 1983; Bolin, 1983, 1986, Post et al., 1990), significant releases of C from the terrestrial biosphere during the transition from current to double-CO_2 conditions would be a strong positive feedback to climate change. We have constructed a simple model to investigate the potential magnitude of this transient C pulse (King et al., 1990; Neilson and King, in press).

2. Methods

Three different C-flux processes are treated in the simple C pulse model; 1) release of stored C to the atmosphere by drought-induced vegetation dieback and wildfire, 2) sequestering of C back into the terrestrial biosphere via vegetation regrowth in areas of dieback, and 3) sequestering of C in the tropics through the expansion of tropical forests (as suggested by the vegetation scenarios of Smith et al. (1991, in press) and K. Prentice and Fung (1990)). Forest systems contain the bulk of the above-ground store of terrestrial C (Olson et al., 1983), and only changes in these systems are considered in the model. Below-ground changes in C are not considered in the model, but would probably change more slowly than above-ground pools. Because tropical forests are being cut at a rapid rate, increases in the tropical forest acreage, as simulated with the Holdridge life-zone model, are not realistic. Thus, a potential tropical C sink is not considered further here.

The rate of C release during vegetation dieback is a function of: 1) the amount of direct combustion of above-ground organic matter; 2) the efficiency of that combustion in terms of CO_2 release; 3) the amount of residual organic matter, either after a fire, or after drought-induced mortality; 4) the rate of decay of that organic matter; and, 5) the amount of C lost to surface waters. Carbon release to the atmosphere is calculated as a combination of the fraction of organic C directly burned, followed by exponential decay of the remaining large fragments (Olson 1963). Model parameters include: the amount of above-ground organic C in the forest vegetation killed by either drought or fire; the fraction of above-ground organic C burned in a fire; and the decay rate of remaining organic matter. The efficiency of combustion to CO_2 is assumed to be 100%, and the organic matter lost to surface waters is assumed to rapidly decay to CO_2.

After vegetation in an area has died, the area becomes available for regrowth. Net above-ground C accumulation is defined as the total accumulation of organic matter yr^{-1} taking into account C losses through autotrophic and heterotrophic respiration (Whittaker, 1975). Organic C accumulation was modeled using a logistic equation, with an upper limit on organic matter density. Model parameters consist of: 1) the initial post-dieback, live organic C (i.e., the amount of resprouting vegetation or seedlings germinating the year following the fire or dieback); 2) the intrinsic capacity of the forest ecosystem to accumulate C (analogous to maximal net ecosystem productivity); and, 3) the maximum C density potential of the region.

Fundamental to the model calculations are estimates of the area of forest dieback and forest regrowth. For the initial analysis, the areal estimates were calculated in the following manner. The current distribution of potential natural vegetation and the future steady-state double-CO_2 vegetation, both estimated using the Holdridge system (Smith et al., 1991, in press), define the beginning and ending points of climatically-induced vegetation change. The area of extra-tropical forest vegetation that changes from one forest type to another is calculated from the two vegetation maps and termed the "change zone" in this paper.

In our initial estimates of vegetation dynamics during climate change, all extra-tropical forest change is assumed to occur through catastrophic dieback and that the dieback is followed by wildfire. The rate of drought-induced mortality and burning is set proportional to the rate of climate change. If the change to double-CO_2 climate conditions is assumed to occur in 100 years, then the model assumes that 1/100 of the change zone dies back and burns in a given model year. Within that parcel of land, regrowth is assumed to begin one year after the dieback and fire. The global yearly flux of C to the atmosphere is then calculated as follows. The current year's C loss from wildfire is added to the amount of C released from the decay of the remaining organic matter from all the areas burned in previous years. Then the C sequestered by the terrestrial biosphere through vegetation regrowth is subtracted from the C released to the atmosphere to get the total yearly C flux.

The pulse model is currently zero dimensional in that all extra-tropical forests are aggregated together in calculating the change zones. An average C density value for the extra-tropical forests was calculated by weighting C density values for different forested biomes by the relative area of the biome. Carbon density values were taken from Olson et al. (1983).

A standard set of average parameter values was used to calculate the pulse for the four different climate-vegetation change scenarios created by Smith et al. (1991, in press). For the standard parameter set, the change to double-CO_2 climate conditions was assumed to occur in 50 years. Other model parameters were estimated from the literature as follows. The fraction of above-ground organic C consumed by a fire was set at 10% (Fahenstock and Agee, 1983; Auclair, 1985), the decay rate of the remaining organic C was set at 0.05 yr^{-1} (Foster and Lang, 1982; Auclair, 1985, Harmon et al., 1986; Edmonds, 1987), the initial post-dieback organic C was estimated to be 0.4 kg m^{-2} (Auclair, 1985), and the accumulation of organic C was set at 2 t ha^{-1} yr^{-1} (Waring and Schlesinger, 1985; Auclair, 1985; Bormann and Likens, 1979).

Because of the uncertainty present in the above values, the pulse was also estimated using two different sets of values for the key model parameters. One set of pulse estimates, here called "low," are based on the combination of plausible parameter values that minimize the magnitude of the pulse, and another set, called "high," are based on the combination of plausible pulse estimates that maximize the pulse. As an additional sensitivity analysis, each of the model parameters was varied independently while the other parameters remained fixed. This allowed an estimate of the importance of the individual parameters on the magnitude of the pulse.

3. Results

An example of model output of potential changes in extra-tropical forest C storage for the GFDL climate scenario is presented in Figure 2. Vegetation dieback first dominates the modeled carbon dynamics, resulting in a flux of carbon to the atmosphere (Figure 2). Later, that carbon is removed from the atmosphere by vegetation regrowth. The time period over which these changes proceed is a direct function of the assumed rate of climate change. If the rate of climate change is assumed to be 50 years, the maximum change in atmospheric carbon stores occurs at about 55 years (Figure 2). The total amount of carbon temporarily shifted from terrestrial vegetation to the atmosphere is called the cumulative carbon pulse.

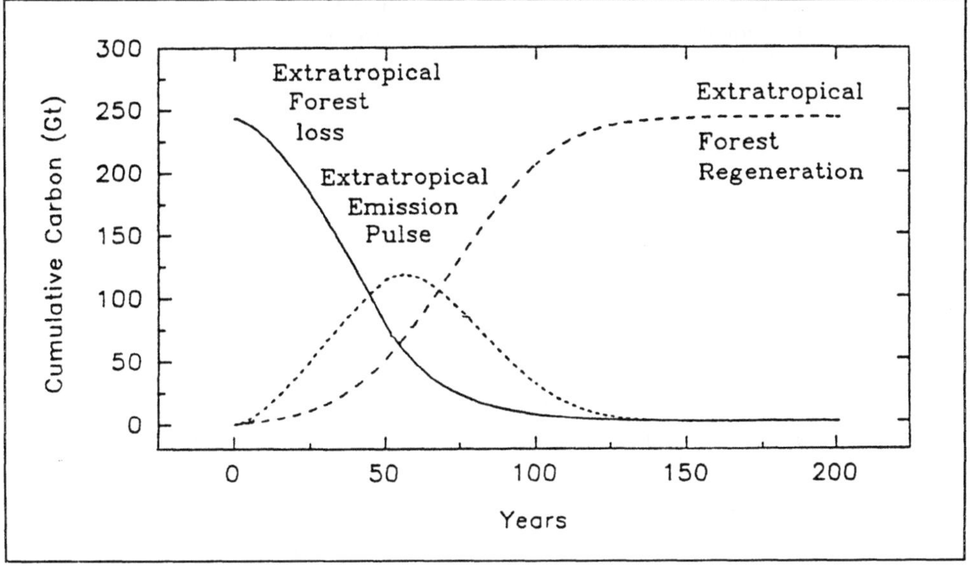

Figure 2. Example of the pulse model calculation for the GFDL climate scenario. Vertical axis represents cumulative changes in C.

The average yearly release of C to the atmosphere calculated using the standard set of model parameters ranges between 1.4 Gt yr^{-1} (OSU scenario) and 2.6 Gt yr^{-1} (UKMO scenario) for a period of about 50 years (Table 3). When the model is run with the "low" parameters, the estimated pulse is considerably less, ranging between 0.1 (OSU) and 0.2 Gt yr^{-1} (UKMO) for about 100 years. When the model is run with the "high" parameters, the pulse is ranges between 1.7 (OSU) and 3.4 Gt yr^{-1} (UKMO) for about 50 years. Given that anthropogenic releases from fossil fuel burning are currently about 6 Gt yr^{-1} (Marland et al., 1989), the transient release of C from forest dieback in extra-tropical regions could increase that amount by 2 to almost 50%. Thus, the potential exists for a significant positive feedback to climate change from vegetation redistribution, but equally clear is the tremendous uncertainty in the pulse estimates. The range of plausible model estimates varies by more than an order of magnitude. The cumulative estimates of the carbon pulse vary by a factor of two, from 70 Gt C (OSU scenario) to 130 Gt C (UKMO scenario; Table 4).

The relative sensitivity of the estimates of the pulse to variation in the value of the specific model parameters is shown in Figure 3. In these analyses, the cumulative flux of C to the atmosphere caused by vegetation dieback and decay is displayed as the dependent variable.

Table 3. Range of annual fluxes of CO_2 (Gt yr^{1}) to the atmosphere calculated by the C pulse model using as input extra-tropical forest redistribution estimates from Smith et al (1991, in press). Data from two different climate scenarios are presented. Low estimates for each climate scenario represent the plausible combination of model parameter values giving the smallest pulse rates, while the high estimates are from the combination of plausible model parameters that yield the highest pulse rates. Pulse estimates are also given for the standard combination of model parameters based on expert judgement of the most likely parameter values.

Type of Parameter Combination	**Climate Scenario**	
	OSU Pulse Estimate (Gt yr^{-1})	UKMO Pulse Estimate (Gt yr^{1})
Low	0.1	0.2
Standard	1.4	2.6
High	1.7	3.4

Table 4. Cumulative total of the flux of C to the atmosphere caused by vegetation dieback as calculated by the pulse model. Pulse estimates for each scenario are based on the standard model parameters.

Climate Scenario	**Cumulative Extra-tropical Carbon Pulse (Gt)**
OSU	70
GISS	80
GFDL	120
UKMO	130

Four of the five model parameters have significant impacts on the total pulse modeled. The percentage of the above-ground C actually consumed in a fire has relatively little impact on the magnitude of the pulse (Figure 3a). The explanation for this is that the remainder of the biomass releases C to the atmosphere according to the exponential decay function. However, if the decay rate were relatively slow, or the rate of regrowth was high, the magnitude of the pulse would be primarily determined by the amount of C consumed by fire.

Variation in the decay rate of remaining C, as would be expected given the exponential function, has a large impact on the pulse (Figure 3b). A variation in the decay rate from 0.03 to 0.05 yr^{1} results in a variation in the pulse of about 20% to 30%. Values for log decay

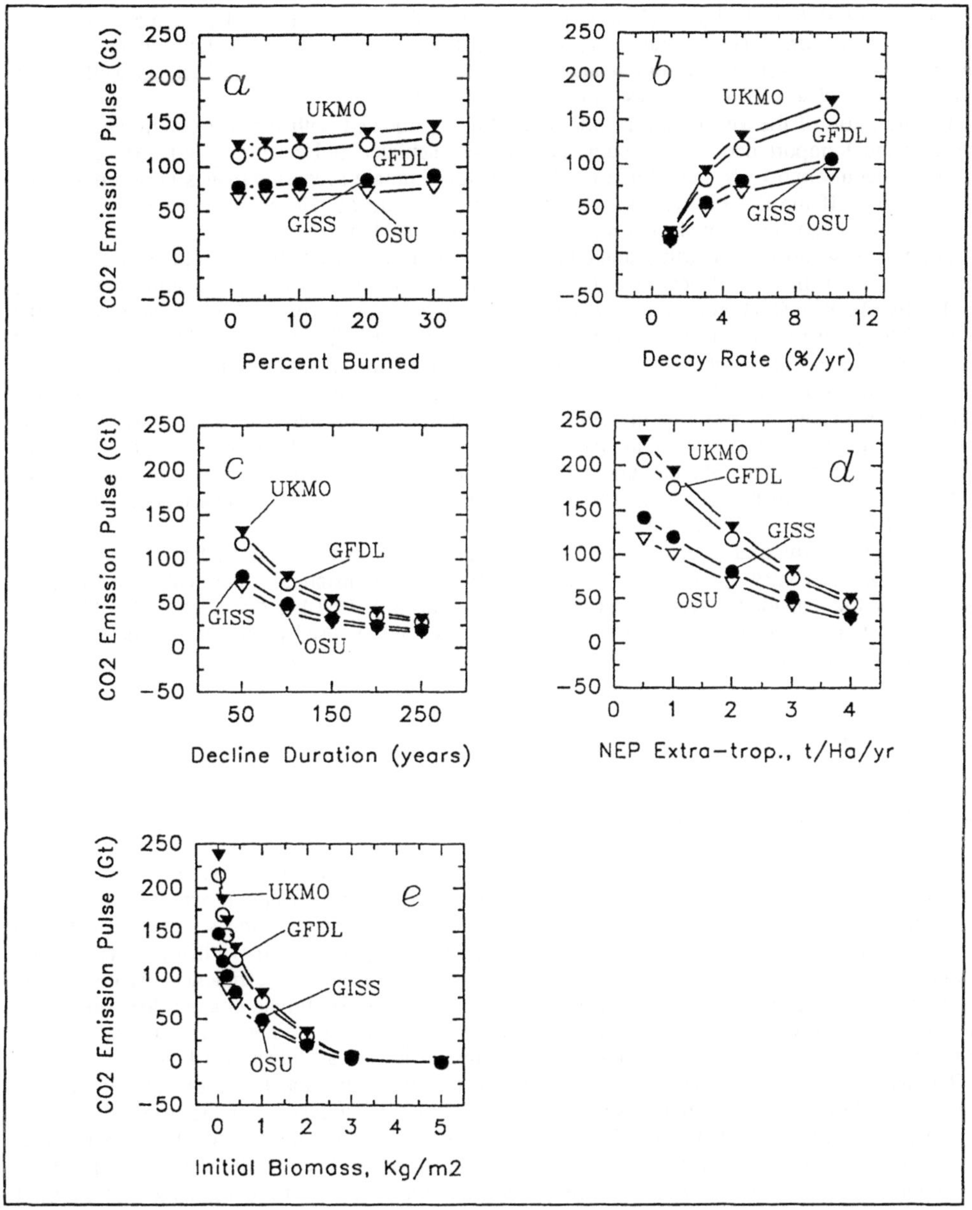

Figure 3a-e. Sensitivity of the extra-tropical pulse to variation of model parameters. Vertical axis is the total pulse of C to the atmosphere. Variable names are listed under the horizontal axis.

range from 0.006 to 0.52 yr^{-1} (Harmon et al., 1986). Decay rates generally decrease poleward from the equator. The average log mineralization rates of 20 temperate deciduous and evergreen trees given in Harmon et al. (1986) is .08 $\pm$ .08 yr^{-1}. Thus 0.05 yr^{-1} is well within the range of observed decay rates, and may be conservative.

The duration of the vegetation decline, here directly equated with the rate of climate change, is of critical importance to the magnitude of the calculated pulse (Figure 3c). If vegetation dieback occurs in 50 yrs, the release of C for all four climate scenarios ranges between 1.4 and 3.3 Gt yr^{-1}. If the decline occurs over 100 years, the rate of release is between 0.4 and 0.8 Gt yr^{-1}, and becomes a much less significant climatic feedback.

The two vegetation regrowth parameters have a strong effect on the magnitude of the pulse. An increase in the rate of net ecosystem productivity from 2 to 3t ha^{-1} yr^{-1} would decrease the pulse by about 30% (Figure 3d). If the amount of live post-dieback C is increased from 0.4 kg m^{-2} to 1 kg m^{-2}, the C pulse decreases by 35% to 50% (Figure 3e). Of the five model parameters, the pulse is most sensitive to the amount of live post-dieback organic C.

The model is also sensitive to two important inputs, the amount of vegetation change (i.e. the size of the change zones) and the amount of C stored in forested biomes. Under the OSU scenario, the Holdridge system indicates that 16% of earth's land surface changes from one biome type to another, whereas under the warmer UKMO scenario, 56% of the land surface undergoes a shift in biome type. The estimated pulse for the UKMO scenario is 85% greater than that calculated for the OSU scenario.

Quantifying terrestrial C storage is a difficult task, and estimates of C density for specific biome types can vary appreciably (Post et al., 1990). For instance, a new estimate of the C density of the Canadian boreal forest (Botkin and Simpson, 1990) is one-third of the Olson et al. (1983) value used in this analysis. When this new value is used in the model, the estimated C pulse is reduced by about 85% using the standard model parameters.

4. Discussion

4.1. MODEL ASSUMPTIONS

Most of the critical model assumptions pertain to the rate and process of vegetation redistribution. The current model assumes that: 1) vegetation redistribution occurs at a constant rate directly proportional to an assumed rate of climate change; 2) the vegetation within a change zone undergoes complete dieback followed by drought-induced wildfire; 3) there is no impact of higher CO_2 concentrations on plant growth and water-use efficiency; 4) no change in biomass occurs in areas outside of the change zones; and, 5) climate change does not affect the rate of regrowth of vegetation nor the amount of carbon stored in extra-tropical forests. The implications of these assumptions on the pulse estimates are discussed below.

The assumption of a constant rate of vegetation redistribution (and hence dieback as defined in the model) is clearly an oversimplification. Trace-gas-induced climate forcing will be imposed on other mechanisms of climate variability and the result will be a non-linear trajectory of climate change. For instance, a sequence of historically unparalleled severe drought years may be followed by several "normal" wet years. The combination of drought years may be sufficient to induce dieback or severe wildfire across broad regions, resulting in a pulse of C to the atmosphere larger than the average calculated by the current model.

The assumption of complete vegetation dieback followed by wildfire is central to the pulse concept as presented above. However, the sensitivity analysis shows that the pulse is actually not that sensitive to changes in the amount of C released by direct incineration. A large C pulse could still occur if vegetation is killed simply by drought stress and then decays.

Certainly, higher fire frequencies would be expected with increased fuel loads, but complete burnover of the change zones is not required to produce a flux of C to the atmosphere.

Nevertheless, to put our assumption of widespread wildfire in the context of current fire patterns, under the UKMO scenario (the most extreme of the four climate change scenarios used) the model assumes 50 x 10^6 ha burn every year globally. In 1987, 30 x 10^6 ha burned in China and Russia alone (Levine, 1990). Thus, the magnitude of the burn required for a given year in the UKMO scenario has almost occurred in the recent past. What is particularly extreme about the wildfire assumption is that a historically rare event is required every year.

The wildfire assumption is less important to the pulse concept than the argument that significant vegetation dieback will take place due to increased drought stress. The amount of dieback that occurs will be primarily a function of the rate of change of moisture availability. If moisture availability decreases slowly, the expected decrease in biomass would occur gradually without high mortality. The net release of C (if any) to the atmosphere would thus be negligible (Figure 4, trajectory 1). However, if moisture availability decreases rapidly, the seasonal demands for water by vegetation could, over a few years, substantially exceed moisture supply resulting in widespread drought-induced mortality (Figure 4, trajectory 2).

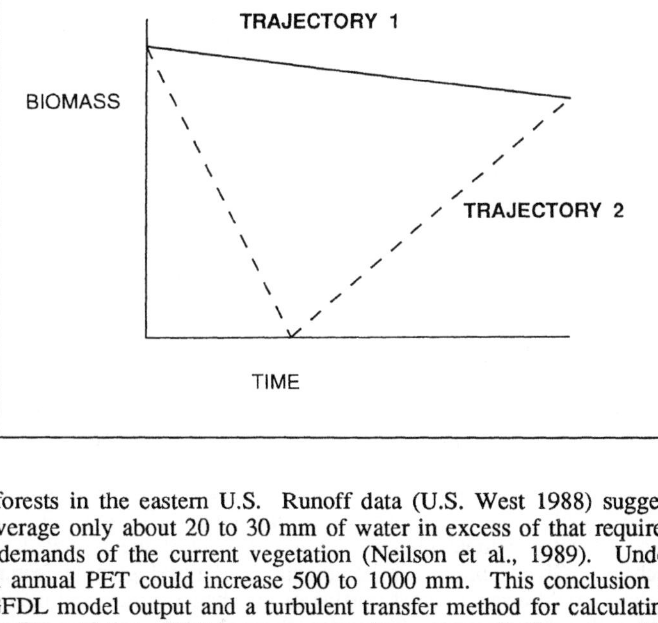

Figure 4. Two different trajectories of biomass change in response to climate change. If climate change is slow, biomass change will proceed along trajectory 1. If climate change is fast, biomass change will proceed along trajectory 2.

As an example, consider forests in the eastern U.S. Runoff data (U.S. West 1988) suggest that in August there is on average only about 20 to 30 mm of water in excess of that required to meet the transpirational demands of the current vegetation (Neilson et al., 1989). Under double-CO_2 conditions, total annual PET could increase 500 to 1000 mm. This conclusion is based upon the GISS and GFDL model output and a turbulent transfer method for calculating PET (Gucinski et al., 1990). Even if these PET increases were to occur gradually over a 100 year period (e.g., 5 to 10 mm yr^{-1}), the excess water would be used up in only a few years, assuming no change in precipitation. Annual precipitation for the eastern U.S. is simulated to range between $\pm$ 10% of current values in the GISS and GFDL double-CO_2 climate scenarios (Manabe and Wetherald, 1987; Hansen et al., 1988). How these changes would affect the seasonal water balance is uncertain at this time.

Two implications of these results are that 1) widespread drought-induced dieback is possible, and that 2) declines could occur relatively early in a transition to double-CO_2 climate conditions. (These arguments are again based on assuming a constant rate of climate change.

As discussed above, this is unlikely, and could result in faster or slower responses in the biosphere than indicated here.) A third implication is that an entire region may undergo decline within a few years rather than beginning at an ecotone and spreading into the core of a biome. Because the amount of excess water at the end of the growing season in U.S. forests is about the same across large areas (Neilson et al., 1989; Neilson et al., 1992), large changes in evaporative demands caused by climate change will deplete the reserves at approximately the same rate, leading to synchronous decline across a region. If these arguments are correct, the resulting C flux to the atmosphere would be greater than calculated here.

One way to test this hypothesis on a landscape to continental scale is to construct a complete water balance model that incorporates water movement through a vegetation canopy and calculates vegetation biomass as a direct function of the amount of water actually transpired (e.g., Dickinson, 1983). Simulating double-CO_2 climate conditions using such a model will show those areas where significant drought stress could be expected to occur and thus the possible magnitude of vegetation dieback. Ongoing modeling efforts show increased drought stress and reduced leaf area index (LAI) in the U.S. under the climate scenarios used here. Drought-induced forest dieback in response to climate change has also been simulated in boreal forest ecosystems by Bonan et al. (1990).

The rate of dieback will also be affected by the amount of drought stress mature trees can withstand before they are killed. There are two aspects of drought stress relevant here, the magnitude of the stress in any given year and its duration over a period of years. These questions are beyond the scope of this paper, but should be a focus of future research.

The third assumption in this modeling work is that the direct effects of CO_2 do not have any effect on plant growth and water-use efficiency. Clearly this will not be the case, but how vegetation across a region will respond to higher CO_2 concentrations is highly uncertain (Allen, 1990). In fact, two opposing scenarios can be proposed. First, higher CO_2 concentrations may increase ecosystem water-use efficiency and thus ameliorate drought stress to some extent. This would decrease the amount of dieback and thus reduce the pulse. However, catastrophic decline could still occur under this scenario, if the rate of climate change still exceeded the rate of increase in water-use efficiency (Neilson et al., 1989). The second scenario begins with the same assumption, that higher CO_2 conditions will increase water-use efficiency. However, the increased water-use efficiency will likely allow plants to put on more biomass using the water balance hypotheses previously discussed. The increased biomass will result in most of the available water being utilized for transpiration. Then, additional climate change to drier conditions could still cause drought stress and vegetation dieback (Neilson et al., 1989). These scenarios need to be tested in ecosystem models that incorporate full calculation of water balance including stomatal conductance, and by experimental studies.

The fourth assumption, complete dieback in the change zones and no change over the remainder of the earth's land surface, was made as a first approximation in order to run the pulse model. Clearly, differences in local topography and soil water holding capacity will act together with regional climate changes to control the spatial distribution and chronology of dieback. Given appropriate combinations of physiographic features and changes in climate, dieback could occur in areas outside of the change zones, and none could occur in certain areas within the change zones. Therefore, the dieback assumption made for this study can be more accurately stated as follows: the area of extra-tropical forest dieback is equivalent to the size of the change zone. A more subtle point is that the Smith et al. (1991, in press) Holdridge scenarios simply suggest where biomes will change from one type to another. It cannot show if vegetation within a biome undergoes decline and then regrows to the same biome class, albeit perhaps to a different biomass or combination of species within that biome class. Consequently, the areas of dieback could conceivably be larger than the change zone area. Again, what is needed to calculate dieback is a fully mechanistic modeling of regional water balance that incorporates canopy control on water flux from the soil.

The fifth and final assumption, no climatically-induced changes in regrowth rates or in C density of extra-tropical forests, was made to simplify the model calculations. Regrowth rates may change, especially with higher CO_2 levels. Higher CO_2 levels could increase plant growth rates, at least under non-water and nutrient-limiting situations. Regrowth rates could lag, however, if fires are very hot and sterilize the forest floor, eliminating seed banks and tree sprouting. Later stages of regrowth could be slowed also if the general poleward migration of trees is slower than the rate of dieback and climate change.

4.2. MODEL LIMITATIONS

A few of the limitations in the scope of the current pulse model are listed here and discussed in light of their implications on the calculated magnitude of the pulse.

1. The Holdridge life-zone classification model is strictly correlational and does not simulate current vegetation patterns well (K. Prentice 1990). Also, the direct effects of CO_2 are not considered in this model or in other currently available global vegetation models. Improving our simulation of vegetation response to climate change using mechanistic water balance models (Neilson et al., in press) that incorporate the direct effects of CO_2 on plant growth is fundamental to improving estimates of biospheric feedbacks.

2. The current and future vegetation scenarios, and thus the C storage values, do not take into account current and future land-use. Inclusion of current land-use would reduce current C storage estimates used in the model, and perhaps would reduce C sequestered under future climate conditions. The first step would decrease the pulse estimate, while the second might increase the pulse estimates.

3. Only changes in above-ground C storage were considered. Soils contain more C than above-ground vegetation (Post et al., 1982; Bolin, 1983, 1986), and changes in vegetation will affect soil C storage, although perhaps at a rate of change too slow to be of significance from the rapid feedback perspective. Under steady-state, double-CO_2 conditions, soils could be either a sink or source of CO_2 depending on the magnitude of the climate and vegetation change (Smith et al., 1991, in press; Dixon and Turner, 1991). Temperature changes alone could cause a release of 61 Gt of C over the next 60 years, assuming no change in the annual addition of plant debris to the soil (Jenkinson et al., 1991). Transient soil C dynamics thus deserve significant attention in future research on biospheric feedbacks to climate change.

4. Only CO_2 fluxes from incineration and decay were included in the pulse model. Other greenhouse gases such as CH_4 are also released during these processes (e.g., Cofer et al., 1990), but were not included in the model. These gases have a higher radiative effect per molecule than CO_2. Thus, even if their release during burning is substantially less than CO_2, they could still produce a relatively strong radiative effect.

5. The model is zero dimensional and thus lumps all extra-tropical forest types together. Disaggregating the pulse simulation may allow more precision in calculating C fluxes to the atmosphere.

6. The model does not include any climate feedbacks of the biospheric response. These feedbacks could range from increased radiative forcing by trace gases to changes in radiative balance caused by increased atmospheric particulates injected by increased fire

frequencies. A strong positive feedback could lead to faster warming and thus to increased rates of vegetation change and C emissions beyond those simulated here.

7. There are significant limitations in the ability of the current generation of GCMs to simulate current and future climate conditions (e.g. IPCC 1990). Since the climate output from these models drives the vegetation models, the GCM limitations further necessitate that the results presented here be considered a sensitivity analysis rather than a prediction of what could occur over the next 100 years.

4.3. PULSE CHRONOLOGY

The results presented in this paper suggest that if climate change is rapid and large enough, an average of 1-3 Gt of C could be released to the atmosphere over a period of 50 to 100 years from vegetation dieback. This additional carbon flux to the atmosphere would affect the rate of radiative warming and should be included in transient simulations of future climate change. In order to do such transient simulations estimates are needed of when vegetation dieback would begin to release CO_2 to the atmosphere.

A simplistic answer can be made based on the modeling framework used here to make climate and vegetation change simulations. Current conditions are simulated by GCMs based on atmospheric CO_2 concentrations in the early 1960s. Most double CO_2 climate and vegetation simulations are made for steady state conditions, and no information is provided on the amount of time required to reach those steady state climate conditions. For this study we have assumed a linear rate of climate change, and thus for vegetation change. In this context, the pulse in our model framework would begin in the 1960s (i.e. the year after the base year for CO_2 concentration in the models) and continue as modeled (Figures 2 and 5, Trajectory A).

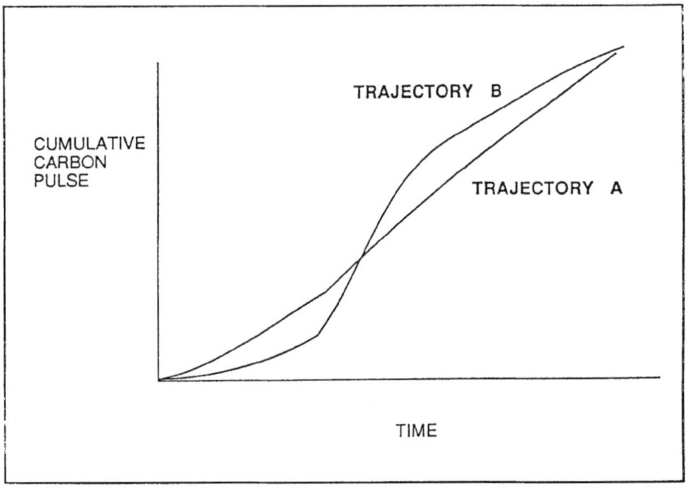

Figure 5. Two different trajectories for the transient carbon pulse. Under trajectory A, the rate of dieback proceeds at a linear rate in response to linear changes in climate. Under trajectory B, dieback lags behind climate change until drought tolerance thresholds are past. At that time the rate of dieback increases, resulting in a faster release of C to the atmosphere.

The major flaw in the above argument is that neither climate nor vegetation change will proceed at a linear rate. Climate forcing caused by increases in trace gas concentrations will be imposed upon other mechanisms of climate variability. Vegetation response in turn will be affected by the inertia of the system and lags in the biological response of trees to climate change. Dieback will not begin until soil moisture is depleted to a lethal level, which is a function of the biology of the tree species and the amount of currently available but unutilized moisture. Dieback and thus the CO_2 pulse will lag behind climate change. When the pulse begins, it will likely proceed at a rate faster than the long-term average calculated here for the entire period of climate change to double CO_2 conditions (Figure 5, Trajectory B).

Estimating when the pulse begins is really a matter of determining how much climate can change before extensive dieback begins. In order to simulate this, one would need to couple a realistic transient climate change scenario with a dynamic simulation of ecosystem response to changes in temperature, water balance, and CO_2 concentrations. Even then, the time course projected may not correspond to the realized time course of change if the climate models do not adequately simulate other mechanisms of climate variability or do not simulate extreme events well.

5. Conclusions

The pulse model described above was developed as a simple approach to estimate whether vegetation redistribution could release enough C to the atmosphere to act as a potential significant positive feedback to climate change. The preliminary results discussed here suggest that vegetation redistribution could result in a negligible yearly flux of C to the atmosphere or a flux equivalent to half of current fossil fuel emissions. The variation in the pulse estimates is extremely high because of uncertainties in the future rate of climate change and in the values of key model parameters. Nevertheless, these results indicate the importance of considering ecosystem dynamics when analyzing the response of the global carbon cycle to climate change. Ecosystem dynamics as shown here could cause terrestrial carbon storage to significantly decrease and then increase as climate changes. This is a much different trajectory of change than if one assumes a strictly linear change in carbon storage to a new steady-state condition.

Whether a significant pulse will occur is primarily a function of the rate of change in the regional water balance. If evaporative demand increases at a rapid enough rate, current moisture supplies could be overdrafted in a relatively short period of time (~5 to 20 years), resulting in severe drought stress and substantial mortality of vegetation. This could lead to a much greater annual release of C and total C pulse than modeled here, where the fastest rate of dieback was calculated as 1/50 of the total change to simulated double-CO_2 conditions. Furthermore, since available moisture appears to be almost fully utilized across a biome (Neilson et al., 1989; Neilson et al., in press), the climatically-induced depletion of available moisture could occur almost simultaneously throughout a region, resulting in extensive areas of dieback distributed across a region rather than situated along ecotones. Thus, instead of 1/50 of the potential decline area suffering dieback each year, an entire region could undergo dieback in just a few years. This would substantially increase the pulse over the levels simulated here. Potential changes in water-use efficiency caused by increases in CO_2 concentrations complicate the above argument. If water-use efficiency on a landscape scale in fact increases and changes faster than climatically-induced drought stress, significant vegetation dieback would be averted.

Another major uncertainty in the current pulse estimates is the amount of C currently stored in the terrestrial biosphere. If the estimates of C density are in fact overestimated by a factor of two or three (Botkin and Simpson, 1990), the amount of C that could be released by vegetation redistribution would be substantially reduced. Finally, changes in PET will be a principal driver of drought-induced vegetation change. Different methods for estimating PET yield substantially different results (Gucinski et al., 1990), leading to corresponding uncertainty in the vegetation redistribution estimates.

These uncertainties suggest a number of research priorities. Of critical importance is the need to improve our simulation of landscape to continental scale water balance by incorporating into models the physiological processes controlling the movement of water through vegetation. This modeling approach also needs to include the simulation of vegetation growth form, as different growth forms have different requirements for moisture and energy (Neilson et al., in press). Model experiments should be conducted on the potential role of enhanced CO_2 concentrations on water-use efficiency at the landscape scale and higher. Validation of PET estimation methods is also required.

Soil processes were not considered in this modeling work, but linkages between above- and below-ground processes need to be built as research on biospheric feedbacks continue. Some of the important linkages include the effect of changes in vegetation distribution and productivity on soil C, the effect of climate change on trace gas emissions from soils, and the effect of soil fertility and structure on vegetation redistribution.

In support of the modeling work, experimental work is needed to characterize the effect of higher CO_2 concentrations on the physiological processes of major tree growth forms (e.g. broadleaf evergreen, needle-leaved evergreen). Important questions include how higher CO_2 levels affect 1) partitioning of above- and below-ground biomass, 2) plant respiration, 3) stomatal conductance, and 4) water-use efficiency. Other experimental, field, and historical research is needed to document the amount of drought stress that particular tree species/growth forms can withstand before mortality occurs.

Finally, various aspects of ecosystem C dynamics also need investigation. More field data are needed on 1) the amount of C stored in ecosystems, 2) biomass decay rates, and 3) biomass accumulation rates following disturbance. Remote sensing techniques to assist in the extrapolation of field data will be invaluable.

6. Acknowledgements

The authors thank Hermann Gucinski, David Tingey, Gregg Marland, Ron Ritschard, and John Weishampel for comments on an earlier draft of the manuscript. The research described in this article has been funded by the U.S. Environmental Protection Agency. This document has been prepared at the EPA Environmental Research Laboratory in Corvallis, Oregon, through Contract No. 68-C8-0006 to ManTech Environmental Technology, Inc. and through Cooperative Agreement No. CR 816257 with Oregon State University. It has been subjected to the Agency's peer and administrative review and approved for publication. Mention of trade names or commercial products does not constitute endorsement or recommendation for use.

7. References

Allen, L.H., Jr. (1990) 'Plant responses to rising carbon dioxide and potential interactions with air pollutants', J. Environ. Qual. 19, 15-34.

Auclair, A.N.D. (1985) 'Postfire regeneration of plant and soil organic pools in a *Picea mariana-Cladonia stellaris* ecosystems', Can. J. For. Res. 15, 279-291.

Bolin, B. (1983) 'The carbon cycle', in The Major Biogeochemical Cycles and Their Interactions, SCOPE Report No. 21. John Wiley, Chichester, pp. 41-45.

Bolin, B. (1986) 'How much CO_2 will remain in the atmosphere', in B. Bolin, B.R. Döös, J. Jäger, and R.A. Warrick (eds.), The Greenhouse Effect, Climatic Change, and Ecosystems, John Wiley & Sons, Chichester, pp. 93-155.

Bonan, G.B. and Shugart, H.H. (1989) 'Environmental factors and ecological processes in boreal forests', Ann. Rev. Ecol. Syst. 20, 1-28.
Bonan, G.B., H.H. Shugart, and D.L. Urban. 1990. 'The sensitivity of some high-latitude boreal forests to climatic parameters', Clim. Change 16, 9-29.
Bormann, F.H. and G.E. Likens. (1979) Pattern and process in a forested ecosystem. Springer-Verlag, New York.
Botkin, D.B. and Simpson, L.G. (1990) 'Biomass of the North American boreal forest', Biogeochem. 9, 161-174.
Cofer, W.R., III, Levine, J.S., Winstead, E.L. and Stocks, B.J. (1990) 'Gaseous emissions from Canadian boreal forest fires', Atmos. Environ. 24A, 1653-1659.
Davis, M.B. (1986) 'Climatic instability, time lags, and community disequilibrium', in J. Diamond and T.J. Case (eds.), Community Ecology, Harper and Row, NY.
Daubenmire, R. (1968) 'Soil moisture in relation to vegetation distribution in the mountains of Northern Idaho', Ecology 49, 431-438.
Dickinson, R.E. (1983) 'Land surface processes and climate - surface albedos and energy balance', Adv. Geophys. 25, 305-353.
Dixon, R. and Turner, D.P. (1991) 'The global carbon cycle and climate change: responses and feedbacks from below-ground systems', Environ. Poll. 73, 245-262.
Dolph, J., D. Marks, and King, G.A. (In press) 'Sensitivity of the regional water balance in the Columbia River Basin to climate variability: Application of a spatially distributed water balance model', in R.J. Naiman (ed.), Watershed Management: Balancing Sustainability with Environmental Change, Springer-Verlag. New York, NY, U.S.A.
Edmonds, R.L. (1987) 'Decomposition rates and nutrient dynamics in small-diameter woody litter in four forest ecosystems in Washington, U.S.A.', Can. J. For. Res. 17, 499-509.
Emanuel, W.R., Shugart, H.H., and Stevenson, M.P. (1985a) 'Climatic change and the broad-scale distribution of terrestrial ecosystem complexes', Clim. Change 7, 29-43
Emanuel, W.R., Shugart, H.H., and Stevenson, M.P. (1985b) 'Response to comment: Climatic change and the broad-scale distribution of terrestrial ecosystem complexes', Clim. Change 7, 457-460.
Fahnestock, G.R. and J.K. Agee. (1983) 'Biomass consumption and smoke production by prehistoric and modern forest fires in Western Washington', J. For. 81, 653-657.
Foster, J.R. and Lang, G.E. (1982) 'Decomposition of red spruce and balsam fir boles in the White Mountains of New Hampshire', Can. J. For. Res. 12, 617-626.
Gleick, P.H. (1987a) 'Regional hydrologic consequences of increases in atmospheric CO_2 and other trace gases', Clim. Change 10, 137-161.
Gleick, P.H. (1987b) 'The development and testing of a water balance model for climate impact assessment: modeling the Sacramento Basin', Water Resources Res. 23, 1049-1061.
Gucinski, H., Marks, D., and Turner, D.P., (editors) (1990) Biospheric Feedbacks to Climate Change: The Sensitivity of Regional Trace Gas Emissions, Evapotranspiration, and Energy Balance to Vegetation Redistribution, Status of Ongoing Research, U.S. Environmental Protection Agency. EPA/600/3-90/078.
Hansen, J., Fung, I., Lacis, A., Rind, D., Lebedeff, S., Ruedy, R., and Russell, G. (1988) 'Global climate changes as forecast by Goddard Institute for Space Studies three-dimensional model', J. Geophys. Res. 93, 9341-9364.
Harmon, M.E., J.F. Franklin, F.J. Swanson, P. Sollins, S.V. Gregory, J.D. Lattin, N.H. Anderson, S.P. Cline, N.G. Aumen, J.R. Sedell, G.W. Lienkaemper, K. Cromack, Jr., and K.W. Cummins. (1986) 'Ecology of coarse woody debris in temperate ecosystems', Adv. Ecol. Res. 15, 133-302.
Holdridge, L.R. (1947) 'Determination of world formulations from simple climatic data', Science 105, 367-368.
Holdridge, L.R. (1967) Life Zone Ecology, Tropical Science Center, San Jose, PR.

Houghton, R.A. 1990. 'The global effects of tropical deforestation', Environ. Sci. Tech. 24, 414-422.

Intergovernmental Panel on Climatic Change (IPCC). (1990) Scientific Assessment of Climate Change. Report for WGI Plenary Meeting. Cambridge University Press, Cambridge, UK.

Jenkinson, D.S., Adams, D.E. and Wild, A. (1991) 'Model estimates of CO_2 emissions from soil in response to global warming', Nature 351, 304-306.

King, G.A., Winjum, J.K., Dixon, R.K., and Arnaut, L.Y, (editors). (1990) Response and Feedbacks of Forest Systems to Global Climate Change, US Environmental Protection Agency, Corvallis, OR, USA, EPA/600/3-90/080.

Lashof, D.A. (1989) 'The dynamic greenhouse: Feedback processes that may influence future concentrations of atmospheric trace gases and climatic change', Clim. Change 14, 213-242.

Lettenmaier, D.P., and Gan, T.Y. (1990) 'Hydrologic sensitivities of the Sacramento-San Joaquin river basin, California, to global warming', Water Resources Res. 26, 69-87.

Levine, J.S. (1990) Global biomass burning: Atmospheric, climatic and biospheric implications. Eos 11, 1075-1077.

Manabe, S., and Wetherald, R.T. (1987) 'Large-scale changes in soil wetness induced by an increase in carbon dioxide', J. Atmos. Sci. 44, 1211-1235.

Marland, G., A. Boden, R.C. Griffin, S.F. Huang, P. Kanciruk, and T.R. Nelson. (1989) Estimates of CO_2 emissions from fossil fuel burning and cement manufacturing, based on the United Nations energy statistics and the U.S. Bureau of Mines cement manufacturing data. ORNL/CDIAC-25. Oak Ridge National Laboratory, Oak Ridge, Tennessee.

Mitchell, J.F.B. (1989) 'The "greenhouse" effect and climate change', Rev. Geophys. 27(1), 115-139.

Neilson, R.P., and King, G.A. (In press) 'Continental scale biome responses to climatic change', in D. McKenzie, E. Hyatt, and J. MacDonald, (eds.), Proceedings of International Symposium: Ecological Indicators. Ft. Lauderdale, FL, Oct. 16-19, 1990, Elsevier Science Publishers, Ltd.

Neilson, R.P., King, G.A., and Koerper, G. (In press) 'Toward a rule-based biome model', Lands. Ecol.

Neilson, R.P., King, G.A., DeVelice, R.L., Lenihan, J., Marks, D., Dolph, J., Campbell, B., and Glick, G. (1989) Sensitivity of Ecological Landscapes and Regions to Global Climate Change. EPA/600/3-89/073, NTIS-PB90-120-072/AS, Washington D.C.

Neilson, R. P., G.A. King, R.L. DeVelice, and J.M. Lenihan. (1992) 'Regional and local vegetation patterns: the responses of vegetation diversity to subcontinental air masses', In: Landscape Boundaries: Consequences for Biotic Diversity and Ecological Flows (Ecological Studies 92). Proceedings of Scientific Committee on Problems of the Environment, International Council of Scientific Unions, Workshop on Ecotones, edited by A. Hansen and F. di Castri. Springer-Verlag, New York.

Olson, J.S. (1963). 'Energy storage and the balance of producers and decomposers in ecological systems', Ecology 44, 322-331.

Olson, J.S., Watts, J.A., and Allison, L.J. (1983) Carbon in Live Vegetation of Major World Ecosystems, ORNL-5862, Oak Ridge National Laboratory, Oak Ridge, TN., USA.

Overpeck, J.T., Rind, D., and Goldberg, R. (1990) 'Climate-induced changes in forest disturbance and vegetation', Nature 343, 51-53.

Post, W.M., Emanuel, W.R., Zinke, P.J., and Stangenberger, A.G. (1982) 'Soil carbon pools and world life zones', Nature 298, 156-159.

Post, W.M., Peng, T-H., Emanuel, W.R., King, A.W., Dale, V.H., and DeAngelis, D.L. (1990) 'The global carbon cycle', Am. Scientist 78, 310-326.

Prentice, K.C. (1990) 'Bioclimatic distribution of vegetation for General Circulation Model studies', J. Geophys. Res. 95, 11811-11830.

Prentice, K.C., and Fung, I.Y. (1990) 'Bioclimatic simulations test the sensitivity of terrestrial C storage to perturbed climates', Nature 346, 48-51.
Prentice, I.C., Sykes, M.T., and Cramer, W. (1991) 'The possible dynamic response of northern forests to global warming', Global Ecol. Biogeog. Lett. 1, 129-135.
Schlesinger, M.E., and Zhao, Z.C. (1989) 'Seasonal climatic change introduced by doubled CO_2 as simulated by the OSU atmospheric GCM/mixed-layer ocean model', J. Climate 2, 429-495.
Smith, J.B., and Tirpak, D. (editors). (1989) The Potential Effects of Global Climate Change on the United States, Report to Congress, EPA-230-05-89-050, U.S. Environmental Protection Agency, Washington, DC, USA.
Smith, T.M., Shugart, H.H., Bonan, G.B., and Smith, J.B. (1991) 'Modeling the potential response of vegetation to global climate change', Adv. Ecol. Res. 22, 93-116.
Smith, T.M., Leemans, R., and Shugart, H.H. (In press) 'Sensitivity of terrestrial carbon storage to CO_2 induced climate change: comparison of four scenarios based on general circulation models', Clim. Change.
Stephenson, N.L. (1990) 'Climatic control of vegetation distribution: the role of the water balance', Amer. Nat. 135, 649-670.
U.S. West. (1988) Hydrodata user's manual: USGS daily and peak values. Version 2.0. US West Optical Publishing, Denver, Colorado.
Waring, R. H. and W.H. Schlesinger. (1985) Forest Ecosystems, Concepts and Management. Academic Press, Inc., Harcourt Brace Jovanovich, Publishers. Orlando.
Whittaker, R.H. (1975) Communities and Ecosystems, 2nd Edition, MacMillan Publishing Co., Inc., New York, NY, USA.
Woodward, F.I. (1987) Climate and Plant Distribution, Cambridge University Press, London, England.

A GEOGRAPHICALLY-BASED ECOSYSTEM MODEL AND ITS APPLICATION TO THE CARBON BALANCE OF THE LUQUILLO FOREST, PUERTO RICO

CHARLES A. S. HALL
SUNY College of Environmental Science and Forestry
Syracuse NY 13210 USA

MARSHALL R. TAYLOR
Resources Planning Associates, Incorporated
Cornell Business and Technology Park
Ithaca NY 14850 USA

EDWIN EVERHAM
SUNY College of Environmental Science and Forestry
Syracuse NY 13210 USA

ABSTRACT. We have developed a geography-based computer model of the Bisley Experimental Watershed ecosystem that simulates basic forest dynamics as a function of meteorological inputs and hydrologic simulation, as influenced by topography, soils, land use, and cover. The model is parameterized based on steady state levels of, and hurricane impacts on, biomass, necromass and rate processes of the tabonuco forest, and is stable over decades. Over a 60 yr simulation (without hurricanes) leaf biomass remains approximately constant and woody biomass of those regions not having severe sunlight, moisture, or nutrient limitations increases slowly in agreement with observations. Necromass decreases slowly. Small quantities of C leave the ecosystem in stream water, especially during large rain events. When topographically-sensitive hurricane impacts are included, leaf and woody biomass are converted to necromass. In the model the recovery of the watersheds' hydrology, leaf and woody biomass, and necromass are consistent with field observations. We used this model to simulate the C dynamics of the forest over centuries using empirical values and found that this forest acted to pump C from the atmosphere to the ocean at a rate of about 90 kg ha^{-1} yr^{-1}.

1. Introduction

In our search for possible sinks of C to balance the global C budget the biota is normally the sink of choice--especially by non biologists, from Revelle and Suess (1957) to Bacastow and Keeling (1973) to Broecker (1979) to Tans et al. (1991). But the mechanisms for this sink are not clear. Revelle and Suess (1957) suggested that the increases in the CO_2 concentration of the atmosphere should be fertilizing the growth (actually an increase in the standing stock is required) of the earth's collective vegetation, but there is little data to support this idea, due mostly to the difficulty of

Water, Air, and Soil Pollution **64**: 385–404, 1992.
© 1992 *Kluwer Academic Publishers.*

the large-scale measurements. Other investigators remain unconvinced, arguing that maximum forest biomass is determined by factors other than CO_2 concentration, and, because it would seem that even the small quantities that would be required to balance the global C budget (10-20 g m^{-2} yr^{-1} for the approximately 70 10^{-6} km^2 of all forested land) would be obvious over the many decades that we have been studying forests. In addition, of course, many forests are being destroyed rapidly (Houghton et al., 1983; Detwiler and Hall, 1988).

Lugo and Brown (1986) have suggested another possible mechanism--that natural forests have always acted as CO_2 sinks relative to the atmosphere. The mechanism is simple: as trees grow they remove CO_2 from the atmosphere, sequester it temporarily in their biomass, pass this C to dead organic material (necromass) and the soil, and some proportion to rivers and eventually the deep sea through riverine transport and sinking processes in the sea. Another proposal of Lugo and Brown (1986) is that most tropical forests have been disturbed prior to 1876 by natural or human factors. These forests sequester C during the short time period simulated by most ecosystem models. Odum (1970) concluded that the biomass of the Luquillo Experimental Forest in Puerto Rico was increasing based on relatively long term measurements from 1943 to the late 1960's, a period with no major hurricanes that followed the cessation of moderate levels of human disturbance. McDowell (personal communication) has found that the C concentration of Luquillo streams increases as a function of stream discharge, so that there is C "flushing" following heavy rainfall events, such as most hurricanes.

Many of the processes that maintain the C balance are sensitive to the frequency and magnitude of disturbance, including treefalls and hurricanes (Walker et al., 1991). Processes moving C to the deep ocean may be accelerated in the tropics due to the abundance there of coccolithophoric algae containing calcium carbonate tests. Obviously in order to resolve whether these mechanisms are important for the global C budget we need to examine them on a global basis. But any such large scale estimates need to be based on detailed knowledge of mechanism at a much finer resolution. The purpose of this paper is to describe a computer model that can test Lugo and Brown's hypothesis at the level of a small ecosystem--say a watershed--to allow extrapolation to much larger forested areas.

2. Modeling Methodology

2.1 Spatial Modeling

Ecosystem processes, such as those affecting C inventories and fluxes, are inherently geographic (Hall et al., 1985; Hall and Uhlig, 1991). Virtually all processes that drive ecosystem response are sensitive to the magnitude of the input from the external environment: for example the growth of plants is critically sensitive to available sunlight, and may be sensitive to soil moisture and other meteorologically-forced variables, including relative humidity. One of the problems faced when attempting to model ecosystems is that critical factors often vary greatly over small spatial scales, that is,

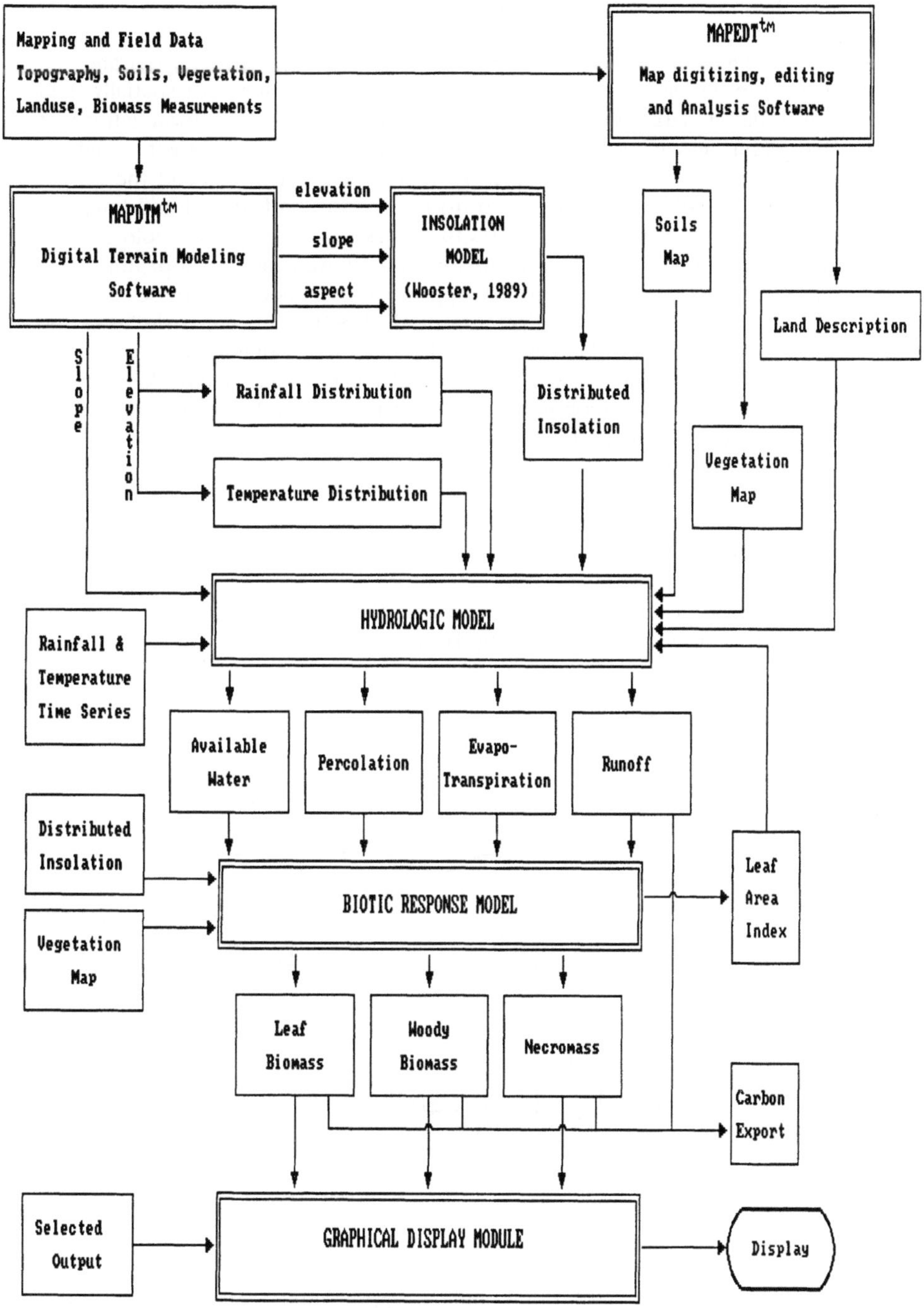

Figure 1. Overview of the Geography-based Ecosystems Model

over tens or hundreds of meters. In addition, many ecologic processes are very sensitive to short term conditions but must be simulated over very long time periods. Until recently it has been difficult to include such variability (in both spatial and temporal scales) in computer simulation models due to the absence of spatially- and temporally-distributed data (observations) and to the limits of the formulations of existing ecosystem modeling algorithms and coefficients --and their solution. Additionally, it has been extremely difficult to represent and comprehend the output from those earlier models which did attempt to incorporate both spatial and temporal variation. During the latter half of the 1980's the necessary computer technology (both hardware and software) matured to a point that has allowed the development of far more comprehensive and comprehensible approaches to modeling these complex variations (French and Taylor, 1982; Costanza et al., 1990). This paper discusses the development of a model of the basic spatially- and temporally-varying ecological processes that influence forest growth. The model incorporates meteorologic inputs and hydrologic modeling in addition to an ecosystem process model (Figure 1). Model parameters are derived using digital terrain modeling and analysis. We present the spatially-varying input data and results using an interactive computer graphic visualization program that aids greatly in the comprehension of the results.

2.2 Application Setting

The initial application of our approach has been to the Luquillo Experimental Forest (LEF) of Puerto Rico, one of the 15 Long Term Ecological Research (LTER) sites.

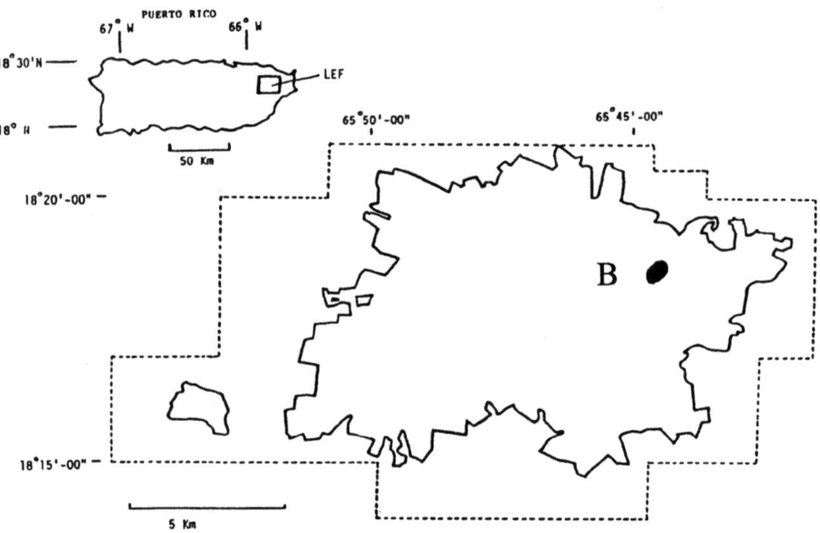

Figure 2. Location and map of the Luquillo experimental forest and the Bisley watershed. B indicates the location of the Bisley watershed.

At the LEF more than 30 principal investigators presently are involved in an assessment of the basic structural and functional properties of this tropical forest. Virtually all of the biotic data are readily convertible to C. One of the study sites which has received intensive investigation is the 13 ha Bisley Experimental Watershed (Figure 2). Both meteorologic and streamflow gauges have been installed at Bisley, and a relatively comprehensive, spatially distributed database of many characterizing variables (vegetation type, damage assessments after Hurricane Hugo, some soil parameters) has been constructed. Basic data collection in the Luquillo forest has occurred intermittently since about 1920 and with increasing frequency over time, and

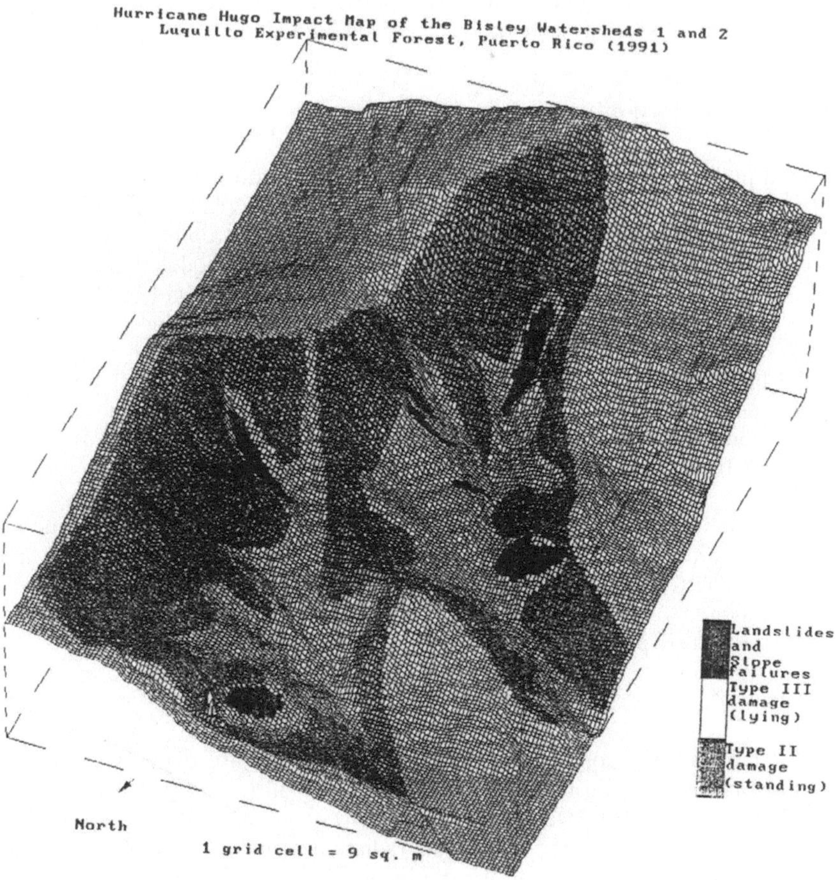

Figure 3. Map of hurricane damage in Bisley watershed (from Scatena, personal communication).

has resulted in a reasonable time series of important system variables (Odum, 1970; Brown et al., 1983). The Bisley watershed has been intensively studied since the mid 1980's.

Much of the research at Luquillo, especially the LTER research, is focused on the study of ecosystem response to, and recovery from, major disturbances, specifically: treefalls, landslides, and hurricanes. Major hurricanes, such as Hugo in 1989, impact the Luquillo region an average of about once in 60 yr, causing severe, nonuniformily distributed damage to the forest (Figure 3). Our initial efforts combined meteorologic input models (Wooster 1988, Everham et al., 1991), a distributed small watershed model and a basic ecosystem-level biotic model derived and parameterized for the Luquillo forest by H.T. Odum based on his extensive experience with the Luquillo forest (Odum 1970) and additional spatially-specific data derived by Scatena et al. (1991).

Overview of the Model Structure

The overall philosophy of our modeling approach was to construct an easy to use and readily accessible spatial and temporal computing structure (called GEOPLOT) into which simple or complex modules can be placed. The model represents a region as a matrix of cells. Each cell represents a parcel of land within a modeled matrix of other cells of identical size but varying attributes representing physical and biotic features.

Since many of the input variables to the model are spatially complex and derivable from maps, the basic model structure is identical for all cells but dependent upon spatial geographic and digital terrain parameters. Most intramodel communications between the diverse component modules are performed through a common spatial data structure and temporal file naming rules.

At this time we have programmed a workable, although somewhat inefficient, modeling superstructure within which we can build various physical and biologic model components. The remainder of this section describes the meteorologic, hydrologic and biologic components which we have incorporated, and then summarizes some of the data preparation and programming concerns we have encountered.

Meteorology

The major meteorologic components of the current model use digital terrain models of the study area to: 1) distribute over space daily rainfall (currently solely a function of elevation), 2) distribute over space mean daily temperature (also a function of elevation), and 3) determine the daily solar radiation incident to each computational cell. Future development plans call for the incorporation of topographic (orographic) factors for rain generation, and stochastic factors.
Currently, mean daily temperature is used in the estimation of daily potential evapotranspiration in the hydrologic model. Our initial application simply distributed mean daily temperature from long term means of recorded data according to a dry adiabatic lapse rate.

The most interesting of the current meteorologic components is the solar insolation model (Wooster, 1989). This model uses a digital terrain analysis to characterize each computational cell by latitude, longitude, slope, and aspect. The model then computes the total potential insolation on a cell for each hour of the day, or can integrate total flux over the day. Mean atmospheric scattering and blockage of sunlight by topographic features are considered. Radiation reduction by clouds is incorporated crudely into the model by using monthly cloud absorption factors.

Hydrology

Figure 4 illustrates the basic components of the hydrologic simulation. Much of the formulation of the model was derived from methods developed by the USDA Soil Conservation Service and others studying the hydrology of small watersheds and agricultural catchments (USDA, 1972). Parts of the formulation have been based on a model called SWRRB (Simulator for Water Resources in Rural Basins), and the criteria for our model development were similar to those criteria applied to SWRRB (Williams et al., 1985; Arnold et al., 1989). These include: 1) the model must be physically-based, using inputs that can be understood and measured in the field or derived by other than trained hydrologists; 2) the model must be robust, that is it must produce reasonable and consistent results when data changes are made to reflect various physical or land management changes; 3) the model must allow simulation of long time periods; and 4) the model must provide a mechanism for any of the spatially-distributed variables (i.e., a hurricane or a management option) to be provided by an external source at any computational time step.

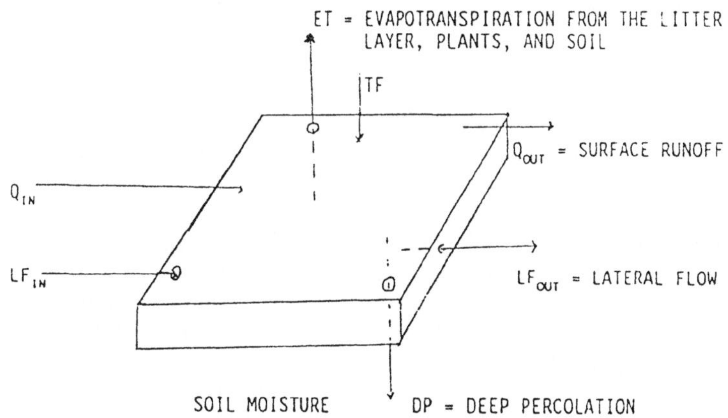

Figure 4. Schematic of hydrological simulation model.

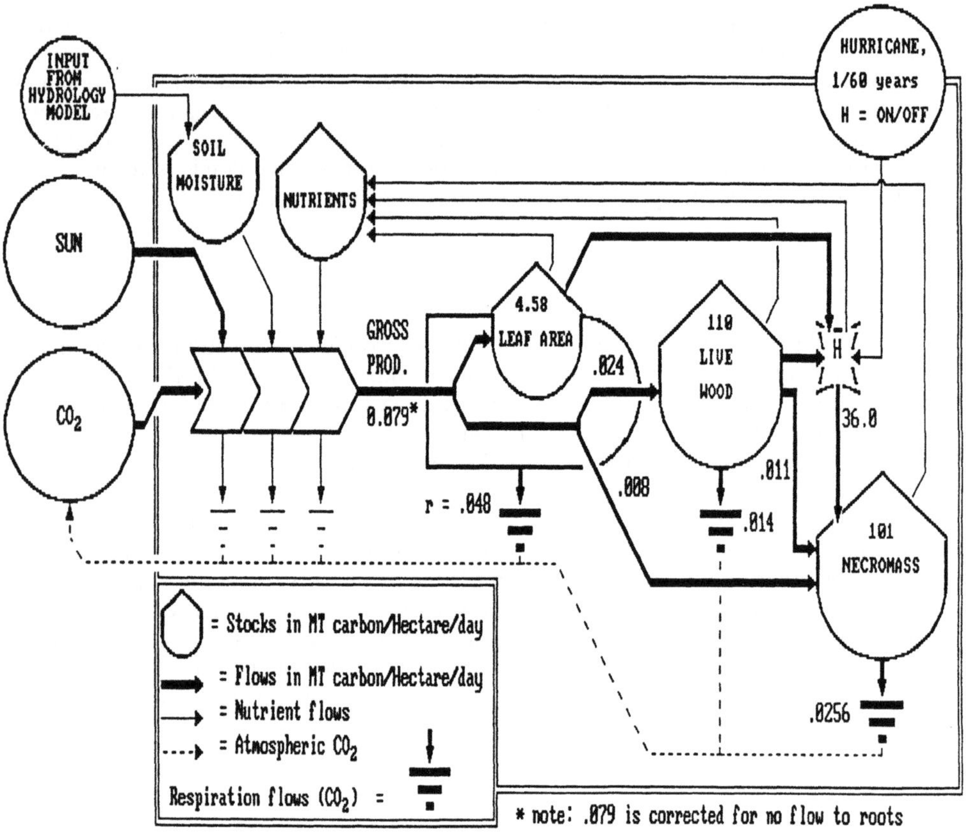

Figure 5. Synthesis model of stocks and flows of C for Bisley watershed. C flows are in heavy lines. Numbers are derived from Odum (1970) by converting to C (divided by 8). Biomass and necromass are adjusted to Scatena's mean values for Bisley.

Our model differs from SWRRB in several respects. First, our model is formulated as a spatially distributed model, so that each computational cell responds to varying meteorological inputs, soil properties, slope and vegetation, and interacts with surrounding cells. SWRRB's components for stochastic weather variable generation have been replaced with the previously-described meteorologic models. Our current model lacks the multiple soil layer capabilities of SWRRB, and does not perform pond and reservoir routing. Since our model improves the spatial description of soils significantly over what can be incorporated into SWRRB, we do not believe that the modeling of soils as a single homogenous layer (bounded by a confining layer) is a cause of significant relative error. Finally, all of SWRRB's crop growth components have been dropped from the current formulation, replaced by the tree-based biotic response model discussed in the next section.

Biotic Response

The biotic model is a conceptually aggregated, but spatially complex, synthesis of the basic components of the Luquillo ecosystem (Odum 1970). The ecosystem model has three basic biotic state variables (leaves, wood and necromass) and a nutrient pool that is diminished by tree growth and incremented by necromass decomposition. Growth limiting functions for sunlight, soil moisture and nutrients were derived from various sources and incorporated into the model. These parameters were weighted to give in aggregate the growth response observed by Odum (1970) as summarized in Figure 5. At this time measurements and experiments are being conducted in the Luquillo Forest that will give us far better estimates for these parameters.

The current model represents the basic biotic processes in each of six community types (ridge or tabonuco-representing the dominant tree on drier ridges, upland slopes, upper riparian, lower riparian, landslides and treefalls) and is initialized from an empirical vegetation map (Scatena, personal communication). For each community type, the model has three major simulated components: leaf biomass, woody biomass, and necromass (dead organic material) which are initialized in part from data given in Scatena et al., 1991. C values for leaves, biomass and necromass are derived by multiplying these dry weight values by 0.5. Although at this time the model represents only 6 broad community types, in the future it is planned to include more species level information which will allow the asking of more specific questions and to see if the inclusion of increasing levels of detail changes the overall behavior of the model. The remainder of this section will present the formulation in more detail and will discuss the interconnections between the various model components.

Gross photosynthesis is represented as a maximum possible rate which is limited by light, available water, and nutrient factors.

In this case we are using values for species thought representative for the entire forest or (in some simulations) the communities listed in Table 1. For example, trees representative of mature forests, such as Dacryodes excelsa (tabonuco) or Sloanea berteriana can live under very low light conditions, but have a relatively low maximum growth rate. Trees representative of clearings (including those caused by hurricane blow downs, landslides and treefalls) cannot survive under low light conditions but grow very rapidly under high light conditions. These species, including the rarer Ochroma sp. and Hampea sp. and the ubiquitous Cecropia schreberiana, can grow at an astonishing rate (up to 8 percent of mass a day--Ned Fetcher, personal communication) in a new forest clearing in the Luquillo forest. A two yr old clearing is often covered by trees two to four meters tall. As discussed above, the model is being developed in conjunction with field measurements that are designed to provide additional data in support of the modeling functions. The present empirical input is limited, and we can use only representative species for the parameter derivations. The best data is for the response of trees to light (Fetcher personal communication). Explicit measurements of tree growth as a function of moisture are not available, so we assumed that trees grow best under conditions where they are most abundant. Thus, we assumed that tabonuco trees grew best under drier conditions, and palms under wetter conditions. Our best guesses for these parameters are given in Table 1.

In the initial applications the nutrient limitation factor has been assigned a constant value (0.66) to allow the model in aggregate to reproduce observed growth rates given the other constraints as in Table 1. Studies are currently underway to derive nutrient limiting functions for additional representative tree species, patterns of soil parameters, and on incorporation of a nutrient transport component into the hydrology model.

Model Component Interaction

In this section we develop some of the interactions between the meteorologic, hydrologic and biotic response models (Figure 1). The meteorologic models currently provide rainfall, insolation and temperature to both the hydrologic and biotic models. The hydrologic model uses geography-specific values for both insolation and temperature to estimate daily total potential evapotranspiration (E_o). E_o serves as a limiter to both potential canopy evaporation and ground level (soil) evapotranspiration. Similarly, the hydrologic model provides estimates of available water to the biotic model. Leaf biomass values from the biotic model are used by the hydrologic model (at the next time step) to estimate the current Leaf Area Index (LAI). LAI is used in both the canopy storage/evaporation and soil evaporation components of the hydrologic model.

The model allocates photosynthate first to leaf respiration (which includes decay of leaves that fell off the tree), which is a function of current leaf biomass, then to leaf production (if required), with any that remains going into wood production. We assume that the leaf respiration includes normal rebuilding of leaf tissue (e.g. that damaged by insects), but not that damaged by large disturbances (e.g. hurricanes). Thus following hurricane damage such photosynthate as is available is first allocated to rebuilding leaves. Once pre-hurricane leaf biomass values are developed all allocation goes to wood production. Although this model does not mimic the reallocation of bole C storage into leaves as a "start up" following defoliation the model output rebuilds leaves in about a half year, as did the real forest.

The photosynthate available for wood production is allocated first to wood respiration, with nutrient release occurring in the process. Similarly, part of the wood biomass goes to necromass and part is respired. Necromass is derived from total biomass (leaves plus wood) at a constant rate slightly less than its decay. The rates for necromass production and decay, with the corresponding nutrient flux, have been adjusted to approximate steady state conditions and to produce the observed slow increase in the forest biomass over time (Odum, 1970). Over many hundreds of years this small increase is balanced by a leakage of C from the ecosystem into rivers and subsequently to the sea. Hurricanes accelerate this process by periodically converting biomass into more mobile particulate and dissolved organic material. Hurricanes convert about

Table 1. Derivation of growth parameters used in initial model simulations.

Many of these values are extremely crude pending further investigations.
Light in 100 W m^{-2} d^{-1}

Community type (limiter)	species used	maximum growth rate g/g/d	minimum value	half saturation
Tabonuco	Dacryodes	0.005		
(light)	excelsa		2	20
(water)			0	20
upland slopes	Sloanea	0.005	2	20
(light)	berteriana		0	20
(water)				
upper riparian	Prestoa	0.002		
(light)	montana		1	10
(water)			0	60
lower riparian	Prestoa	0.002		
(light)	montana		1	10
(water)			0	60
landslide	Cecropia	0.08		
(light)	schreberiana		5	30
(water)	or			
treefalls	Ochroma	0.08		
(light)			5	30
(water)				

one-third of the biomass to necromass, depending on location (Figure 3) about once per 60 yr.

Additions to the available nutrient pool are, as outlined above, a function of the respiration rates of each component (leaves, woody material, necromass) times the component biomass and nutrient to C ratios. Nutrient pool reductions are simply a function of gross photosynthesis and, again, nutrient to C ratios.

Data Preparation

The above discussion illustrates how many of the model parameters are dependent upon gross physical characteristics (elevation, slope, aspect, soil type, plant community type) of the elemental parcels of land (cells). Two geographic information tools (MAPEDT^{tm} and MAPDTM^{tm} were used for spatial data digitizing and analysis RPA, 1991). Topography of the watershed and mapping of vegetation coverage and other physical characteristics were digitized or derived with these tools.

Much of the data preparation for the various component models was accomplished by digital terrain analysis. Examples include: 1) the spatial distributions of both the average daily temperature and rainfall used in the hydrologic model, which are computed as a function of elevation, 2) the use of two characterizing variables for the insolation model (slope and aspect), and 3) the use of slope analyses to provide estimates of soil depths for each cell. These values are used in several segments of the hydrologic model.

3. Verification and Validation

Model verification is the process of determining that a simulation procedure is producing results consistent with the theoretical formulation of that model. That is, confirming that the numerical procedures are in fact producing output which represents the authors' conceptual and quantitative understanding and mathematical description of the system. Model validation is the process of determining that the model's output is in fact relevant to, and consistent with, the behavior of the real system which is being modeled. Both model verification and validation are difficult to address when a model has been designed to simulate a complex recovery from a low frequency event. The addition of spatial and temporal variation adds to the complexity of this process, but also specifies more rigorous validation criteria, since the model must agree with nature over space as well as time. In our study we have approached verification and validation in two ways: 1) examination of simulated hydrologic and biologic conditions (at as small as a daily time step) throughout calendar yr 1990, and 2) application of the model to a 5 yr simulation which spans hurricane Hugo. No formal validation criteria have been established for the model yet, but a number of comparisons and heuristic analyses have been made, which is summarized below.

Hydrologic Model

The first item examined in the verification and validation of the hydrologic model was how well the model reproduced the field-observed annual hydrologic budget for calendar year 1990. A summary of that comparison is presented in Table 3 where column A presents the field estimates, column B presents the primary model results and

Table 3.
1990 Hydrolologic Budgets
Bisley Experimental Watersheds 1 and 2

	A*	B	C	D	E	F
Rainfall and Condensation	12.3	12.52	12.52	12.36	12.52	13.38
Throughfall	7.87	7.81	6.65	7.62	7.81	7.59
Evapotranspiration	7.8	4.89	6.13	4.93	4.82	5.83
Runoff	5.9	6.56	5.34	6.38	6.69	6.44
Soil Moisture Change	0.0	0.08	0.08	0.08	0.07	0.10
Deep Percolation	0.0	0.98	0.97	0.97	0.94	0.98

*Rainfall throughfall and runoff were measured (August 1987 - December 1990); Condensation estimated from Odum and Pigeon, 1970. Soil moisture change and deep percolation were not measured but assumed in evapotranspiration, which was calculated by difference.

columns C through F present results from the specified sensitivity analyses.
Run Descriptions:

A) Measured values
B) Primary simulation run
C) Higher insolation due to cloud factor = 0.0
D) Adjusted rainfall
E) Soil depths decreased by 50%
F) LAI = constant 6.0

Of particular interest to the authors was an examination of the spatial response of the model. Rather than accepting the model solely based on good agreement with the overall watershed budget, we needed to verify that the behavior of smaller computational cells were consistent with observations and logic. These comparisons were conducted by examination of simulation time series for a number of selected cells. With limited field data available for comparison, this particular task essentially involved verification of "appropriate" behavior. Still examining calendar year 1990, the simulated basin outflow (overland flow calculated by the SCS runoff equation plus return flow) was plotted against field observations and flows at the nearest downstream US Geological Survey stream gaging station. Care must be taken when making comparisons with flows at the gaging station because the Bisley Watershed represents less than 1% of the drainage area at the gage and is located in the mountainous extremes of that drainage area. Nonetheless there appears to be a basic agreement between our model and the field data.

Validation of results for the 5 yr simulation is more difficult than for the period where field observations exist. In this case our analyses were limited to a search for computational anomalies in the model. Of particular interest were 1) the canopy storage and evaporation component of the model during the period immediately after hurricane Hugo, and 2) whether long term levels of state variables (soil moisture, return flows, evapotranspiration) were stable and reasonable. No significant errors have been found, but since we are in the early stages of constructing this model our analyses continue.

Biotic Model

The biological model was validated in the sense that when we varied the nutrient limiter (which was unknown) to generate observed values of forest regrowth in the 60 yr between hurricanes the mean resultant photosynthesis was within 10 % of the value given in Figure 2 of Odum (1970). Validation of the spatial patterns will require more spatial data, but the model does maintain the coarse-scale pattern of community biomass values observed by Scatena et al. (1991).

4. Results for Carbon

When we applied this model to examining the C balance through time, and the role of forests in the proposed C sink, we found that in the absence of hurricanes the Bisley forest was a small C sink relative to the atmosphere (about 10 g m^{-2} or 100 kg ha^{-2} yr^{-1}) over time (Figure 6a). When a hurricane hits there is a large transfer from the biomass pool to the necromass pool, although no net change in the ecosystem C (Figure 6b). Over longer periods of time the total C pool can increase or decrease depending upon the relative magnitude of the biomass-to-necromass vs the necromass-decomposition coefficients. In the cyclical process of ecosystem disturbance and recovery from disturbance the net effect of the coefficients that we used is a small accumulation of refractory C in the ecosystem, especially for those ecosystems growing in favorable sites. A summary of the long term pattern of the Bisley Forest using a stochastic rainfall generator, a stochastic hurricane generator with an average of one hurricane every 60 yr and a flow-weighted mean C concentration for streams of 4.0 mg 1^{-1} is presented as Figure 8. During periodic extreme rainfalls accumulated material is flushed from the forested ecosystem, and some unknown portion of that is probably sequestered in the deep sea. Thus over time these disturbed and recovering ecosystems pull C out of the atmosphere, sequester it in biomass and then necromass following disturbance, and then export some small part of this, about 95 ($\pm$ 15) kg ha^{-1} y^{-1}, mostly in extreme hydrological events. In theory this value should be equal to measured C exports at the bottom of the watershed, but the field data measuring this process are not yet summarized.

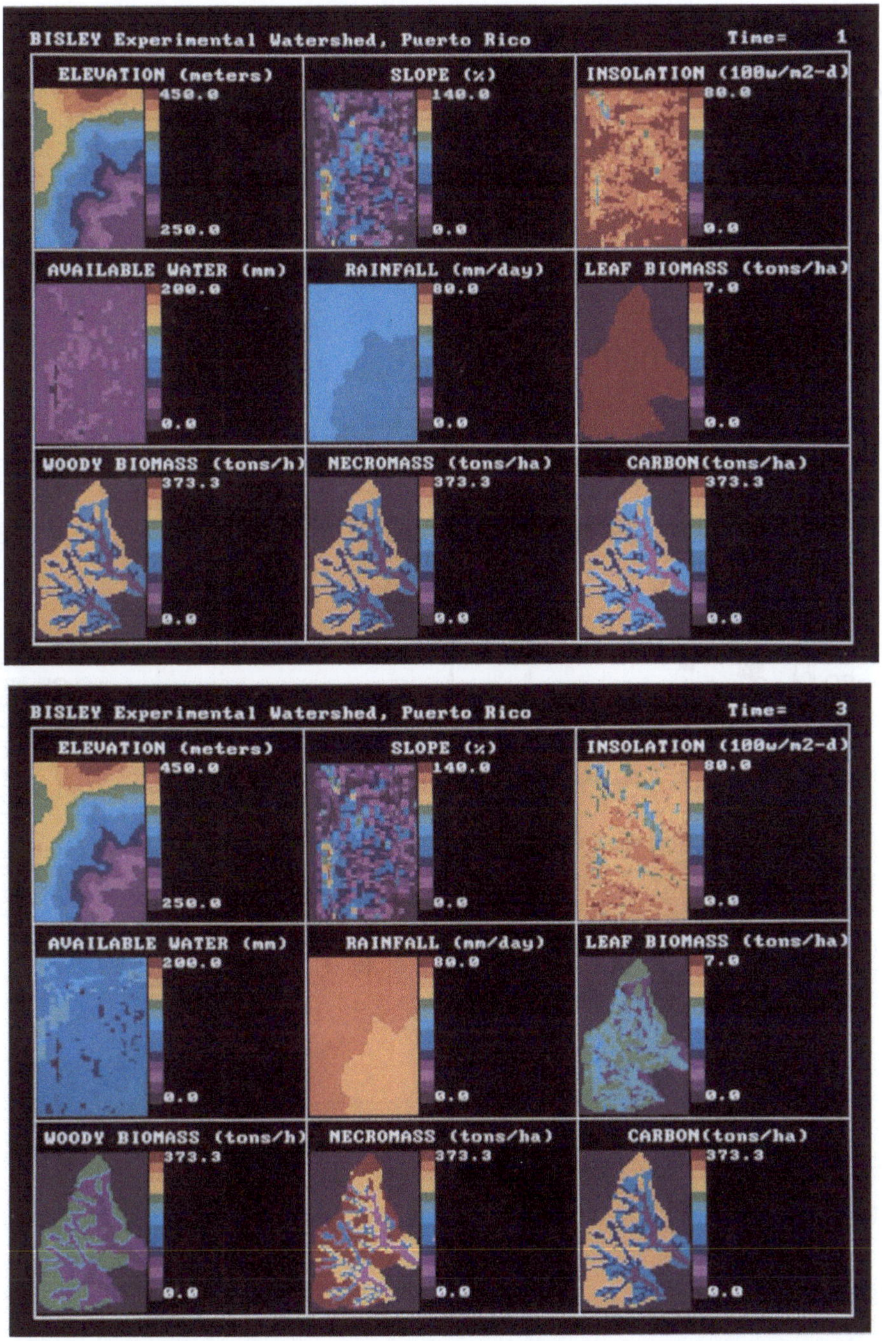

Figure 6. Geographical representation of major foraging functions and biotic response for Bisley watershed before (top-August) and after (bottom-September) Hurricane Hugo. Dark areas in lower panels represent areas where vegetation was not sampled. Values are weekly averages. Dendritic patterns in lower panel represent different community types as in Table 1. Note transition of biomass to necromass following hurricane.

5. Discussion

The quality of our results is obviously a function of the quality of the input data, which at this time is crude. In particular our results to date do not take advantage of the great sophistication of the spatial data that we have and will be synthesizing in future years. Over time we will develop much more sophisticated models, although it is not clear to what degree this will change or increase the accuracy of our analysis. Nevertheless, even this crude model supports the concept that tropical forests can operate to pump CO_2 from the atmosphere into biomass and necromass and eventually

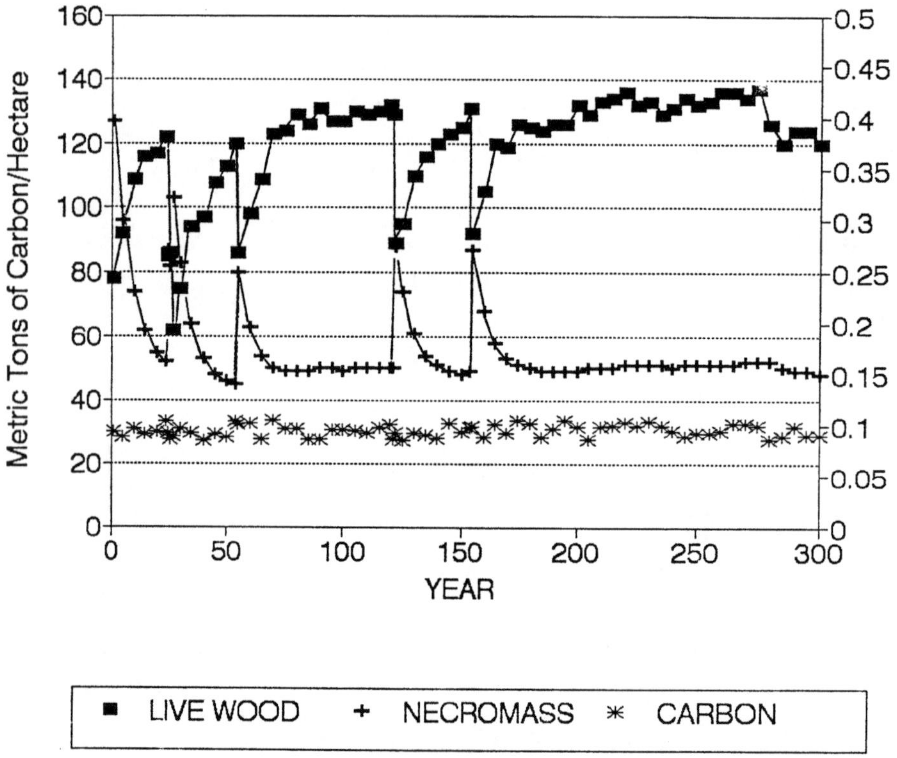

Figure 7. Simulated long term C storage in biomass and necromass of Bisley watershed, along with annual discharge of C in streams. Simulation starts just after a hurricane, and then runs for 300 years with random hurricanes (at a frequency of one per 60 years (although this run received many!) and random annual rainfall.

into the sea. Richey (personal communication) believes that his earlier estimates (Richey 1982; Degens et al. 1991) of 0.4 Gt transported per year from rivers to the sea was too low by perhaps a factor of two. This proposed "C pump" is nearly sufficient to explain the sink required by Tans et al. (1990) if the C is sequestered in the sea. Our discharge rate of 95 kg/ha/yr would be 0.68 Gt yr^{-1} extrapolated to all of the world's 7.2 10^{9} ha of forests and savannas. This is within the range of Richey's estimates of C discharged in rivers. Thus Lugo and Brown's proposed mechanisms for forests acting as sink is supported by this analysis.

6. Conclusions

We have outlined a geography-based ecosystems model which incorporates meteorologic, hydrologic and biotic response models into an integrated structure. Several geographic information and interactive graphic tools were used for data development and presentation of model results, as exemplified by Figure 1. Our approach involved attention to interdisciplinary model integration and to improving the modeling of both spatial and temporal variation. We believe that the approach has considerable merit.

An important gauge of the value of our efforts is whether or not the model, as formulated, can be reasonably applied to different areas. The meteorologic models are simple in theory and should be applicable to any region for which adequate data exists. Development of the hydrologic model emphasized parameters based on fundamental physical characteristics rather than modeling parameters which must be derived by analysis of an historical record. As such, the hydrologic model should be generally applicable in any region, however, the lack of a "below the root zone" groundwater model certainly limits its applicability in many areas. Also, improvements are necessary in the data connections between the hydrologic and biotic response models. The evapotranspiration estimates of the hydrologic model (simple functions of temperature, radiation and leaf area index) need to be replaced by more accurate, biotic-response specific, functions. Finally, at the current time, the authors can give few guidelines on the estimation of biotic response rate constants through means other than extensive field observation. Our continuing efforts will be placed in the following areas:

1) development of improved verification and validation procedures,

2) application to other areas of the Luquillo mountains and elsewhere,

3) preparation of better guidelines for both the obtaining data for both the hydrologic and biotic response models,

4) improvements in the computational efficiencies of the model superstructure,

5) development of better data bases for the C model,

6) finally, and most important, we need much better information on how tree growth responds to environmental variables. We are planning to do this based on Hall et al. (in press) which gives procedures now for estimating biotic response to meteorological variables.

In the meantime, these early model results support the concept that natural forests can act to generate sinks for atmospheric C on a scale of the right order of magnitude to help solve the "missing C" question.

Acknowledgments

We thank Nadia Iwachiw for manuscript and final copy production, Will Ravenscroft for help with figures and three anonymous reviewers. Funding provided in part by the National Science Foundation (85R-8811902), The University of Puerto Rico (Center for Energy and Environment Research), the United States Department of Agriculture (Southern Forest Research Station, Institute of Tropical Forestry) and United States Department of Energy (DEFG02-90ER61080).

References

Arnold, J.G., Williams, J.R., Nicks, A.D. and Sammons, N.B. (1989) 'SWWRB, a basin scale simulation model for soil and water resources management', Texas A & M University Press.

Bacastow, R. and Keeling, C. (1973) 'Changes from A.D. 1700 to 2070 as deduced from a geochemical model' in G.M. Woodwell and E.V. Pecan (eds.), Carbon and the Biosphere, 24th Brookhaven Symposium in Biology, Upton NY, pp. 86-135.

Broecker, W.S., Takahashi, T., Simpson, H.J. and Peng, T.-H. (1979) 'Fate of fossil-fuel carbon dioxide and the global carbon budget', Science 206, 409.

Brown, S., Lugo, A.E., Silander, S. and Liegel, L. (1983) 'Research History and Opportunities in the Luquillo Experimental Forest', USDA/USFS: New Orleans LA.

Costanza, R., Sklar, F.H. and White, M.L. (1990) 'Modeling coastal landscape dynamics', BioScience 40, 91-107.

Degens, E.T., Kempe, S. and Richey, J.E. (eds.) (1991) Biogeochemistry of Major World Rivers. John Wiley and Sons, 356 p.

Detwiler, R.P. and Hall, C.A.S. (1988) 'Tropical forests and the global carbon budget', Science 239, 42-47.

Everham, E.M., Wooster, K.B. and Hall, C.A.S. (1991) 'Forest landscape climate modeling', in M.A. Buford (ed.). Proceedings of the 1991 Symposium on Systems Analysis in Forest Resources USFS. Southeastern Forest Exp. Station. Technical paper SE-74. Ashville NC.

French, P.N. and Taylor, M.R. (1982) 'Interactive management and updating of spatial data bases', Proceedings of the National Conference on Energy Resource Management, Baltimore MD.

Loucks, D.P., French, P.N. and Taylor, M.R. (1986) 'Interactive modeling and display of spatial and time-varying phenomena using CAPLIB', Proc. 4th Annual ASCE Conference on Computing in Civil Engineering, ASCE, Boston.

Lugo, A.E. and Brown, S. (1986) 'Steady state terrestrial ecosystems and the global carbon cycle', Vegetatio 68, 83-90.

Hall, C.A.S., Detwiler, R.P., Bogdonoff and Underhill, P.S. (1985) 'Land use change and carbon exchange in the tropics: I. Detailed estimates for Costa Rica, Panama, Peru, and Bolivia', Environmental Management. 9, 313-334.

Hall, C.A.S. and Uhlig, J. (1991) 'Refining estimates of carbon released from tropical land-use change', Can. J. For. Res. 21, 118-131.

Hall, C.A.S., Stanford, J.S. and Hauer, R. (in press) 'The distribution and abundance of organisms as a consequence of energy balances along multiple environmental gradients', Oikos.

Houghton, R.A., Hobbie, J.E., Melillo, J.M., Moore, B., Peterson, B.J., Shaver, G.R. and Woodwell, G.M. (1983) 'Changes in the carbon content of terrestrial biota and soils between 1860 and 1980: A net release of CO_2 to the atmosphere', Ecological Monographs 53, 235-262.

Odum, H.T. Summary, an emerging view of the ecological system at El Verde in Odum, H.T. and Pigeon, R.F. (1970) 'A tropical rain forest: A study of irradiation and ecology at El Verde, Puerto Rico', U.S. Atomic Energy Commission, Oak Ridge TN.

Revelle, R. and Suess, H.E. (1957) 'Carbon dioxide exchange between atmospheric and ocean, and the question of an increase of atmospheric C02 during the past decades', Tellus 9, 18-27.

Richey, J.E. (1981) 'Fluxes of organic matter in rivers relative to the global carbon cycle. in G. Likens, F. Mackenzie, J. Richey, J. Sedell, and K. Turekian, ed., Flux of Organic Carbon by Rivers to the Oceans. National Research Council/DOE Conf. 8009140.

RPA (Resources Planning Associates, Incorporated) (1991) *MPA Series Users's Guide. Ithaca NY.

Scatena, F.N., Silver, W., Siccima, T. and Sanchez, M.J. (unpublished manuscript) Biomass and nutrient content of the Bisley experimental watersheds, Luquillo Experimental Forest, Puerto Rico.

Taylor M.R., French, P.N. and Loucks, D.P. (1983) 'Digital color mapping and resource planning', Proceedings of the First National Conference on Microcomputers in Engineering, ASCE, Orlando FL.

Tans, P.P., Fung, I.Y., Takahashi, T. (1990) 'Observational constraints on the global atmospheric CO2 budget', Science (Washington DC) 247, 1431-1438.

U.S. Department of Agriculture. (1972) Soil Conservation Service. National Engineering Handbook, Hydrology Section 4.

Walker, L.R., Lodge, D.J., Brokaw, N.V.L. and Waide, R.B. (1991) 'Introduction to hurricanes in the Caribbean', Biotropica 23, 313-316.

Williams, J.R., Nicks, A.D. and Arnold, J.G. (1985) 'Simulator for water resources in rural basins', J. Hydraulic Engineering, ASCE 111, 970-986.

Wooster, K. (1989). 'A geographically-based microclimatological computer model for mountainous terrain with application to the Luquillo Experimental Forest, Puerto Rico'. MS Thesis, SUNY ESF, Syracuse NY.

POTENTIAL OF MARINE MACROALGAE AS A SINK FOR CO_2: CONSTRAINTS FROM A 3-D GENERAL CIRCULATION MODEL OF THE GLOBAL OCEAN

James C. Orr and Jorge L. Sarmiento
Program in Atmospheric and Oceanic Sciences
Princeton University
Princeton, NJ 08544-0710

ABSTRACT. A 3-D global ocean model used previously to determine natural oceanic uptake of anthropogenic CO_2 is used here to evaluate another proposed strategy for mitigation of rising atmospheric CO_2. As a reference, this study bases itself on previous efforts with the same model to evaluate the potential of Fe fertilization as a means to enhance oceanic CO_2 uptake. From that base, we test the feasibility of slowing the rise in atmospheric CO_2 by enhancing growth of seaweed, a proposal resurrected from previous efforts considering it as a means to grow marine biomass as fuel for energy production. To determine its maximum potential, logistical and financial constraints are ignored. An enhanced growth of 1 GT C yr^{-1} is prescribed to be evenly distributed over a large ocean area such as the equatorial band from 18°S to 18°N and the northern and southern subtropics from 18° to 49° latitude. Results from these simulations clearly demonstrate that the CO_2 invasion from the atmosphere is substantially less than C removed from the surface via enhanced growth. When enhanced growth is supported only by naturally available nutrients, the enhancement to the air to sea CO_2 flux averages 0.2 GT C yr^{-1} for the first 100 yr. When nutrients are supplied artificially to support the enhanced growth, the mean enhanced air to sea flux is more (for the first 100 yr it averages 0.72 GT C yr^{-1} when all enhanced growth is harvested but only 0.44 GT C yr^{-1} without harvesting); however, generating enhanced marine growth at 1 GT C yr^{-1} requires an unreasonably large supply of nutrient—close to the world's current rate of fertilizer production for P and substantially more than that for N. Less nutrient is needed if the enhanced algal growth is not harvested and thus respired, but respiration increases demand for oxygen so that significant anoxia develops. We conclude that growth of macroalgae is an inefficient mechanism for sequestering anthropogenic CO_2 and that the use of macroalgae as an additional fuel source will actually result in a net transfer of CO_2 from ocean to atmosphere; however, there would be a reduction in the atmospheric CO_2 increase rate if macroalgae were used as a partial replacement for fossil fuel.

1. Introduction

Since the beginning of the industrial revolution, more than half the fossil CO_2 emitted to the atmosphere has stayed there, with much of the rest having been transferred to the ocean. Transfer occurs by gas exchange at the air-sea interface and is enhanced by hydrolysis of CO_2. This reaction of CO_2 with water produces both HCO_3^- and CO_3^{-2} to the extent that in seawater, out of all three dissolved species, less than 1% remains as CO_2. The relative abundance and the equilibria between these inorganic C species is largely responsible for the unusual behavior of CO_2 with respect to its partitioning between air and sea (98.5% resides in the ocean); most of the other gases in the atmosphere, such as N_2, O_2, and Ar, are significantly more abundant in the atmosphere (~1% of each is found in the ocean).

Water, Air, and Soil Pollution **64**: 405–421, 1992.
© 1992 *Kluwer Academic Publishers.*

Thus in the global C cycle, the ocean naturally acts as a significant sink. The ocean's role as a sink for CO_2 has been investigated with various ocean models ranging in complexity from those such as the box-diffusion model (Oeschger et al., 1975) and the outcrop-diffusion model (Siegenthaler, 1983) to three-dimensional general circulation models (GCM's) of the world's oceans (Maier-Reimer and Hasselman, 1987; Bacastow and Maier-Reimer, 1990; Sarmiento et al., 1992). These ocean models all agree that oceanic CO_2 uptake is significant: the Intergovernmental Panel on Climate Change (IPCC) summary (Houghton et al., 1990) shows the range in estimates as 2.0 ± 0.8 GT C yr^{-1}.

Recently, Tans et al. (1990) used an atmospheric model, the measured latitudinal gradient in atmospheric CO_2, and measurements of the difference between atmospheric and oceanic pCO_2, to suggest that oceanic uptake was less than 1.0 GT C yr^{-1}. Although the lower estimate by Tans et al. has generated considerable activity to account for their necessarily larger missing sink for CO_2 that must be land based, it now appears that the discrepancy between estimates may not be significant after some corrections are applied. Sarmiento and Sundquist (1992) suggest that the Tans et al. estimate should be raised to range from 1.1 to 2.4 GT C yr^{-1} based on several corrections: the ocean's cooler surface skin temperature, CO production, its atmospheric transport, and its subsequent oxidation, as well as river transport of both organic and inorganic C. Further discussion regarding other differences between estimates of oceanic CO_2 uptake is provided by Sarmiento et al. (1992).

Prompted by the ever-increasing concentration of CO_2 in our atmosphere, it is only natural that many look at the ocean now with the hope of enhancing its already significant role as a CO_2 sink. Proposed enhancement strategies have taken on various forms, but all rely on the physics of the ocean to either actively supply essential nutrients or to passively sequester additional CO_2.

1.1. DEEP-SEA INJECTION OF CO_2

One strategy involves injection of anthropogenically produced CO_2 directly into the deep sea (Marchetti, 1977), whereby it would be insulated from the air-sea interface for an extended period because of the density structure and circulation pattern of the oceans; if this CO_2-enriched water then contacted calcite-rich sediments, the capacity for uptake would be further enhanced. Modeling efforts (Maier-Reimer, 1991a; Bacastow and Stegen, 1991; Stegen et al., 1991) have shown such a scenario to be viable even when up to one third of of all anthropogenic CO_2 expected to be produced during the next century is injected into the deep ocean.

1.2. IRON FERTILIZATION

Another strategy that has recently received much attention is based on the hypothesis of Martin et al. (1990) that states that the large nutrient-rich regions of the surface ocean, namely the Southern Ocean, and the North and equatorial Pacific, are not depleted of nutrients because they lack Fe as an essential micronutrient. This hypothesis also presumes that during the past, increased supply of Fe (via increased atmospheric dust supply) has been linked to increased biological productivity and thus greater export of C to the deep. Martin (1990) supports his hypothesis with evidence found in ice cores.

Almost since its inception, the Fe limitation hypothesis has been the subject of considerable controversy because these nutrient-rich regions of the surface ocean have long been an enigma and may just as well be explained by alternative hypotheses (e.g., light limitation, zooplankton grazing, and temperature effects). However, if substantial Fe limitation is prevalent, then large regions of the ocean could be fertilized with Fe as a means to reduce surface nutrients, presuming the process could be made cost effective and logistically feasible. This reduction in surface nutrients would be accompanied by a corresponding reduction in the partial pressure of CO_2 (pCO_2) in surface seawater, increasing the CO_2 difference between air and sea, and hence enhancing the flux of CO_2 into the ocean by gas exchange.

Before going further, a simple yet important concept should be made clear: when a given amount of organic C is exported from the surface to deep ocean, an equivalent amount of C will not invade the surface ocean from the atmosphere. As mentioned previously, one must also consider the result of equilibrium-driven shifts between all three organic C species (CO_2, HCO_3^-, and CO_3^{-2}). To do this, we employ a perturbation model for the carbonate chemistry of seawater that relates surface ocean changes in pCO_2 to changes in ΣCO_2, .i.e., the sum of all three inorganic C species (Sarmiento et al., 1992; Sarmiento and Orr, 1991). Proper consideration of the potential for Fe fertilization and its effect upon uptake of CO_2 must therefore consider the ocean's chemistry as well as its biology.

Also, one must consider ocean physics, which determines the rate at which deep waters with additional capacity for taking up CO_2 are exposed at the surface. For this purpose, Peng and Broecker (1991a,b) and Joos et al. (1991a,b) have employed box models while Sarmiento and Orr (1991) and Maier-Reimer (1991b) have performed analogous simulations with 3-D GCM's.

After 100 yr of successful Fe fertilization, Joos et al. find an increased atmospheric CO_2 perturbation due to fertilization of -107 ppm in the IPCC "business-as-usual" CO_2 emission scenario and -90 ppm in a "constant-emission" scenario; Sarmiento and Orr (1991) find -72 ppm and -61 ppm for the same scenarios, respectively. Over the next century then, with the same two scenarios but without nutrient depletion, the total increase in atmospheric pCO_2 (i.e., 431 ppm for business-as-usual and 148 ppm for constant emission), would be reduced by ~20% and ~50%, respectively, if Fe fertilization were completely effective. Note that the greatest reduction in the increase is that from 431 ppm to 148 ppm which occurs by controls of emissions.

Sarmiento and Orr (1991) show that one side effect to any successful nutrient depletion in the Southern Ocean is the corresponding reduction in mean oceanic oxygen by 16 μmol kg^{-1}, a drop of nearly 10%, due to increased respiration at depth combined with loss of O_2 from the surface ocean to the atmosphere through gas exchange. Loss of oxygen results in anoxia at 1600 m in the southwestern Indian Ocean.

Sarmiento and Orr also discuss possible another side effect, i.e., production of N_2O gas during remineralization by the process of nitrification. N_2O is itself a greenhouse gas, so its enhanced production in the ocean and escape to the atmosphere is counterproductive to a process designed to reduce atmospheric concentrations of greenhouse gases. Furthermore, the radiative forcing of N_2O is 206 times more potent per unit mass than for CO_2 (Houghton et al., 1990). The perturbation atmospheric N_2O produced by nutrient depletion in the Southern Ocean would represent a reduction in the benefit derived by removing atmospheric CO_2 via the same nutrient depletion scenario.

Given these side effects, and others such as the fact that the total oceanic biological production would double but with all the effect concentrated in a small area of the ocean, and that it seems unlikely to be able to achieve such large uptakes as estimated by these extreme scenarios of nutrient depletion, all the groups who did the iron fertilization studies concluded that the iron fertilization scenario was not likely to play a role in strategies for controlling future growth of atmospheric CO_2.

1.3. ENHANCED ALGAL GROWTH

Like Fe fertilization, enhanced algal growth has been proposed as a means to mitigate rising atmospheric CO_2. In principle, the mechanism for enhanced algal growth is similar to that for Fe fertilization because both invoke enhanced growth of plants in the sea, and both, as proposed, are fueled by the abundance of nutrients found at the surface in some areas of the ocean. The idea of enhancing growth of marine algae is not new, in fact, it has been previously considered as a means to grow biomass to serve as fuel for energy production (e.g., Bird, 1987; McKay, 1983).

For the most part, proposals for enhanced algal growth only call on macroalgae. Enhanced growth of marine macroalgae has been considered most thoroughly in coastal and near-coastal regions, i.e., in waters up to 200 m in depth. Deeper waters are not as favorable because of logistics for tending growth, in kelp-like beds or in the water column with the macroalgae held in some confining structure but moored in a fixed position to the sediment. Adey (1984) has floated mats made of plastic mesh at sea, to which seaweed spores, present naturally in seawater, attached and grew at rates many times higher than found for primary productivity in the open ocean. North (1987) has also done experiments with a floating open ocean kelp farm. Reviews on this subject can be found in Glenn et al. (1991) and Ritschard (1992).

Although practical experiments have been confined to ocean depths less than 200 m because of the mooring requirement, seaweed farms in the open ocean are considered feasible (Ryther, 1984). If moorings would prove unnecessary, and if other logistics could be worked out and made cost effective, maximum potential for growth could be realized by floating seaweed farms over much of the open ocean.

As with Fe fertilization, a realistic model considering such enhanced growth of macroalgae in the ocean must include not only biological uptake but also ocean physics and chemistry. We report here development of such a model to determine the maximum effect for the role of enhanced algal growth as a mechanism for mitigating rising levels of atmospheric CO_2. To determine the maximum potential, financial and logistical constraints are ignored.

Whether or not potential exists for mitigating rising atmospheric CO_2, it is important to emphasize once more (as mentioned for Fe fertilization) that these simulations of enhanced growth allow clear demonstration of one basic principle that seems often overlooked: all C lost from the surface layer via enhanced growth (or enhanced new production) is not replaced by an equivalent invasion of CO_2 from the atmosphere. One must also consider oceanic carbonate chemistry, ocean physics, and to a lesser degree, gas exchange. The model described herein does all this.

Furthermore, the model determines the extent to which enhanced algal growth is possible given the natural availability of nutrients in the surface ocean. If nutrient supply is insufficient, nutrients must then be added artificially if substantial growth is to be achieved. Also, the model determines the oxygen demand for scenarios where the biomass produced from the enhanced growth is not harvested. It seems likely that harvesting would not be feasible in the open ocean; hence, the biomass from the enhanced growth would be respired either deeper in the water column or after reaching the sediments. Therefore, the model determines the extent of oxygen depletion for scenarios with no harvesting.

2. Model Description

This study employs a primitive-equation, world-ocean GCM based on the model developed by Bryan (1969) at the Geophysical Fluid Dynamics Laboratory (GFDL) in Princeton. It consists of 12 vertical levels with resolution of 4.5 degrees latitudinally and 3.75 degrees longitudinally. Its maximum depth is 5000 m with bathymetry as reasonable as allowed by the grid resolution. The model is forced by annual mean observed temperature and salinity at the surface (Levitus, 1982) and annual mean wind stress from Hellerman and Rosenstein (1983). More of its specifics as applicable to this study can be found in Sarmiento et al. (1992), Sarmiento and Orr (1991), Najjar(1990), and Toggweiler et al. (1989a,b).

As with our 3-D modeling study for Fe fertilization (Sarmiento and Orr, 1991), we use a perturbation approach developed for CO_2 by Sarmiento et al. (1992) and first extended for use with nutrients by Joos et al. (1991a,b). For the Fe fertilization study, we adapted the perturbation method to the basic modeling approach developed by Najjar (1990) for work in the 3-D GCM, then expanded it to include the N_2O tracer and to include additional effects accounting for anoxia. Unlike the simulations for Fe fertilization, all simulations reported herein necessarily assume that either Fe is not a limiting micronutrient, or if so, that it is supplied artificially in sufficient quantity to the region of enhanced growth so as not to limit growth.

To determine the potential for enhanced growth of seaweed in the ocean with naturally available nutrients, this work employs a model identical to that used for nutrient depletion studies, including efforts to evaluate the feasibility of Fe fertilization (Sarmiento and Orr, 1991). For this case, as with Fe fertilization, surface nutrients are forced to zero in the region of enhanced productivity and to observations elsewhere. Other simulations in this study address the feasibility of enhancing growth further by adding nutrients at the surface in the region of enhanced growth. In this case, the total enhanced growth rate is specified a priori (as 1 GT C yr^{-1}) and is spread evenly over large areas of the ocean (e.g., over the entire equatorial band from 17.78°S to 17.78°N).

3. Results of Enhanced Algal Growth in a 3-D Ocean GCM

3.1. GROWTH WITH NATURALLY AVAILABLE NUTRIENTS

3.1.1. *At the Equator.* Insight into the capacity of the ocean for enhanced algal growth is gained by first considering enhanced growth with only naturally available nutrients (the same as for the change in new production due to Fe fertilization). When enhanced growth is carried out with naturally available nutrients over all ocean area in the equatorial band, the net growth over the whole ocean is reduced, −0.2 GT C yr^{-1} after 100 yr (Table 1), a rather surprising result. This reduction in growth contrasts sharply with the analogous simulation (Fe fertilization) carried out in the Southern Ocean (south of 31.12°S) where, after 100 yr, the mean-ocean enhancement was 13.2 GT C yr^{-1} (Sarmiento and Orr, 1991).

To understand the difference, one must look beyond the main similarity between both regions—high nutrients in surface waters. Instead this difference is partially explained by differences in the rate of nutrient supply from nutrient-rich waters below. In fact, much of the Southern Ocean sits atop convective plumes which provide rapid communication between the surface and nutrient-rich waters at depth; the equatorial surface waters are relatively light and the density structure there isolates these waters from most of the rest of the nutrient-rich, deeper ocean. Furthermore, one must consider the relative amount of nutrient depleted water that escapes the region of enhanced growth and later makes its way to surface waters either in the same region or elsewhere. This negative new production is discussed below.

Figure 1 exhibits the history of enhanced growth for the simulation of nutrient depletion at the equator. At initialization, enhanced growth quickly peaks to very high values of 40 GT C yr^{-1}, but

Table 1. Changes in average atmospheric pCO_2, oceanic O_2, and oceanic PO_4^{-3} for enhanced growth supported only by naturally available nutrients after 100 yr with atmospheric CO_2 forced by the IPCC business-as-usual scenario.

	Atm. pCO_2 Perturbation (ppm)	Oceanic O_2 Perturbation (μmol Kg^{-1})	Oceanic PO_4^{-3} Perturbation (μmol Kg^{-1})	Enhanced Growth (GT C yr^{-1})
Not Harvested				
Northern Subtropics (17.78°N–48.90°N)	−8.8	−0.80	–	0.47
Equator (17.78°S–17.78°N)	−2.8	+2.22	–	0.21
Southern Subtropics (48.90°S–17.78°S)	−28.1	−5.41	–	2.04
Harvested				
Northern Subtropics (17.78°N–48.90°N)	−8.8	+0.04	−0.01	0.10
Equator (17.78°S–17.78°N)	−2.2	+1.49	−0.01	0.05
Southern Subtropics (48.90°S–17.78°S)	−31.5	+0.02	−0.04	0.37

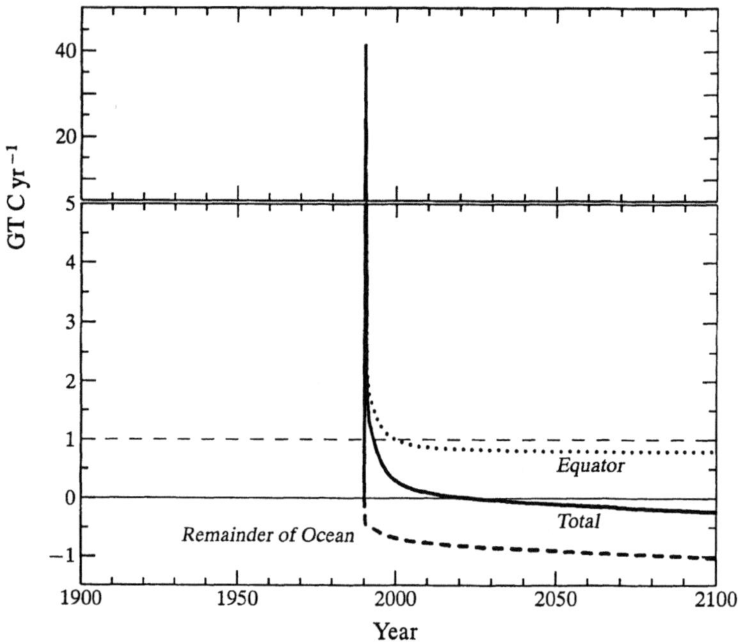

Figure 1. Time history for the rate of change of enhanced growth at the equator (dots), perturbation new production elsewhere (dashes), and the whole ocean total (solid) for the simulation of enhanced growth at the equator (17.78°S to 17.78°N) with only naturally available nutrients. Enhanced growth at the equator is more than matched by reduced new production elsewhere to the extent that the net enhancement over the whole ocean decreases.

it falls off rapidly on the time scale of months as surface nutrient concentrations drop, a function of their being forced towards zero with a 30-day restoring time constant (Sarmiento and Orr, 1991). After the nearly instantaneous rise and rapid fall, changes are more gradual: growth at the equator falls from 1.0 GT C yr^{-1} after about 10 yr to about 0.8 GT C yr^{-1} after 100 yr. Outside the equatorial region of enhanced growth, the mean perturbation to the new production is negative, an effect attributable to a general loss in the supply of surface nutrients to these outer regions as forced by the ongoing depletion at the equator and subsequent migration of these nutrient-poor waters from the equator to surrounding waters. After ~35 yr, the net effect results in lower new production over the whole ocean, partially a consequence of sequestering significant nutrients below the regions which resupply surface waters to the area of enhanced growth, a particularly important phenomenon for the equatorial Pacific.

The result of enhanced growth is some enhanced uptake of atmospheric CO_2 (an additional 2.8 ppm after 100 yr) as shown in Figure 2; enhanced oceanic uptake of atmospheric CO_2 for the analogous simulation of Fe fertilization in the Southern Ocean (Sarmiento and Orr, 1991) is much greater (72 ppm after 100 yr).

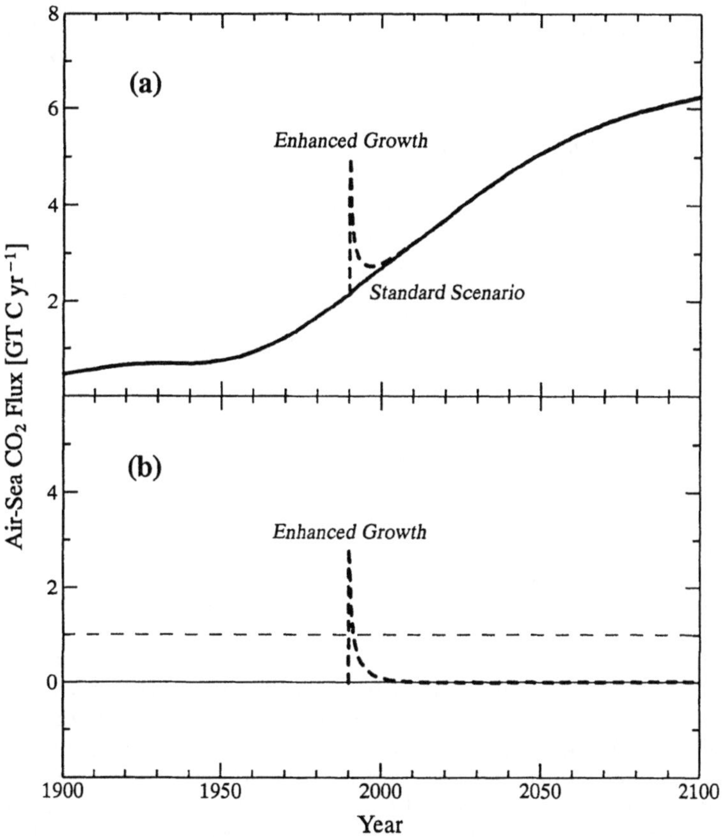

Figure 2. Total ocean perturbation flux of CO_2 from atmosphere to ocean for enhanced growth at the equator as supported only by naturally available nutrients. Both (a) the total ocean perturbation flux and (b) the flux due only to enhanced growth indicate that significant flux occurs only within the first 10 yr.

For the simulation with enhanced growth at the equator, the enhanced oceanic uptake of atmospheric CO_2 is only significant within the first 10 yr of fertilization; afterwards the net effect is negligible. Note that the mean air-sea CO_2 flux does not match the mean enhanced growth (i.e., invasion of atmospheric CO_2 is substantially less). Again, the discrepancy is explained by the buffer capacity of seawater for CO_2 (e.g., oceanic pCO_2 changes only ~10% per 100% change in ΣCO_2) and low mixing between the surface and deep ocean at the equator.

From Figure 1 it is clear that the supply of nutrients to the surface of the equatorial region is insufficient to support a substantial increase in the mean global-ocean new production, partially because waters depleted in nutrients (and deriving from the region of enhanced growth) eventually make their way into surrounding waters and substantially reduce new production there. After the first few years of enhancement at the equator, one sees that any net growth over the whole-ocean requires equivalent input of nutrients artificially.

3.1.2. *In the Northern Subtropics.* Figure 3 shows the same analysis but for the scenario where growth of macroalgae is enhanced over all the northern subtropical ocean (17.78°N to 48.90°N). Net enhancement is greater in this simulation than for that with enhanced growth at the equator (see Figure 1). That is, the total of the enhancement (forced by the nutrient supply into the growing region) minus the degradation (forced by depleted supplies of surface nutrients for surrounding regions—a product of enhanced growth) is more positive when growth occurs in the northern subtropics. In fact, unlike with growth at the equator, growth in the northern subtropics produces a net enhancement over the whole ocean that remains positive throughout the 110 yr simulation; however, despite greater net enhancement, it remains substantially less than 1 GT C yr^{-1}. To reach this higher target, at least the residual must be supplied artificially.

Figure 4 shows the resulting enhancement to the air-sea CO_2 flux. It is larger for this simulation with growth in the northern subtropics relative to the that with growth at the equator, but the flux is only substantial over the first 20 yr; subsequently, it declines gradually until after 100 yr it is only ~0.1 GT C yr^{-1}.

3.1.3. *In the Southern Subtropics.* As a reference, Table 1 also provides results for a simulation of enhanced growth in the southern subtropics (48.90°S to 17.78°S). Resulting net enhancement over the whole ocean (2.0 GT C yr^{-1} after 100 yr) is much higher relative to previously mentioned simulations where growth is enhanced in the northern subtropics or at the equator. However, the total net enhancement for this simulation of enhanced growth in the southern subtropics is still much lower than for the simulation where growth is enhanced throughout the entire Southern Ocean south of 31.12°S (13.2 GT C yr^{-1} after 100 yr), as detailed for Fe fertilization by Sarmiento and Orr (1991). These differences are largely explained by the relative proportion of the Southern Ocean contained within the region of enhanced growth. Within the Southern Ocean, the indigenous nutrient-rich surface waters and high frequency of convection (the mechanism by which surface waters communicate rapidly with the large pool of nutrient-rich water at depth) allows for maximum enhanced production. The southern subtropics—as defined for this simulation—overlaps much of the nutrient-rich, highly-convective Southern Ocean; hence, net enhanced growth there easily surpasses the target rate of ~1 GT C yr^{-1}, at least for the first 110 yr, i.e., the duration of these simulations. However, as will be discussed later, the vicinity of the Southern Ocean is perhaps the most implausible for successful implementation of enhanced growth of macroalgae.

3.1.4. *With Harvesting.* Previously mentioned results derive from simulations which consider that biomass produced from enhanced growth degrades deeper in the water with the same scaling as

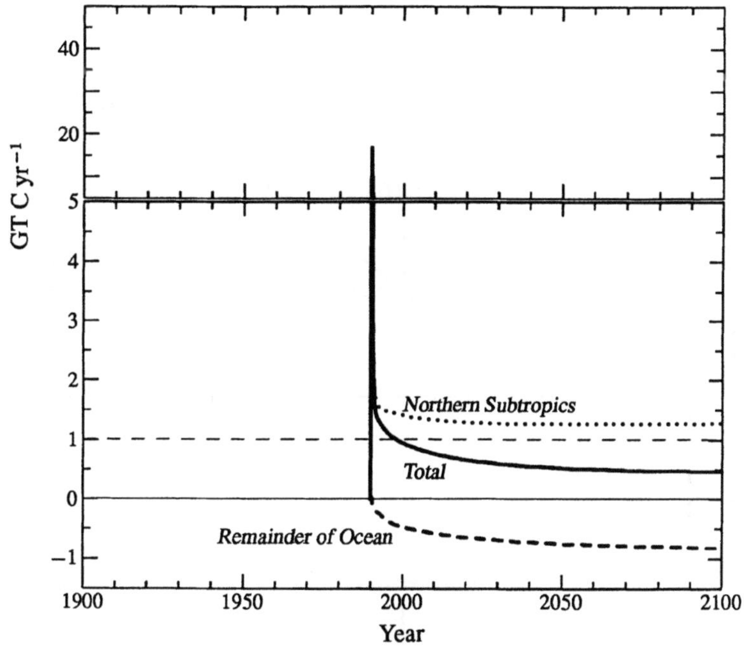

Figure 3. Time history for the rate of change in enhanced growth in the northern subtropics (dots), perturbation new production elsewhere (dashes), and the whole ocean total (solid) for the simulation of enhanced growth in the northern subtropics (17.78°N to 48.90°N) with only naturally available nutrients. The net enhancement over the whole ocean is substantial but still much less than 1 GT C yr^{-1}.

described by Sarmiento and Orr (1991) for Fe fertilization. If biomass were degraded in place within the surface layer, the corresponding production of CO_2 would counter any CO_2 reduction due to enhanced growth; therefore, in this scenario, any effort towards enhanced growth would not be worthwhile. Alternatively, the biomass from enhanced production could theoretically be harvested. Table 1 also reports results from simulations when growth is harvested, i.e., not respired. One advantage to this approach is that without respiration, the process of enhanced growth does not produce CO_2 at depth; hence, there would be no CO_2 produced from the enhanced growth that could eventually make it to the surface and escape to the atmosphere. However, a disadvantage also results because harvesting of marine organic matter represents a net loss of nutrient from the ocean.

Results show that when all macroalgae is harvested, the net enhanced growth over the whole ocean is substantially less for the simulations in the northern and southern subtropics relative to simulations without harvesting; for harvesting at the equator, net growth remains low but is no longer negative. Despite such relatively large changes in net enhanced growth, Table 1 reveals that the enhanced flux of invading atmospheric CO_2 remains nearly constant whether or not macroalgae is harvested. Again enhanced growth cannot be tied too closely to the enhanced air-sea CO_2 flux. Note that if the harvested macroalgae is burned as a supplemental fuel, the result would be addition of CO_2 to the atmosphere; as an alternative fuel, there would be some enhanced CO_2 removal from the atmosphere.

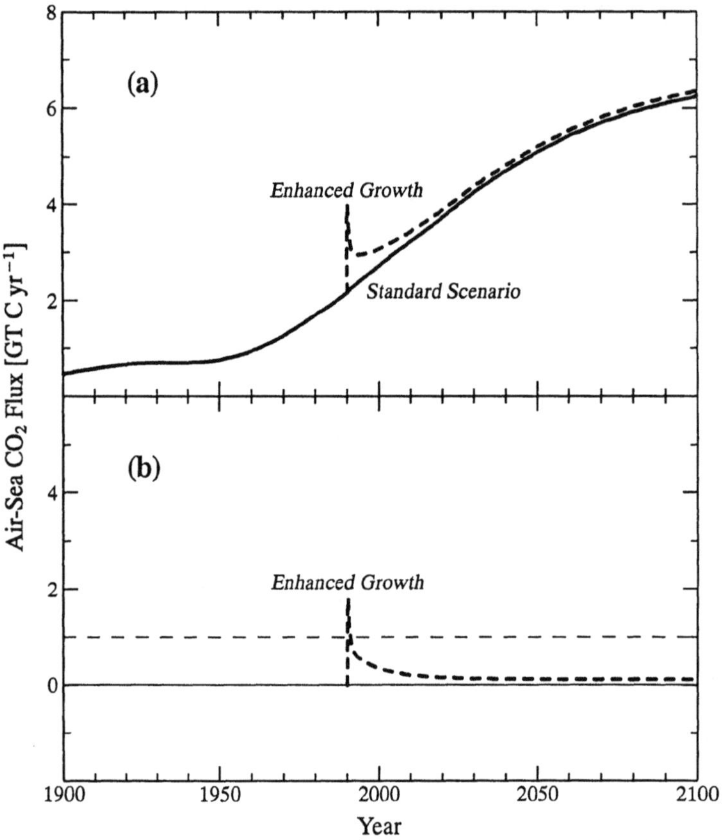

Figure 4. Total ocean perturbation flux of CO_2 from atmosphere to ocean for enhanced growth in the northern subtropics as supported only by naturally available nutrients. Both (a) the total ocean perturbation flux and (b) the flux due only to enhanced growth evidence significance but remain small after the first 20 yr.

3.2. GROWTH WHEN NUTRIENTS ARE ADDED ARTIFICIALLY

With but one exception cited above, if enhanced growth is to occur over the ocean as a whole, nutrients must be added artificially. Thus it is instructive to look at results from model scenarios of enhanced algal growth where the necessary nutrients are supplied artificially.

3.2.1. *With Harvesting.* Table 2 shows results for simulations where all biomass from enhanced growth is harvested. For these and later simulations, total enhanced growth is set to 1 GT C yr^{-1} and is evenly distributed over large bands of the world's oceans. Resulting oceanic uptake of CO_2 (33.8 ± 1.2 ppm) depends little upon the region selected for enhanced growth. After 100 yr, on average 72% of the organic C lost from the surface ocean through enhanced growth has been replenished by an equivalent amount of CO_2 invading the surface ocean from the atmosphere. As clearly demonstrated in Figure 5, even when growth is harvested, the enhanced flux of atmospheric CO_2 invading the ocean never matches the equivalent 1 GT C yr^{-1} removed from the surface by enhanced growth.

Table 2. Changes in average atmospheric pCO_2, oceanic O_2, and oceanic PO_4^{-3} for total enhanced growth at 1 GT C yr^{-1} as supported by artificially supplied nutrients. Atmospheric CO_2 is forced as described in Table 1.

	Atm. pCO_2 Perturbation (ppm)	Oceanic O_2 Perturbation (μmol Kg^{-1})	Oceanic PO_4^{-3} Perturbation (μmol Kg^{-1})
Not Harvested			
Northern Subtropics (17.78°N–48.90°N)	−20.4	−3.91	–
Equator (17.78°S–17.78°N)	−23.0	−4.51	–
Southern Subtropics (48.90°S–17.78°S)	−19.2	−3.87	–
Harvested			
Northern Subtropics (17.78°N–48.90°N)	−33.7	+0.25	−0.04
Equator (17.78°S–17.78°N)	−35.1	+0.15	−0.04
Southern Subtropics (48.90°S–17.78°S)	−32.8	+0.33	−0.03

As for O_2, there is little impact when growth is harvested because no O_2 is respired and most of that produced at the surface is lost to the atmosphere. Although this approach with harvesting produces a larger enhanced air-sea flux of CO_2 with no oxygen deficiencies, it requires more nutrients be added artificially (see Discussion) because the equivalent of 1 GT C yr^{-1} is lost from the ocean. Demand for artificially-supplied nutrients is less if growth is not harvested.

3.2.2. *Without Harvesting.* Also shown in Table 2 are results from simulations where the enhanced growth is not harvested. For these simulations, the biomass from enhanced growth is respired deeper in the water column according to the same scaling as with Fe fertilization (Sarmiento and Orr, 1991). Again the resulting air-sea flux of CO_2 (20.9 ± 1.9 ppm) depends little upon the region of enhanced growth. With 1 GT C yr^{-1} of enhanced growth for 100 yr, on average 0.44 GT C yr^{-1} will be replenished by CO_2 invading the ocean from the atmosphere. Although this flux of CO_2 is lower because growth is not harvested, there is less demand for artificial nutrient supply. For instance, Tables 1 shows, that after 100 yr of enhancement in the northern subtropics, the natural nutrient supply will support 47% of the 1 GT C yr^{-1} growth when remineralization is allowed; without remineralization, the natural nutrient supply can only support about 10%.

However, despite lower nutrient demand, two undesirable side effects result when growth is not harvested, i.e, when respiration is allowed. First, the oxygen demand produces substantial anoxia. Figure 6a shows that for the simulation with enhanced growth at the equator, anoxia extends from the Central America towards the middle of the Pacific and spans several vertical layers of the model from 119 to 914 m; significant anoxia is also found in the Indian Ocean. For enhanced growth in the northern subtropics, anoxia extends over much of the North Pacific spanning vertically from 371 to 1347 m (Figure 6b). Second, during remineralization, this model produces N_2O, another greenhouse gas. It is formed during remineralization by the process of nitrification. Some of this N_2O formed in the ocean escapes to the atmosphere, and once there, acts to increase radiative forcing (as per previous discussion, the radiative forcing of N_2O is $\sim$200 times more potent per unit mass than is CO_2). The model produces a perturbation atmospheric N_2O concentration of

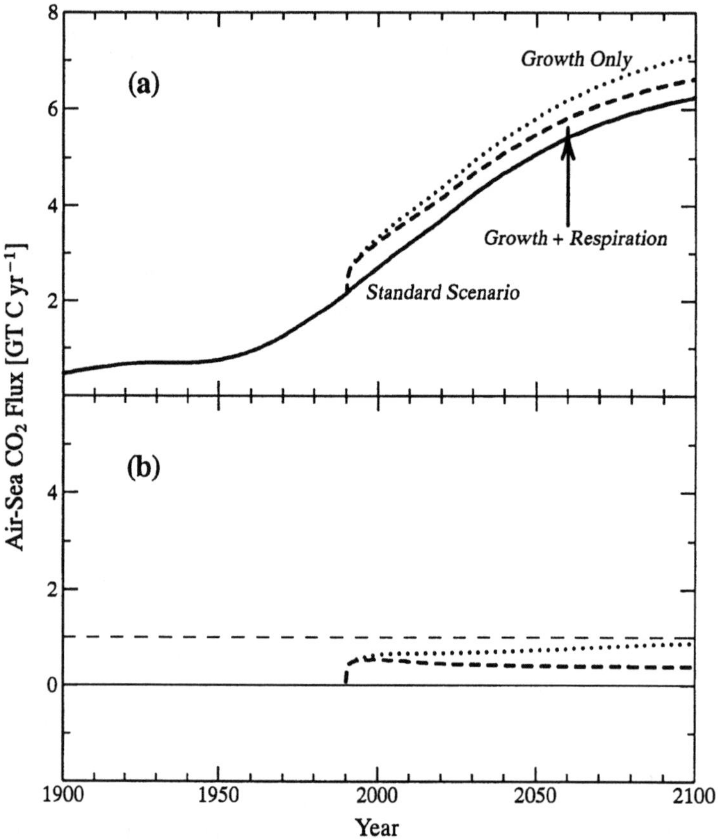

Figure 5. Fluxes of CO_2 from atmosphere to ocean for enhanced growth in the northern subtropics when nutrients are supplied artificially to support growth at a constant 1 Gt C yr^{-1} for cases when macroalgae is harvested (dots) and when it is not (dashes). Both (a) the total flux and (b) the portion of the flux due to enhanced growth show that invasion of atmospheric CO_2 cannot match the rate of enhanced growth. When enhanced growth is not harvested (allowed to be respired) the enhanced air-sea CO_2 flux is lower.

1 to 2 ppb after 100 yr of enhanced growth when nutrients are supplied artificially and growth is not harvested. This represents about a 1 to 2% reduction in the benefit achieved by lowering atmospheric pCO_2 about 20 ppm during the same simulations. For comparison, simulations using only naturally available nutrients produce less N_2O, unless they contain a portion of the Southern Ocean; simulations with harvesting produce no N_2O.

4. Discussion

From Figure 7 we see that the ratio of the cumulative enhanced air-sea CO_2 flux to the cumulative new production never reaches unity for any simulation in the northern subtropics. In fact, it never

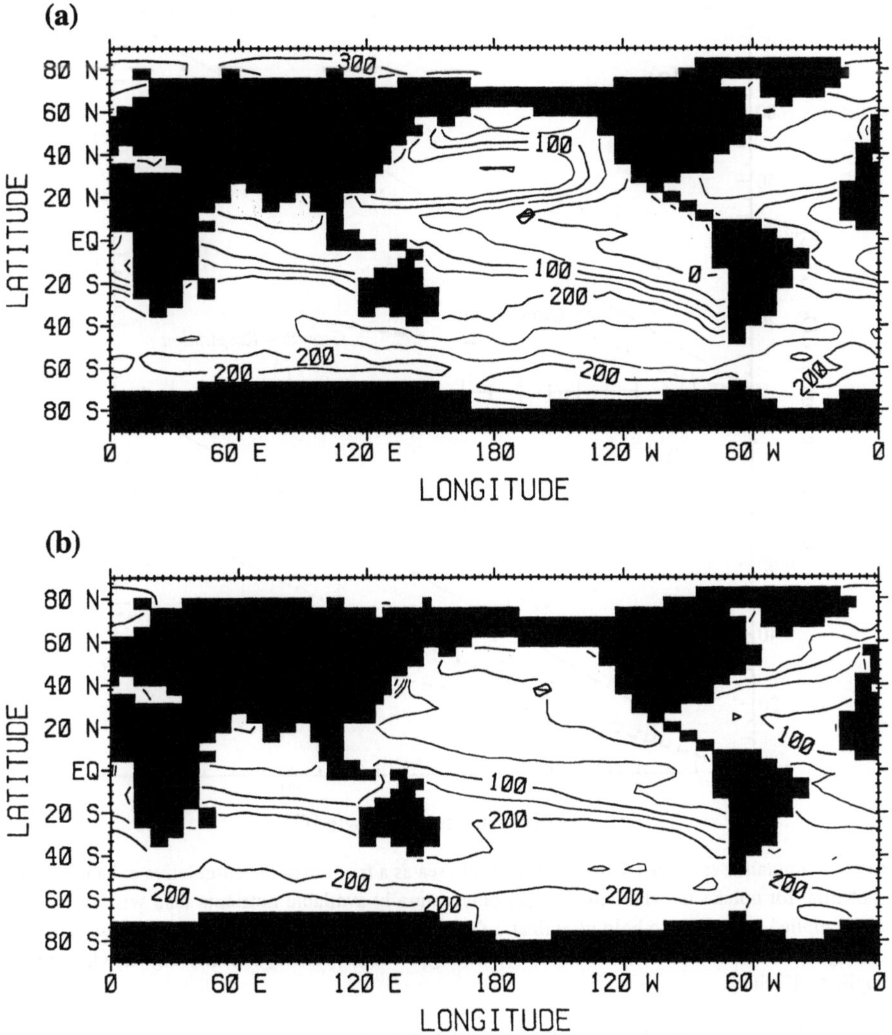

Figure 6. Map of oxygen perturbation + observations for simulations of no harvesting with enhanced growth at 1 GT C yr^{-1} evenly distributed over (a) all equatorial ocean (map at 483 m) and (b) all northern subtropical ocean (map at 755 m). Stippled areas denote anoxic conditions.

reaches unity in any simulation. This ratio falls with time during simulations without harvesting (with respiration) but rises slightly in simulations with harvesting. For enhanced growth in the northern subtropics with only naturally available nutrients, this fraction of the cumulative enhanced air-sea flux to the cumulative enhanced growth is 0.61 (after 100 yr) when growth is harvested, but only 0.27 without harvesting. When nutrients are added artificially to enhance growth to 1 GT C yr^{-1}, the same fraction is higher: 0.71 when growth is harvested, but only 0.43 without

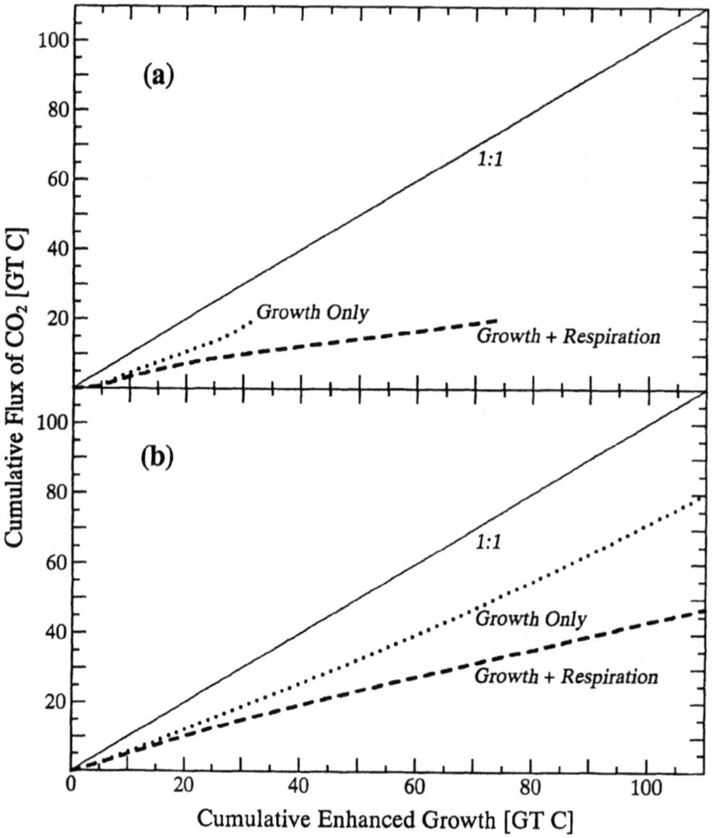

Figure 7. Cumulative enhanced CO_2 flux from air to sea as a fraction of the cumulative enhanced growth. Less potential for uptake is exhibited with (a) only naturally available nutrients than with (b) sufficient nutrients supplied artificially to hold growth at 1 GT C yr^{-1}. The enhanced CO_2 flux is always less than enhanced growth, and the enhanced-flux-to-enhanced-growth fraction is generally higher when nutrients are supplied artificially.

harvesting. All cases show then, that the enhanced flux of CO_2 from atmosphere to ocean never equals the enhanced export of organic C from surface to deep via enhanced biological productivity.

When only naturally available nutrients are used for enhanced growth, simulations show that the mitigating effect upon atmospheric CO_2 is relatively modest. The only exception is when the region of enhanced growth contains a significant portion of the Southern Ocean whose surface is rich in nutrients and is rapidly resupplied with nutrient-rich waters from below, largely by convection. Unfortunately, the logistics for deploying and maintaining open ocean aquacultures are perhaps most implausible in the Southern Ocean, if only because of its high winds, stormy seas, and extreme temperatures. For these reasons, and because the Southern Ocean has already been thoroughly studied with the same model, albeit for Fe fertilization, we have focused here on milder and warmer regimes, specifically the subtropics and the equatorial region. These simulations show that if naturally available nutrients are depleted in the region of enhanced growth, new production suffers elsewhere because the region of enhanced growth exports relatively nutrient-depleted waters.

Also, there is no reason that nutrient demand would diminish even if enhanced growth were more localized and still supplied only with naturally available nutrients. Extensive localized growth centers would still export relatively nutrient-poor waters, thereby causing significant reduction of new production in surrounding waters.

To enhance growth of macroalgae then, one must supply nutrients artificially to regions of enhanced growth. The supply of nutrients to surface waters could be enhanced by pumping up nutrient-rich deep waters or by addition of fertilizer to the surface. Unfortunately, pumping deeper waters to the surface is not plausible for at least two reasons. First, the cost of pumping deeper water to the surface is estimated to be 6–7 times greater than any return from energy production that uses seaweed as a biomass fuel (see Ritschard, 1992). Such high cost derives from the large volume of water that must be pumped to the surface. For instance, production of 1 GT C yr^{-1} of enhanced algal growth would require at least all the nutrients found in deep water pumped at the rate of 6.6×10^6 m^3 s^{-1}. For comparison, the mean annual rate of discharge for the Amazon river is 0.175×10^6 m^3 s^{-1} (Oltman, 1968). Second and more importantly, pumping nutrient-rich deeper water to the surface to enhance biological production as a means of ultimately reducing atmospheric CO_2 would not make sense, because deeper waters are also enriched in CO_2; if undertaken, such pumping would not result in enhanced CO_2 uptake. With this option confuted, let us now turn to the remaining alternative.

Can we add fertilizer to the surface ocean? At the current annual rates for world production of nitrogen fertilizer at ~100 Tg N yr^{-1} and phosphorus fertilizer ~50 Tg P_2O_5 yr^{-1} (FAO, 1988), and with the relatively constant proportion of C:N:P (130:16:1) in seawater, one can determine the relative proportion of the world's annual production of fertilizer which is necessary to maintain enhanced algal growth at 1 GT C yr^{-1}. If we use all the N fertilizer produced today, there is only enough to support enhanced growth at 0.7 GT C yr^{-1}; for P fertilizer, it is better, but still, 90% of the world's current fertilizer production would then go towards enhanced growth of macroalgae.

5. Conclusions

Although it appears there are no outspoken proponents firmly backing enhanced growth of marine macroalgae as a viable means to help limit rising atmospheric CO_2, this idea does derive from former studies that have evaluated it as a means to fuel energy production and it is now being reconsidered in this new sense. The simulations discussed herein investigate its maximum potential, i.e., if enhanced growth could be achieved on a large scale. Moreover, these simulations demonstrate a key principle that all too often seems neglected in considerations of this nature.

Not all organic carbon removed from the ocean's surface layer by enhanced biological productivity is replaced by an equivalent amount of CO_2 invading the surface ocean from the atmosphere. These simulations show that at best (i.e., if enough nutrients could be added to support enhanced biological productivity at 1 GT C yr^{-1} and all growth is harvested) enhanced influx of atmospheric CO_2 is only 71% of the enhanced productivity, as averaged over the first 100 yr. The relative amount is substantially less if growth is not harvested: 43% when nutrients are supplied artificially, but only 27% when just naturally supplied nutrients are available. One consequence of the fact that the atmospheric CO_2 uptake is less than the algal growth is that if the seaweed were used as an additional source of fuel, the net effect would be to remove CO_2 from the ocean and add it to the atmosphere; however, if it were used as an alternative source, i.e. as a partial replacement for fossil fuel, there would be some net reduction in the increase rate of atmospheric CO_2.

Other conclusions can now be drawn regarding the feasibility of enhanced growth and inherent side effects. Enhancing algal production on order $\sim$1 GT C yr^{-1} increases local demand for nutrients beyond what is naturally available. Furthermore, naturally available nutrients, as fuel for enhanced growth, are depleted in the growing region and these depleted waters are exported elsewhere where new production eventually drops, i.e., when nutrient depleted waters make it to the surface. Therefore, for enhanced algal growth to be effective, nutrients must be supplied artificially.

As previously mentioned, if nutrients could be supplied artificially, the atmospheric flux of of CO_2 would be most pronounced if macroalgae could be harvested. Unfortunately, with harvesting, nutrient demand would be high. The nutrient demand could not be satisfied by pumping deep water to the surface because (1) it is too costly, and (2) it would not reduce atmospheric CO_2. The other alternative, adding fertilizer to surface waters, is also not feasible: nutrient demand would then require close to the world's production rate for P fertilizer and substantially more than that for N fertilizer.

Less nutrient is required when enhanced growth is not harvested; however, nutrient demand remains prohibitive, partially because not all nutrients regenerated at depth return to the surface of the region of enhanced growth. Furthermore, regeneration of enhanced growth depletes oxygen to the extent that substantial anoxia results.

ACKNOWLEDGEMENTS. This work has only been possible because of the long-established collaboration with NOAA GFDL. This manuscript has benefited from discussions with E. Glen and R. Ritschard. S. Mateu-Orr assisted in preparation of the manuscript and tables. The longterm support of the U.S. Dept. of Energy under contract DEFG 02-90ER61052 is gratefully acknowledged.

References

Adey, W. H., 1984. Biomass from the sea: a new approach to obtain optimal solar energy conversion in the ocean, Bioenergy council, Washington, D.C., 21 pp.

Bacastow, R., and E. Maier-Reimer, 1990. Ocean-circulation model of the carbon cycle, *Climate Dynamics, 4*, 95-125.

Bacastow, R., and G. R. Stegen, 1991. *Proc. OCEANS 91, 3*, 1654–1657, Honolulu.

Bird, K. T., 1987. Cost analyses of energy from marine biomass, in K. T. Bird and P. H. Benson (eds.), *Seaweed Cultivation for renewable resources*, Elsevier, New York, pp. 327–350.

Bryan, K. 1969. A numerical method for the study of the circulation of the world ocean, *J. Comput. Phys., 4*, 347–376.

FAO, 1988. *FAO Fertilizer Yearbook, FAO Statistics Series No. 89*, Food and Agriculture Organization of the United Nations, Rome.

Glen, E. P., K. J. Kent, T. L. Thompson, and R. J. Frye, 1991, Seaweeds and halophytes to remove carbon from the atmosphere, *Electric Power Res. Inst., ER/EN-7177*, Palo Alto, CA, 71 pp.

Hellerman, S. and M. Rosenstein, 1983. Normal monthly wind stress over the world ocean with error estimates, *J. Phys. Oceanogr., 13*, 1093–1104.

Houghton, J. T., G. J. Jenkins, and J. J. Ephraums, 1990. *Climate Change, The IPCC Scientific Assessment*, Cambridge U. Press, Cambridge, 365 pp.

Joos, F., J. L. Sarmiento, and U. Siegenthaler, 1991a. Estimates of the effect of Southern Ocean iron fertilization on atmospheric CO_2 concentrations, *Nature, 349*, 772–775.

Joos, F., U. Siegenthaler, and J. L. Sarmiento, 1991b. Possible effects of iron fertilization in the Southern Ocean on atmospheric CO_2 concentrations, *Global Biogeochem. Cycles, 5*, 135–150.

Levitus, S., 1982. Climatological atlas of the World Ocean, *NOAA Prof. Pap. 13*, U.S. GPO., Washington, D.C., 173 pp.

Maier-Reimer, E., 1991a. in M. Latif (ed.), *Strategies for future climate research*, Max-Planck-Institut fur meteorologie, Hamburg, pp. 319–339.

Maier-Reimer, E., 1991b, private communication.

Maier-Reimer, E., and K. Hasselmann, 1987. Transport and storage of CO_2 in the ocean – an inorganic ocean-circulation cycle model, *Climate Dynamics, 2*, 63–90.

Marchetti, C., 1977. On geoengineering and the CO_2 problem, *Climatic change, 1*, 59–68.

Martin, J. H., 1990. Glacial-interglacial CO_2 change: the iron hypothesis, *Paleoceanogr., 5*, 1-13.

Martin, J. H., S. E. Fitzwater, and R. M. Gordon, 1990. Iron deficiency limits phytoplankton growth in Antarctic waters, *Global Biogeochem. Cycles, 4*, 5–12.

McKay, L. B., 1983. Seaweed raft and farm design in the United States and China, *NTIS Report PB84-163864*, NOAA, Sea Grant, Rockville, MD, 108 pp.

Najjar, R. G., 1990. Simulations of the phosphorus and oxygen cycles in the world ocean using a general circulation model, Ph.D. thesis, Princeton Univ., 190 pp.

North, W., 1987, Ocean farming of *Macrocyctis*, the problems and non-problems, in K. T. Bird and P. H. Benson (eds.), *Seaweed cultivation for renewable resources*, Elsevier, New York, pp. 39–68.

Oeschger, H., U. Siegenthaler, U. Schotterer, and A. Gugelmann, 1975. A box diffusion model to study the carbon dioxide exchange in nature, *Tellus, 27*, 168–192.

Oltman, R.E., 1968, Reconnaissance investigations of the discharge and water quality of the Amazon River, *Geol. Surv. Circ. U.S., 552*, 16pp.

Peng, T.-H., and W. S. Broecker, 1991a. Dynamic limitations on the Antarctic iron fertilization strategy, *Nature, 349*, 227–229.

Peng, T.-H., and W. S. Broecker, 1991b. Factors limiting the reduction of atmospheric CO_2 by iron fertilization, *Limnol. Oceanogr.*, in press.

Ritschard, R. 1992, Marine Algae as a CO_2 sink, *Water, Air, and Soil Pollution*, this issue.

Ryther, J. H., 1984, Technology for the commercial production of macroalgae, in M. Z. Lowenstein (ed.), *Energy applications of biomass*, Elsevier, London, pp. 39–68.

Sarmiento, J. L., and J. C. Orr, 1991. Three dimensional ocean model simulations of the impact of Southern Ocean nutrient depletion on atmospheric CO_2 and ocean chemistry, *Limnol. Oceanogr.*, in press.

Sarmiento, J. L., and E. Sundquist, 1991. Oceanic uptake of anthropogenic CO_2: a new budget, *Nature*, submitted.

Sarmiento, J. L., J. C. Orr, and U. Siegenthaler, 1992. A perturbation simulation of CO_2 uptake in an ocean general circulation model, *J. Geophys. Res., 97*, in press.

Siegenthaler, U., 1983. Uptake of excess CO_2 by an outcrop-diffusion model of the ocean, *J. Geophys. Res., 88*, 3599–3608.

Stegen, G. R., K. H. Cole, and R. Bacastow, 1991. in Proc. IEA international conference on technology responses to global environmental challenges, Kyoto, Japan.

Tans, P. P., I. Y. Fung, and T. Takahashi, 1990. Observational constraints on the global atmospheric CO_2 budget, *Science, 247*, 1431–1438.

Toggweiler, J. R., K. Dixon, and K. Bryan. 1989a. Simulations of radiocarbon in a course-resolution world ocean model. 1. Steady state pre-bomb distributions, *J. Geophys. Res., 94*, 8217–8242.

Toggweiler, J. R., K. Dixon, and K. Bryan. 1989b. Simulations of radiocarbon in a course-resolution world ocean model. 2. Distributions of bomb-produced ^{14}C, *J. Geophys. Res., 94*, 8243–8264.

AN APPROACH TO ASSESSMENT OF MANAGEMENT IMPACTS ON AGRICULTURAL SOIL CARBON

Thomas O. Barnwell, Jr., Robert B. Jackson, IV
U.S. EPA, Athens, GA, 30613-0801, USA

Edward T. Elliott, Ph.D., Ingrid C. Burke, Ph. D.
Colorado State University, Ft. Collins, CO, 80523, USA

C. Vernon Cole, Ph.D.
USDA/ARS, Ft. Collins, CO, 80523, USA

Keith Paustian, Ph.D., Eldor A. Paul, Ph.D.
Michigan State University, East Lansing, MI, 49060, USA

Anthony S. Donigian, Avinash S. Patwardhan
Aqua Terra Consultants, Mountain View, CA, 94043, USA

Allen Rowell, Kevin Weinrich
Computer Sciences Corporation, Athens, GA, 30613-0801, USA

ABSTRACT. Agroecosystems contain about 12% of the terrestrial soil C and play an important role in the global C cycle. We describe a project to evaluate the degree to which management practices can affect soil C in agroecosystems. The objectives of the project are to determine if agricultural systems can be managed to conserve and sequester C and thereby reduce the accumulation of CO_2 in the atmosphere, and to provide reference datasets and methodologies for agricultural assessments.

1. Introduction

Carbon movement between the atmosphere, oceans, and terrestrial systems is in a dynamic equilibrium. One recent generally accepted estimate of the sizes of these pools, and the fluxes among them, is shown in Figure 1, redrawn from the Intergovernmental Panel on Climate Change's (IPCC) report Climate Change (Houghton et al. 1990).

The atmospheric pool, at 750 Gt C (1Gt = 10^{12}kg) is the smallest of the major C pools and is reliably measured to be increasing at a rate of 3 Gt C yr^{-1}. Many recent studies, summarized in the IPCC report, have projected that this increase will cause worldwide climate change. This increase is attributed to deforestation (~2 Gt C yr^{-1}) and fossil fuel combustion (~5 Gt C yr^{-1}). A mass balance argument estimates that about 4 Gt C

Water, Air, and Soil Pollution **64**: 423–435, 1992.
© 1992 *Kluwer Academic Publishers.*

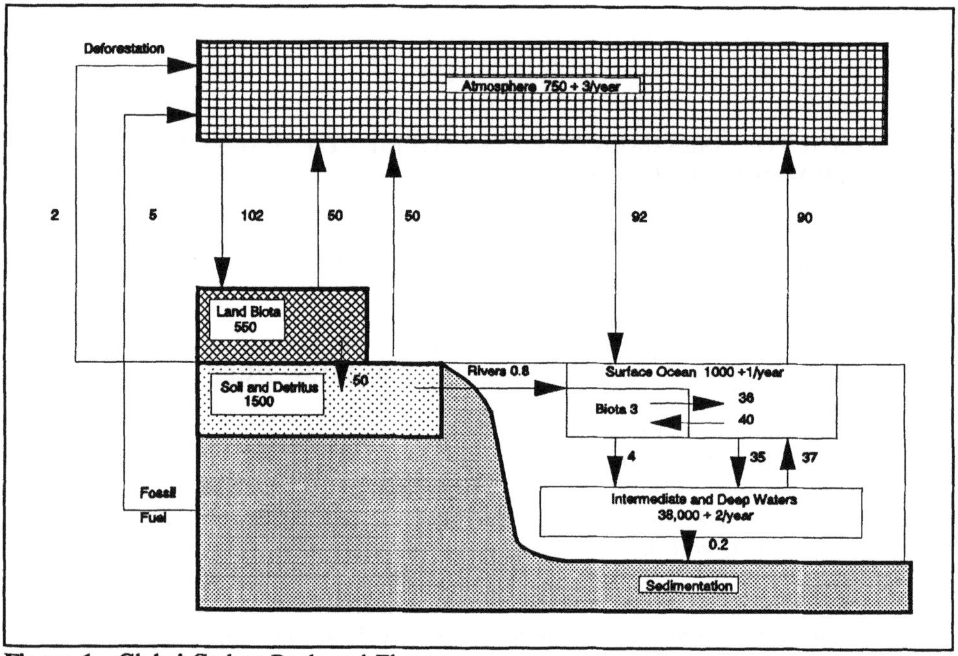

Figure 1 - Global Carbon Pools and Fluxes

yr^{-1} of these anthropogenic emissions are recycled back into the oceanic and terrestrial pools. The IPCC report allocates this recycled C equally between terrestrial and oceanic systems, but a recent paper (Tans et al. 1990) argues that the terrestrial sink is larger than the oceanic sink, and is between 2.0 and 3.4 Gt C yr^{-1}.

The surface ocean pool, at 1000 Gt C, is a large C reservoir, and some proposals for active management have been made. These proposals involve stimulation of biological activity in the southern oceans through addition of Fe, one of the limiting nutrients in these systems. There is, however, considerable scientific discussion with respect to the validity of this proposal. The deep ocean, at 38,000 Gt C, is the largest C reservoir, but is relatively inaccessible and generally not believed to be subject to active management.

Just over 2000 Gt C are contained in terrestrial pools, about 550 Gt C in living vegetation and 1500 Gt C in soil organic matter (SOM). Because this C pool is three times the size of the atmospheric pool, changes in terrestrial systems are important in affecting the sensitive atmospheric pool. It is interesting to examine the time history of net terrestrial C emissions (Figure 2, Houghton et al. 1983). For much of the last 100 yr, C emissions from land use change (degradation of biota and soils) have exceeded emissions from fossil fuel combustion. Fossil fuel emissions of C surpassed those from biota and soils in the 1950s and have increased dramatically since, although there has been a short-term trend in the 60s and 70s indicating that the biota/soils emissions have been constant, or at least decreasing less rapidly.

The U. S. Environmental Protection Agency is evaluating the potential for active management of terrestrial C, both through reforestation and good stewardship of agricultural resources. This paper deals with the approach being developed for agricultural systems. Clearly, some agricultural management practices can induce dramatic changes in soil organic C. The classic experiments at Rothamsted, England (Figure 3, Jenkinson 1991) demonstrate a three-fold increase in soil C storage through application of farm yard manure. It is probably not possible to achieve these results on a large scale, but the data do demonstrate potential. An interesting note concerning these data is that soil C still has an upward trend after 140 yr of active management. While reforestation can probably sequester great quantities of C in the short-term, steady-state will likely be reached in 15 to 30 yr. Soils, on the other hand, have the potential for much longer accumulation.

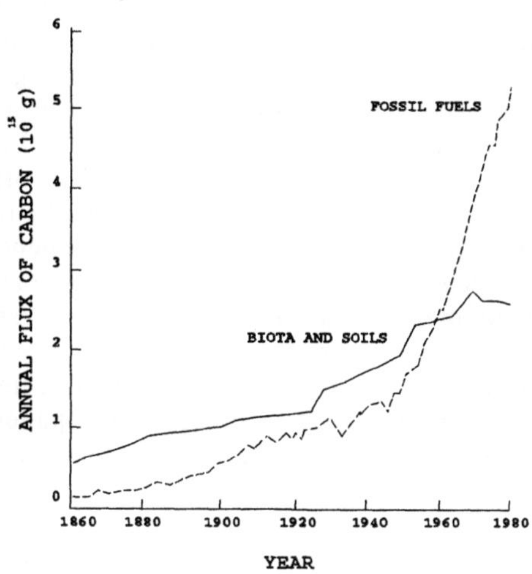

Figure 2 - Time history of terrestrial carbon emissions.

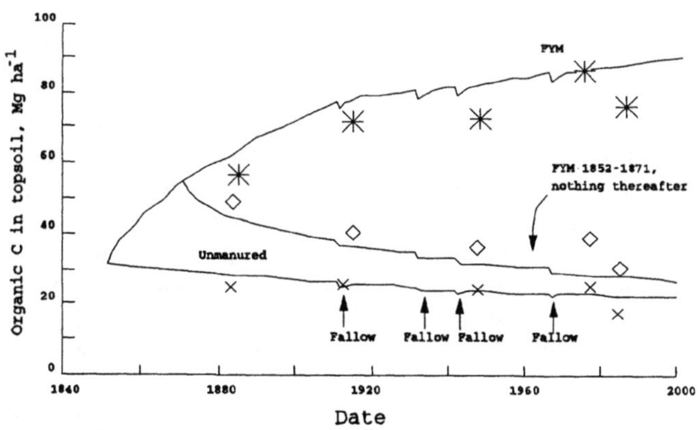

Figure 3 - Organic carbon in the topsoil of a plot at Rothamsted, England. The symbols show experimental observations; the solid line is a model fit. FYM is farm yard manure.

2. Assessment of soil C sequestration potential for U. S. agroecosystems

The objectives of the project are to determine if agricultural systems can be managed to conserve C and thereby reduce the accumulation of CO_2 in the atmosphere, and to provide reference datasets and methodologies for agricultural assessments. Application of agroecosystem management practices can increase the terrestrial C pool, thereby reducing the build-up of atmospheric C. In particular, many environmentally beneficial agricultural practices increase soil C content. The program will assess the potential agroecosystem contribution to global C conservation and sequestration.

The impact of alternative management practices on agricultural soil C is being estimated by a soil C mass balance model and analysis of agricultural production cycles, development of data bases for agroecosystem C pools and dynamics, development of spatial data bases for assessment of agricultural C sequestration poten-tial, and development of cost factors. A brief review of the Agroecosystems Carbon Pools and Dynamics project is included in this paper, as well as a review of soil C models. A strategy for incorporating policy considerations in the analysis is also described.

3. Agroecosystem Carbon Pools and Dynamics

The objective of this project element is to conduct field and laboratory investigations to develop defensible data to evaluate C pools and dynamics for agroecosystems, to identify critical soil and climatic parameters that might affect those dynamics, and to validate the models developed to assess the impact of alternative agricultural management practices on C storage and fluxes in agroecosystems.

The conversion of native ecosystems to agricultural lands in the Great Plains and Corn Belt regions of the Central United States was perhaps the most extensive ecological disturbance known in North America during the past 150 yr (Wilson 1978). A major consequence of this land conversion has been the substantial loss of organic matter from soils across the regions (Haas et al. 1957, Campbell 1978). These soil C losses have been a primary anthropogenic source of CO_2, second only to fossil fuel combustion in contributing to historical increases of global CO_2 concentrations (Houghton et al. 1983, Post et al. 1990).

Since soil C levels in most Great Plains and Corn Belt agricultural areas have been reduced to levels well below those existing prior to the establishment of agriculture, the potential should exist to increase present soil C levels, thereby providing a sink for atmospheric CO_2. At present there is considerable debate whether increases in atmospheric CO_2 concentrations and global warming will result, independently, in significant changes in soil C storage (e.g., Prentice and Fung 1990, Schlesinger 1990). While not discounting the importance of these factors, increasing agricultural soil C levels may be more dependent on agricultural management practices and land use changes.

Many of the agricultural practices that have been traditionally applied in the Great Plains and Corn Belt (e.g., moldboard plowing, fallow with no plant cover, removal or burning of crop residues and monoculture cropping (Cole et al. 1989)) represent a "worst-case" for soil C maintenance by minimizing organic matter inputs and promoting a high degree of soil organic matter decomposition. Currently, agricultural management in the United States is undergoing rapid change in response to a variety of pressures, including high costs of energy and chemical inputs, environmental concerns about nutrient and pesticide pollution and soil erosion, and product quality demands by consumers. It is encouraging to note that many of the techniques designed to address these issues have also demonstrated a potential for increasing soil C levels above those maintained under conventional management. Such practices include reduced or no-till management (Blevins et al. 1983, Dick 1983, Lamb et al. 1985); reduction in the proportion of fallow land relative to crop in semiarid regions (Wood et al. 1991); and use of cover crops, green manure, and animal manure (Paustian et al. 1992, Jenkinson 1991, Vitosh et al. 1973). However, the extent to which soil C levels can be regulated through the application of various combinations of these practices for different cropping systems, soil types, and climatic regions is not yet clear.

In addition to changing management trends on production land, land use changes may have a considerable impact on region-wide C storage potential. Over the past 50 yr, significant acreage in cropland has been abandoned or converted back into perennial grassland and forest. For example, more than 100,000 ha of farmland in northeast Colorado alone were abandoned during the 1930s. In more recent years, large areas of Great Plains cropland have been converted to grassland under the Conservation Reserve Program (Joyce and Skold 1988). Similar changes have occurred in regions of the Corn Belt, in particular in the Great Lakes region and in eastern areas bordering on the Appalachians. For example, from 1950 to 1990, the agricultural land base of Michigan declined by $3x10^6$ ha, much of which was previously in corn, soybean and wheat production (Michigan Dept. Agric. 1990). To date, there has been little systematic evaluation of how these shifts in land use have affected regional soil C balances or what effect future changes may have.

While data from long-term field studies has provided much information about the possible rates and directions of change in soil C under various management regimes, our current knowledge base is very fragmented. Most of what we know concerning soil organic matter (SOM) dynamics has been obtained by studying SOM losses. We understand less about controls on SOM accumulation and how they vary across soil types, climatic regions and management regimes. Such information is crucial for estimating the potential for C sequestration in agricultural soils of the Great Plains and Corn Belt regions.

Studies are focusing on corn, soybeans, wheat and rangelands. Priority is being given to sites with historical data of the effects of management practices on SOM and supporting data pertaining to climatic conditions. Data will be used to parameterize and evaluate C budget models developed in other tasks of the project. A network of field sites (Figure 4) has been organized through a cooperative agreement with Colorado State University and Michigan State University. Emphasis is on collection, preparation, and evaluation of C budget data. Sites where comparisons of management practices are possible will be given highest priority. Because of their spatial and temporal variability, CO_2 and other greenhouse gas fluxes will be inferred from soil storages rather than measured. Outputs of the studies will be useful in the quantification of soil and plant organic C pools and dynamics for the dominant agricultural crops, including rangelands, in the two regions.

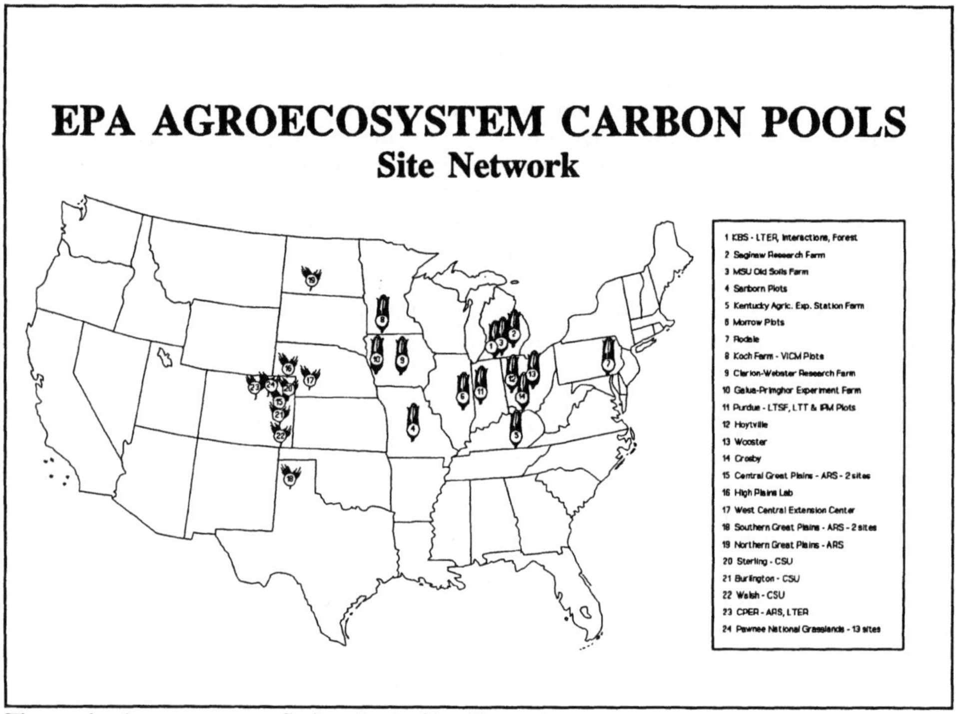

Figure 4 - Agroecosystem Carbon Pools Site Network

4. Soil Carbon Models

A literature review of soil C modeling, CO_2 emissions and climate change, and soil C budgets and impacts of soil and agricultural practices has been completed (Barnwell et al. 1991). In addition, a summary of major C computer models identified in the literature survey are included. Based on the preliminary review, a list of evaluation criteria and capabilities for comparison of soil C models has been compiled, and these criteria have been used to select models for evaluation of agricultural soil sequestration capabilities. These models will be incorporated into a comprehensive evaluation framework as described in section 5.

The focus of the literature search has been on the identification and documentation of available soil C models; investigation of soil C budgets, transformation, and pathways in agroecosystems; and determination of the impacts of tillage and cropping practices on the soil C cycle. The preliminary evaluation of available soil C models listed in Figure 5 reduced the number of 'candidate models' to four or five primary models that were evaluated further. The existing models are primarily 'research models' in that few have undergone more than limited independent field testing and validation. One of the key issues is to provide a level of model complexity (and associated data requirements) that is consistent with the project needs <u>and</u> is also compatible with the national level

Model Name	Reference
CENTURY Model	Parton et al., 1988
DNDC Model	Li et al., 1991a; Li et al., 1991b; Li et al., 1991c
Jenkinson and Rayner Model	Jenkinson and Rayner, 1977; Jenkinson 1990, Jenkinson et al., 1991
Frissel and van Veen	Frissel and van Veen, 1981; van Veen and Paul, 1981; van Veen et al., 1984
OBM Model	Esser, 1990
NCSWAP Model	Clay et al., 1985
TEM Model	Raich et al., 1991
Box Model	Box, 1988
Gildea et al. Model	Gildea et al., 1986
Moore et al. Model	Moore et al., 1981
Smith Model	Smith, 1979
PHOENIX Model	McGill et al., 1981

Figure 5 - List of soil C models reviewed

databases available for providing the model input.

The models selected for use and/or modification to meet the needs of representing soil C cycles in agroecosystems and impacts of management practices are CENTURY and DNDC. These models share a common ability to examine the impacts of alternative management practices on soil organic C, and are readily accessible for these studies. A brief review of the capabilities and application experience with these models follow.

4.1 CENTURY Model

Parton et al. (1988) developed the CENTURY model to simulate the dynamics of C, N, P, and S in cultivated and uncultivated grassland soils. The model uses monthly time steps for simulation, and runs can be performed for time periods ranging from 100 to 10,000 yr. The model simulates three soil organic matter fractions: 1) active fraction (turnover time of 1 to 5 yr); 2) a protected fraction that is resistant to decomposition (turnover time 20 to 40 yr); and 3) a physically protected and chemically resistant fraction (turnover time of 200 to 1500 yr). Model simulations yielded satisfactory results in terms of long-term soil formation, weathering of P and S from soil parent material and the formation of organic P and S in different SOM fractions.

Schimel et al. (1990) studied grassland biogeochemistry and its link to atmospheric processes. Model simulations obtained using the CENTURY model adequately represented the observed geography of C and N. Cole et al. (1989) modeled the effects of land use on SOM dynamics in the North American Great Plains using the CENTURY model. The model results were compared with a 30- to 40-year observed data set. The study found that "the effects of a range of soil textures on the original amounts and forms of C, N and P in the soils and changes upon cultivation correspond well with observations of detailed field studies". Parton et al. (1989) studied the regional pattern dynamics of C, N, and P in the U.S. central grassland region using the CENTURY model. A comparison of simulated P and C results with observed data showed good correlation. Paustian et al. (1991) performed model analyses using the CENTURY model for studying the influence of organic amendments and N-fertilization on SOM in long-term plots. Changes in soil C and N after 30 yr were predicted within $\pm$ 30% of observed values.

A geographic information system (GIS) was used with the CENTURY model for regional modeling of grassland geochemistry (Burke et al. 1990). The driving variables for the model consisted of soil texture, monthly precipitation, and monthly minimum and maximum temperatures, which were included in the GIS. Net primary production was affected primarily by climatic factors, and it closely followed spatial variation in precipitation. However, it was reported that SOM is controlled by soil texture.

4.2 DNDC Model

Li et al. (1991a) developed the physically based model DNDC (Denitrification Decomposition Model) for estimating N_2O and N_2 evolution from soils. The submodels included in the model are thermal-hydraulic flow, decomposition and denitrification. The model calculates dynamic soil temperature and moisture profiles, and estimates the shift of aerobic-anaerobic conditions in soils using climatic data of rainfall events and air temperature. The dominant processes between rainfall events are the decomposition of organic matter and the oxidation reactions. These processes affect levels of total organic C, soluble C and nitrate in soils between rainfall events. However, during rainfall the dominant process is that of denitrification resulting in production of N_2O and N_2. The major input data requirements for the DNDC model consist of the climatic scenario, soil texture and its properties, and initial organic matter and nitrate in soil. Li et al. (1991b) have also conducted sensitivity and simulation analysis for the DNDC model. Li et al. (1991c) applied the DNDC model for estimating N_2O emissions from fallow, grass and sugarcane fields at Belle Glade, FL. Good agreement was reported between simulated and observed results of N_2O emissions, N mineralization rates, and changes in nitrate and ammonium at the soil surface.

5. Assessment Framework

An important aspect of this effort is the development of the modeling framework and methodology that will define the agricultural production systems and scenarios (i.e., crop-soil-climate combinations) to be assessed in terms of national and international policy, the integration of the model needs with available databases, and the operational mechanics of evaluating C sequestration potential with the integrated model/database system. Figure 6 represents our operational strategy at this time.

The objective of the assessment is to estimate the greenhouse gas emissions from agriculture and to estimate the relative impact of various policy alternatives on these emissions. We are working closely with the U. S. EPA's Office of Policy and Program Evaluation (OPPE) and Iowa State University's Center for Agricultural and Rural Development (CARD) to define a reasonable set of policy alternatives for this assessment. An example set of policy alternatives is shown in the box labeled "OPPE" in Figure 6. Included are national policy alternatives such as the impact of a Global Agreement on Trade and Tariffs (GATT) and their possible effect on national agricultural production levels, as well as agriculture policy alternatives that might be affected through a revised Farm Bill, such as incentives to selectively promote conservation tillage, sustainable agriculture, or good stewardship of the conservation reserve.

As shown in the box labeled "CARD", these policies will be used to drive economic models that predict national production levels using models such as the Basic Linked System (BLS), a set of national and regional models linked by policy and a world price determination process. CARD's Comprehensive Environmental Economic Policy Evaluation System (CEEPES) will be used to translate policy alternatives into basic data for use in soil C models. These data, including such elements as agricultural practices, fertilization rates, and erosion rates will be used in the soil C models to produce net greenhouse gas emissions on a per unit area basis.

Data will be produced for a range of "Production Areas," which are hydrologic areas defined by the Water Resources Council that are used for aggregated economic reporting. The economic models also will provide statistics on crop acres, acreage in the conservation reserve, and other rural land usage. The unit-area emissions will be combined with areal-extent data in a GIS to produce an estimate of total greenhouse gas emissions and to calculate an aggregate greenhouse gas warming index. The greenhouse gas warming index will represent the combined impact of CO_2 and N_2O emissions. We plan to produce a table (Figure 6) that, we hope, will communicate the aggregate impact of alternative policies in a simple, effective manner.

6. Conclusions

It is clear that agricultural soil C has been depleted over the last century. This study will assess the current size of this C pool and examine the potential for alternative agricultural management practices to alter its size. Many of the agricultural management techniques that favor accumulation of soil organic C also have other positive environmental benefits, so their use is being encouraged for other reasons. While it is not clear at present the extent to which soil C levels can be managed using these techniques, we hope through this study to provide a better estimate than currently available.

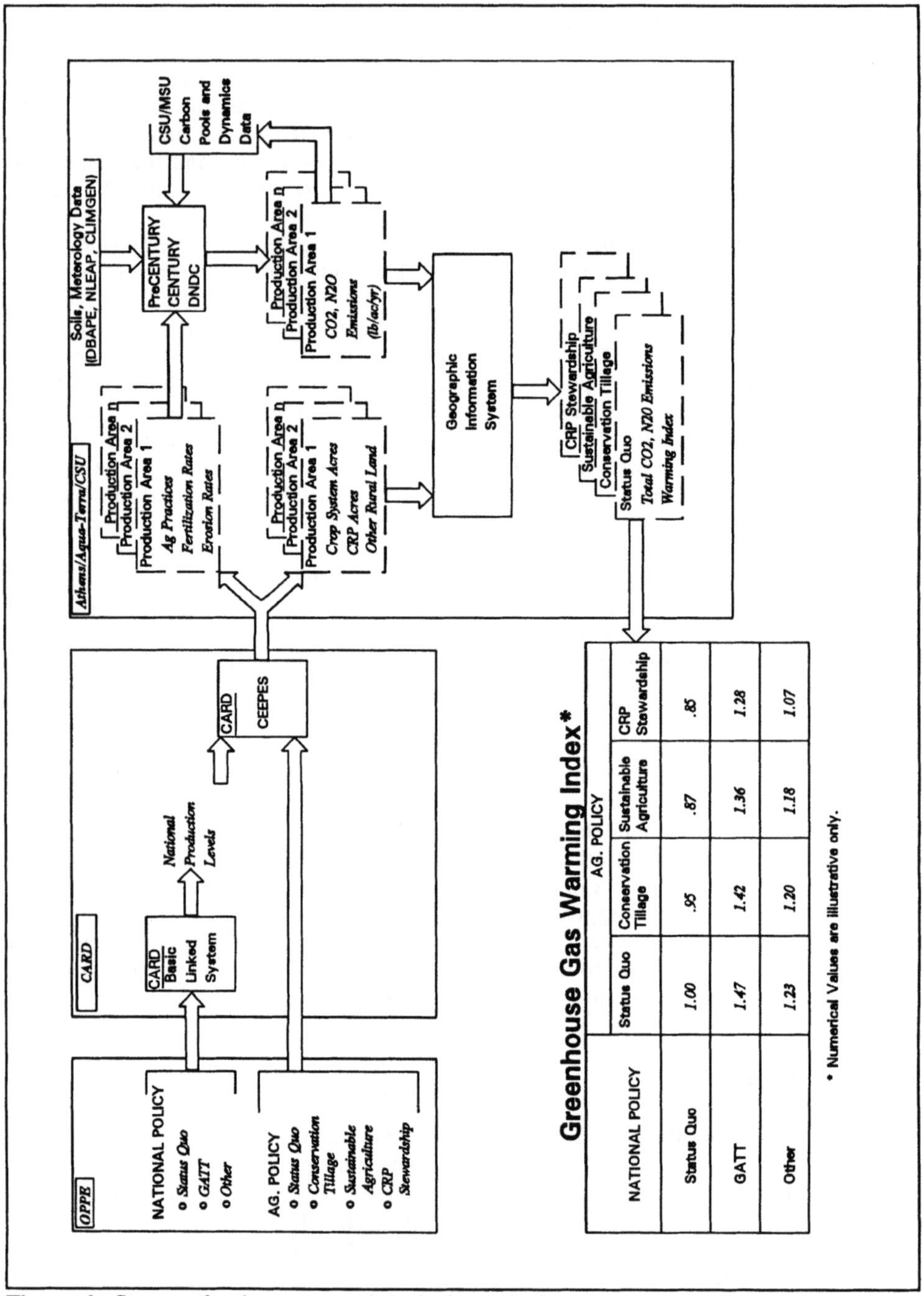

Figure 6 - Strategy for Assessment of Agricultural Carbon Sequestration Potential

7. References

Barnwell, T. O. Jr., R. B. Jackson, IV, E. T. Elliott, E. A. Paul, K. Paustian, A. S. Donigian, A. S. Patwardhan, A. Rowell, and K. Weinrich. 1991. Assessment of Methods, Models, and Databases for Soil Carbon Sequestration Potential for U. S. Agroecosystems. (Internal Report). Environmental Research Laboratory, U S EPA, Athens, GA, September 1991.

Blevins, R.L, G.W. Thomas, M.S. Smith, W.W. Frye and P.L. Cornelius. 1983. Changes in soil properties after 10 years continuous non-tilled and conventionally tilled corn. Soil & Tillage Res. 3:135-146.

Box, E.O. 1988. Estimating the seasonal carbon source-sink geography of a natural, steady-state terrestrial biosphere. J. of Applied Meteorology. 27(10):1109-1124

Burke, I.C., D.S. Schimel, C.M. Yonker, W.J. Parton, L.A. Joyce and W.K. Lauenroth. 1990. Regional modeling of grassland biogeochemistry using GIS. Landscape Ecology 4(1):45-54.

Campbell, C.A. 1978. Soil organic carbon, nitrogen and fertility. In: Schnitzer, M. and S.U. Khan (eds) Soil Organic Matter. Developments in Soil Science 8, Elsevier Scien. Pub. Co., Amsterdam. pp. 173-271.

Clay, D.E., C.E. Clapp, J.A.E. Molina, and D.R. Linden. 1985. Nitrogen-tillage-residue management. I. Simulating soil and plant behavior by the model NCSWAP. Plant and Soil. 84:67-77.

Cole, C.V., J.W.B. Stewart, D.S. Ojima, W.J. Parton and D.S. Schimel. 1989. modeling land use effects on soil organic matter in the North American Great Plains. In: Clarholm, M. and L. Bergstrom (eds.). Ecology of Arable Land. Kluwer Academic Publishers, Dordrecht, The Netherlands. 89:98.

Dick, W.A. 1983. Organic carbon, nitrogen, and phosphorus concentrations and pH in soil profiles as affected by tillage intensity. Soil Sci. Soc. Am. J. 47:102-107.

Esser, G. 1990. Modeling global terrestrial sources and sinks of carbon dioxide with special reference to soil organic matter, Chapter 10. In: A.F Bouman (ed.) Soils and the Greenhouse Effect. Proc. of the International Conference Soils and the Greenhouse Effect. John Wiley and Sons, Chichester, U.K. 225-261.

Frissel, M.J., and J.A. van Veen. 1981. Simulation model for nitrogen immobilization and mineralization. In: I.K. Iskandar (ed). Modeling Wastewater Renovation by Land Disposal. John Wiley and Sons, New York, 359-381.

Gildea, M.P., B. Moore, C.J. Vorosmarty, B. Bergquist, J.M. Melillo, K. Nadelhoffer, and B.J. Peterson. 1986. A global model of nutrient cycling: I. Introduction, model structure and terrestrial mobilization of nutrients. In: D. Correll (ed.). Watershed Research Perspectives. Smithsonian Institution Press, Washington, D.C. 1-36.

Haas, H.J., C.E. Evans and E.R. Miles. 1957. Nitrogen and carbon changes in Great Plains soils as influenced by cropping and soil treatments. USDA Tech. Bull. 1164. United States Department of Agriculture, Washington, DC.

Houghton, J. T., G. J. Jenkins and J. J. Ephraums. 1990. Climate Change. The IPCC Scientific Assessment. Cambridge University Press, Cambridge, UK.

Houghton, R. A., J. E. Hobbie, J. M. Melillo, B. Moore, B. J. Peterson, G. R. Shaver, and G. M. Woodwell. 1983. Changes in the carbon content of terrestrial biota and soils between 1860 and 1980: A net release of CO_2 to the atmosphere. Ecological Monographs. 53(3), pp. 235-262.

Jenkinson, D. S. 1991. The Rothamsted long-term experiments: are they still of use? Agronomy Journal, 83(1), pp 2-10.

Jenkinson, D.S, D.E. Adams, and A. Wild. 1991. Model estimates of CO_2 emissions from soil in response to global warming. Nature. 351:304-306.

Jenkinson, D.S. 1990. The turnover of organic matter and nitrogen in soil. Phil. Trans. R. Soc. Lond. B. 329:361-368.

Jenkinson, D.S. and J.H. Rayner. 1977. The turnover of soil organic matter in some of the Rothamsted classical experiments. Soil Sci. 123:298-305.

Joyce, L.A. and M.D. Skold. 1988. Implications of changes in the regional ecology of the Great Plains. In: Impacts of the Conservation Reserve Program in the Great Plains. Symposium Proceedings, Denver, CO Sept. 16-18, 1987. USDA Forest Service Gen. Tech. Report RM-158. April 1988.

Lamb, J.A., G.A. Peterson, and C.R. Fenster. 1985. Wheat fallow tillage systems' effect on newly cultivated grassland soils' nitrogen budget. Soil Sci. Soc. Am. J. 49:352-356.

Li, C.S., S.E. Frolking, and T. Frolking. 1991a. DNDC, A model of nitrous oxide evolution from soil driven by rainfall events: I. Model development. J. of Geophysical Research (in press)

Li, C.S., S.E. Frolking, and T. Frolking. 1991b. DNDC, A model of nitrous oxide evolution from soil driven by rainfall events: II. Sensitivity and simulation. J. of Geophysical Research, (in press).

Li, C.S., S.E. Frolking, R.C. Harriss, and R.E. Terry. 1991c. DNDC model simulation: A case study on N_2O emissions at Belle Grade, Florida. J. of Geophysical Research. (in press).

McGill, W.B., H.W. Hunt, R.G. Woodmansee, and J.O. Reuss. 1981. Phoenix, A model of the dynamics of carbon and nitrogen in grassland soils. In: F.E. Clark and T. Rosswall (eds.). Terrestrial Nitrogen Cycles, Processes, Ecosystem Strategies and Management Inputs. Ecol. Bull. (Stockholm) 33:49-115.

Michigan Department of Agriculture. 1990. Michigan agricultural statistics 85p.

Moore, B., R.D. Boone, J.E. Hobbie, R.A. Houghton, J.M. Melillo, B.J. Peterson, G.R. Shaver, C.J. Vorosmarty, and G.M. Woodwell. 1981. A simple model for analysis of the role of terrestrial ecosystems in the global carbon budget. In: B. Bolin (ed.). Modeling the Global Carbon Cycle. SCOPE 16. John Wiley and Sons. New York, New York, USA, 365-385.

Parton, W.J., C.V. Cole, J.W.B. Stewart, D.S. Ojima and D.S. Schimel. 1989. Simulating regional patterns of soil C, N, and P dynamics in the U.S. Central Grasslands Region. In: M. Clarholm and L. Bergstrom (eds.). Ecology of Arable Land. Kluwer Academic Press. 99-108.

Parton, W.J., J.W.B. Stewart and C.V. Cole. 1988. Dynamics of C, N, P and S in grassland soils: A model. Biogeochemistry. 5:109-131.

Paustian, K., W.J. Parton and J. Persson. 1992. Influence of organic amendments and N-fertilization on soil organic matter in long-term plots: Model analyses. Soil Sci. Soc. Am. J. (in press).

Post, W.M, T.H. Peng, W.R. Emanuel, A.W. King, V.H. Dale and D.L. DeAngelis. 1990. The global carbon cycle. Amer. Sci. 78:310-326.

Prentice, K.C. and I.Y. Fung. 1990. The sensitivity of terrestrial carbon storage to climate change. Nature, 346:48-51.

Raich, J.W., E.B. Rastetter, J.M. Melillo, D.W. Kicklighter, P.A. Steudler, B.J. Peterson, A.L. Grace, B. Moore III, and C.J. Vorosmarty. 1991. Potential net primary productivity in South America: Application of a global model. Ecological Applications, 1:399-429.

Schimel, D.S., W.J. Parton, T.G.F. Kittel, D.S. Ojima, and C.V. Cole. 1990. Grassland biogeochemistry: Links to atmospheric processes. Climatic Change, 17:13-25.

Schlesinger, W.H. 1990. Evidence from chronosequence studies for a low carbon-storage potential of soils. Nature, 348:232-234.

Smith, O.L. 1979. An analytical model of the decomposition of soil organic matter. Soil Biol. Biochem. 11:585-606.

Tans, P. P., I. Y. Fung, and T. Takahashi. 1990. Observational constraints on the global atmospheric CO_2 budget. Science, 247, 23 March 1990, pp 1431-1438.

van Veen, J.A. and E.A. Paul. 1981. Organic carbon dynamics in grassland soils. 1. Background information and computer simulation. Can. J. Soil Sci. 61:185-201.

van Veen, J.A., J.N. Ladd, and M.J. Frissel. 1984. Modeling C and N turnover through the microbial biomass in soil. Plant and Soil, 76:257-274.

Vitosh, M.L., J.F. Davis and B.D. Knezek. 1973. Long-term effects of manure, fertilizer, and plow depth on chemical properties of soils and nutrient movement in a monoculture corn system. J. Envir. Qual. 2:296-299.

Wilson, A.T. 1978. The explosion of pioneer agriculture: contribution to the global CO_2 increase. Nature 273:40-42.

Wood, C.W., D.G. Westfall, and G.A. Peterson. 1991. Soil carbon and nitrogen changes on initiation of no-till cropping systems. Soil Sci. Soc. Am. J. 55(2):470-476.

NATURAL SINKS OF CO_2 WORKSHOP SUMMARY – TECHNICAL SYNTHESIS, CONCLUSIONS, KEY FINDINGS AND RESEARCH RECOMMENDATIONS

PART V

Natural Sinks of CO_2: Technical Synthesis from the Palmas Del Mar Workshop

JOHN P. DOWNING
Pacific Northwest Laboratories
Marine Sciences Laboratory
439 West Sequim Bay Road
Sequim, WA 98382 USA

DOMINIC A. CATALDO
Pacific Northwest Laboratories
Environmental Sciences Department
900 Battelle Boulevard
Richland, WA 99352 USA

ABSTRACT. Natural CO_2 sinks in terrestrial and marine environments are important components of the global carbon cycle, yet the sign and magnitudes of key fluxes among them are unknown. The results of the Palmas Del Mar Workshop - Natural Sinks of CO_2 presented in this special issue and its companion hard-bound volume of Water, Air, & Soil Pollution, provide a synthesis of current research on the carbon cycle, CO_2 sinks and associated processes and fluxes, and critical research needs to assess the potential role of forest and land-use management in carbon sequestration. The papers in this volume present data, observations, and model simulations that demonstrate: 1) the existence of natural CO_2 sinks that could mitigate a significant amount of CO_2 emissions from fossil-fuel combustion; 2) probable, human-caused imbalances in C exchanges among vegetation, soils, and the atmosphere; 3) enhanced C storage in vegetation in response to excess atmospheric CO_2; 4) strong interactions among carbon, nutrient and hydrological cycles; and 5) an excess of carbon production over consumption in several, large managed forests. Although it appears unlikely that the search for the "missing" C sink required to balance the C budget will end in the open ocean, new estimates of C storage in mangrove wood and peat, suggest that coastal ecosystems have the capacity to store significant amounts of carbon in vegetation and sediments. Convincing analyses are also presented indicating the technical and economical feasibility of managing existing lands to sequester additional carbon. Long-term field studies of CO_2 fertilization effects and carbon cycling by plants and soils in geographically important systems, native forests, and coastal ecosystems will go a long way toward meeting the research needs identified at the workshop.

Water, Air, and Soil Pollution **64**: 439–453, 1992.
© 1992 *Kluwer Academic Publishers.*

1. Background

Future warming of the atmosphere, alteration of our climate, and degradation of the global environment are potential consequences of human activities that have increased the concentration of CO_2 in the atmosphere nearly 30% since about 1800. The principal causes of increased atmospheric CO_2 are fossil fuel combustion and changes in land use, primarily deforestation (USDOE 1985, CES 1989, and IPCC 1990). The consequences of increasing atmospheric CO_2 are potentially serious enough that control of CO_2 emissions and other remedial actions are being considered by some developed nations.

Major scientific efforts have been focused on the global carbon cycle in the past several decades. These are well documented in IEA (1982), USDOE (1985), SCOPE 29 (1986), and IPCC (1990). According to the Intergovernmental Panel on Climate Change (IPCC 1990), in the 1980's CO_2 emissions from fossil fuel combustion (5.4 ± 0.5 Pg $C \cdot yr^{-1}$) and land-use change (1.6 ± 1.0 Pg $C \cdot yr^{-1}$) totaled 7.0 ± 1.5 Pg $C \cdot yr^{-1}$. Of these emissions, 3.4 ± 0.2 Pg $C \cdot yr^{-1}$ remained in the atmosphere and about 2.0 ± 0.8 Pg $C \cdot yr^{-1}$ was taken up by the ocean. The fate of the remaining CO_2, 1.6 ± 1.4 Pg $C \cdot yr^{-1}$ of "missing carbon" (USDOE 1985), is intensely debated in political arenas and the scientific community. A serious implication of the apparent imbalance among known energy emissions, estimates of the C flux from land-use changes, and natural sinks of CO_2 is that model predictions of future atmospheric CO_2 concentrations lack credibility (Edmonds 1992). The identification and quantification of natural CO_2 sinks is a critical gap in our current understanding of the global carbon cycle and was the main focus of a workshop convened at Palmas Del Mar, Puerto Rico in February 1992.

2. Introduction

The workshop was attended by 30 leading scientists, representing U.S. and European government agencies, research institutions, academia, and private-sector interests. It built on previous assessments by the U.S. Department of Energy (USDOE 1985) and the IPCC (1990). The objectives were to evaluate the global carbon cycle, assess apparent imbalances among known anthropogenic CO_2 sources and natural sinks, and to explore the potential for managing natural sinks, both terrestrial and marine, to sequester carbon from energy sources. In this paper, we provide a technical synthesis of presentations made at the workshop and the papers in this volume.

There were four key findings of the workshop. In addition, the participants made 26 recommendations for future research on C reservoirs and fluxes, CO_2 fertilization effects, and the development of models and management practices to exploit natural CO_2 sinks. The key findings, research recommendations, and a consensus statement for the workshop are summarized in Chapter 25 (Lugo and Wisniewski 1992).

3. Carbon Cycles in Terrestrial and Marine Systems

Terrestrial carbon reservoirs and fluxes involve diverse ecosystems including forests, grasslands, agroecosystems, tundra, rivers, lakes, and freshwater wetlands. For the purposes of this workshop, the marine carbon reservoir was partitioned into a coastal component and an open-ocean component. The coastal ocean includes saline wetlands, estuaries, bay-lagoon complexes, and continental shelves out to the 200-meter isobath. Marine waters beyond the shelf edge are included in the open ocean.

Figure 1 is a simple schematic of major sources, sinks, and fluxes of the global carbon cycle. It is included only to provide a general framework for the synthesis presented here. Because the size and direction of some C fluxes are still the object of scientific debate, their values are not shown on the schematic. Similar diagrams appear in previous treatments of the global carbon cycle. For example, detailed schematics appear in Sundquist (1985), USDOE (1985), and several papers in this volume.

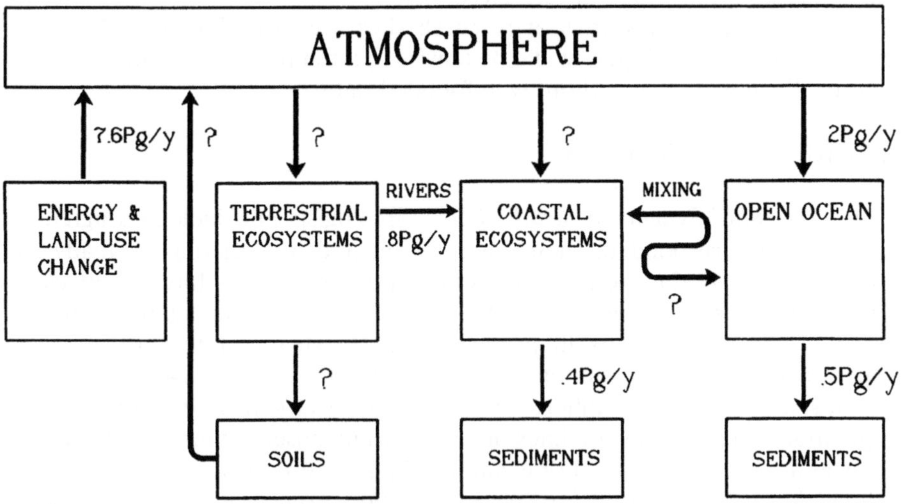

Figure 1. Carbon Cycle Schematic (? = contested direction and/or amount).

In evaluating the global carbon cycle, several spatially and functionally separate, yet interrelated, compartments must be considered. To date, most analyses of the carbon cycle have focused on the atmospheric and oceanic system while treating terrestrial systems as a neutral component with respect to carbon. It now appears important to add a separate component in conceptual models to represent coastal systems; see Twilley et al. (1992). As we discuss below, coastal systems may have the capacity to store significant amounts of carbon in sediments and ecosystems, and they can transform the chemical form of carbon discharged by rivers and thereby affect its storage elsewhere in the system. The concept that the terrestrial system is a "neutral biosphere" with respect to carbon arises from a poor understanding of carbon dynamics in individual components. From an operational standpoint, the terrestrial system is composed of a variety of soil/plant ecosystem types, and these should be considered individually. This should result in greater accuracy in determining net carbon fluxes for the terrestrial system.

Each carbon reservoir has capacities for storing and transporting carbon in various forms, and for exchanging CO_2 with the atmosphere. The ecosystems which mediate chemical transformations and exchange of carbon have unique assemblages of biota, soils, microclimates, and stressors that influence their capacity to exchange carbon in terrestrial and marine systems. Thus, it is critical that geography, as well as biotic and abiotic variability, be considered in the selection of key representative ecosystems when quantifying carbon fluxes (Brown et al. 1992). The research and observations of Huettl and Zoettl (1992) similarly indicate that studies to resolve carbon dynamics and storage in terrestrial systems should be site specific and that thoroughly documented histories of research sites are essential. Based on these observations and approaches, it would be unwise to attempt generalizations from any one site to the global scale without well-founded data.

3.1. TERRESTRIAL SYSTEMS

3.1.1. *Are Terrestrial Systems in Steady State?* Ecologists and biologists agree that biological systems respond to their environment and are seldom in steady state (Lugo and Brown 1992). Some ecosystem responses, such as forest dieback, are obvious while others may be very subtle. Until recently it was assumed that the terrestrial system, and forests in particular, are in steady state with respect to atmospheric CO_2. This assumption implies that the flux of carbon from the terrestrial system to the atmosphere equals the flux of CO_2 from the atmosphere to the vegetation and soils. Several lines of evidence and observations now suggest that these fluxes are not in balance. For example: (1) mature stands accumulate biomass (see discussion below); (2) all forests leach carbon from soils that is replaced by atmospheric CO_2; (3) human activities are constantly modifying forests, and in many cases, resetting them to the log phase of growth; (4) all studies of carbon budgets of mature forest stands show an excess of carbon production over consumption; and (5) soils of mature forests appear to accumulate carbon over long time periods. Discussion of the steady-state assumption at the workshop highlighted the need to factor non-steady state terrestrial processes into future analyses of the global carbon cycle.

Data presented at the workshop indicate that large areas of managed tropical and temperate forests are not in steady state with respect to carbon storage. Recent changes in forest areas on Peninsular Malaysia for the period 1972 to 1982, documented by Brown et al. (1992), indicate that forest area decreased by 17% while forest biomass was reduced by 28%. These forests are degrading with respect to biomass and are not in steady state. An assessment of recent trends in forest dieback by Mueller-Dombois (1992) provides further evidence that forests in many regions are in a state of constant change. While direct cause-and-effect links could not be assigned to many cases of dieback, Mueller-Dombois shows that periodic diebacks probably result from an interaction of several factors. In monospecific stands, trees tend to segregate spatially by maturity and environmental stress appears to act on specific age groups. Dieback in these situations is triggered by nutrient deficiency, pollution, severe wind, etc., and contributing factors such as natural pathogens and disease. For example, Gregor (1992) reports that productivity may be declining in damaged-mature or nutrient-limited forests of Germany as a result of air pollution.

In contrast with the deforested areas discussed above, Brown et al. (1992) conclude that the biomass of inaccessible and previously logged Malaysian forests increased by 7 to 14 % and 7 to 152%, respectively, from 1972 to 1982. Huettl and Zoettl (1992) reported that growth of managed forests in Germany (formerly West Germany) has significantly increased in the last four decades, and that standing-wood volume has increased by about 50 to 60%. Increased standing wood volume in Finland, Sweden, Austria, Switzerland, and France over the last 30 years has also been reported. The causes of these increases are only partially known. In Germany, changes in management (e.g. thinning and establishment of more productive species) and age structure (replacement of older stands with younger, fast-growing ones) have caused some of the increased forest growth. Other probable causes are enhanced N deposition (15 to 50 kg $N \cdot ha^{-1} \cdot yr^{-1}$), regional changes in weather patterns (specifically increased precipitation in southwestern Germany in the last 100 yrs), and CO_2 fertilization. Further evidence for increased biomass in temperate forests is presented by Sampson (1992). Harvest and biomass surveys of North American timberlands indicate a 0.13 Pg $C \cdot yr^{-1}$ increase in tree biomass. Sampson also estimates that 33% of harvested wood (0.05 Pg $C \cdot yr^{-1}$) is stored in-use, in landfills, or burned as fuel. Thus, North American timberlands appear from this analysis to be a sink of at least 0.18 Pg $C \cdot yr^{-1}$.

3.1.2. *CO_2 Fertilization*. There was considerable discussion about the potential for global CO_2 fertilization to offset emissions of CO_2 from fossil fuel use. Recent results indicate that CO_2, N, and other nutrient ions can affect plant growth and increase plant biomass. Drake (1992) shows that photosynthesis in some C_3 plants, growing in a coastal wetland and exposed to elevated atmospheric CO_2, is dramatically increased. In the same plants, excess CO_2 increased rooting depth, biomass production, and soil nitrogen fixation, and decreased respiration and transpiration. These fertilization effects were not observed in the C_4 plants studied by Drake. It is yet to be determined if this fertilization effect is characteristic of all C_3 plants or is limited to specific plant families and/or species. Although Drake observed sustained CO_2 effects for five growth cycles, it is not known if fertilization effects persist for the entire life cycle of long-lived plants without adverse physiological/ecological consequences.

Rastetter et al. (1992) discuss model experiments on the effects of elevated temperature and CO_2 fertilization on vegetation and soils in temperate-forest and arctic-tundra ecosystems. Their simulations suggest that soil N and plant C increase in response to excess atmospheric CO_2, while plant nitrogen decreases slightly. They also show that a 5° C increase in soil temperature decreases soil and plant nitrogen, and increases soil and plant carbon. These responses alters soil C:N ratios. Using a forest gap model, Post et al. (1992) show that a model forest consisting initially of aspen and paper birch, and succeeded by black and white spruce, has considerable potential for increased C storage in response to elevated CO_2. The amount of increased storage depends on species composition, soil conditions, and N dynamics. There was little change in the model soil organic matter (SOM) pools with time or between control and $2xCO_2$ simulations. Both the Rastetter et al. and Post et al. models indicate that enhanced biomass growth in a warmer atmosphere with elevated CO_2 could sequester significant quantities of atmospheric CO_2 on a global basis. The importance of this potential sink underscores the need to evaluate critical model assumptions by continued field research on effects of CO_2 fertilization in geographically important terrestrial systems and native tree species.

3.1.3. *Carbon and Nutrient Cycle Linkages.* Field studies reviewed by Strain and Thomas (1992) and Drake (1992) demonstrate important linkages among carbon, nutrient, and water cycles. For example, CO_2 fertilization can improve plant water status and ameliorate drought stress. In nutrient-limited systems, however, these effects are reduced. European research (Gregor 1992; Huettl and Zoettl 1992) clearly shows that net primary productivity increases in forests with long-term management and that balanced-mineral (K, Mg, and P) supplements appears critical to sustain forest growth in some stressed European forests. The model simulations of Rastetter et al. (1992) suggest that the major cause of enhanced C storage by elevated CO_2 and temperature is a transfer of N from soils to vegetation. Their results also show that "open" model ecosystems, with enhanced flow of N through the inorganic N pool, store more C than "closed" model systems.

Johnson (1992) presented data indicating that harvesting forests, in most cases, does not effect soil-C inventories. In addition, he showed that increasing soil fertility, especially by adding N fixers, lime, or P, can substantially increase soil-C content. Charcoal from combustion of forest debris adds significant refractory C to soils (Johnson 1992). Refractory C in soils could have formed at a slow rate (0.2-0.3 $Pg \cdot yr^{-1}$, according to Schlesinger 1990). The complexity of carbon and nutrient

3. Carbon Cycles in Terrestrial and Marine Systems

Terrestrial carbon reservoirs and fluxes involve diverse ecosystems including forests, grasslands, agroecosystems, tundra, rivers, lakes, and freshwater wetlands. For the purposes of this workshop, the marine carbon reservoir was partitioned into a coastal component and an open-ocean component. The coastal ocean includes saline wetlands, estuaries, bay-lagoon complexes, and continental shelves out to the 200-meter isobath. Marine waters beyond the shelf edge are included in the open ocean.

Figure 1 is a simple schematic of major sources, sinks, and fluxes of the global carbon cycle. It is included only to provide a general framework for the synthesis presented here. Because the size and direction of some C fluxes are still the object of scientific debate, their values are not shown on the schematic. Similar diagrams appear in previous treatments of the global carbon cycle. For example, detailed schematics appear in Sundquist (1985), USDOE (1985), and several papers in this volume.

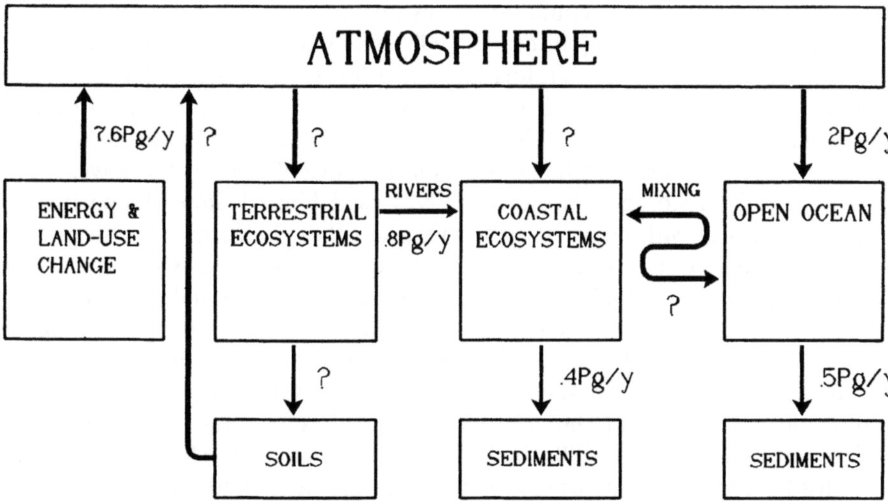

Figure 1. Carbon Cycle Schematic (? = contested direction and/or amount).

In evaluating the global carbon cycle, several spatially and functionally separate, yet interrelated, compartments must be considered. To date, most analyses of the carbon cycle have focused on the atmospheric and oceanic system while treating terrestrial systems as a neutral component with respect to carbon. It now appears important to add a separate component in conceptual models to represent coastal systems; see Twilley et al. (1992). As we discuss below, coastal systems may have interactions reported at the workshop was viewed by most participants as one of the critical areas for future research.

3.1.4. *Management of Carbon Storage.* Using geographic information systems (GIS) technology, and assuming reduced human pressure, Brown et al. (1992) estimate the potential for sequestering carbon in tropical forests. Their analysis, which includes C in below-ground biomass and coarse woody debris, indicates that the C storage potential is 2.5 Pg $C \cdot yr^{-1}$. Management of marginal lands and forests can increase carbon sequestration significantly, according to results reported by Johnson et al. (1992). Their data clearly show that management can reduce C loss from soils following disturbance. After drastic physical disturbances and forest fires, C may accumulate in soils. Cultivation of forest soils, on the other hand, can cause a loss of SOM. N fertilization can reduce such losses. Jackson's (1992) estimate of C fluxes from U.S. Agroecosystems in 1987 include an uptake of 0.367 Pg $C \cdot yr^{-1}$ in plant crops and a release of 0.018 Pg $C \cdot yr^{-1}$ from live-animal respiration and CH_4 fermentation, for a net of about 0.35 Pg $C \cdot yr^{-1}$. However, as Jackson points out, this estimate "is incomplete because rice CH_4, plant and animal waste CH_4 and CO_2, and soil-atmosphere C fluxes could not be estimated from the census data".

Using published data, Winjum et al. (1992) evaluate the potential contributions of five management practices (reforestation, afforestation, natural regeneration, silviculture, and agroforestry) to C sequestration. Their estimates for all practices combined are in the range 50 to 100 Pg C over a fifty-year period at an estimated cost of \$3.40 per t C. Of the practices evaluated, silviculture has the lowest values for potential carbon storage. Sampson (1992) used United States Department of Agriculture statistics with conversion factors and other data from the literature to estimate potential U.S. mitigation of CO_2 emissions by a variety of forest management practices, including conversion of marginal crop and pasture lands to forests, planting trees for windbreaks and in urban areas, and growing wood for fuel. The potential reduction of C emissions that could be achieved by these practices ranges from 0.14 to 0.38 Pg $C \cdot yr^{-1}$. According to Glenn et al. (1992), global utilization of available arid lands for halophyte cultivation, assuming 5.5 Mt $C \cdot ha^{-1}$ is stored annually by 130×10^6 ha, might remove as much as 0.7 Pg $C \cdot yr^{-1}$. Options for long-term storage of C in halophyte biomass include wet and dry burial in local soils, conversion to fodder, oil, meal, and biomass fuels. Marland and Marland (1992) use a growth-function-based model to explore strategies to store C in trees. Based on the model simulations, they conclude that the most effective strategy for forests with large standing biomass is to protect them. Lands with little biomass and low productivity should be managed for forest growth; highly productive lands would be most effectively used for growing biomass fuels and wood for making long-lived products.

3.1.5. *Model Calculations.* Hall et al. (1992) apply a geographically based ecosystem model to simulate long-term carbon storage in a forest disturbed by hurricanes at an average frequency of once per 60 years. The model accumulates a small amount of refractory carbon after a 300-yr simulation of cyclic disturbance and recovery. Smith et al. (1992) used a forest gap model, driven by $2xCO_2$ climates from general circulation models (GCM), to simulate changes in species composition for an Alaskan site. The predicted stand characteristics vary widely from one GCM projection to another. When forced with the relatively mild, $2xCO_2$ climates from the GISS and OSU GCM's, forest biomass changes gradually over a period of about 100 years. For the severe (hot and wet) GFDL and UKMO $2xCO_2$ climates, the onset of dieback of some species is sudden and they disappear from the landscape in less than fifty years. King and Neilson (1992) examine the potential pulse of C from drought-induced forest dieback and wildfire resulting from global warming predicted by four GCM's with a simple forest dieback/regrowth model. Depending on the GCM scenario and model input parameters assumed, the average-annual C release during a pulse (duration ~100 yr) ranged from 0.1 to 3.4 Pg $C \cdot yr^{-1}$. A method for analyzing the potential C storage in agroecosystems, using existing soil-carbon models and GIS data bases, is described in Barnwell et al. (1992). Many of the simplifying assumptions made in the above model studies warrant further investigation.

3.2. FLUVIAL CARBON

The atmosphere and rivers link the terrestrial and oceanic carbon reservoirs. Rivers deliver carbon to the ocean as dissolved inorganic carbon (DIC), mainly carbonate ions; carbonate sediments and particulate organic matter (PIC and POC); and dissolved organic matter (DOC). Sundquist (1985) gives the following ranges for fluvial carbon inputs to the oceans in Pg $C \cdot yr^{-1}$: DIC - 0.37 to 0.45, DOC - 0.08 to 0.32, POC - 0.04 to 0.3, PIC - 0.02 to 0.2 (Twilley et al.'s 1992 estimate is 0.35 Pg $C \cdot yr^{-1}$ for DOC plus POC). Despite strong buffering by sea water alkalinity, about 17% of the carbonate carbon discharged by rivers, escapes to the atmosphere as CO_2. The current estimate of total fluvial carbon input to the ocean is about 0.8 Pg $C \cdot yr^{-1}$ (SCOPE 42 1991) of which Wollast and MacKenzie (1989) estimate that as much as 0.6 Pg $C \cdot yr^{-1}$ may result from human activities, mainly enhanced erosion.

The fate of fluvial carbon in the ocean is unclear. Some PIC and POC may be stored temporarily (100's to 1,000's yr) in estuarine and continental shelf sediments. DOC is an energy source for marine bacteria which decomposes to DIC; the residual DOC is added to the large reservoir of oceanic DOC (1,600 Pg C). Human activities on developed coasts could influence the chemical and biological transformation of carbon and where it is stored in coastal systems. Although not a major focus of the Palmas Del Mar Workshop, participants recognized that the fate of fluvial carbon may affect the global balance of carbon fluxes significantly. For these reasons, more realistic treatment of fluvial carbon in models of coastal systems was a major research need that was identified at the workshop.

4. Marine Systems

Marine systems comprise major sinks for atmospheric CO_2 and carbon from terrestrial sources. The coastal-ocean reservoir is connected by carbon fluxes, directly by mixing and indirectly by the atmosphere, to an open-ocean reservoir. Carbon in marine systems occurs as DIC, PIC, DOC, and POC. Bicarbonate and carbonate ions comprise 99% of the DIC in the ocean (36,600 Pg C, calculated from 1972-77 GEOSECS data, Sundquist 1985 and Takahashi et al. 1981). Surface DIC concentrations (< 1,000 m) are about 15% lower than at depth. The vertical DIC gradient is maintained by settling of biologically produced PIC and POC, convection in cold regions, and replenishment of DIC to the surface by mixing and upwelling. The turnover time of DIC between surface and deep reservoirs is on the order of centuries. Of the plausible causes for a disequilibrium between atmospheric and surface-ocean pCO_2, the oceanic circulation and inorganic carbon chemistry are thought by most to control the uptake of atmospheric CO_2 by the ocean. Major uncertainties, however, remain about the amount and dynamics of the DOC (Toggweiler 1989) and the role of new production in sequestering carbon in the ocean (ASLO 1991).

The net flux of atmospheric CO_2 into the open ocean has been estimated by various models to be in the range, 2 ± 0.8 Pg $C \cdot yr^{-1}$, in the 1980's (Houghton et al. 1990). These estimates have come from box- and general-circulation model calculations (Oeschger et al. 1975, Peng 1984, Volk and Hoffert 1985, Siegenthaler and Oeschger 1987, Bacastow and Maier-Reimer 1990, and most recently Sarmiento et al. 1992). Others have estimated the oceanic uptake of CO_2 from surface partial pressures (pCO_2) and wind fields. Although most estimates of CO_2 uptake cluster about 2 Pg $C \cdot yr^{-1}$, the range of individual estimates is 0.3 (Tans et al. 1990) to 2.6 (Takahashi et al. 1986) Pg $C \cdot yr^{-1}$. Recent adjustments for fluvial carbon inputs, carbon monoxide oxidation and transport, and underestimates of skin temperature suggest that the values of Tans et al. (1990) are too low and that the lower end of the range of estimates has shifted back toward a canonical value of 2 Pg $C \cdot yr^{-1}$. Workshop participants were left with the perception that estimates of C uptake by the open ocean have not changed drastically in the past two decades, and that the search of new carbon sinks will probably not end in the ocean.

Efforts to quantify the inventory and fluxes of carbon in the coastal ocean have been less conclusive than for the open ocean. Twilley et al. (1992) estimated that about 0.16 $Pg \cdot yr^{-1}$ of C are stored in mangrove wood, 0.02 Pg $C \cdot yr^{-1}$ accumulates in mangrove peat, and as much as 0.05 Pg $C \cdot yr^{-1}$ is exported to the coastal ocean. Their carbon budget for coastal ecosystems has other new source estimates, including 0.36 and 0.08 Pg $C \cdot yr^{-1}$ of total organic carbon from rivers and wetlands, respectively. Estimates of C sinks in coastal wetlands (0.025 Pg $C \cdot yr^{-1}$), coral reefs, shelf and estuarine sediments (0.41 Pg $C \cdot yr^{-1}$), give a total potential C storage of as much as 0.6 Pg $C \cdot yr^{-1}$.

To date, no global carbon-cycle model for the coastal ocean with dynamics (advection and diffusion) and realistic biogeochemical processes exists. Carbon-cycle models and observations for specific regions have also met with very limited success in balancing changes in the carbon inventory in a coastal system with net inputs and outputs. It was acknowledged that geographical limitations in the data base, infrequent sampling, and inadequate methodologies are largely responsible for the lack of progress in such investigations. This is a critical gap that warrants immediate attention and a high priority in carbon cycle research planning.

4.1. MODEL ESTIMATES OF OCEANIC CO_2 UPTAKE

All models of oceanic CO_2 uptake assume that the pre-industrial carbon cycle was in steady state. This assumption is based on ice core records with nearly constant atmospheric CO_2 ($\pm$ 5 ppm) for about 1000 years prior to 1800. During this period, gas exchange, ocean circulation, new production, and sea water carbonate chemistry are assumed to be in steady state. It is also commonly assumed that CO_2 uptake by the ocean may be explained by physiochemical processes alone and that the coastal carbon reservoirs and fluxes (Figure 1) can be ignored.

Early box-ocean models included rudimentary biological processes to convert surface DIC to particulate organic matter (POM) which settles to deeper layers (Hoffert et al. 1981, Volk and Hoffert 1985). Model conversion of DIC to POM depletes nutrients in the surface layer and removes carbon in accordance with the Redfield ratio (C:N:P), a process commonly called the "biological pump". Nutrients removed with POM from the surface are replenished by vertical diffusion in the models. In the late 1980's, speculation about the benefits of stimulating new production with micronutrients (ASLO 1991) prompted more realistic biology in ocean models. A multi-component biology model has been incorporated in two general circulation models (GCM). The biology model includes multiple trophic levels, parameterized by region-specific coefficients.

Orr and Sarmiento (1992) report the results of iron-fertilization experiments with a GCM-based carbon-cycle model. Their experiments showed that, after 100 years of nutrient depletion, the organic-matter flux in the model increased 6 to 30 Pg $C \cdot yr^{-1}$ (about twice the present new production in the ocean). During the same period, the pCO_2 in the model atmosphere decreased by 46 to 85 ppm indicating a 98-to-181 Gg C uptake by the model ocean. Although these appear to be large amounts of C, they are small compared to the projected CO_2 emissions for the next century of fossil fuel combustion. Another interesting result, was that when the model ocean was run with 1 Pg $C \cdot yr^{-1}$ of enhanced new production, it only took up 0.4 (without harvesting) to 0.7 (with harvesting) Pg $C \cdot yr^{-1}$ from the model atmosphere. This suggests that creating a new sink for CO_2 by manipulating oceanic biology may be very inefficient. Ritschard's (1992) reassessment of iron fertilization and macroalgae farming, using available ocean engineering technologies, suggests that logistical and unknown environmental consequences render the prospects of successful and economical enhanced CO_2 uptake very dim.

The model results of Orr and Sarmiento (1992) and Manabe and Stouffer (1988), paleoceanographic data (Broecker et al. 1985), and long time series of chlorophyll measurements (Venrick et al. 1987) all suggest that ocean biota may not be in steady state, and that ocean circulation could have more than one mode of operation. Simulations with non-steady-state ocean circulations and new production, unconstrained by present-day temperature, salinity, and nutrient distribution will be very difficult. This capability will be valuable for understanding how the carbon cycle might respond if the ocean were to operate differently in a warmer climate than it does now. Models to investigate such changes are under development.

5. Conclusions

The papers in this volume present data, observations, and model simulations that demonstrate: 1) the existence of natural CO_2 sinks that could mitigate a significant amount of CO_2 emissions from fossil-fuel combustion (6 ± 0.9 Pg C in 1991, Edmonds 1992); 2) probable, human-caused imbalances in C exchanges among vegetation, soils, and the atmosphere; 3) enhanced C storage in vegetation in response to excess atmospheric CO_2; 4) strong interactions among carbon, nutrient and hydrological cycles; and 5) an excess of carbon production over consumption in several, large managed forests. Convincing analyses are also presented indicating that it is technically and economically feasible to manage existing lands to sequester additional carbon.

The importance of natural C sinks, apart from concerns about global climate change, and uncertainty about the size and dynamics of C sinks underscore the need to critically examine assumptions and data gaps in model calculations of terrestrial C dynamics. Long-term field research on the effects of CO_2 fertilization and carbon cycling by vegetation and soils in geographically important systems, native forests, and coastal ecosystems will go a long way toward meeting this need. Even with such research, the complexities of carbon- and nutrient-cycle interactions were viewed by most participants as one of the critical areas for future research.

Based on the few papers presented at the workshop and subsequent discussions of model estimates of open-ocean C uptake, it appears unlikely that the search for the "missing" C sink required to balance the C budget will end in the open ocean. New estimates of C storage in mangrove wood and peat, however, suggest that coastal ecosystems may have the capacity to store significant amounts of carbon in vegetation and sediments. Considerable uncertainty exists about the dynamics of C from rivers and exchanges of C between the coastal and open ocean. In addition, geographical limitations of existing data and infrequent sampling of rivers, coastal ecosystems, and sediments will likely continue to limit progress in balancing the C budget and modeling C exchange in the coastal ocean. These are critical problems that warrant immediate attention and a high priority in research planning.

ACKNOWLEDGMENTS

The authors wish to thank the Global Change Program of Pacific Northwest Laboratory (PNL) and the U.S. Department of Energy (USDOE) for financial support while this paper was in preparation. PNL is operated for DOE under contract DE-AC06-76RLO-1830 by Battelle Memorial Institute. The comments of Roger Dahlman and John Kinsman greatly improved the paper. We thank the workshop participants who provided summaries of their results and reviewed an early version of the paper. Finally, we thank Eileen Stoppani for turning the manuscript into photo-ready copy.

REFERENCES

ASLO (American Society of Limnology and Oceanography Symposium) (1991) 'What Controls Phytoplankton Production in Nutrient-Rich Areas of the Open Sea?', February 22-24, 1991, San Marco, California.

Bacastow, R.B. and Maier-Reimer, E. (1990) 'Ocean-circulation model of the carbon cycle', Clim. Dyn. 4, 95-125.

Barnwell, Thomas O. Jr., Jackson Robert B. IV, Elliott, Edward T. (1992) (this volume).

Broecker, Wallace S., Peteet, Dorothy M. and Rind, David (1985) 'Does the ocean-atmosphere system have more than one stable mode of operation?', Nature 315 (6014), 21-25.

Brown et al. (1992), this volume.

CES (Committee on Earth Sciences) (1989) 'The U.S. Global Change Research Program', in Our Changing Planet: The FY 1989 Research Plan.

Drake (1992), this volume.

Edmonds (1992), this volume.

Glenn, et al. (1992), this volume.

Gregor (1992), this volume.

Hall, et al. (1992), this volume.

Hoffert, M.I., Callegari, A.J. and Hsieh, C.T. (1981) 'A box-diffusion carbon cycle model with upwelling, polar bottom water formation and a marine biosphere.' in B. Bolin (ed.), Carbon Cycle Modeling SCOPE 16, John Wiley, New York, pp. 287-305.

Houghton, J.T., Jenkins, G.J. and Ephraums, J.J. (1990) 'Climate Change, The IPCC Scientific Assessment', Cambridge University Press, New York, pp. 365.

Huettl and Zoettl (1992), this volume.
IEA (Institute for Energy Analysis) (1982) 'Carbon Dioxide Review 1982', William C. Clark (ed.), Oxford University Press, New York.
IPCC (Intergovernmental Panel on Climate Change) (1990) 'Scientific Assessment of Climate Change', World Meteorological Organization/United Nations Environmental Program, Cambridge University Press, New York.
Jackson (1992), this volume.
Johnson (1992), this volume.
King and Neilson (1992), this volume.
Lugo and Brown (1992), this volume.
Lugo and Wisniewski (1992), this volume.
Manabe, S. and Stouffer, R.J. (1988) 'Two Stable Equilibria of Coupled Ocean-Atmosphere Model', J. of Climate 1, 841-866.
Marland and Marland (1992), this volume.
Mueller-Dombois (1992), this volume.
Oeschger, H., Siegenthaler, U., Schotterer, U. and Gugelmann, A. (1975) 'A box diffusion model to study the carbon dioxide exchange in nature', Tellus 27, 168-192.
Orr and Sarmiento (1992), this volume.
Peng, T.-H. (1984) 'Invasion of Fossil Fuel CO_2 into the Ocean', in W. Brutsaert and G.H. Jirka (eds.), Gas Transfer at Water Surfaces, D. Reudel Publishing Company, pp. 515-523.
Post et al. (1992), this volume.
Rastetter et al. (1992), this volume.
Ritschard (1992), this volume.
Sampson (1992), this volume.
Sarmiento, J. L., Orr, J. C. and Siegenthaler, U. (1992) 'A Perturbation Simulation of CO_2 Uptake in an Ocean General Circulation Model', J. Geophys. Res.
Schlesinger, W. (1990) 'Evidence from Chronsequence Studies for a Low Carbon Storage Potential of Soils', Nature 348, 232-239.
SCOPE 42 (Scientific Committee on Problems of the Environment) of the International Council of Scientific Unions (ICSU) (1991) 'Biogeochemistry of Major World Rivers', E.T. Degens, S. Kempe and J.E. Richey (eds.), John Wiley & Sons, New York.
SCOPE 29 (1986) 'The Greenhouse Effect, Climatic Change, and Ecosystems', Bert Bolin, Bo R. Doos, Jill Jager and Richard A. Warrick (eds.), John Wiley & Sons, New York.
Siegenthaler, U. and Oeschger, H. (1987) 'Biospheric CO_2 emissions during the past 200 years reconstructed by deconvolution of ice core data', Tellus 39(B), 140-154.
Smith et al. (1992), this volume.
Strain and Thomas (1992), this volume.

Sundquist, E. T. (1985) 'Geological Perspectives on Carbon Dioxide and the Carbon Cycle', in E.T. Sundquist and W.S. Broecker (eds.), The Carbon Cycle and Atmospheric CO_2: Natural Variations Archean to Present, Geophysical Monograph 32, American Geophysical Union, Washington, D.C.

Takahashi, T., Goddard, J., Sutherland, S., Chipman, D.W. and Breeze, C.S. (1986) 'Seasonal and Geographic Variability of Carbon Dioxide Sink/Source in the Oceanic Areas: Observations in the North and Equatorial Pacific Ocean, 1984-1986 and Global Survey', Technical Report to U.S. DOE Carbon Dioxide Research Division, Washington, D.C., pp. 65.

Takahashi, T., Broecker, W.S. and Bainbridge, A.E. (1981) 'The alkalinity and total carbon dioxide concentration in the world oceans', in B. Bolin (ed.), Carbon Cycle Modeling, SCOPE 16, John Wiley, New York, pp. 271-286.

Tans, P.P., Fung, I.Y. and Takahashi, T. (1990) 'Observational constraints on the global atmospheric CO_2 budget', Science 247, 1431-1438.

Toggweiler, J.R. (1989) 'Is the downward dissolved organic matter (DOM) flux important in carbon transport?', in W.H. Berger, W.S. Smetacek, and G. Wefer (eds.), Productivity of the Ocean: Present and Past, John Wiley & Sons, Chichester, U.K., pp. 65-83.

Twilley et al. (1992), this volume.

USDOE (U.S. Department of Energy) (1985) 'Atmospheric Carbon Dioxide and the Global Carbon Cycle', John R. Trabalka (ed.), DOE/ER-0239.

Venrick, E.L., McGowan, J.A., Cayan, D.R. and Hayward, T.L. (1987) 'Climate and Chlorophyll a: Long-Term Trends in the Central North Pacific Ocean', Science 238, 70-72.

Volk, T., and Hoffert, M. I. (1985) 'Ocean Carbon Pumps: Analysis of Relative Strengths and Efficiencies in Ocean-Driven Atmospheric CO_2 Changes', in The Carbon Cycle and Atmospheric CO_2: Natural Variations Archean to Present, AGU Monograph 32, pp. 99-110.

Winjum, et al. (1992), this volume.

Wollast, R. and MacKenzie, F.T. (1989) 'Global Biogeochemical Cycles and Climate', in A. Berger (ed.), Climate and Geo-Sciences, Kluver Acad. Publ., Hingham, pp. 453-473.

NATURAL SINKS OF CO_2 CONCLUSIONS, KEY FINDINGS AND RESEARCH RECOMMENDATIONS FROM THE PALMAS DEL MAR WORKSHOP

ARIEL E. LUGO

Institute of Tropical Forestry USDA Forest Service, Southern Forest Experimental Station, Call Box 25000, Rio Piedras, Puerto Rico, 00928–2500, USA

JOE WISNIEWSKI

Wisniewsli & Associates, Inc, 6862 McLean Province Circle, Falls Church, Virginia, 22043, USA

Abstract. This paper documents the conclusion, key findings and research recommendations from the Natural Sinks of CO_2 Workshop held in Palmas Del Mar, Puerto Rico during 24–27 February 1992. Also presented are some preliminary estimates of actual and potential carbon sinks in terrestrial and aquatic ecosystems.

1. Introduction

Environmental, energy, and socioeconomic issues associated with carbon dioxide (CO_2) and other "greenhouse" gases have been debated intensely in political arenas and the scientific community due to recent concerns over potential for human-induced changes in the Earth's climate. Because of the complexity of the Earth's systems, the current and long-term interactions between atmospheric CO_2 and the complete carbon cycle have yet to be examined in an integrated manner. To evaluate the global carbon budget and the potential for managing the Earth's carbon sinks, a workshop of leading scientists was convened in Palmas Del Mar, Puerto Rico during 24–27 February 1992.

The workshop was motivated by needs to: (1) understand properties, processes and magnitudes of natural sinks of atmospheric CO_2; (2) expand the scope of measurements and models to provide a truly interdisciplinary analysis of the global carbon cycle; (3) encourage broad participation (academic, government and private sector) in research that offers solutions to increasing atmospheric CO_2; and (4) balance the carbon budget as a basis for better policy making.

This paper is a synopsis of findings presented at the workshop. New information was presented at the workshop indicating that substantial potential exists for capturing and sequestering carbon in land and water environments. Carbon is being absorbed by natural sinks and it is economically feasible to manage additional land, soil and coastal ecosystems with the objective of building carbon stocks.

Water, Air, and Soil Pollution **64:** 455–459, 1992.
© 1992 *Kluwer Academic Publishers.*

2. Conclusions

The following five points represent the consensus agreement of all workshop participants:

- The world's carbon cycle is important apart from climate change and is poorly understood.
- New information, particularly from research in tropical forests, demonstrates that exchanges of CO_2 between the biota and the atmosphere have not been in steady state for several centuries. A vast number of natural and managed ecosystems are currently accreting carbon, and the quantity may be large enough to account for the so-called "missing carbon".
- Rising CO_2 in the atmosphere accelerates net primary productivity and probably increases carbon storage in terrestrial ecosystems.
- New information demonstrates that the carbon cycle is inextricably linked to nutrient and hydrological cycles.
- Appropriate ecosystem management can enhance net carbon storage and is compatible with conservation of biodiversity, sustainable land use, energy conservation, and economic development.

3. Other Key Findings

The following statements are based on presentations and discussions at the workshop:

- There is a clear need to add the coastal system to the terrestrial, oceanic, and atmospheric systems in models of the carbon cycle because this sector of the biosphere (particularly in the tropics) has a high rate of carbon sequestration that has not been accounted for in terrestrial and oceanic carbon models.
- Natural recovery of ecosystems from disturbance and through succession sequesters large amounts of carbon and this process does not require substantial investments in management.
- It is economically feasible to scientifically manage existing degraded, damaged or saline lands to sequester carbon.
- Management of forest lands for multiple uses, including carbon sequestration, should reduce deforestation elsewhere by reducing pressure on the land caused by human activities.

Systems and processes that have the potential to or are sequestering carbon are shown in the following table. These estimates, many of which are first-order, were presented by various speakers at the workshop.

SOME PRELIMINARY ESTIMATES OF ACTUAL AND POTENTIAL CARBON SINKS IN ECOSYSTEMS AS REPORTED AT THE PALMAS DEL MAR WORKSHOP ON NATURAL SINKS OF CO_2 HELD DURING 24–27 FEBRUARY 1992.

Some of the estimates are for potential sinks only, which require new management regimes (e.g., temperate marginal lands and saline lands). Some are actual sinks (e.g., coastal systems and growth of temperate forests). Others are a mixture of potential and actual (e.g., tropical forests and forest soils), as the present state of knowledge does not allow separation of actual from potential sinks.

System		Approximate Rate of Annual Carbon Accumulation (Pg)
Temperate Forests [1]		1.4–4.3
Forest growth	(1.0–2.0)	
Marginal lands	(0.15–0.25)	
Forest soils	(0.25–2.0)	
Tropical Forests [2]		1.5–3.2
Saline Lands [3]		0.5–1.0
Coastal Systems [4]		0.3–0.6
Oceans [5]		1.2–2.8
CO_2 stimulation of production limited by total nitrogen now in terrestrial ecosystems [6]		1–6% of net primary productivity

1. Post *et al.* (1992), Sampson (1992) and Winjum *et al.* (1992).
2. Brown *et al.* (1992).
3. Glenn *et al.* (1992).
4. Twilley *et al.* (1992).
5. Orr and Sarimento (1992).
6. Rastetter *et al.* (1992).

4. Research Recommendations

The following recommendations are made to bridge gaps in definition and understanding of the global carbon cycle:

Carbon Reservoirs

- Continued measurement of carbon reservoirs and turnover is needed for key components of terrestrial ecosystems, including soils and vegetation.
- Techniques are needed, using satellite data, to improve estimates of total forest biomass and its change.
- Many forests that are experiencing dieback are also experiencing vigorous regrowth among standing dead trees. It is critical to determine to what degree these transition forest states act as sources/sinks of carbon and provide an opportunity to sequester carbon.
- The size and dynamics of carbon reservoirs for coastal ecosystems need to be quantified.
- Estimates of annual and interannual variability in the inorganic carbon inventory of the oceans are needed.

Carbon Fluxes

- Eddy-correlation measurements of CO_2 fluxes in the atmosphere over different terrestrial ecosystems are needed to identify and quantify geographical locations of CO_2 sources or sinks.
- There is a need to measure and understand rate-controlling factors for carbon fluxes between different ecosystem compartments and the atmosphere.
- The transport and fate of riverine carbon are poorly known, except for a few temperate river deltas and estuaries. The processes associated with these fluxes need to be characterized in tropical and arctic regimes.
- In addition to carbon fluxes, the fate of riverine nutrient inputs must be determined because they affect present and future coastal system net primary productivity.
- The role of the coastal ecosystem as a sink for atmospheric carbon needs to be determined, and the rate-limiting processes involved with CO_2 uptake defined.
- Using paleoceanographic records, research is needed on whether or not the coastal carbon inventory is in steady state.

CO_2 Fertilization

- Carbon dioxide enrichment studies under field conditions for whole forest ecosystems are needed over long time periods for wet, moist and dry conditions in boreal, temperate and tropical life zones, because the potential acceleration of net ecosystem production by higher concentrations of atmospheric CO_2 could result in a large carbon sink.
- Additional field measurements on the response of autotrophic and heterotrophic respiration to elevated CO_2 concentration are needed.
- Field research centers should be established with international cooperation to study all aspects of the carbon budget for key ecosystem types in different locations.
- Most of the vegetation of the Earth is in some state of disturbance and it is critical to understand how this affects vegetation response to CO_2 enrichment.

Approaches to Management of Carbon

- Identify degraded, damaged and saline lands available for carbon sequestration and develop prescriptions for their management.
- Quantify above- and below- ground carbon sequestration and conservation opportunities through sustainable forestry and agriculture management options.
- Further analyses of economic and environmental cost/benefit data for enhancing carbon sequestration through management actions are needed. Preliminary economic analyses in the United States suggest feasibility but these analyses need to address all costs and benefits, including environmental considerations, and must be expanded to other regions.

Modeling

- Global carbon cycle analyses should be carried out with coupled atmosphere, ocean, terrestrial and coastal components of the global carbon system. New advances in the modeling of terrestrial carbon dynamics now make comprehensive carbon cycle modeling more feasible, although models of coastal processes still need further development.
- The excellent data bases for temperate and boreal zone forests are now ready for geographically-referenced analysis without steady state assumptions. Such analyses will determine the size of the carbon sink/source in these ecosystems.
- Carbon sinks in coastal systems need to be factored into the global carbon budget in ocean model calculations.
- Ocean model calculations need to be run without the constraints of present-day temperatures and salinities and steady-state biology to examine the effects of feedbacks in oceanic CO_2 uptake.
- Ecosystem models should be improved so that they simulate the interactive effects of vegetation, nutrients and regional water balance; and the direct effects of CO_2 on plant growth.
- Representation of processes such as clouds, atmospheric particles, and hydrology should be improved in general circulation models (GCMs) so that model outputs better meet the accuracy and resolution requirements of ecological modelers. Caution is required when using current GCM outputs.

Acknowledgments

We acknowledge the input from *all* workshop participants who were instrumental in assembling these Conclusions, Key Findings and Research Recommendations. We particularly cite the extensive review and input provided by Roger Dahlman and John Kinsman. Special thanks are offered to Sandra Brown and Bert Drake for assembling the table.

References

Brown et al. (1992) This volume.
Glenn et al. (1992) This volume.
Orr and Sarimento (1992) This volume.
Post et al. (1992) This volume.
Sampson (1992) This volume and personal communication.
Twilley, et al. (1992) This volume.
Winjum et al. (1992) This volume.

LIST OF PARTICIPANTS

NATURAL SINKS OF CO_2 WORKSHOP
24–27 February 1992 – Palmas Del Mar, Puerto Rico

Sandra Brown
Department of Forestry
University of Illinois
110 Mumford Hall
1301 W. Gregory Drive
Urbana, IL 61801

Dominic Cataldo
Environmental Sciences
Battelle Pacific Northwest Laboratory
900 Battelle Blvd.
Richland, WA 99352

Roger Dahlman
Environmental Sciences Division
Office of Energy Research
Department of Energy
Washington, D.C. 20585

John Downing
Marine Sciences Laboratory
Battelle Pacific Northwest Laboratory
439 West Sequim Road
Sequim, WA 98382

Bert Drake
Smithsonian Environmental Research
Center
P.O. Box 28
Edgewater, MD 21037

Jae Edmonds
Battelle Pacific Northwest Laboratory
901 D Street., S.W., Suite 900
Washington, D.C. 20024-2115

Heinz D. Gregor
Federal Environmental Agency
(Umweltbundesamt)
Bismarckplatz 1
1000 Berlin 33
Germany

Edward Glenn
Environmental Research Laboratory
2601 East Airport Drive
Tucson, AZ 85706-6985

Charles A.S. Hall
SUNY College and Forestry
Science and Forestry
Syracuse, NY 13210

Reinhard F. Huettl
Forestry Department
Kali and Salz, AG
P.O. Box 102029
D-3500 Kassel
Germany

Robert B. Jackson, IV
U.S. EPA Environmental Research
Laboratory
960 College Station Road
Athens, GA 30613-7799

Dale W. Johnson
Desert Research Institute
P.O. Box 60220
Reno, NV 89506

Mark G. Johnson
ManTech Environmental Technology,
Inc.
U.S. EPA Environmental Research
Laboratory
200 S.W. 35th Street
Corvallis, OR 97333

George King
ManTech Environmental Technology,
Inc.
U.S. EPA Environmental Research
Laboratory
200 S.W. 35th Street
Corvallis, OR 97333

Water, Air, and Soil Pollution **64**: 461–462, 1992.

John Kinsman
Environmental Affairs
Edison Electric Institute
701 Pennsylvania Avenue, N.W.
Washington, D.C. 20004–2696

Ariel E. Lugo
U.S.D.A. Forest Service
Call Box 25000
Rio Piedras, Puerto Rico 00928-2500

Gregg Marland
Environmental Research Laboratory, Inc.
Oak Ridge National Laboratory
Oak Ridge, TN 37831-6335

Billy M. McCormac
Editor, Water, Air and Soil Pollution
12861 Alta Tierra Road
Los Altos Hills, CA 94022

Dieter Mueller-Dombois
Department of Botony
University of Hawaii at Manoa
3190 Maile Way
Honolulu, HI 96822

James C. Orr
Program in Atmospheric and Oceanic Science
Princeton University
Princeton, NJ 08544-0710

Wilfred M. Post
Environmental Sciences Division
Oak Ridge National Laboratory
P.O. Box 2008, Bldg. 1000
Oak Ridge, TN 37831-6335

E. B. Rastetter
The Ecosystems Center
Marine Biological Laboratory
Woods Hole, MA 02543

Ronald L. Ritschard
Center for Atmospheric & Biospheric Effects of Technology
Lawrence Berkeley Laboratories
University of California
Berkeley, CA 94720

R. Neil Sampson
American Forestry Association
P.O. Box 2000
Washington, D. C. 20013

Boyd R. Strain
Department of Botany
Duke University
Durham, NC 27706

Robert R. Twilley
Department of Biology
Universtiy of Southwestern Louisiana
P.O. Box 42451
Lafayette, LA 70504-2451

Peter Van Voris
Battelle Pacific Northwest Laboratories
900 Battelle Blvd.
Richland, WA 99332

John F. Weishampel
Department of Environmental Sciences
University of Virginia
Charlottesville, VA 22903

Jack Winjum
National Council for Air and Stream Improvement
U.S. EPA Environmental Research Laboratory
200 S.W. 35th Street
Corvallis, OR 97333

Joe Wisniewski
Wisniewski and Associates, Inc.
6862 McLean Province Circle
Falls Church, VA 22043

AUTHOR INDEX

SUBJECT INDEX

Environmental and Resource Economics

Published in cooperation with the European Association of Environmental and Resource Economists (EAERE)

Managing Editorial Board:

J. B. Opschoor (Chairman), *Free University Amsterdam, The Netherlands; Advisory Council for Research on Nature and Environment, Rijswijk, The Netherlands;* **H. Folmer (President EAERE, Vice-Chairman ex officio),** *Agricultural University Wageningen, The Netherlands;* **H. M. A. Jansen (Executive Managing Editor),** *Free University Amsterdam, Institute for Environmental Studies, The Netherlands*

The primary concern of *Environmental and Resource Economics* (ERE) is the application of economic theory and economic methods to those environmental issues that require rational economic analysis. The current debate includes a number of relatively unexplored topics and *ERE* provides an international forum for the advancement of discussing them. ERE's content should be directly influential on policy making.

Areas of particular interest include:
* evaluation and development of instruments of environmental policy
* cost-benefit and cost effectiveness analysis
* sectoral environmental policy impact analysis
* modelling and simulation
* institutional arrangements
* resource pricing and the valuation of environmental goods
* environmental quality indicators.

Subscription Information **ISSN** 0924-6460
1992, Volume 2 (6 issues)
Subscription rate: Dfl.436.00/US$222.00
incl. postage and handling
Private rate: Dfl.246.00/US$125.00
incl. postage and handling
Special rate for EAERE individual members available upon request

P.O. Box 322, 3300 AH Dordrecht, The Netherlands
P.O. Box 358, Accord Station, Hingham, MA 02018-0358, U.S.A.

Journal Highlight

KLUWER
ACADEMIC
PUBLISHERS

Contaminated Soil '90

Proceedings of the Third International KfK/TNO Conference on Contaminated Soil, December 10—14, 1990, Karlsruhe, FRG

edited by **F. Arendt, M. Hinsenveld, W. J. van den Brink**

Contaminated soil is very much a problem of our times: the theoretical fundamentals underpinning standards have been greatly improved, many more countries are taking inventories of their contaminated sites, legislation is being established in an increasing number of countries. Technical developments, too, are on the move: in-situ techniques are being developed, especially biological and air-stripping ones, and immobilization techniques, such as electroosmosis, are receiving an increasing amount of attention.
These volumes contain altogether more than 300 contributions from about 20 countries. The first volume contains papers related to policies, risk assessment, behaviour and the analysis of contaminants, and the second volume deals with remediation and safeguarding techniques, as well as polluted sediments. The coverage is complete, and the volumes will be required reading for soil scientists, engineers and technologists, pollution scientists, and all others who will have to deal with contaminated soil in the future.

1991, 1620 pp. ISBN 0–7923–1058–6
Hardbound $425.00/Dfl. 650.00/£225.00

P.O. Box 322, 3300 AH Dordrecht, The Netherlands
P.O. Box 358, Accord Station, Hingham, MA 02018-0358, U.S.A.

KLUWER ACADEMIC PUBLISHERS

MIX
Papier aus verantwortungsvollen Quellen
Paper from responsible sources
FSC® C105338

If you have any concerns about our products,
you can contact us on
ProductSafety@springernature.com

In case Publisher is established outside the EU,
the EU authorized representative is:
Springer Nature Customer Service Center GmbH
Europaplatz 3, 69115 Heidelberg, Germany

Printed by Libri Plureos GmbH
in Hamburg, Germany